Atomic Processes in Electron–Ion and Ion–Ion Collisions

NATO ASI Series

Advanced Science Institutes Series

A series presenting the results of activities sponsored by the NATO Science Committee, which aims at the dissemination of advanced scientific and technological knowledge, with a view to strengthening links between scientific communities.

The series is published by an international board of publishers in conjunction with the NATO Scientific Affairs Division

A Life Sciences **B Physics**	Plenum Publishing Corporation New York and London
C Mathematical and Physical Sciences	D. Reidel Publishing Company Dordrecht, Boston, and Lancaster
D Behavioral and Social Sciences **E Engineering and Materials Sciences**	Martinus Nijhoff Publishers The Hague, Boston, Dordrecht, and Lancaster
F Computer and Systems Sciences **G Ecological Sciences** **H Cell Biology**	Springer-Verlag Berlin, Heidelberg, New York, London, Paris, and Tokyo

Recent Volumes in this Series

Volume 139—New Vistas in Nuclear Dynamics
edited by P. J. Brussaard and J. H. Koch

Volume 140—Lattice Gauge Theory: A Challenge in Large-Scale Computing
edited by B. Bunk, K. H. Mütter, and K. Schilling

Volume 141—Fundamental Problems of Gauge Field Theory
edited by G. Velo and A. S. Wightman

Volume 142—New Vistas in Electro-Nuclear Physics
edited by E. L. Tomusiak, H. S. Caplan, and E. T. Dressler

Volume 143—Atoms in Unusual Situations
edited by Jean Pierre Briand

Volume 144—Fundamental Aspects of Quantum Theory
edited by Vittorio Gorini and Alberto Frigerio

Volume 145—Atomic Processes in Electron-Ion and Ion-Ion Collisions
edited by F. Brouillard

Volume 146—Geophysics of Sea Ice
edited by Norbert Untersteiner

Series B: Physics

ME: NATO Advanced Study Institute on the Physics of Electron-ion and Ion-ion Collisions (1985: Han-sur-Lesse, Belgium)

Atomic Processes in Electron–Ion and Ion–Ion Collisions

Edited by

F. Brouillard

Catholic University of Louvain
Louvain-la-Neuve, Belgium

Plenum Press
New York and London
Published in cooperation with NATO Scientific Affairs Division

Proceedings of a NATO Advanced Study Institute on
the Physics of Electron-Ion and Ion-Ion Collisions,
held September 30-October 12, 1985,
in Han'Lesse, Belgium

Library of Congress Cataloging in Publication Data

NATO Advanced Study Institute on the Physics of Electron-ion and Ion-ion Collisions (1985: Han-sur-Lesse, Belgium)
Atomic processes in electron-ion and ion-ion collisions.

(NATO ASI series. Series B, Physics; vol. 145)
"Published in cooperation with NATO Scientific Affairs Division."
"Proceedings of a NATO Advanced Study Institute on the Physics of Electron-Ion and Ion-Ion Collisions held September 30-October 12, 1985, in Han'Lesse, Belgium"—T.p. verso.
Includes bibliographical references and index.
1. Electron-ion collisions—Congresses. 2. Ion-ion collisions—Congresses. I. Brouillard, F., 1937- . II. North Atlantic Treaty Organization. Scientific Affairs Division. III. Title. IV. Series: NATO ASI series. Series B, Physics; v. 145.
QC794.6.C6N44 1985 539.7 86-22699
ISBn 0-306-42413-4

A Division of Plenum Publishing Corporation
233 Spring Street, New York, N.Y. 10013

Printed in the United States of America

PREFACE

Four years after a first meeting in BADDECK, Canada, on the Physics of Ion-Ion and Electron-Ion collisions, a second Nato Advanced Study Institute, in HAN/Lesse, Belgium, reexamined the subject which had become almost a new one, in consideration of the many important developments that had occured in the mean time.

The developments have been particularly impressive in two areas : the di-electronic recombination of electrons with ions and the collisional processes of multiply charged ions.

For dielectronic recombination, a major event was the obtainment, in 1983, of the first experimental data. This provided, at last, a non speculative basis for the study of that intricate and subtle process and strongly stimulated the theoretical activities.

Multiply charged ions, on the other hand, have become popular, thanks to the development of powerful ion sources. This circumstance, together with a pressing demand from thermonuclear research for ionisation and charge exchange cross sections, has triggered systematic experimental investigations and new theoretical studies, which have contributed to considerably enlarge, over the last five years, our understanding of the collisional processes of multiply charged ions.

Dielectronic recombination and multiply charged ions were therefore central points in the programme of the A.S.I. in HAN/Lesse and are given a corresponding emphasis in the present book.

Dielectronic recombination is treated extensively in the lectures by Gordon DUNN (experimental) and Yukap HAHN (theoretical) but is also present in the general theory developped by David MOORES and Michaël PINDZOLA. In addition, dissociative recombination, a companion process of di-electronic recombination of fundamental importance to molecular dynamics is discussed in the lectures of Brian MITCHELL and Annick GIUSTI.

Multiply charged ions are dealt with both in the lectures on electron-ion and ion-ion collisions.
Experiments on electron impact ionisation and excitation are reviewed by Ronald PHANEUF while the related techniques are discussed by Pierre DEFRANCE.

The theory of charge exchange and ionisation in heavy particle collisions is presented in the lectures by Michel BARAT and Ratko JANEV. Experiments and techniques are reported by Ken DOLDER and Ken DUNN. Transfer ionisation phenomena receive the special attention they deserve in specific lectures by Erhard SALZBORN.

The physics of electron-ion and ion-ion collisions is strongly linked to plasmas. It was therefore natural to include in HAN/Lesse some related contributions. Specific collisional problems relevant to astrophysical, Tokamak- and dense, laser produced plasmas have been discussed in three lectures by Jacques DUBAU, Mike HARRISON and Pierre JAEGLE, respectively. Unfortunately, the lecture of Pierre Jaegle has not been available for publishing here.

Finally, it was felt useful to also include in this book two of the excellent seminars delevered in HAN/Lesse, one by Bernard PIRAUX on electron impact ionisation of simple atoms and an other one by Anders BARANY presenting the main features of the ion storage ring to be built in Stockholm and principally dedicated to atomic and molecular physics.

On the whole, this book intends to give a self-consistent but non-exhaustive account of the current investigations in the field of electronic and ionic collisions.
A useful complement to it is the series of lectures published in 1983 (Plenum) under the title : "Physics of Ion-Ion and Electron-Ion Collisions" in the wake of the BADDECK Advanced Study Institute, the first on the subject.
It would be unfair not to mention at this place the name of J.W. Mc GOWAN, who got that first Institute to start and injected in it a lot of his enthusiastic energy. The success of that first Institute was an encouragment to organise a second one and this was possible four years later, thanks to the dynamic collaboration of Michel BARAT, Gordon DUNN and Pierre DEFRANCE.

Or course, the organisation of such meetings would not be possible without financial support. One can never too much emphasise the impressive service that NATO does Scientific research by intensively funding it through its Institutes and Workshops.
In particular, I like to thank Dr Craig SINCLAIR, presently in charge of the NATO Scientific Affairs for the simplicity and efficiency of his managment.

The Institute is also much indebted to the "Fonds National de la Recherche Scientifique" and his general secretary Paul LEVAUX for his substantial support.

The Belgian ministery of Education also deserves our gratitude for making the "Domaine des Masures", in HAN/Lesse, available to the Institute The personal involvement of Mr. HERMAN and Mr. BREUSKIN must be here specially acknowledged.

Finally, it is just fair to point out the discrete but decisive contribution of Nicole COISMAN, who kindly took care of all the administration work.

F. Brouillard

Louvain-la-Neuve, May 1986

CONTENTS

LECTURES ON COLLISIONAL PROCESSES IN PLASMAS

SEMINARS

THEORY OF ELECTRON-ION COLLISIONS

D.L. Moores

Departement of Physics and Astronomy
University College London, Gower street
GB London WC1E6BT Great Britain

BASIC THEORY

We consider scattering of electrons by an ion of nuclear charge Z and with N electrons. The Hamiltonian for the N+1 electron system is

$$H_{N+1} = \sum_{i=1}^{N+1} \left(-\frac{1}{2}\nabla_i^2 - \frac{Z}{r_i}\right) + \sum_{i>j=1}^{N+1} \frac{1}{|\underline{r}_i - \underline{r}_j|} \tag{1}$$

and the Schrödinger equation for the complete system is

$$[H_{N+1} - E]\,\Psi = 0 \tag{2}$$

We introduce the ion eigenstates Φ_i which satisfy

$$[H_N - E_i]\,\Phi_i = 0 \tag{3}$$

and let

$$E = E_i + \frac{1}{2}k_i^2. \tag{4}$$

We look for solutions of (2) with asymptotic form

$$\begin{aligned}\Psi_\gamma \underset{r_{N+1}\to\infty}{\sim} & \ \Phi_\gamma \chi_{m\gamma}(\sigma_{N+1})\, e^{ik_\gamma z_{N+1}} \\ & + \sum_{\gamma'} \Phi_{\gamma'}\chi_{m\gamma'}(\sigma_{N+1}) f_{\gamma\gamma'}(\hat{r}_{N+1}) \frac{e^{ik_{\gamma'} r_{N+1}}}{r_{N+1}} \\ & + \sum_{\gamma''} \int \Phi_{\chi\gamma''}\chi_{m\gamma''}(\sigma_{N+1}) \frac{f_{\underline{\chi}\gamma}(\hat{r}_{N+1})}{r_{N+1}}\, e^{i(k_{\gamma''} r_{N+1} + \eta(\underline{\chi}, \underline{r}_{N+1}))}\, d\underline{\chi}\end{aligned} \tag{5}$$

In equation (5) $\chi_m(\sigma)$ are spin eigenfunctions; the first summation on the right hand side is over all energetically accessible discrete target states γ and $f_{\gamma\gamma'}(\hat{r}_{N+1})$ is the excitation amplitude for the transition $\gamma \rightarrow \gamma'$. If the total energy is such that the ion can be ionised, one must add the second sum and integral over continuum states $\Phi_{\underline{\chi}\gamma''}$ where $\underline{\chi}$ is the wave vector of the ejected electron and $f_{\underline{\chi}\gamma}(\hat{r}_{N+1})$ is

the direct ionisation amplitude, $\eta(\underline{\chi},\underline{r})$ being a phase factor allowing for dynamic screening of the continuum electrons. In the case of ionisation, it is necessary to consider in addition the asymptotic forms of Ψ_γ when $r_i \to \infty$, $i < N+1$.
More will be said about ionisation later, but for the present the discussion will be confined to excitation and we shall ignore the third term on the right hand side of (5).
The differential cross section for the transition $\gamma \to \gamma'$ is given by

$$\frac{d\sigma_{\gamma'\gamma}}{d\Omega_{N+1}} = \frac{k_{\gamma'}}{k_\gamma} \quad |f_{\gamma'\gamma}(\hat{r}_{N+1})|^2 \tag{6}$$

EXPANSION OF THE TOTAL WAVE FUNCTION

In order to express $f_{\gamma'\gamma}(\hat{r}_{N+1})$ in terms of quantities that are convenient to calculate, we write the solution of (2) in the form of an eigenfunction expansion, in a coupled angular momentum representation in LS coupling. Let the space plus spin coordinates of the pth electron be $\underline{x}_p$, and the target quantum numbers be $\alpha_i S_i L_i M_{S_i} M_{L_i}$ where α_i stands for the atomic configuration and let us work in a representation $\alpha_i S_i L_i M_{S_i} M_{L_i} k_{\alpha_i S_i L_i}$ 1/2 $l_i \Gamma$, where $\Gamma = SLM_S M_L \pi$, π being the parity and the quantum numbers S L M_S and M_L referring to the complete system. We form vector coupled functions of all coordinates except r_{N+1}, the radial coordinate of the $N+1^{th}$ electron, of the form

$$\psi_i^\Gamma = \sum_{\substack{M_{S_i} M_{L_i} \\ m_{s_i} m_{l_i}}} C^{S_i\ 1/2\ S}_{M_{S_i} m_{s_i} M_S} C^{L_i l_i L}_{M_{L_i} m_{l_i} M_L} \Phi_{\alpha_i S_i L_i M_{S_i} M_{L_i}}(\underline{x}_1 \ldots \underline{x}_N) \chi_{m_{s_i}}(\sigma_{N+1}) Y_{l_i m_{l_i}}(\hat{r}_{N+1}) \tag{7}$$

The eigenfunction expansion then takes the form

$$\Psi_j^\Gamma(\underline{x}_1 \ldots \underline{x}_{N+1}) = \sum_i \frac{\psi_i^\Gamma F_{ij}(r_{N+1})}{r_{N+1}} + \sum \chi_k^\Gamma(\underline{x}_1 \ldots \underline{x}_{N+1}) a_{kj}^\Gamma \tag{8}$$

or

$$\Psi_j^\Gamma = \sum_i \Theta_i + \sum_k \chi_k a_{kj}. \tag{9}$$

The set of quantum numbers $\alpha_i S_i L_i k_{\alpha S_i L_i} l_i$ constitutes a free channel; if $k^2_{\alpha_i S_i L_i} > 0$ the state $\alpha_i S_i L_i$ may be excited and we say the channel is open; if $k^2_{\alpha_i S_i L_i} < 0$ the channel is closed. The functions χ_k^Γ are single configuration antisymmetrised N+1 - electron bound type wave functions made up from the target orbitals and are known as bound channel or correlation functions. They go to zero exponentially as $r_i \to \infty$. Their inclusion not only allows for extra electron short-range correlation, but also is required in order to satisfy certain orthogonality conditions needed to ensure uniqueness requirements.
In equation (8) the subscript j refers to a particular solution of the Schrödinger equation, that is, corresponding to particular boundary conditions. One possible such set is

$$F_{ij}(o) = 0$$

$$F_{ij}(r) \underset{r\to\infty}{\sim} k_i^{-1/2}[\sin\theta_i \delta_{ij} + \cos\theta_i K_{ij}], \quad k_i^2 > 0$$

$$F_{ij}(r) \underset{r\to\infty}{\sim} 0, \quad k_i^2 < 0 \tag{10}$$

where

$$\theta_i = k_i r - \frac{1}{2}\ell_i\pi + \frac{Z-N}{k_i}\,\ell n 2k_i r + \arg\Gamma(\ell_i+1+i\,\frac{Z-N}{k_i}).$$

The quantities K_{ij} form the elements of the reactance matrix, $\underline{K}$. Alternatively, we could impose complex boundary conditions, replacing F_{ij} in (8) by G_{ij} where

$$\begin{aligned} G_{ij}(0) &= 0 \\ G_{ij}(r) &\underset{r\to\infty}{\sim} k_i^{-1/2}[e^{-i\theta_i}\delta_{ij} - e^{i\theta_i}S_{ij}]\ , \quad k_i^2 > 0 \end{aligned} \tag{11}$$

S_{ij} form the elements of the S matrix. It is easy to see that the K matrix and the S matrix functions are related by the transformation

$$\underline{G} = -2i\ \underline{F}\ (1-i\underline{K})^{-1} \tag{12}$$

and hence

$$\underline{S} = (1-i\underline{K})^{-1}(1+i\underline{K}). \tag{13}$$

Expanding the plane wave term in (5) in partial waves and then equating the incoming and outgoing wave terms in (5) and (8), using (11), one obtains f_{ij} in terms of the S matrix elements :

$$f_{ij}(\hat{r}) = \frac{i\pi}{(k_i k_j)^{1/2}} \sum_{\substack{L\,S\,\pi \\ \ell_i \ell_j m_{\ell_j}}} i^{\ell_j-\ell_i}(2\ell_j+1)^{1/2}[\delta_{ij}-S^{\Gamma}_{ij}] \tag{14}$$

$$Y_{\ell_i m_{\ell_i}}(\hat{r})\ C^{L_i \ell_i L}_{M_{L_i} 0 M_L}\ C^{L_i \ell_j L}_{M_{L_j} m_{\ell_j} M_L}\ C^{S_i \frac{1}{2} S}_{M_{S_i} m_{S_i} M_S}\ C^{S_j \frac{1}{2} S}_{M_{S_j} m_{S_j} M_S}$$

The differential cross section is obtained on substituting (14) into (6) and the total cross section obtained by integrating over all scattered electron directions and averaging over initial and summing over final spins, giving

$$\sigma(\alpha_i S_i L_i \to \alpha_j S_j L_j) = \frac{\pi}{k_i^2} \sum_{\substack{\ell_i \ell_j \\ LS\pi}} \frac{(2L+1)(2S+1)}{2(2L_i+1)(2S_i+1)} |\delta_{ji}-S_{ji}|^2 \tag{15}$$

VARIATIONAL PRINCIPLE AND THE COUPLED EQUATIONS

We consider the functional

$$I_{pq} = \langle \Psi_p | H-E | \Psi_q \rangle \tag{16}$$

where Ψ_p and Ψ_q have asymptotic form defined by (8) and (10). We consider variations $\delta\Psi$ about the exact functions Ψ due to variations δF in F such that

$$\delta F_{ij} \underset{r\to\infty}{\sim} k_i^{-1/2}\cos\theta_i\ \delta K_{ij} \tag{17}$$

we have

$$\delta I_{pq} = < \delta\Psi_p|H-E|\Psi_q > + < \Psi_p|H-E|\delta\Psi_q > + <\delta\Psi_p|H-E|\delta\Psi_q > \quad (18)$$

and also

$$< \Psi_p|H-E|\delta\Psi_q > - < \delta\Psi_q|H-E|\Psi_p >^{\star} = \delta K_{pq} \quad (19)$$

hence

$$\delta(I_{pq}-K_{pq}) = <\delta\Psi_p|H-E|\psi_q> + <\delta\Psi_q|H-E|\Psi_p>^{\star} + <\delta\Psi_q|H-E|\delta\Psi_p> \quad (20)$$

and hence for small variations about the exact functions the following condition must hold :

$$\delta\ [<\underline{\Psi}|H-E|\underline{\Psi}> - \underline{K}] = 0 \quad (21)$$

where $\underline{\Psi}$ is written for a vector whose components are Ψ_i. We impose the condition (21) for small variations about some approximate functions. This is equivalent to, from (20),

$$< \delta\Psi|H-E|\Psi > = 0 \quad (22)$$

if $\quad \Delta\Psi = \Psi_{approx} - \Psi_{exact}$

then we get

$$\underline{K}^{exact} = \underline{K}^{calc} - < \Psi|H-E|\Psi >^{calc} + < \Delta\Psi|H-E|\Delta\Psi > \quad (23)$$

if $< \Psi|H-E|\Psi >^{calc} = 0$ then the error in $\underline{K}$ is quadratic in the error in the wave function.
It may be shown that in order to define the functions F_{ij} uniquely it is necessary to impose the orthogonality constraints

$$< F_{ij}|P_\alpha > = 0 \qquad \text{if} \quad \ell_i = \ell_\alpha \quad (24)$$

where P_α are the target radial wave functions and that this may be effected without imposing any constraint on the total wave function provided a suitable set of correlation functions is included. To take account of (24) we replace (22) by

$$< \delta\Psi|H-E|\Psi > + \sum_{i,\alpha} \lambda_{i\alpha} < \delta F_i|P_\alpha > \quad (25)$$

where $\lambda_{i\alpha}$ are Lagrange multipliers. This condition, together with the expansion (8) reduces after some lengthy algebra to the basic equations of the theory, the set of coupled integro-differential equations[1] for the unknown functions F_{ij} and the constants a_{kj} and $\lambda_{i\alpha}$:

$$\left[\frac{d^2}{dr^2} - \frac{\ell_i(\ell_i+1)}{r^2} + \frac{2Z}{r} + k_i^2\right] F_{ij}(r) + \sum_p [V_{ip}(r)-W_{ip}(r)]\, F_{pj}(r)$$

$$+ \sum_k U_{ik}\, a_{kj} + \sum_\alpha \lambda_{i\alpha}\, P_\alpha(r) = 0 \quad (26)$$

$$\underline{U} = r < \underline{\Psi}|H|\underline{\chi} > \quad (27)$$

$$< \underline{U}|\underline{F} > + (H-E)\underline{a} = 0 \quad (28)$$

$$< P_\alpha|F_i > = 0 \quad (29)$$

The direct potentials $V_{ip}(r)$ for $i \neq p$ fall off at large r faster than r^{-1}; the exchange terms $W_{ip}(r)F_{pj}(r)$ fall off exponentially. These equations must be solved with boundary conditions (10) to give the K matrix. It is clearly necessary to truncate the expansion in some way in order to make the calculation numericallv tractible, and in practice the expansion is restricted to just a few atomic states. Any integral over target continuum states is awkward to deal with and is usually omitted. The truncation of the expansion may be compensated for to some extent by the introduction of pseudo-states into the expansion, which if chosen judiciously may be used to mock the effects of a larger number of coupled channels. As well as for scattering problems, these equations may be used for bound states of the N+1 electron system by solving at energies for which all channels are closed.

SOLUTION OF THE EQUATIONS

Three principal methods for solving the coupled integro-differential equations are in current use.
In the R matrix method[2], the space surrounding the ion is separated into two regions, an inner one, $r < a$ and an outer one, $r > a$. In the inner region, the wave function is expanded in a set of square-integrable functions Ψ_k of the form (9) chosen to diagonalise the Hamiltonian for $r < a$;

$$< \Psi_k | H | \Psi_{k'} > = E_k \, \delta_{kk'} \qquad (30)$$

The functions Ψ_k are assumed to form a basis for expansion of the wave function at any energy E;

$$\Psi_E = \sum_k A_{Ek} \, \Psi_k . \qquad (31)$$

The logarithmic derivative of the channel functions at the boundary r=a may be expressed in terms of the R matrix

$$R_{ij} = \frac{1}{a} \sum_k \frac{G_{ik}(a) G_{jk}(a)}{E_k - E} \qquad (32)$$

Where $G_{ij}(r)$ are the radial parts of Ψ_k . The K matrix is obtained by matching the logarithmic derivative of the inner solution with that of the solution in the asymptotic region $r > a$. Although the asymptotic solution must be obtained at each energy, the R matrix is obtained at any energy from a single diagonalisation. A correction (the Buttle correction) may be made for the fact that a finite basis set is used in the inner region. Standard methods are available for obtaining the solutions in the asymptotic region.
In the linear algebraic method[3], the coupled integro-differential equations are reduced to a set of linear lagebraic equations by means of appropriate finite difference formulae. The integration region is divided up into intervals and the values of the radial functions at each mesch point are expressed as linear combinations of each other. The resulting set of linear equations in which the unknowns are the function values at the mesh points and the K matrix elements are solved by sophisticated matrix inversion techniques.
In the third technique, developed by Henry and collaborators[4], the equations are converted into integral equations, which are solved by a non-iterative procedure.

RESONANCES

A distinctive feature of the close coupling method is the automatic inclusion of resonance structure in the K matrix and thus in the cross sections if either closed or bound channels are included. To investigate formally the resonant structure of equations (26) let us for simplicity omit the bound channels. Following Feshbach[5], we introduce projection operators P, Q and R. The operator R projects on to the subspace spanned by the terms retained in the eigenfunction expansion; P projects out the open channel subspace of R and Q the closed channel subspace. Thus

$$\begin{aligned} P + Q &= R \\ P^2 &= P \\ Q^2 &= Q \end{aligned} \tag{33}$$

We then have

$$P(H_{N+1}-E)R\Psi^R = 0 \tag{34}$$

$$Q(H_{N+1}-E)R\Psi^R = 0 \tag{35}$$

Hence

$$P[H_{N+1}+H_{N+1}Q\frac{1}{Q(E-H_{N+1})Q}QN_{N+1} - E]\,P\Psi^R = 0. \tag{36}$$

The second term here corresponds to virtual transitions via the closed channels and is a non local optical potential. If we introduce the eigenfunctions $\xi_n(r)$ of $QH_{N+1}Q$ satisfying

$$(QH_{N+1}Q)\ \xi_n(r) = \varepsilon_n\xi_n(r) \tag{37}$$

this set of equations corresponds to the closed channel subset, neglecting coupling to the open channels. If E is in the neighbourhood of one of the eigenstates ε_n we can write (36) in the form

$$[\mathcal{H}_{N+1}-E]P\Psi^R = -\frac{PH_{N+1}Q\xi_n\rangle\langle\xi_n QH_{N+1}P\Psi^R}{E-\varepsilon_n} \tag{38}$$

where $\mathcal{H}_{N+1}$ is the sum of the slowly-varying terms in (36). The formal solution to this equation is

$$P\Psi_o^R\rangle = |P\Psi^R\rangle + \frac{\mathcal{G}PH_{N+1}Q\xi_n\rangle\langle\xi_n QH_{N+1}P\Psi_o^R\rangle}{E-\varepsilon_n-\Delta_n} \tag{39}$$

$$\Delta_n = \langle\ \xi_n QH_{N+1}P\,\mathcal{G}PH_{N+1}Q\xi_n\ \rangle \tag{40}$$

This shows how the discrete spectrum of $QH_{N+1}Q$ introduces resonant behaviour into the open channel functions at energies in the vicinity of the eigenvalues ε_n. The level shift Δ_n arises from coupling to the continuum. Each resonance is associated with the formation and decay of a quasi-bound state of the N+1 electron system. In the case of electron-ion scattering the existence of the long range Coulomb potential means that infinite Rydberg series of resonance occur. Similar arguments can be produced showing that bound channels likewise introduce resonant behaviour.

MANY-CHANNEL QUANTUM DEFECT THEORY

It is plain that, in a problem in which a number of channels has been included, some open, some closed, the resonance structure in the cross sections can become very complex indeed, with for example overlapping Rydberg series of resonances converging on to different thresholds. In order to obtain usable results, the resonance structure must be averaged over in some way. One useful tool for doing this is provided by many channel quantum defect theory (MQDT)[6]. This theory starts from the coupled equations (26) and first makes two major approximations : the integral operators (exchange terms) W_{ij} are dropped and the direct potentials V_{ij} are assumed to be short range,

$$V_{ij}(r) = 0, \quad r > r_0. \tag{41}$$

Starting from these two assumptions a rigorous theory may be developed relating the asymptotic form of the solutions of the equations in one energy region with those in another. Thus MQDT provides a link between bound or quasi-bound states and the continuum. Suppose we consider a problem in which I channels are included; when they are all open, solution of (26) with (10) yields a reactance matrix of dimension I, $\mathcal{K}$. Suppose the energy is now such that these are I_0 open channels ($I_0<I$). For closed channels we put

$$k_i^2 = -(Z-N)^2/\nu_i^2 \quad , \qquad i > I_0 \tag{42}$$

Solution of the coupled equations yields a reactance matrix of dimension I_0., $\underline{K}$. If is the analytic continuation of the open channel K-matrix to this energy, we partition it in the form

$$\mathcal{K} = \begin{pmatrix} \mathcal{K}_{oo} & \mathcal{K}_{oc} \\ \mathcal{K}_{co} & \mathcal{K}_{cc} \end{pmatrix} \tag{43}$$

where o and c refer to open and closed submatrices respectively. From MQDT one derives the result that

$$\underline{K} = \mathcal{K}_{oo} - \mathcal{K}_{oc}(\mathcal{K}_{cc} + \tan\pi\nu_c)^{-1}\mathcal{K}_{co} \tag{44}$$

and

$$\underline{S} = \mathcal{S}_{oo} - \mathcal{S}_{oc}(\mathcal{S}_{cc} - e^{-2\pi i\nu_c})^{-1}\mathcal{S}_{co} \tag{45}$$

These formulae clearly show up the resonance structure in $\underline{K}$ or $\underline{S}$; $\underline{S}$ has apole at each zero of $\mathcal{S}_{cc} - e^{-2\pi i\nu_c}$.
($\underline{S}$ and $\mathcal{S}$ are the corresponding scattering matrices related to $\underline{K}$ and $\mathcal{K}$ respectively by expression (13)).
Since $\mathcal{S}_{cc}$ is slowly varying, an extra resonance is introduced each time ν_c increases by about unity.
MQDT has many applications, of which two will be mentioned here. First, by calculating the slowly varying matrix $\mathcal{K}$ above all thresholds and extrapolating to lower energy, $\underline{K}$ may be determined from (44) at any energy. Alternatively, $\underline{K}$ may be calculated at a few points and fitted to the form (44).
MQDT may also be used for averaging over resonance structure just below a threshold. We consider the case when all the bound channels have the same energy. Just below threshold there will be an infinite series of resonances. We require the average over a resonance

$$< |S_{ij}|^2 > = \int_{\nu_c=p}^{p+1} |S_{ij}|^2 d\nu_c \tag{46}$$

with S given by (45).

By transforming to a representation in which $\mathcal{S}_{cc}$ is diagonal, Gailitis showed that if i,j are open and a,b closed,

$$< |S_{ij}|^2 > = |\mathcal{S}_{ij}|^2 + \sum_{a,b} \frac{\mathcal{S}_{ia}\mathcal{S}_{aj}\mathcal{S}^{\star}_{ib}\mathcal{S}^{\star}_{bj}}{I - \mathcal{S}_{aa}\mathcal{S}_{bb}^{\star}} \tag{47}$$

For one closed channel c this reduces to

$$< |S_{ij}|^2 > = |\mathcal{S}_{ij}|^2 + \frac{|\mathcal{S}_{ic}|^2 |\mathcal{S}_{cj}|^2}{\Sigma\ |\mathcal{S}_{ck}|^2} \tag{48}$$

The physical interpretation of (48) is that the second term describes formation and decay of the resonance in the scattering. $|\mathcal{S}_{ic}|^2$ is the probability of forming the resonance state from i while

$$|\mathcal{S}_{cj}|^2 \ / \ \sum_k |\mathcal{S}_{ck}|^2$$

is the probability that the resonance breaks up into j.

The smoothly-varying averaged cross section obtained from (47) or (48) (the Gailitis[7] average) is not continuous with the non-resonant cross section above threshold. There is a sudden discontinuous drop due to loss of flux into new channels. Pradhan and Seaton[8] have investigated the effects of radiative decay of the resonances on the excitation cross section. As the principal quantum number increases, radiative decay starts to compete with autoionisation, and thus reduces the resonant contribution. Eventually as the threshold is approached the resonant contribution vanishes and the cross section is continuous at threshold. In this case (47) becomes

$$< |S_{ij}|^2 > \quad |\mathcal{S}_{ij}|^2 + \sum_{a,b} \frac{\mathcal{S}_{ia}\mathcal{S}^{\star}_{ib}\mathcal{S}^{\star}_{aj}\mathcal{S}^{\star}_{jb}}{e^{2\pi\nu^3 r/z^2} - \mathcal{S}_{aa}\mathcal{S}^{\star}_{bb}} \tag{49}$$

r being the radiative transition probability.

BOUND STATES

MQDT may also be used to study bound states. When all channels are closed, the condition for an eigenstate is

$$|\ \underline{K} + \tan \pi \nu\ | = 0 \tag{50}$$

The energies at which the zeros of this determinant occur give the bound states of the system of target ion plus electron.(50) may be used to check the accuracy of a scattering calculation by comparing the zeros of (50) with energies obtained from spectroscopic methods.

FINE STRUCTURE TRANSITIONS AND RELATIVISTIC EFFECTS

If relativistic effects are small, they may be neglected during the collision and cross sections between fine structure states may be obtained by a unitary transformation of the K matrix from one angular momentum coupling scheme to another[9] :

$$K^{J\pi}_{\alpha\beta} = \sum_{LSij} A^{SLJ}_{\alpha i} \, K^{SL\pi}_{ij} \, A^{SLJ}_{j\beta} \tag{51}$$

If relativistic effects are not small it may be necessary to do a relativistic scattering calculation, with a Breit-Pauli or a Dirac Hamiltonian[10,11].

THE DISTORTED WAVE AND COULOMB-BORN METHODS

The distorted wave method is a simpler approximation to the close coupling method described above. In it, approximate radial wave functions are calculated from central potentials, ignoring exchange terms and channel coupling. In this section we describe the form of the method as used by Saraph, Seaton and Shemming[12], although many variations exist. A suitable central potential $V_i(r)$ with asymptotic form $2(Z-N)/r$ is constructed. Radial wave functions F_i, orthogonal to the target orbitals, of the form

$$F_i = f_i - \sum_\alpha < f_i | P_\alpha > P_\alpha \tag{52}$$

where

$$\left(\frac{d^2}{dr^2} - \frac{\ell_i(\ell_i+1)}{r^2} + V_i(r) + k_i^2\right) f_i(r) = 0 \tag{53}$$

$$f_i(r) \underset{r\to\infty}{\sim} k_i^{-1/2} \sin\left(k_i r - \frac{Z-N}{k_i} \ln r + \tau_i\right) \tag{54}$$

are calculated and the total wave functions of the form

$$\Psi_i = \theta_i + \sum_j \phi_j C_{ji} \tag{55}$$

where θ_i are constructed using F_i. With these choices, the DWK matrix is of the form

$$\underline{K}^{DW} = -<\underline{\theta}|H-E|\underline{\theta}>+<\underline{\theta}|H-E|\underline{\phi}><\underline{\phi}|H-E|\underline{\phi}>^{-1}<\underline{\phi}|H-E|\theta> \tag{56}$$

The method can give reasonable accuracy provided $K^{DW} << 1$. A special case of the DW approximation is the Coulomb-Born method in which

$$V_i(r) = 2(Z-N)/r \qquad \text{all } r.$$

The functions $F_i(r)$ are thus Coulomb functions.

IONISATION

Although a precise theory of ionisation would not make any separation between them it is customary to distinguish three different mechanisms leading to ionisation of positive ions by electrons. The first is the direct process

$$A^{+n} + e^- \rightarrow A^{+n+1} + e^- + e^-.$$

Then we have inner shell excitation followed by autoionisation

$$A^{+n} + e^- \rightarrow (A^{+n})^{**} + e^-$$

$$(A^{+n})^{**} \xrightarrow{h\nu} (A^{+n})^{*} + e^-$$

$$(A^{+n})^{**} \xrightarrow{\text{radiationless}} A^{+n+1} + e^- + e^-$$

an thirdly, what has come to be known as resonant capture double auto-ionisation

$$A^{+n} + e^- \rightarrow (A^{+(n-1)})^{**} \rightarrow (A^{+n})^* + e^-$$

$$\rightarrow A^{+n+1} + e^- + e^- .$$

In fact, the third process is just a special case of the second, in which the intermediate stage $A^{+(n-1)**}$ is a resonant state in the scattering of an electron by A^{+n**}.
We shall first discuss the direct process.

The formulation of the basic theory of electron impact ionisation in a fashion suitable for the calculation of cross sections remains one of the outstanding problems of atomic physics. The fundamental difficulty lies in the correct description of the double continuum state, in which two electrons interact with each other and with the nucleus out to infinity via the long range Coulomb interaction. For calculation of cross sections we really still only have the 20 year old theory of Rudge and Seaton[13] and Peterkop[14]. For simplicity we consider the case of a hydrogenic ion with charge z. The Hamiltonian is

$$H = -\frac{1}{2}\nabla_1^2 - \frac{1}{2}\nabla_2^2 - \frac{z}{r_1} - \frac{z}{r_2} + \frac{1}{r_{12}} \qquad (54)$$

and the Schrödinger equation

$$[H-E]\,\Psi(\underline{r}_1,\underline{r}_2) = 0 \qquad (55)$$

Let the ionisation energy be I the wave vector of the incident electron be $\underline{k}_0$ and the wave vectors of the two continuum electrons after ionisation be $\underline{k}_2$ and $\underline{k}_1$. By energy conservation

$$\frac{1}{2}k_0^2 - I = \frac{1}{2}(k_1^2 + k_2^2) = E \geqslant 0 \qquad (56)$$

The asymptotic forms of the wave function are

$$\Psi(\underline{r}_1,\underline{r}_2) \underset{r_1\to\infty}{\sim} \phi_0(\underline{r}_2)\, e^{i\underline{k}_0\cdot\underline{r}_1} + \frac{1}{r_1}\sum_a \phi_a(\underline{r}_2) f_a(\hat{r}_1) e^{ik_a r_1}$$
$$+ \frac{1}{r_1}\int \phi(z,\underline{k}_2,\underline{r}_2) f(\underline{k}_1,\underline{k}_2) e^{i(k_1 r_1+\eta_1)}\, d\underline{\psi} \qquad (57)$$

$$\Psi(r_1,r_2) \underset{r_2\to\infty}{\sim} \frac{1}{r_2}\sum_a \phi_a(\underline{r}_1) g_a(\hat{r}_2) e^{ik_a r_2} + \frac{1}{r_2}\int \phi(z,\underline{k}_2,\underline{r}_1)$$
$$g(\underline{k}_1,\underline{k}_2) e^{i(k_1 r_2+\eta_2)}\, d\underline{\psi} \qquad (59)$$

where

$$(\frac{1}{2}\nabla^2 + \frac{z}{r} - \varepsilon_a)\,\phi_a(\underline{r}) = 0 \qquad (59)$$

$$(\frac{1}{2}\nabla^2 + \frac{z}{r} + \frac{1}{2}k_2^2)\phi(z,\underline{k}_2,\underline{r}) = 0 \qquad (60)$$

Rudge and Seaton, and Peterkop have derived an integral expression for the ionisation amplitude $f(\underline{k}_1,\underline{k}_2)$ in the form

$$f(\underline{k}_1,\underline{k}_2) = (2\pi)^{-5/2} e^{i\Delta(\underline{k}_1,\underline{k}_2)} \int \Psi(\underline{r}_1,\underline{r}_2)[H-E]\Phi^{\star}(\underline{r}_1,\underline{r}_2)\, d\tau \qquad (61)$$

where $\Psi(\underline{r}_1,\underline{r}_2)$ is an exact solution of the Schrödinger equation with asymptotic form (57/58) which may be expressed as an outgoing six dimensional Coulomb - distroted spherical wave

$$\Psi(\underline{r}_1,\underline{r}_2) \underset{r_1,r_2\to\infty}{\sim} \left(\frac{K}{r^5}\right)^{1/2} f(K\sin\alpha\ \hat{r}_1,\ K\cos\alpha\ \hat{r}_2)\, e^{i(Kr + \frac{Z(\hat{\Omega})}{K} \ln 2Kr)} \qquad (62)$$

where we have introduced hyperspherical coordinates

$$r = \sqrt{r_1^2 + r_2^2} \qquad a = \arc\tan r_2/r_1 \qquad K = \sqrt{k_1^2 + k_2^2} \qquad (63)$$

Let $\hat{\Omega}_r$ stand for the set of angles $\alpha\ ,\theta_1\phi_1\theta_2\phi_2$ in r-space and $\hat{\Omega}_k$ for the equivalent in momentum space. The potential

$$V = -\frac{z}{r_1} - \frac{z}{r_2} + \frac{1}{r_{12}} \qquad (64)$$

may be written

$$V = -\frac{Z(\hat{\Omega})}{r} = -\frac{1}{r}\left(\frac{1}{\cos\alpha} + \frac{1}{\sin\alpha} - \frac{1}{(1-\sin 2\alpha\ \hat{r}_1\cdot\hat{r}_2)^{1/2}}\right) \qquad (65)$$

The function $\Phi(\underline{r}_1,\underline{r}_2)$ in (61) is the final state wave function and has asymptotic form

$$\Phi \sim \delta(\hat{\Omega}_k-\hat{\Omega}_r)\left(\frac{2\pi}{iKr}\right)^{5/2} e^{i(Kr + \bar{\omega} \ln Kr + \beta)} + \text{ingoing wave} \qquad (66)$$

(The presence of the ingoing wave ensures non-singularity at r=0 but it does not contribute to the integral). In order that the integral (61) (after having been reduced to a surface integral) converge to a finite limit, it is necessary that

$$\bar{\omega} = Z(\hat{\Omega}_k)/K \qquad (67)$$

where

$$Z(\hat{\Omega}_k) = K\left(\frac{z}{k_1} + \frac{z}{k_2} - \frac{1}{|\underline{k}_1-\underline{k}_2|}\right) \qquad (68)$$

Various choices are possible, consistent with (66) for $\Phi(\underline{r}_1,\underline{r}_2)$.

BORN APPROXIMATION

In the Born approximation one takes

$$\Phi = e^{i\underline{k}_1\cdot\underline{r}_1} \phi(z,\underline{k}_2,\underline{r}_2) \qquad (69)$$

However, (68) is not satisfied : the integral oscillates logarithmically and its phase remains indeterminate.

PRODUCT OF COULOMB FUNCTIONS

If $\Phi(\underline{r}_1,\underline{r}_2)$is taken to have the form of a product of Coulomb functions with effective charges z_1 and z_2,

$$\Phi(\underline{r}_1,\underline{r}_2) = \phi(z_1,\underline{k}_1,\underline{r}_1)\phi(z_2,\underline{k}_2,\underline{r}_2) \tag{70}$$

it may then be shown that in order that the expression (61) contain no divergent phase factor asymptotically, z_1 and z_2 must satisfy the relation

$$\frac{z_1}{k_1} + \frac{z_2}{k_2} = \frac{z}{k_1} + \frac{z}{k_2} - \frac{1}{|k_1-k_2|} = \frac{Z(\hat{\Omega}_k)}{K} \tag{71}$$

where

$$k_1 = k \sin\alpha_k \; , \quad k_2 \cos\alpha_k \tag{72}$$

and the phase factor Δ is given by

$$\Delta(\underline{k}_1,\underline{k}_2) = 2\left(\frac{z_1}{k_1} \ln \frac{k_1}{k} + \frac{z_2}{k_2} \ln \frac{k_2}{k}\right) \tag{73}$$

Equation (71) takes full account of the asymptotic Coulomb potentials. It may be further shown that with appropriate normalisation and phase factors the exchange amplitude $g(\underline{k}_1,\underline{k}_2)$ is gi⁻⁻en by

$$g(\underline{k}_1,\underline{k}_2) = f(\underline{k}_2,\underline{k}_1) \tag{74}$$

and the ionisation cross section is then

$$Q(E) - \frac{1}{4\pi k_0} \sum_{S=0,1} (2S+1) \int_0^{E/2} k_1 k_2 \sigma(S|k_1,k_2) d(\tfrac{1}{2} k_2^2) \tag{75}$$

where

$$\sigma(S|k_1,k_2) = \frac{1}{4\pi} \int d\hat{k}_0 d\hat{k}_1 d\hat{k}_2 |f_S(\underline{k}_1,\underline{k}_2) + (-1)^S g_S(\underline{k}_1,\underline{k}_2)|^2 \tag{76}$$

$\underline{k}_0$ being the incident electron momentum and S the total spin.

$$Q(E) = \frac{1}{4} \sum_{S=0,1} (2S+1)\; Q^S(E). \tag{77}$$

Expanding the integrand in (76) $Q^S(E)$ may be expressed as the sum of three terms, the direct

$$Q_\alpha^S(E) \frac{1}{4\pi^2 k_0} = \int_0^{E/2} d(\tfrac{1}{2}k_2^2) \int k_1 k_2 d\hat{k}_0 d\hat{k}_1 d\hat{k}_2 |f_S(\underline{k}_1,\underline{k}_2)|^2 \tag{78}$$

the exchange

$$Q_{ex}^S(E) = \frac{1}{4\pi^2 k_0} \int_0^{E/2} d(\tfrac{1}{2}k_2^2) \int k_1 k_2 d\hat{k}_0 d\hat{k}_1 d\hat{k}_2 |g_S(\underline{k}_1,\underline{k}_2)|^2 \tag{79}$$

and the interference term

$$Q_{int}^S(E) = \frac{(-1)^S}{2\pi^2 k_0} \int_0^{E/2} d(\tfrac{1}{2}k_2^2) \int k_1 k_2 d\hat{k}_0 d\hat{k}_1 d\hat{k}_2 \{\mathrm{Re}\; f_S^\star(\underline{k}_1,\underline{k}_2) g_S(\underline{k}_1,\underline{k}_2)\} \tag{80}$$

It should be noted that the interference term depends on the phase factor $\Delta(\underline{k}_1,\underline{k}_2)$. The superscript S in equations (78-80) is to indicate that in general the radial wave functions will depend on the total spin S. In practical calculations, two "no-exchange" approximations have been used. In the first, only the interference term is dropped, and then, because of (74), the cross section is given by the direct term alone, but with the upper limit of integration over k_2^2 replaced by E. In the second, $g(\underline{k}_1,\underline{k}_2)$ is set equal to zero. In exchange approximations, the interference term is included.

In applications of this theory the amplitude is determined from (61) with suitable approximate functions Ψ and Φ and the cross section obtained by evaluating (75). If Φ is of the form (70) and z_1 and z_2 satisfy (71) then combining (71) and (73) one may obtain a variety of different but equivalent expressions for the phase factor $\Delta(k_1,k_2)$. If the effective charges are chosen such that (71) is not satisfied however then these expressions are not equivalent, and the phase factor, and hence the interference terms in (76) must be regarded as arbitrary. The uncertainty in what to use for the phase factor is sometimes referred to as the "phase problem"; ionisation calculations in which (71) is not satisfied, such as the usual first or second Born approximations, must be regarded as containing an adjustable parameter. Most calculations that have been carried out to date fall into this category.

Whereas methods based on the above theory, with some arbitrary choice of phase, are capable of giving cross sections which are in reasonable agreement with experiment even at quite low energy, their deficiencies are clearly revealed when comparison is made with the results of experiments designed to give more stringent tests of theory. The experiment of Alguard et al[15] uses beams of spin-polarised electrons and of H atoms and measures the asymmetry A between singlet and triplet cross sections

$$A = \frac{Q^{\circ} - 3Q^{1}}{Q^{\circ} + 3Q^{1}} \tag{81}$$

This quantity depends strongly on the interference terms (80) and is equal to the ratio of interference to total cross sections if $f_0(\underline{k}_1,\underline{k}_2)=f_1(\underline{k}_1,\underline{k}_2)$, that is, if the radial wave functions are independent of the total spin of the system. Out of twelve different calculations with which comparison is made, not one gives results which agree in magnitude or in shape with the measurements.

Calculations in which z_1 and z_2 have been chosen deliberately to satisfy (71) (Rudge and Schwartz[16], Luffman[17]) have also been carried out but since they arbitrarily require that one charge be fixed at a constant value and the other determined from (71) these choices involve unrealistic physical assumptions and, not surprizingly, do not give any better results. The problem with the use of (71) is that although it takes full account of the three-body asymptotic potentials it provides only one relation between two parameters, the angle-dependent charges z_1 and z_2. In the absence of another independent relation between them, one of them must be indeterminate.
It is plain that an attempt ought to be made to develop an alternative formulation, and we now proceed to discuss various possibilities.

USE OF HYPERSPHERICAL COULOMB FUNCTIONS

In this section we shall show that it is possible to retain the integral approach but at the same time eliminate any arbitrary parameters. Suppose that instead of (70) one considers

$$\Phi(\underline{r}_1,\underline{r}_2) = \Phi_6(Z, -\underline{K}|\underline{r}) \tag{82}$$

where Φ_6 is a six-dimensional Coulomb function satisfying the Coulomb wave equation in hyperspherical coordinates

$$(\nabla_6^2 + \frac{2Z}{r} + K^2)\ \Phi_6\ (Z, -\underline{K}\ \underline{r}) = 0 \qquad (83)$$

where Z is an as-yet undetermined "charge" and where $\underline{K} = (\underline{k}_1, \underline{k}_2)$ and $\underline{r} = (\underline{r}_1, \underline{r}_2)$ are six-dimensional momentum and position vectors respectively. The properties of Φ_6 are discussed by Peterkop[14]. It has asymptotic form

$$\Phi_6^\star \sim \delta(\hat{\Omega}_k - \hat{\Omega}_r)(\frac{2\pi}{iKr})^{5/2}\ e^{i(Kr + \frac{Z}{K}\ \ell n 2Kr)} \qquad (84)$$

+ ingoing wave in hyperspherical space.

(Peterkop[14], equation 16.51 with n=6; our $\Phi_6^\star = \psi_6^-$ of Peterkop). In (26), $\hat{\Omega}_r = (\hat{r}_1, \hat{r}_2, \alpha)$ and $\hat{\Omega}_k = (\hat{k}_1, \hat{k}_2, \beta)$.

Following the procedure of Peterkop and Rudge and Seaton[13] it is easily shown that for functions $X(\underline{r}_1, \underline{r}_2)$ and $Y(\underline{r}_1, \underline{r}_2)$ which are everywhere bounded and continuous such that $(\overline{H}-\overline{E})X = 0$, one has the identity

$$\mathcal{L} \equiv \int X(H-E)Y d\underline{r}_1 d\underline{r}_2 = -\frac{1}{2} \lim_{r\to\infty} r^5 \int \{ X\frac{\partial Y}{\partial r} - Y\frac{\partial X}{\partial r} \}\ d\hat{\Omega}_r \qquad (85)$$

$$d\hat{\Omega}_r = d\hat{r}_1\ d\hat{r}_2\ \sin^2\alpha\ \cos^2\alpha\ d\alpha$$

We take now $X = \Psi$ and $Y = \Phi_6$ substitute the asymptotic forms (62) and on the right hand side of (85) and integrate over $d\hat{\Omega}_r$ to get

$$\mathcal{L} = \lim_{r\to\infty}\ i(2\pi)^{5/2}(2Kr)^{i(\zeta(\hat{\Omega}_k)-Z)}\ f(\hat{\Omega}_k) \qquad (86)$$

Note that any converging wave component such as the second term in (84) gives zero contribution to the integral on the right hand side of (85). In order that the integral in (86) converges in the limit $r \to \infty$, there should be no divergent phase factor and we must have

$$\frac{\zeta(\hat{\Omega}_k)}{K} = \frac{1}{k_1} + \frac{1}{k_2} - \frac{1}{|k_1 - k_2|} = \frac{Z}{K} \qquad (87)$$

(where we have assumed z=1).

From (85), (86), (87) we obtain the exact expression for the ionisation amplitude

$$f(\underline{k}_1, \underline{k}_2) = -\ i(2\pi)^{-5/2} \int \Psi(H-E)\Phi_6\ d\underline{r}_1 d\underline{r}_2 \qquad (88)$$

Relation (87), which was first obtained by Peterkop[14] is the equivalent of (71) and defines an angle-dependent asymptotic charge of the three-body system. The important fact to notice about (87) is that it contains only one parameter Z which is uniquely defined. Hence when (88) is used to determine the ionisation amplitude with Z given by (87) then the full effects of the three-body asymptotic Coulomb potentials are taken into account and there is no phase ambiguity to worry about. We also observe that all ionisation amplitudes which differ by (and up to) a constant phase were the latter is a symmetric function of $\underline{k}_1$ and $\underline{k}_2$ are physically equivalent. If one uses (82) and (87) in (88) then in view of (83) one may replace $(H-E)\Phi_6$ in the integral (88) by

$$\frac{1}{r}\ [\zeta(\hat{\Omega}_r) - \zeta(\hat{\Omega}_k)]\Phi_6 \qquad (89)$$

The cross section is given by

$$Q(E) = \frac{1}{4} \sum_{S=0,1} (2S+1) Q^S(E) \tag{90}$$

where

$$Q^S(E) = \frac{K^4}{k_0} \int |f_S(\hat{\Omega}_k) + (-1)^S f_S(\bar{\Omega}_k)|^2 d\hat{\Omega}_k \tag{91}$$

where $d\bar{\Omega}_k$ has α_k replaced by $\pi/2 - \alpha_k$.

We thus have a form of the integral formulation in which the dynamic screening is described by a single, angle-dependent charge parameter instead of two. In (88), Ψ is an exact solution. In practical calculations, (88) could be used with approximate choices for Ψ of increasing accuracy. A first order approximation to Ψ is

$$\Psi(\underline{r}_1, \underline{r}_2) = e^{i\underline{k}_0 \cdot \underline{r}_1} P(\underline{r}_2) \tag{92}$$

where $P(\underline{r}_1)$ is the wave function for hydrogen in its ground state. A better approximation to Ψ would be a close coupling expansion of the form

$$\Psi_S(\underline{r}_1, \underline{r}_2) = \sum_{LM_L n} \frac{2\pi i^{\ell n+1}}{k_0^{1/2}} Y^{\star}_{\ell_n m_n}(\hat{k}_0) C^{L_n \ell_n L}_{M_{L_n} m_n M} \Psi^S_n(\underline{r}_1, \underline{r}_2) \tag{93}$$

where $\Psi^S_n(\underline{r}_1, \underline{r}_2)$ is of the form (9) and is obtained from the solution of a system of equations of the form (26). Such functions may be systematically improved by including more and more channels, and pseudo-states to mimic the effects of the continuum. Given modern computing capabilities, the method described in this section is a potentially powerful technique for obtaining accurate electron-hydrogen ionisation cross sections.

CLOSE COUPLING EXPANSION FOR IONISATION

It would clearly be desirable if a method for treating ionisation by the close coupling method, which has proved so successful in dealing with excitation problesm, could be developed. It is thus instructive to investigate the form the equations would take for ionisation[13]. For simplicity, we take the case of electron plus hydrogen. Let $\phi_\gamma(\underline{r}_2)$ be the bound state eigenfunctions for hydrogen and $\phi_p(r_2) = \phi(1, \chi_p, \underline{r}_2)$ represent the continuum states of the atom. We have

$$E = \frac{1}{2} (k_p^2 + \chi_p^2) \tag{94}$$

Let us expand the total wave function in the form

$$\Psi(\underline{r}_1, \underline{r}_2) = \sum_\gamma \phi_\gamma(\underline{r}_2) F_\gamma(\underline{r}_1) + \int \phi_p(\underline{r}_2) F_{\underline{\chi}_p}(\underline{k}_p | \underline{r}_1) d\underline{\chi}_p \tag{95}$$

Substitution into the Schrödinger equation gives

$$(\nabla_1^2 + k_\gamma^2) F_\gamma(\underline{r}_1) = 2 \sum_{\gamma''} V_{\gamma\gamma''}(\underline{r}_1) F_{\gamma''}(\underline{r}_1) + 2 \int V_{\gamma_p}(\underline{r}_1) F_{\underline{\chi}_p}(\underline{k}_p | r_1) d\underline{\chi}_p \tag{96}$$

where

$$V_{\alpha\beta}(\underline{r}_1) = \int \phi^{\star}_\gamma(\underline{r}_2) \left[\frac{1}{r_{12}} - \frac{1}{r_2}\right] \phi_\beta(\underline{r}_2) d\underline{r}_2 \tag{97}$$

and

$$(\nabla_1^2 + k_p^2)F_{\underline{\chi}_p}(k_p|r_1) = 2\sum_{\gamma''} V_{p\gamma''}(\underline{r}_1)F_{\gamma''}(\underline{r}_1) + 2\int V_{pp''}(\underline{r}_1)F_{\chi_p''}(\underline{k}_p''|\underline{r}_1)dx_p'' \tag{98}$$

with

$$V_{pp''}(\underline{r}_1) = \int \phi_p^\star(\underline{r}_2)\,[\frac{1}{r_{12}} - \frac{1}{r_1}]\,\phi_{p''}(\underline{r}_2)d\underline{r}_2 \tag{99}$$

If either α and β is a bound state, the potential $V_{\alpha\beta}(\underline{r}_1)$ falls off faster than r_1^{-1} as $r_1 \to \infty$ therefore asymptotically

$$(\nabla_1^2 + k_p^2)F_{\underline{\chi}_p}(k_p|\underline{r}_1) \approx 2\int V_{pp''}(\underline{r}_1)F_{\underline{\chi}_{p''}}(\underline{k}_{p''}|\underline{r}_1)d_{\underline{\chi}_{p''}} \tag{100}$$

It may be shown that

$$\int V_{pp''}(\underline{r}_1)F_{\underline{\chi}_{p''}}(\underline{k}_{p''}|\underline{r}_1)d\underline{\chi}_{p''} \underset{r_1\to\infty}{\sim} \frac{k_p\Gamma(\chi_p,k_p\hat{r}_1)}{r_1} F_{\underline{\chi}_p}(k_p|\underline{r}_1) \tag{101}$$

where

$$\Gamma(\underline{\chi},\ \underline{k}) = \frac{1}{k} - \frac{1}{|\underline{\chi}-\underline{k}|} \tag{102}$$

so that

$$(\nabla_1^2 + k_p^2)F_{\underline{\chi}_p}(\underline{k}_p|\underline{r}_1) \sim \frac{1}{r_1}\,(1 - \frac{k_p}{|\underline{\chi}_p-\underline{k}_p|})\,F_{\underline{\chi}_p}(\underline{k}_p|\underline{r}_1) \tag{103}$$

and hence

$$F_{\underline{\chi}_p}(\underline{k}_p|\underline{r}_1) \sim \frac{1}{r_1} f(\underline{k}_p,\ \underline{\chi}_p\hat{r}_1)\ e^{i\Phi} \tag{104}$$

$$\Phi = k_p r_1 + \Gamma(\chi_p, k_p\hat{r}_1)\ell n2k_p r_1 - \frac{2}{\chi_p}\ell n\frac{\chi_p}{X} - 2(\frac{1}{k_p} - \frac{1}{|\underline{\chi}_p - k_p\hat{r}_1|})\ \ell n\frac{k_p}{X} \tag{105}$$

K-HARMONICS

Peterkop[14] has suggested treating the integral over the continuum by means of a discrete expansion in K-Harmonics $Y_{K\gamma}(\hat{\Omega})$:

$$\Psi = \sum_{\gamma'} \phi_{\gamma'}(\underline{r}_2)F_{\gamma'}(\underline{r}_1) + r^{-5/2}\sum_{K,\gamma} Y_{K\gamma}(\hat{\Omega})F_{K\gamma}(r) \tag{106}$$

where $Y_{K\gamma}(\hat{\Omega})$ are hyperspherical harmonics, eigenfunctions of the Laplace operator on the unit hypersphere in six dimensional space;

$$[\Delta_6 + K(K+4)]\ Y_{K\gamma}(\hat{\Omega}) = 0 \tag{107}$$

Owing to non-orthogonality between terms in the first and second expansions the $F_\gamma(\underline{r}_1)$ and the $F_{K\gamma}(r)$ are not unique and supplementary conditions would have to be introduced. Asymptotically,

$$(\frac{d^2}{dr^2} - (K_i+\frac{3}{2})(K_i+\frac{5}{2})/r^2 + k^2)\ F_j(r) = -\frac{2}{r}\sum_{i=1}^{\infty} Z_{ij}F_j(r) \tag{108}$$

The $F_j(r)$ are coupled through off-diagonal Coulomb potentials.

THE R-MATRIX METHOD

Burke[18] has investigated the application of the R-Matrix method to ionisation. In the inner region ($r_{N+1}<a$, $r_{N+2}<a$) an expansion isadopted of the form

$$\Psi_\ell = \sum_{ijk} a_{ijk\ell}\, \Phi_i(\underline{x}_1 \ldots \underline{x}_N \hat{r}_{N+1}\sigma_{N+1}\hat{r}_{N+2}\sigma_{N+2}) u_j(r_{N+1}) v_k(r_{N+2})$$

$$+ \sum_j \chi_j(\underline{x}_1 \ldots \underline{x}_{N+2})\, b_{j\ell} \qquad (109)$$

where $\underline{x}_j$ denotes the space and spin coordinate of electron j.($\underline{x}_j=\underline{r}_j$, σ_j)

In the first expansion N electrons are in orbitals which vanish on the boundary while the remaining two are represented by continuum orbitals u_i and v_k that are non-zero on the boundary.
The continuum orbitals are then matched on the boundary to an expansion involving just two electrons in the outer region. When r_{N+1} and r_{N+2} are both large, expansions in hyperspherical coordinates may be used, but these do not give good representation of the wave function when one of r_{N+1} and r_{N+2} is large but the other small.
Asymptotic forms which go to the correct limit when either r_{N+1} or r_{N+2} is small are being developed by Altick[19].

THRESHOLD LAWS

A great deal of the effort in recent work has been concerned with studying the ionisation threshold law. Following the early, purely classical studies by Wannier[20], work by Peterkop; Rau[21] Greene[22,23] and Klar and Schlecht[24] has led to a good qualitative understanding of the physics of the near-threshold ionisation process. (But see also Temkin[25]). This work will be discussed in the talk of B. Piraux at this meeting and so will be omitted here. Dr. Piraux will also discuss calculations[30] of triple-differential cross sections for which a lot of experimental data are available.

AUTOIONISATION

We now turn to the subject of inner shell excitation followed by autoionisation. Dramatic examples of this effect have been observed in a considerable number of experimental investigations. To illustrate a typical example of the kind of system susceptible to the process, we consider Na-like ions. In these, the transitions responsible are

$$A^{+n}(2p^6 3s) + e^- \rightarrow A^{+n}(2p^5 3snl) + e^-$$

$$\rightarrow A^{+n+1}(2p^6) + e^- + e^-$$

At each $2p^5 3snl$ excitation threshold an abrupt jump appears in the ionisation cross section, due to the onset of excitation which in a positive ion has a finite threshold cross section. Observations in these ions have shown that the effect is distinctly observable. However, in other ions, notably Ca^+, Sr^+, Ba^+, Ti^{2+} and Ti^{3+}, the effect dominates over direct ionisation above the first autoionisation threshold.
There have been two principal theoretical approaches to the problem, each having its associated advantages and disadvantages. In the first which we shall consider, direct ionisation and excitation-autoionisation are considered independently and the ionisation cross section is expressed in the form

$$Q = Q_{ion}(E_i) + \sum_{i=i_o}^{i_m} Q_{ex}(i) \frac{A_a(i)}{A_a(i) + \sum_j A_r(i \to j)} \tag{110}$$

where $Q_{ion}(E_i)$ is the direct ionisation cross section for incident electron energy E_i, $Q_{ex}(i)$ is the excitation cross section for the state i, $A_a(i)$ the autoionisation rate of state i and $A_r(i \to j)$ the radiative decay rate from state i to true bound state j. For highly excited states i and for very highly charged ions, $A_a(i) << A_r(i \to j)$ and the indirect term vanishes; all states i decay radiatively. On the other hand, for low excitation and low to moderate degree of ionisation, $A_a(i) >> A_r(i)$ and we have

$$Q = Q_{ion}(E_i) + \sum_{i=i_o}^{i_m} Q_{ex}(i) \tag{111}$$

In this case the evaluation of the indirect contribution reduces to the evaluation of all the excitation cross sections contributing to the sum. This approach has been adopted by Bely[26] and more recently by Lagattuta and Hahn[27] and by Griffin et al[28]. The calculations for Mg^+ of Moores and Nussbaumer[29] following the earlier work of Bely, made use of inner shell ionisation cross sections extrapolated to negative "ejected" electron energies and gave an overestimate of the effect. More recent work by Griffin et al for Mg^+, Al^{2+} and Si^{3+} in which excitation was treated in an exchange-distorted wave approximation, also appeared to give overestimates.

In calculations in which the incident and scattered electrons are assumed to interact with the ion only via the long-range Coulomb interaction or at best an exchange distorted wave interaction, coupling effects between autoionising states and between them and the continuum are ignored. No coupling between the electron and the target ion is allowed for. Thus processes such as resonant capture cannot be taken into account. In the approach of Henry[31] (Li like ions) and Burke[32]et al (1983) the excitation autoionisation contribution is calculated in a close coupling approximation, including the ground state, important excited states and a few autoionising states, regarded as pure bound states, in the expansion. This inclusion of coupling is found significantly to decrease the discrepancy with experiment as the results for Ti^{3+} show.

For Ti^{3+} Burke et al used the R-matrix method and included the ground state $3p^63d$ plus 9 terms out of the 19 arising from the $3p^53d^2$ configuration. This calculation makes allowance for the resonant-capture autoionisation effect via the resonance $3p^53d^2nl$ obtained in the excitation cross sections. These are important near threshold. This approach neglects post-collision interaction, interference effects and can only cope with a small number of autoionising levels at a time. In the calculation, an infinite number of open channels must be neglected and the autoionising levels regarded as being pure bound (no interaction with the continuum). Nevertheless, in including strong coupling with the incident and scattered electrons it seems to allow for the most important effect.

Burke et al[32] have also performed R-matrix calculations for Ca^+. In this ion too the resonant capture process is very important, just above the ionisation threshold. As well as

$$\begin{aligned} Ca^+(3p^64s) + e^- &\to Ca^{+\star}(3p^54snd) + e^- \\ &\to Ca^{2+}(3p^6) + 2e^- \end{aligned}$$

one may have

$$Ca^{+}(3p^{6}4s) + e^{-} \rightarrow Ca^{**}(3p^{5}3d^{2}4s) \rightarrow Ca^{+*}(3p^{5}3d4s) + e^{-}$$
$$\rightarrow Ca^{2+}(3p^{6}) + 2e^{-}$$

The intermediate state $Ca^{**}(3p^{5}3d^{2}4s)$ may be regarded as a resonance in the excitation process $Ca^{+}(3p^{6}4s) \rightarrow Ca^{+*}(3p^{5}3d4s)$. The theoretical results predict an abrupt rise in the cross section at approximately the energy of the observed one, but the peak value of the cross section is nearly twice as big as that observed.

The second approach is the method developed by Moores and Jakubowicz[33]. This starts from the generalisation of equation (61) to ionisation of an ion with N+1 electrons :

$$f(\underline{k},\underline{\chi}) = (2\pi)^{-5/2}\, e^{i\delta(k,\chi)} \int \Psi(N+2)[H-E]\Phi(N+2)\,d\tau \qquad (112)$$

The function Ψ in (112) is taken to be of the form

$$\Psi = \Psi_o(N+1)\ \phi(\underline{z}_o,\underline{k}_o,\underline{r}_{N+2}) \qquad (113)$$

where

$$z_o = Z - N - 1 \qquad (114)$$

and

$$\Phi = \Psi_f(N+1,\chi^2)\ \phi(z_o,\underline{k},\underline{r}_{N+2}) \qquad (115)$$

with

$$\Psi_f = \sum_{LSM_LM_S\gamma} C(\gamma LSM_LM_S)\ Y^{\star}_{\ell_\gamma m_\gamma}(\hat{\chi}_\gamma)\ \Psi_\gamma(N+1) \qquad (116)$$

where $C(\gamma LSM_LM_S)$ is an angular coefficient.
Both Ψ_o and Ψ_γ are of the form (9)

$$\Psi_j = \sum_i \Theta_{ij} + \sum_k \chi_k\, a_{kj} \qquad (117)$$

and are solutions of equations of type (26), the close coupling equations for an N-electron ion plus electron. Ψ_o corresponds to all channels closed and represents the initial bound state of the N+1 - electron ion. Ψ_γ is a solution with one or more channels open. Ψ_o and Ψ_γ being solutions of the same problem, are hence orthogonal. The functions $\phi(z,\underline{k},\underline{r})$ are Coulomb waves or distorted waves satisfying equations of the form (53). In this method the exchange amplitude is chosen to satisfy

$$|g(\underline{k},\underline{\chi})| = |f(\underline{\chi},\underline{k})| \qquad (118)$$

and the phase of the interference term is chosen according to the prescription of Burgess and Rudge[34], which is such as to give near-maximum interference. The cross section is then calculated from (76). The whole method may be referred to as the Coulomb-Born Exchange or the Distorted-Wave Born Exchange approximation.

The expression for the amplitude may be rearranged as

$$f(\underline{k},\underline{\chi}) = \int \Psi_o(N+1)\ V(N+1)\ \Psi_f(N+1)\ d\tau_{N+1} \qquad (119)$$

where

$$V(N+1) = \int \phi^{\star}(z_1,\underline{k}_0,\underline{r}_{N+2}) \sum_{i=1}^{N+1} \frac{1}{r_{i,N+2}}\ \phi(z_1,\underline{k},\underline{r}_{N+2})\,d\underline{r}_{N+2} \qquad (200)$$

V(N+1) is a multipolar operator and (119) resembles the form of the

matrix element for a radiative transition, V replacing the radiative dipole operator. Ionisation is thus described in this approach as a bound-free transition of the N+1 electron ion provoked by electron impact, the impacting electron being little affected in the process. Coupling effects between incident and/or scattered electrons and the ion, except for perhaps a weak, spherically symmetric interaction are ignored. It is thus a high energy approximation.
The use of functions Ψ_0 and Ψ_f satisfying close coupling equations has a number of important advantages. Accurate initial state functions, including correlation, may be used. The possibility of including bound channels and closed channels in Ψ_f means that excitation-autoionisation may be included directly in the calculation of the ionisation process, rather than treating it as an independent excitation contribution. The autoionisation contribution takes the form of resonant behaviour in Ψ_γ.

At energies corresponding to resonances in scattering of electrons by the N-electron ion, sharp peaks will appear in the single-differential cross section, (equation (76)), that is the cross section for ejection of an electron with energy χ^2. Integration over ejected electron energy will give rise to corresponding jumps in the total cross section. Whole Rydberg series of resonances may be included by this means.
It must be admitted however that the method can involve considerable computational complexity if more than a few channels are included.

The method is furthermore not capable of making allowance for the so-called "resonant capture double autoionisation" effect, in which the incident electron is captured into a short lived quasi bound state of the N+2 electron system which subsequently breaks up with the ejection of two electrons. To describe such a process, strong coupling effects between the incident electron and the target would have to be included.

REFERENCES

P.G. Burke and M.J. Seaton, Methods Comput. Phys. 10, 1, 1971.
P.G. Burke , A. Hibbert and W.D. Robb, J. Phys. B, 4, 153, 1971.
M.A. Crees, M.J. Seaton and P.M.H. Wilson, Comput. Phys. Commun., 15, 23, 1978.
E.R. Smith and R.J.W. Henry, Phys. Rev. A7, 1585; A8, 572, 1973.
H. Feshbach, Ann. Phys., 19, 287, 1962.
M.J. Seaton, Proc. Phys. Soc., 88, 801, 1966.
M. Gailitis, Sov. Phys. JETP 17, 1328, 1963.
A.H. Pradhan and M.J. Seaton, J. Phys. B, 18, 1631, 1985.
H. Saraph, Comp. Phys. Commun. 3, 256, 1972.
N.S. Scott and P.G. Burke, J. Phys. B 13, 4299, 1980.
P.H. Norrington and I.P. Grant, J. Phys. B 14, L261, 1981.
H. Saraph, M.J. Seaton and J. Shemming, Phil. Trans. A264, 77, 1969.
M.R.H. Rudge and M.J. Seaton, Proc. Roy. Soc. A 283, 262, 1965.
R.K. Peterkop, Opt. Spektrosk. 13, 153 (Translation Opt. Spectrosc. 13, 87), 1962; "Theory of ionization of atoms by electron impact", Colorado Assoc. Univ. Press, 1977.
M.J. Alguard, V.W. Hughes, M.S. Lubell and P.F. Wainwright, Phys. Rev. Letters 39, 334, 1977.
M.R.H. Rudge and S.B. Schwartz, Proc. Phys. Soc. 88, 567, 1966.
B.L. Luffman, J. Phys. B 2, 162, 1969.
P.G. Burke, "Electron Atom Scattering Theory", Nato Advanced Study Institute, Santa Flavia, Sicily, Sept. 1984.
P.L. Altick, J. Phys. B 16, 3543, 1983.
G.H. Wannier, Phys. Rev. 90, 817, 1953.
A.R.P. Rau, Phys. Rev. A4, 207, 1971.
C.H. Greene, Phys. Rev. A23, 661, 1981.
C.H. Greene and A.R.P. Rau, Phys. Rev. Letters 48, 533, 1982.

H. Klar and W. Schlecht, J. Phys. B 9, 1699, 1976.
A. Temkin, Phys. Rev. Letters 49, 365, 1982.
O. Bely, J. Phys. B 1, 23, 1968.
Y. Hahn and K. Lagattuta, Phys. Rev. A26, 1378, 1982.
K. Griffin, C. Bottcher and M. Pindzola, Phys. Rev. A25, 154, 1982.
D.L. Moores and H. Nussbaumer, J. Phys. B 3, 161, 1970.
F.W. Byron, C.J. Joachain and B. Piraux, J. Phys. B 15, L293, 1982.
R.J.W. Henry, J. Phys. B 12, L309, 1979.
P.G. Burke, A.E. Kingston and A. Thompson, J. Phys. B 16, L385, 1983.
H. Jakubowicz and D.L. Moores, J. Phys. B14, 3733, 1981.
A. Burgess and M.R.H. Rudge, Proc. Roy. Soc. A 273, 372, 1963.

RADIATIVE CAPTURE PROCESSES IN HOT PLASMAS

Yukap Hahn

Physics Department
University of Connecticut
Storrs, CT

ABSTRACT

The theory of electron capture by ionic targets is discussed and a calculational procedure for the evaluation of capture rates and cross sections is outlined. Scaling properties of the radiative and Auger transiton probabilities are examined in detail. Sample calculations of dielectronic recombination (DR) rates and cross sections are given. Refinements of the basic calculation which are described include configuration interaction, intermediate coupling, orders of coupling, overlapping resonances, cascade effects, and the fine-structure effect. The effect of extrinsic electric fields on the DR cross section is considered. Field-induced l-mixing, field mapping of the captured states, field ionization, and possible field-induced shifts of the captured high Rydberg states by time-dependent electric fields are all considered. Finally, atomic processes which are related to DR are briefly summarized.

INTRODUCTION

Atomic collision cross sections and reaction rates are used to analyze spectral lines emitted by astrophysical plasmas[1-3] and laboratory plasmas[4-6]. Line intensities, wave lengths and shifts, widths and shapes all carry information on the system that emits radiation and the effect of its environment. Except for imploded plasmas, electron density is 10^{13}-10^{14} cm^{-3} or lower, so that such plasmas are optically thin and the radiation escapes easily from the system. However, it has become increasingly clear in recent years that the influence of even a small applied electric field can cause a serious perturbation on the system. It is therefore important to incorporate density and field effects into the atomic data one generates.

Atomic processes which produce radiation are:

(i) e-I collisional excitation and subsequent radiative decay

(ii) I-I collisional excitation and subsequent radiative decay

(iii) radiative capture in e-I collisions; both radiative recombination and dielectronic recombination

(iv) bremsstrahlung in e-I collisions

(v) charge exchange to excited states in I-I collisions and subsequent radiative decay

(vi) collisional energy exchange in I-I collisions and decay.

In the following, we consider in detail only the process (iii), which, together with (i) and (v), are the dominant cooling mechanisms in laboratory plasmas at keV temperatures. As is well-known, the radiative capture of electrons by ionic targets proceeds in two distinct modes; (a) direct radiative recombination (RR), and (b) resonant dielectronic recombination (DR), which can often be the dominant mode. Study of DR should be accompanied by the study of other resonant processes in e-I collisions[7-8]; i.e., resonant excitation (RE), Auger ionization (AI) and photo-Auger ionization (PAI). All of these higher-order processes have cross sections which are mutually related. Thus, we have

$$\sigma^{DR} \sim V_a^{capt} \cdot \omega \cdot \tilde{\delta} \quad , \tag{1}$$

where V_a^{capt} is the radiationless capture-excitation probability, ω is the fluorescence yield, and $\tilde{\delta}$ is a Lorentzian shape factor. This is to be compared with

$$\sigma^{RE} \sim V_a^{capt} \cdot (1 - \omega) \cdot \tilde{\delta} \tag{2}$$

$$\sigma^{AI} \sim V_a^{excit} \cdot (1 - \omega) \cdot \tilde{\theta} \tag{3}$$

$$\sigma^{PAI} \sim V_r^{excit} \cdot (1 - \omega) \cdot \tilde{\delta}, \tag{4}$$

where V_a^{excit} is simply the collisional excitation probability, V_r^{excit} is the photo-excitation probability, and $\tilde{\theta}$ is a step function. Some of these processes are being discussed by other lecturers at this workshop.

RADIATIVE CAPTURE

A rigorous theoretical formulation of the e-I collision process was given in refs. 8 and 9, where the coupling of electron and photon fields was properly taken into account. The amplitudes for the processes

$$\begin{array}{lll} e + A^{z+} \longrightarrow (A^{(z-1)+})^{*} + \gamma, & & (RR) \\ e + A^{z+} \searrow (A^{(z-1)+})^{**} \nearrow (A^{(z-1)+})^{*} + \gamma & & (DR) \end{array} \tag{5}$$

are given by[9]

$$T_{fi}^{RR} = \langle f \mid D \mid i \rangle \tag{6}$$

$$T_{fi}^{DR} = \langle f \mid D\, G^{\Gamma}\, V \mid i \rangle \quad , \tag{7}$$

where, with q= electron index,

$$\begin{aligned} D &= -ie \sum_{q} (2\pi\hbar\, \omega_{\vec{k}})^{1/2}\, \vec{r} \cdot \hat{\varepsilon} \\ V &= \sum_{q>q'} \frac{e^2}{r_{qq'}} - \sum_{q} \frac{Z_c\, e^2}{r_q} \\ G &= (E - H + i\, \Gamma^{op}/2)^{-1} \quad . \end{aligned} \tag{8}$$

The initial and final states $|i\rangle$ and $|f\rangle$ are distorted states. The intermediate state Green's function G^{Γ} contains the width operator Γ^{op}. This arises from summing higher-order terms that couple the resonant-closed channel with all of the open channels at a given energy E. Here, the open channels include both the electron-ion channels and the radiation channel. Therefore, T_{fi}^{DR} is not the usual second-order perturbation amplitude; it contains an infinite number of sub-diagrams from the exact perturbation series. Diagrammatically, we have

$$(T_{fi}^{RR})^+ =$$

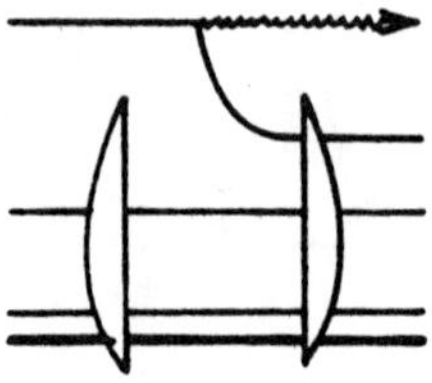

$$= \langle i \mid D \mid f \rangle \tag{9}$$

$$(T_{fi}^{DR})^+ =$$

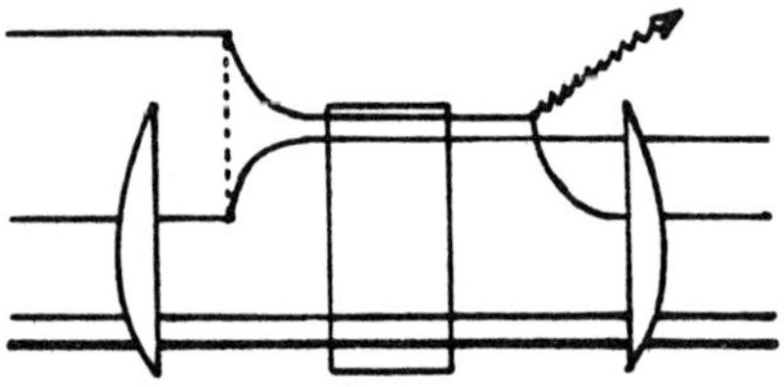

$$= \langle i \mid V \, G^{\Gamma} \, D \mid f \rangle \quad , \tag{10}$$

where

$$G^{\Gamma} = G - \frac{i}{2} G \, \Gamma^{op} G + (\frac{i}{2})^2 G \, \Gamma^{op} G \, \Gamma^{op} G + \ldots \tag{11}$$

$$\Gamma^{op} \cong \pi V \, \delta(E-H_{PP}) \, V + \pi D \, \delta(E-H_{RR}) \, D \quad . \tag{12}$$

H_{pp} is the Hamiltonian for those channels in which only one electron is in the continuum and no photons exist, while H_{RR} is for channels with one photon but no continuum electrons. The total capture cross section is given by $\sim |T_{fi}{}^{RR} + T_{fi}{}^{DR}|^2$; the intereference cross term will be discussed in a later section.

RR Amplitude and Cross Section

By definition, the differential capture cross section is given by

$$d\sigma_{fi} = W_{fi} / F_i \tag{13}$$

where

$$W_{fi} = \frac{2\pi}{\hbar} \left| T_{fi} \right|^2 \rho_f$$

$$F_i = p_c / m_e \tag{14}$$

$$\rho_f = \frac{\hbar \omega^2 d\Omega_{\hat{k}}}{(2\pi\hbar c)^3} .$$

The wave functions for the initial and final states are given as

$$\Psi_i = 4\pi \sum_{\ell_c m_c} \frac{1}{r} e^{i\ell_c} e^{-i\delta_{\ell_c}} R_{p_c \ell_c}(r) Y^*_{\ell_c m_c}(\hat{p}_c) Y_{\ell_c m_c}(\hat{r})$$

$$\Psi_f = \frac{1}{r} R_{n\ell}(r) Y_{\ell m}(\hat{r}) \tag{15}$$

and

$$R_{p_c \ell_c}(r) \longrightarrow \frac{1}{p_c} \sin\left(p_c r + \frac{z}{p_c} \ln(2p_c r) - \frac{\ell_c \pi}{2} + \sigma_{\ell_c} + \delta_{\ell_c}\right) \tag{16}$$

Both $R_{p_c \ell_c}$ and $R_{n\ell}$ are usually evaluated in HF approximation. Thus;

$$\frac{d\sigma^{RR}_{fi}}{d\Omega_{\hat{k}}} = \frac{1}{(2\pi)} \frac{e^2}{\hbar c} \left(\frac{\omega_{fi}}{c}\right)^3 \frac{m_e c}{p_c} \left| \int \Psi_f \, \hat{\varepsilon} \cdot \vec{r} \, \Psi_i d^3 r \right|^2 \tag{17}$$

and

$$\sigma_{n\ell}^{RR} = \sum_{m} \sum_{pol} \int d\Omega_{\hat{k}} \frac{d\sigma_{fi}^{RR}}{d\Omega_{\hat{k}}}$$

$$= \frac{2}{3} \alpha_o^3 \left(\frac{\hbar \omega_{fi}}{Ry}\right)^3 \left(\frac{Ry}{e_c}\right)^{1/2} \left[\ell (R_n^{\ell-1})^2 + (\ell+1)(R_n^{\ell+1})^2\right] (\pi a_o^2). \tag{18}$$

For a purely Coulombic case, $\sigma_{n\ell}$ scales as $n' = z / p_c$; this property is useful in tabulating the cross section for different ions. To evaluate the total capture cross section, contributions from all of the available Rydberg states have to be summed,

$$\sigma_i(\, i{=}n_o \ell_o) = \sum_{n > n_o} \sum_{\ell > \ell_o} \sigma_{n\ell} \quad , \tag{19}$$

where $(n_o \ell_o)$ are the quantum numbers of the outer-most occupied orbital of the target ion before capture. The values of $\sigma_{n\ell}$ as well as σ_i were tabulated in ref. 10. It was advocated there that the effective charge $Z_{eff} = (Z_C+Z_I)/2$ gives good estimates of σ , where Z_C =nuclear core charge and Z_I=degree of ionization before capture. The Kramer's formula[11] was also studied with the above Z_{eff}, and good agreement was obtained with a more careful relativistic calculation[12]. The Kramer's formula is given as

$$\sigma_{tot}^K = \omega_{n_o} \cdot \sigma_{n_o}^K + \sum_{n \geqslant n_o+1} \sigma_n^K \tag{20}$$

where

$$\sigma_n^K = \frac{8\pi}{3\sqrt{3}} \frac{\alpha_o^5}{\bar{\nu}^3} \frac{Z_{eff}^4}{e_c \,(e_c - Z_{eff}^2 \alpha_o^2/2\bar{\nu}^2)} \quad . \tag{21}$$

Here Z_{eff} = effective charge

$\bar{\nu} = n - \bar{\mu}$, $\bar{\mu}$ = average quantum defect

e_c = kinetic energy of the incoming electron

n_o = outer-most occupied orbital before capture

ω_{n_o} = ratio between the number of the unoccupied states and number of the available states in the n_o shell.

Incidentally, note that the radial integrals $R_{n\ell}^{\ell_a = \ell+1}$ in eqn 18 can be evaluated analytically in the pure Coulombic case[10,13]. Table 1 contains the total σ_i for pure Coulombic states.

Table 1. The total direct radiative recombination (RR) cross sections are given, in units of πa_o^2. The first row with $n_o \ell_o$ is for the completely stripped ion, with $Z_C = Z_I$. The numbers after the comma denote the power of 10. For details, see ref. 10.

$n_o l_o$ \ n'	0·25	0·50	0·75	1·0	2·0	3·0	4·0	5·0
0	9·18, −9	1·49, −7	6·23, −7	1·56, −6	1·13, −5	3·25, −5	6·70, −5	1·16, −4
1s	1·63, −9	3·03, −8	1·49, −7	4·38, −7	4·75, −6	1·65, −5	3·77, −5	6·95, −5
2s	6·71, −10	1·46, −8	8·42, −8	2·78, −7	3·78, −6	1·41, −5	3·34, −5	6·26, −5
2p	6·26, −10	1·19, −8	6·09, −8	1·87, −7	2·41, −6	9·56, −6	2·37, −5	4·63, −5
3s	3·42, −10	7·25, −9	4·13, −8	1·38, −7	2·10, −6	8·77, −6	2·23, −5	4·40, −5
3p	3·26, −10	6·29, −9	3·29, −8	1·05, −7	1·59, −6	7·03, −6	1·86, −5	3·77, −5
3d	3·26, −10	6·26, −9	3·24, −8	1·01, −7	1·41, −6	6·02, −6	1·59, −5	3·24, −5
4s	2·06, −10	4·28, −9	2·39, −8	7·98, −8	1·27, −6	5·67, −6	1·52, −5	3·14, −5
4p	1·99, −10	3·85, −9	2·03, −8	6·49, −8	1·04, −6	4·86, −6	1·35, −5	2·84, −5
4d	1·99, −10	3·84, −9	1·99, −8	6·26, −8	9·25, −7	4·23, −6	1·18, −5	2·51, −5
4f	1·99, −10	3·84, −9	1·99, −8	6·25, −8	9·09, −7	4·06, −6	1·12, −5	2·35, −5
5s	1·38, −10	2·82, −9	1·56, −8	5·17, −8	8·36, −7	3·87, −6	1·08, −5	2·30, −5
5p	1·34, −10	2·59, −9	1·37, −8	4·38, −8	7·11, −7	3·43, −6	9·86, −6	2·13, −5
5d	1·34, −10	2·58, −9	1·35, −8	4·25, −8	6·46, −7	3·06, −6	8·83, −6	1·93, −5
5f	1·34, −10	2·58, −9	1·35, −8	4·24, −8	6·30, −7	2·91, −6	8·27, −6	1·80, −5
5g	1·34, −10	2·58, −9	1·35, −8	4·24, −8	6·29, −7	2·89, −6	8·15, −6	1·76, −5
6	9·63, −11	1·91, −9	9·69, −9	3·05, −8	4·59, −7	2·14, −6	6·15, −6	1·35, −5
7	7·25, −11	1·40, −9	7·30, −9	2·30, −8	3·49, −7	1·64, −6	4·76, −6	1·06, −5
8	5·65, −11	1·09, −9	5·69, −9	1·80, −8	2·73, −7	1·29, −6	3·77, −6	8·43, −6
9	4·52, −11	8·70, −10	4·56, −9	1·44, −8	2·19, −7	1·04, −6	3·04, −6	6·82, −6
10	3·70, −11	7·13, −10	3·73, −9	1·18, −8	1·80, −7	8·52, −7	2·49, −6	5·58, −6

DR Amplitude, Cross Section and Rate Coefficient

From eqns 7 and 14, we have

$$\sum_m \sum_{pol} \frac{2\pi}{\hbar}\int |T^{DR}_{fi}|^2 \rho_f/F_i = \sum_{m,pol}\frac{2\pi}{\hbar}\int\left|\sum_d \langle f|D|d\rangle \frac{1}{E-E_d+i\Gamma/2}\langle d|V|i\rangle\right|^2 \rho_f/F_i$$

$$\approx \sum_d \left[\sum_m \sum_{pol}\frac{2\pi}{\hbar}|\langle f|D|d\rangle|^2 \frac{\rho_f}{\Gamma}\right]\frac{\Gamma(d)/2}{(E-E_d)^2+\Gamma^2/4}\left[2\pi|\langle d|V|i\rangle|^2 \frac{1}{F_i}\right]$$

$$\equiv \sum_d \frac{A_r(d\to f)}{\Gamma(d)}\cdot\tilde{\delta}\cdot\left(\frac{4\pi^2}{p_c^2}V_a\right) \equiv \sum_n \sigma^{DR}(i\to d\to f), \qquad (22)$$

where

$$\sigma^{DR}(i\rightarrow d\rightarrow f) = \frac{4\pi\, Ry}{p_c^2\, a_0^2}\, V_a(i\rightarrow d)\, \tau_0\, \tilde{\delta}(e_c)\, \omega(d\rightarrow f)(\pi a_0^2)$$

$$\omega(d\rightarrow f) = \frac{A_r(d\rightarrow f)}{\Gamma_r(d)+\Gamma_a(d)}\ , \qquad \tau_0 = 2.42 \times 10^{-17}\ \text{sec}, \tag{23}$$

$$\tilde{\delta}(e_c) = \frac{\Gamma/2}{(E-E_d)^2+\Gamma^2/4}\ , \qquad E = e_c + E_i\ , \qquad \int \tilde{\delta}\, dE = 1\ .$$

Furthermore,

$$V_a(i\rightarrow d) = (\,g_d/2g_i)\, A_a(d\rightarrow i)$$

$$\Gamma(d) = \Gamma_a(d) + \Gamma_r(d)\ , \qquad \omega(d) = \sum_{f'} \omega(d\rightarrow f')$$

$$\Gamma_a(d) = \sum_{i'} A_a(d\rightarrow i'), \qquad A_a(d\rightarrow i') = 2\pi\, |\langle d|V|i'\rangle|^2 \tag{24}$$

$$\Gamma_r(d) = \sum_{f'} A_r(d\rightarrow f'), \qquad A_r(d\rightarrow f') = 2\pi |\langle d|D|f'\rangle|^2\ .$$

Since V_a and ω consist of A_a and A_r, the evaluation of σ^{DR} amounts to explicit calculation of A_a and A_r. In eqn 17 we took only the diagonal terms in the square of the d-sum. This is valid if the resonance states d do not overlap. Typically, $\Gamma(d) \sim (10^{-2}\, n^{-3})$Ry, while the spacings between resonances are $\sim 2/n^3$Ry. Exceptions to this are discussed in the next section.

The DR rate is defined as a thermal average of $v_c \cdot \sigma^{DR}$, assuming that the plasma, in which DR is taking place, is in local thermodynamic equilibrium. Thus,

$$\alpha^{DR}(i\rightarrow d) = \int d\varphi(e_c)\, v_c\, \sigma^{DR}(i\rightarrow d\ ;\ e_c)$$

$$\sigma^{DR}(i\rightarrow d;\ e_c) = \sum_f \sigma^{DR}(i\rightarrow d\rightarrow f) \tag{25}$$

with

$$d\varphi(e_c) = \frac{4\pi}{(2\pi k_B T_e)^{3/2}} e^{-e_c/k_B T_e} v_c \, de_c \qquad (26)$$

and $e_c = m_e v_c^2/2$. Using σ^{DR} of (2.14) in the isolated resonance approximation (IRA), we then have

$$\alpha^{DR}(i\to d) = \left(\frac{4\pi\ Ry}{k_B T_e}\right)^{3/2} a_0^3 \ V_a(i\to d)\ \omega(d)\ e^{-e_c/k_B T_e} \quad . \qquad (27)$$

Since the resonance widths $\Gamma(d)$ are quite small, as compared to a typical experimental beam width, it is convenient to present the theoretical σ^{DR} in an energy-averaged form, as

$$\bar{\sigma}^{DR}(i\to d) \equiv \frac{1}{\Delta e_c} \int_{e_c - \Delta e_c/2}^{e_c + \Delta e_c/2} \sigma^{DR}(i\to d;\ e_c')\ de_c'$$

$$= \frac{4\pi}{(p_c a_0)^2} \frac{Ry}{\Delta e_c} \left[V_a(i\to d)\ \tau_0\right] \omega(d)\ (\pi a_0^2) \qquad (28)$$

where Δe_c is an energy-bin, chosen arbitrarily. So long as Δe_c is smaller than the typical beam width, $\bar{\sigma}^{DR}$ contains the same information as the original σ^{DR}. Note that $\bar{\sigma}^{DR}$ is essentially of the same structure as α^{DR}; this similarity facilitates comparison between data on $\bar{\sigma}^{DR}$ and α^{DR}

Scaling Properties

Before discussing the DR process in more detail in the next section, we consider the scaling properties of σ^{DR} and α^{DR} in the variables $Z=Z_{eff}$, n and ℓ . Such scaling properties are useful in simplifying the calculation of σ and α for many ionic targets with different degrees of ionization, and provide checks on various quantities, such as A_a and A_r for highly charged target ions. The scaling is valid for pure Coulombic orbitals, and thus breaks down as the correlation effect (V_{ee}) becomes more important; i.e., when $V_{ee} \gtrsim V_{ez}$=nuclear potentials.

Z-scaling. (a) $\Delta n_t \neq 0$:$A_a(ab\to sc)$ contains three bound orbitals a,b,s and one continuum orbital c. Consider the radial equation that these orbitals satisfy. For a,b and s and, neglecting V_{HF}, ($\|V_{HF}\| \ll \|z/r\|$),

$$\left[-\frac{d^2}{dr^2}+\frac{\ell(\ell+1)}{r^2}-\frac{z}{r}-e_n\right]R_{n\ell}(r)=0 \tag{29}$$

with

$$\begin{aligned} R_{n\ell} &\longrightarrow 0 \text{ as } r^{\ell+1} \text{ for } r\rightarrow 0 \\ &\longrightarrow 0 \text{ exponentially for } r\rightarrow\infty \ . \end{aligned} \tag{30}$$

If we divide this equation by Z^2 and let $rZ=t$, then eqn 29 becomes independent of Z, and $e_n = -Z^2/n^2$ Ry . Explicit Z dependence then appears only in the normalization of the wave functions; i.e., through a factor $Z^{3/2}$ for each wave function, following the change $d^3r \dashrightarrow dt^3=z^3d^3r$. The continuum function is (energy) normalized such that

$$R_{p_c\ell_c}/r \longrightarrow \sqrt{\frac{2}{\pi p_c}}\,\frac{1}{r}\sin(\,p_c r+\cdots\,)\sim Z^{1/2}\ . \tag{31}$$

Therefore,

$$\begin{aligned} A_a &\sim \left|\left\langle ab\left|\frac{1}{r_{ij}}\right|sc\right\rangle\right|^2 \\ &\sim \left|Z^{3/2}\,Z^{3/2}\,Z\,Z^{3/2}\,Z^{1/2}\cdot Z^{-6}\right|^2 \sim Z^0=1\ . \end{aligned} \tag{32}$$

Similarly

$$\begin{aligned} A_r &\sim (\hbar\omega_{fd})^3\left|\langle b\,|\,r\,|\,s\rangle\right|^2 \\ &\sim Z^6\cdot\left|Z^{3/2}\,Z^{-1}\,Z^{3/2}\cdot Z^{-3}\right|^2 \sim Z^4\ . \end{aligned} \tag{33}$$

Incidentally, note that

$$A_a^{\text{excit}} \sim \left|\left\langle ac'\,\middle|\,1/r_{ij}\,\middle|\,sc\right\rangle\right|^2 \sim Z^{-2}\ , \tag{34}$$

which is the main reason why collisional excitation to the continuum is negligible at large Z. We also have

$$A_r^{brems} \sim (\hbar\omega_{fi})^3 \,|\langle c \,|\, r \,|\, c\rangle|^2 \sim Z^0 = 1 . \qquad (35)$$

Now, using the above scaling properties, the Z-dependence of ω and α may be studied:

$$\omega = \Gamma_r / \Gamma \sim Z^0 \;, \text{ where } \Gamma_r \gg \Gamma_a \;, \; \omega \approx 1$$
$$\sim Z^4 \;, \text{ where } \Gamma_r \ll \Gamma_a \;, \; \omega \ll 1 \qquad (36)$$

$$\alpha^{DR} \sim Z^{-3} \text{ and } \sigma^{DR} \sim Z^{-4} \;, \text{ for } \omega \approx 1$$
$$\sim Z^{+1} \text{ and } \sigma^{DR} \sim Z^0 \;, \text{ for } \omega \ll 1 \;. \qquad (37)$$

(b) $\Delta n_t = 0$: By definition, this transition involves $|\Delta\ell| \neq 0$ and the same n shell, so that a non-Coulombic behavior is assumed. However, the wave functions involved may still be nearly Coulombic. Thus, the transition energy $\Delta e = e_{n_a\ell_a} - e_{n_a'\ell_a'} \approx \langle a \,|\langle 1/r_{ij}\rangle_i \,| a'\rangle_j \sim Z$. Consequently,

$$A_r \sim (\Delta e)^3 \,|\langle a \,|\, r \,|\, a'\rangle|^2$$
$$\sim Z^3 \cdot |Z^{3/2} Z^{-1} Z^{3/2} \cdot Z^{-3}|^2 \sim Z^1 \qquad (38)$$

$$A_a \sim |\langle ab \,|\, 1/r_{ij} \,|\, sc\rangle|^2$$
$$\sim |Z^{3/2} Z^{3/2} Z \, Z^{3/2} Z^{3/4} \cdot Z^{-6}|^2 \sim Z^{1/2} \;, \qquad (39)$$

where we made use of the fact that ($n_b \gg 1$)

$$|c\rangle \sim \frac{1}{\sqrt{p_c}\; r} \sin(p_c r + \cdots) \sim Z^{-1/4} Z = Z^{3/4} . \quad (40)$$

Following from the above approximate scaling, we have

$$\begin{aligned} \omega &\sim Z^0 \quad \text{for } \omega \approx 1 \\ &\sim Z^{1/2} \quad \text{for } \omega \ll 1 \end{aligned} \quad (41)$$

and

$$\begin{aligned} \alpha^{DR} &\sim Z^{-1} \text{ and } \bar{\sigma}^{DR} \sim Z^{-3/2} \quad \text{for } \omega \approx 1 \\ &\sim Z^{-1/2} \text{ and } \bar{\sigma}^{DR} \sim Z^{-1} \quad \text{for } \omega \ll 1 . \end{aligned} \quad (42)$$

n_b-scaling. For large $n_b = n$, Coulomb wave functions of different n have the same shape in the overlapping region (at small r) but their magnitudes are reduced by $n^{-3/2}$ due to the normalization factors. Thus,

$$\begin{aligned} A_a (ab \to sc) &\sim n^{-3} \\ A_r (b \to t) &\sim n^{-3} . \end{aligned} \quad (43)$$

On the other hand, when the HRS electron b is a spectator, then to a good approximation,

$$\begin{aligned} A_a (abt \to sbc) &\sim n^0 = 1 \\ A_r (ab \to tb) &\sim n^0 = 1 . \end{aligned} \quad (44)$$

The n-dependence in eqns 43 and 44 is important in estimating the HRS contribution to σ and α , as will be seen in a later section.

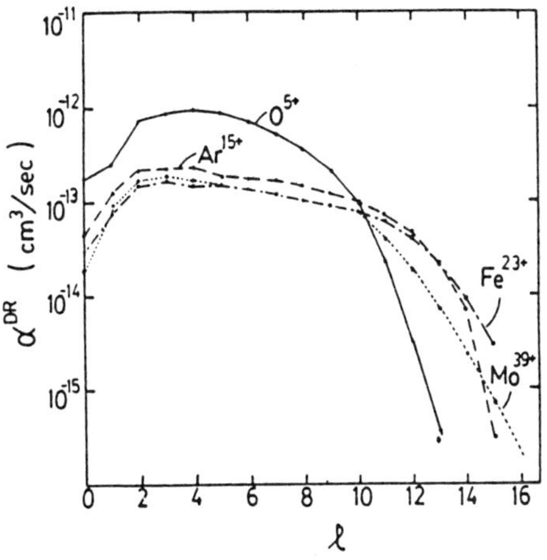

Figure 1. The DR rate coefficients for 2s→2p excitation-capture. The temperatures chosen are 5.17 Ry (O^{5+}), 33.3 Ry (Ar^{15+}), 73.5 Ry (Fe^{23+}) and 201 Ry (Mo^{39+}).

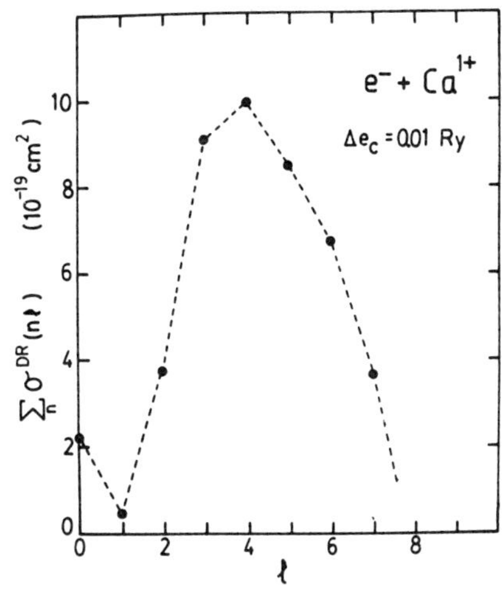

Figure 2. The l-dependence of the DR cross sections for Ca^{1+}, summed over all $n \leqslant n_F = 80$. The large dip in the $\ell = 1$ contribution is due to an accidental cancellation in the matrix element of A_a.

l_b-scaling. The dependence of A_a and A_r on the ℓ_b value of the HRS electron is rather complicated, because any change in ℓ_b produces changes in the shape of the wave function. We present in Figs. 1 and 2 examples of typical ℓ -dependence obtained from explicit calculations. The ℓ^{-3} or ℓ^{-4} type behavior often used at large ℓ is completely misleading. In most cases studied, we found that the ℓ -dependence is as shown in Fig. 3. As is apparent from Figs 1-3, the largest ℓ-value that is important for A_a is

$$
\begin{aligned}
\ell_{max} &\lesssim 12 \quad \text{for high } Z_C, Z_I \text{ and low } n \\
\ell_{max} &\lesssim 5 \quad \text{for high } n .
\end{aligned}
\tag{45}
$$

In addition to the ℓ-dependence of A_a, we note the behavior of $\bar{\sigma}^{DR}$ in two extreme limits that follows as

$$
\begin{aligned}
\omega \approx 1, \quad \Gamma_r \gg \Gamma_a &\longrightarrow \bar{\sigma}^{DR} \sim V_a \sim n^{-3} \\
\omega \ll 1, \quad \Gamma_r \ll \Gamma_a &\longrightarrow \bar{\sigma}^{DR} \sim A_r' \sim \text{constant.}
\end{aligned}
\tag{46}
$$

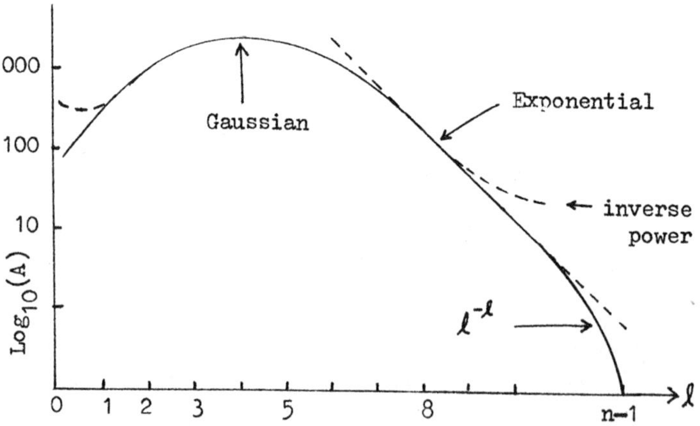

Figure 3. The ℓ-dependence of typical A_a and A_r for the dipole-allowed transitions. For small values of ℓ, there can be a sudden dip, as shown in Fig. 2. The behavior in the dominant region is usually Gaussian. and inverse power law gives an overestimate.

DIELECTRONIC RECOMBINATION

In this section, we illustrate the calculation of α^{DR} and σ^{DR} by specific examples, and summarize the theoretical data accumulated thus far. Comparison with existing experimental measurements will be deferred to a later section. Roughly, atomic data of two different qualities are needed in applications. For the analysis of specific spectral lines, a detailed and careful calculation of the energies and transition probabilities should be performed. The accuracy required is high, but the number of intermediate states to be treated is small. On the other hand, for plasma modelling and for input to rate equations, many energy levels and a complete set of transition probabilities and rate coefficients are needed. Obviously, calculations of the rates for this purpose require an immense computational effort and so have to be simpler, and as a result less accurate. So far a few bench mark calculations have been carried out and the data obtained were used to improve the existing empirical formulas. Calculations are continually upgraded, as some of the effects to be discussed in later sections are incorporated. The scaling properties discussed previously are very useful for these purposes.

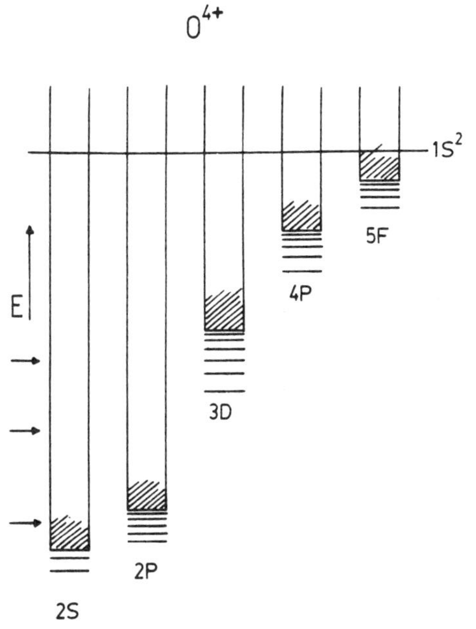

Figure 4. Energy diagram for $O^{4+} = O^{5+} + e_c$ below the ionization threshold $(1s^2)$ of O^{5+}. The 3D bound states are degenerate with the 2S and 2P continua, thus show up as resonance states. The corresponding decay widths are $A_a(3Dn\ell \rightarrow 2Sk_c\ell_c)$ and $A_a(3Dn\ell \rightarrow 2Pk'_c\ell'_c)$ respectively.

In general, there are three classes of DR process: Class 1 involves the excitation of an outer-shell electron to a state in the same shell, $\Delta n_t = 0$; Class 2 covers the excitation of an outer-shell electron to a state of higher n, $\Delta n_t \neq 0$. Finally, Class 3 includes all the inner-shell electron excitations, with $\Delta n_t \neq 0$. The excitation energies involved are the smallest for Class 1 and the largest for Class 3. The above distinctions become less clear-cut as the number of electrons become larger and the energy separations between different sub-shells become comparable in magnitude to inter-shell energy gaps. Figure 4 shows a simple energy diagram. We chose O^{5+}, a lithium-like ion, as our example throughout this section.

Class 1. $\Delta n_t = 0$, outer-shell excitation.

$$1s^2 2s + k_c \ell_c \underset{A_a}{\overset{V_a}{\rightleftharpoons}} 1s^2 2pn\ell \xrightarrow{A_r'} 1s^2 2sn\ell + \gamma \qquad (47)$$

$$1s^2 2pn\ell \longrightarrow 1s^2 2pn'\ell' + \gamma'$$

(i) (d) (f)

$$\ell_c = \ell \pm 1 \qquad n' \leq n \qquad \ell' = \ell \pm 1$$

The resonance condition gives

$$E_i + e_c = E_d , \quad e_c \geqslant 0 . \qquad (48)$$

With the approximations $E_i \approx e_{2s} + 2e_{1s}$ and $E_d \approx e_{2p} + e_n + 2e_{1s}$, we have

$$e_{2s} + e_c \approx e_{2p} + e_n , \text{ with } e_{2s}, e_{2p}; e_{n\ell} < 0 \qquad (49)$$

or

$$\Delta e \equiv e_{2p} - e_{2s} = e_c - e_{n\ell} . \qquad (50)$$

Therefore, for given excitation energy Δe ,

$$e_c = \Delta e + e_{n\ell} \longrightarrow \text{maximum for } n \rightarrow \infty , \; e_c = \Delta e . \tag{51}$$

On the other hand, the lowest allowed state (n, ℓ) for capture is given by the condition that $e_c \geq 0$; i.e.,

$$\Delta e = - e_{n_o \ell_o} \longrightarrow \text{lowest } n_o \ \ell_o \quad . \tag{52}$$

For O^{5+}, $\Delta e = 0.89$ Ry, which gives $n_o \approx 6$. For example, with n=6 and ℓ=0, we have in the AMA (denoted with bars),

(d)	(f)			$\bar{A}_r(\mathrm{sec}^{-1})$	
$1s^2 2p6s \longrightarrow$	$1s^2 2s6s$	+	γ_1	4.18(+8)	← n independent
	$1s^2 2p^2$	+	γ_2	9.92	← n dependent
	$1s^2 2p3p$	+	γ_3	6.71	
	$1s^2 2p4p$	+	γ_4	4.63	
	$1s^2 2p5p$	+	γ_5	3.54	

$$\text{Total} \qquad \bar{\Gamma}_r = 29.0(+8) \ \mathrm{sec}^{-1} \quad . \tag{53}$$

A similar analysis can be carried out in LS coupling, but we forego such cases for simplicity. LS coupling is considered for class 2 examples. The AMA calculation is the simplest, and usually gives a good rough estimate. For coupling of HRS, the jK coupling scheme may be more appropriate. For this (d), there is only one Auger channel, for which

$$\bar{A}_a(d \rightarrow i, \ell_c) = \bar{\Gamma}_a = 4.00(+13) \ \mathrm{sec}^{-1} \tag{54}$$

and thus

$$\bar{\omega}(d) = \bar{\Gamma}_r / (\bar{\Gamma}_r + \bar{\Gamma}_a) = 7.25(-5) \quad . \tag{55}$$

To obtain the contribution from all $n \geq n_o = 6$, we fit $\bar{\Gamma}_a$ and $\bar{\Gamma}_r$ as

$$\bar{\Gamma}_r(d{=}1s^2 2pn\ell) = a + bn^{-3}, \quad a = 4.2(+8) \text{ and } b = 5.36(+11)$$

$$\bar{\Gamma}_a(d) = c + dn^{-3}, \quad c{=}0 \text{ and } d = 8.64(+15). \tag{56}$$

α^{DR} then becomes, for each ℓ,

$$\bar{\alpha}^{DR}(n\ell) = e^{-e_c/k_BT_e}\left(\frac{4\pi\, Ry}{k_BT_e}\right)^{3/2} a_o^{\,3}\, V_a(n{=}n_o)\left(\frac{n_o}{n}\right)^3 \left[\frac{a + b/n^3}{(c+a)+(b+d)/n^3}\right]$$

$$\equiv \bar{\alpha}_o \left(\frac{n_o}{n}\right)^3 \left[\frac{a + b/n^3}{(c+a) + (b+d)/n^3}\right] \quad . \tag{57}$$

For the present case, d>>a so that the denominator behaves as $(b+d)/n^3$ for low n, until n reaches the value $n_c=((b+d)/(c+a))^{1/3}$. For $n>n_c$, the square bracket is constant and α decays as n^{-3}; i.e., (See Fig. 5). In the example above, The total contribution from all n is, with $n_c=(2.07\times10^7)=274$, with

$$\bar{\alpha}^{DR}(n) \sim \bar{\alpha}_o \left(\frac{a}{b+d}\right) n_o^{\,3} \sim \text{constant}, \quad n \lesssim n_c$$

$$\sim \bar{\alpha}_o \left(\frac{a}{c+a}\right)\left(\frac{n_o}{n}\right)^3 \sim n^{-3}, \quad n \gtrsim n_c ,$$

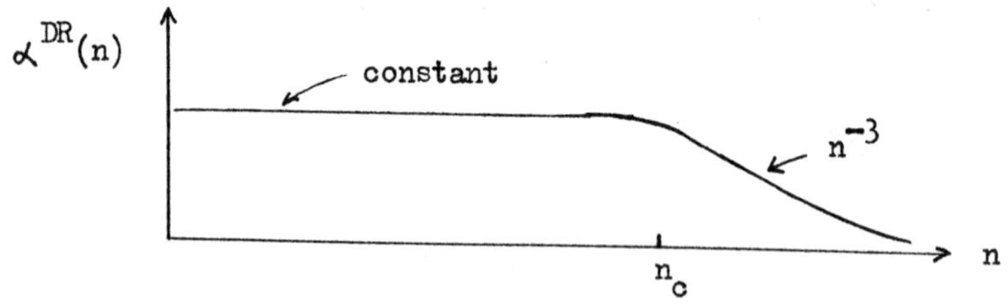

Figure 5. The n-dependence of $\alpha^{DR}(n)$ for the $\Delta n_t = 0$ transitions in light ions with $Z_I \lesssim 5$. For $n < n_c$, $A_a >> A_r$, but at higher n A_r(independent of n) begins to dominate.

$$\sum_{n=n_o}^{\infty} \bar{\alpha}^{DR}(n,o) = \frac{\bar{\alpha}_o\, b}{2(b+d)n_c^{2}}\,\frac{1}{x_o^{2}} + \bar{\alpha}_o n_c \left[\frac{a}{b+d} - \frac{b(a+c)}{(b+d)^2}\,(1.21 - x_o)\right]$$

$$\approx 2.04(-13)\ \mathrm{cm^3/sec},\ \text{at}\ k_B T_e = 5.17\ \mathrm{Ry},\ \ell=0, \qquad (58)$$

where $x_o \equiv (n_o - \frac{1}{2})/n_c$. This is to be compared with the $n=n_o=6$ contribution

$$\alpha^{DR}(n=n_o=6,\ \ell=0) \approx 5.68(-15)\ \mathrm{cm^3/sec}\ . \qquad (59)$$

Similar calculations should be repeated for each ℓ. Finally, the sum over all the ℓ contributions give the final total α^{DR}. The Class 1 transitions are absent for ions of the He- and Ne- sequences.

Class 2

$\Delta n_t \neq 0$, outer-shell electron excitation: again we consider the O^{5+} target. In this case

$$\underset{(i)}{1s^2 2s + k_c \ell_c} \ \underset{A_a}{\overset{V_a}{\rightleftarrows}}\ \underset{(d)}{1s^2 n_a \ell_a n_b \ell_b} \xrightarrow{A_r} (f) + \gamma. \qquad (60)$$

There are a double infinity of intermediate states to consider, where n_a, n_b 3. As an example, we take (d) = $1s^2 3d4d$. This state can radiatively decay as

(d)	(f)		$A_r(\mathrm{sec}^{-1})$	
$1s^2 3d4d$ ——	$1s^2 2p4d$	$+\ \gamma_1$	8.77(+10)	
	$1s^2 2p3d$	$+\ \gamma_2$	2.42(+10)	
	$1s^2 3p3d$	$+\ \gamma_3$	1.56(+10)	← Auger unstable
	$1s^2 3p4d$	$+\ \gamma_4$	(+8)	

$$\Gamma_r = 1.18(+11)\ \mathrm{sec}^{-1} \qquad (61)$$

while the cascade corrected value is, from f_1 and f_2, $\Gamma_r = 1.02(+11)\ \mathrm{sec}^{-1}$, since $\omega(f_3) \approx 0$ and thus the state $1s^2 3p3d$ does not contribute.

On the other hand, the Γ_a are more involved; in the (ab) coupling scheme, we have for the initial state $i=1s^2 2s$, with $e_c=4.47$ Ry and $k_B T_e=5.17$ Ry,

ℓ_c	L_{ab}	S_{ab}	$A_a(s^{-1})$	$\Gamma_a(s^{-1})$	ω	$g_d/2g_i$	$\alpha^{DR}(cm^3/s)$
0	0	0	3.35(+13)	2.18(+14)	4.68(−4)	1/4	0.93(−15)
0	0	1	3.95(+11)	3.06(+12)	3.23(−2)	3/4	2.26
2	2	0	1.28(+13)	1.58(+14)	6.46(−4)	5/4	2.45
2	2	1	1.16(+11)	1.26(+12)	7.51(−2)	15/4	7.73
4	4	0	5.39(+13)	4.64(+14)	2.20(−4)	9/4	6.32
4	4	1	3.29(+10)	3.84(+12)	2.59(−2)	27/4	1.36

Total α^{DR} = 21.0(−15) (62)

In evaluating Γ_a, we have

$$\Gamma_a = A_a(d \rightarrow i=1s^2 2s,\ \ell_c) + A_a'(d \rightarrow i'=1s^2 2p,\ \ell_c') \quad (63)$$

so that, for $i'=1s^2 2p$ and $e_c=3.59$ Ry, the following values of A_a' are needed in the evaluation of Γ_a:

ℓ_c'	L_{ab}	S_{ab}	$A_a'(sec^{-1})$
1	0	0	1.85(+14)
1	0	1	2.67(+12)
1	2	0	3.74(+13)
1	2	1	6.52(+11)
3	2	0	1.08(+14)
3	2	1	4.91(+11)
3	4	0	4.08(+14)
3	4	1	3.25(+12)
5	4	0	2.42(+12)
5	4	1	5.55(+11)

(64)

Thus, for example, for $\ell_c=2$, $L_{ab}=2$, $S_{ab}=1$,

$$\Gamma_a = 1.16(+11) + 6.52(+11) + 4.91(+11) = 1.26(+12)\text{ , etc.} \quad (65)$$

The AMA calculation is much simpler, while other coupling schemes are also possible. This is discussed in the next section.

Class 3

$\Delta n_t \neq 0$, inner-shell electron excitation:

For the O^{5+} example, we have

$$1s^2 2s \;+\; k_c \ell_c \;\underset{A_a}{\overset{V_a}{\rightleftharpoons}}\; 1s2sn_a \ell_a n_b \ell_b \xrightarrow{A_r} (f) + \gamma, \qquad (66)$$

(i) (d)

where n_a and $n_b \geq 2$ in this case. Again the number of states to be included are very large. In general, because of the large excitation energies required for this class of capture, as compared with the other two classes, the Class 3 contribution is usually small except at very high temperature. The DR cross section for this class of capture occurs (as resonances) at much higher energies than that of Classes 1 and 2. In summary, α^{DR} in units of 10^{-13} cm^3/sec for O^{5+} at $k_B T_e$ = 5.17 Ry and 20.7 Ry is:

$k_B T_e$ (Ry) = 5.17

Transition	α^{DR}(cm^3/sec)
2s, $\Delta n_t = 0$	64(−13)
2s, $\Delta n_t \neq 0$	24
1s, $\Delta n_t \neq 0$	0.12

Total α^{DR} = 88(−13) . (67)

Summary of theoretical results for α^{DR}

The most-studied ions are those of the He-isoelectronic sequence, for which very detailed data are now available[14,17]. In addition, some work has been done on the H-sequence[18] and Ne-sequences[19,20] More recently, rather complete calculations have been reported for the Li- and Na- sequences[21,22], and some earlier calculations for the Be- and Mg- sequences are being improved [23,24]. A partial result on Fe^8, C^+ and Ca^+ is also available. More work needs to be done, however, to cover several other important isoelectronic sequences and also to improve on the existing data and the empirical rate formulas. We show here a few sample data for the Li and Na sequences, Figs. 6-9.

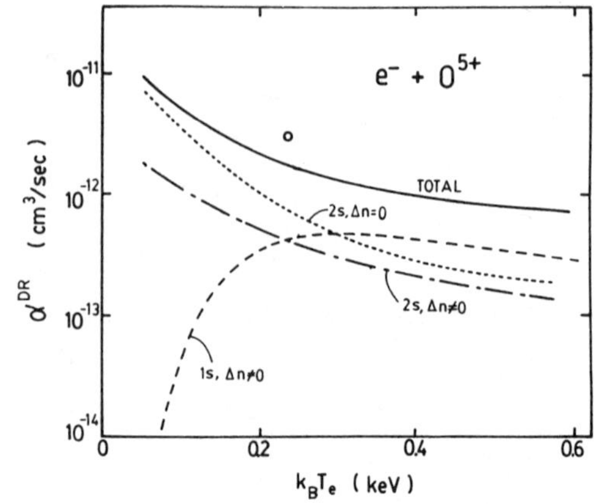

Figure 6. LS coupled, cascade corrected DR rate coefficients for the O^{5+} target ion, as functions of temperature. (Ref. 21) The circle gives the value obtained by the Burgess-Merts formula.

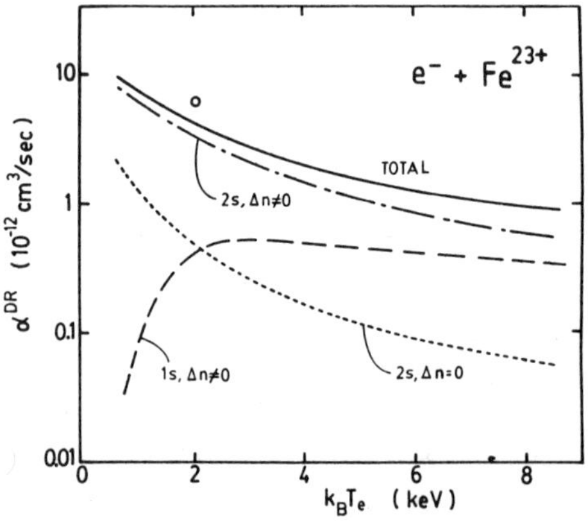

Figure 7. Same as Fig. 6, for the $e + Fe^{23+}$ system. Note the dominance of the 2s, $\Delta n \neq 0$ transitions at all temperatures, and also the importance of the 1s electron excitation contribution.

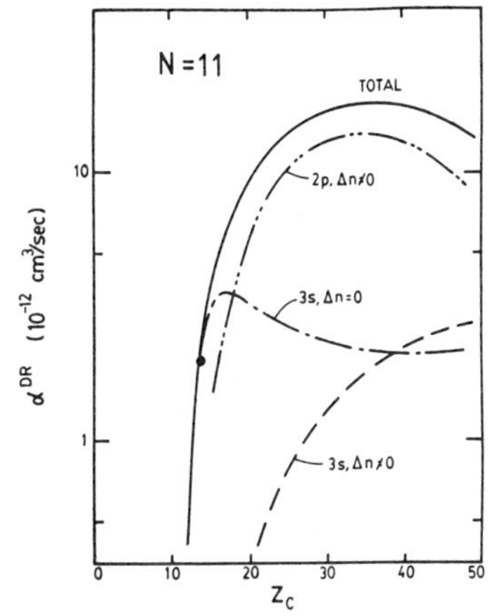

Figure 8. Values of the DR rate coefficients α^{DR} vs. Z_C for ions of the Na isoelectronic sequence. The temperatures used are scaled as Z_{eff}^2. Note the dominance of the 2p electron excitation, as compared to the 3s contribution. Ref. 22.

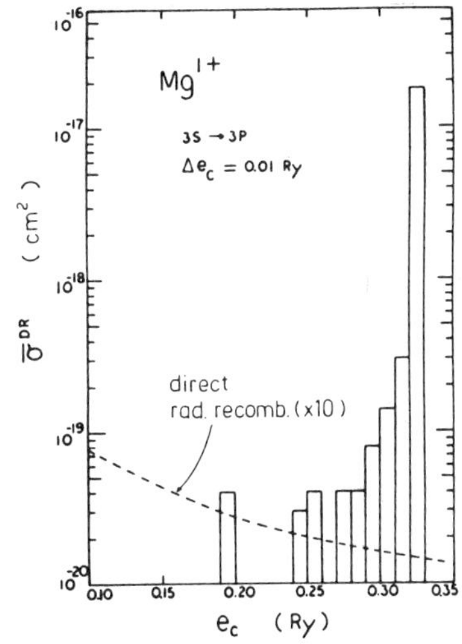

Figure 9. DR cross section, averaged over an energy bin size Δe_c, for the e + M_g^+ system, (histogram). The direct RR contribution is also shown (dashed curve). The large peak near e_c = 0.33Ry comes from the HRS. Ref. 32.

Configuration Interaction

The configuration interaction (CI) effect arises from the fact that the basis functions $\{\Phi_\alpha\}$ one has constructed for the many-electron system are not exact eigenstates of the Hamiltonian, so that residual interaction can mix these states. Probably the best known example of CI is the mixing of the ^1S states of $2s^2$ and $2p^2$, or $3s^2$ and $3p^2$ configurations, where the mixing can be as high as 30 percent. If we assume that the basis set is originally generated by a model Hamiltonian H^M, then the residual interaction is given by $V_e = H - H^M$. Improvement of the wave function is then carried out by diagonalization of the energy matrix obtained with V_e and a subset of $\{\Phi_\alpha\}$, as

$$\mathbb{B}^{-1}\,\mathbb{V}\,\mathbb{B} = \mathbb{V}^D = \text{diagonal}, \tag{68}$$

where

$$V_{\alpha\beta} = \langle \Phi_\alpha | V_e | \Phi_\beta \rangle \tag{69}$$

and thus

$$\Phi_\beta^{CI} = \sum_\alpha b_{\beta\alpha}\,\Phi_\alpha \quad . \tag{70}$$

Some examples are given in Figs. 10 and 11, and also in Table 2. Larger basis sets were used by Roszman and Weiss[56] in their study of the CI effect on α^{DR}. It was generally found that the CI effect on individual configurations can be very large, but the net sum of all contributions from all configurations involved in the mixing is often not very sensitive to CI. A partial explanation for this insensitivity may be that CI is a unitary transformation, so that its effect on the total α^{DR} is small whenever the fluorescence yield $\omega \approx 1$ or $\omega \ll 1$. Only for the intermediate values of ω, which change substantially from one configuration to another, does the CI effect become important.

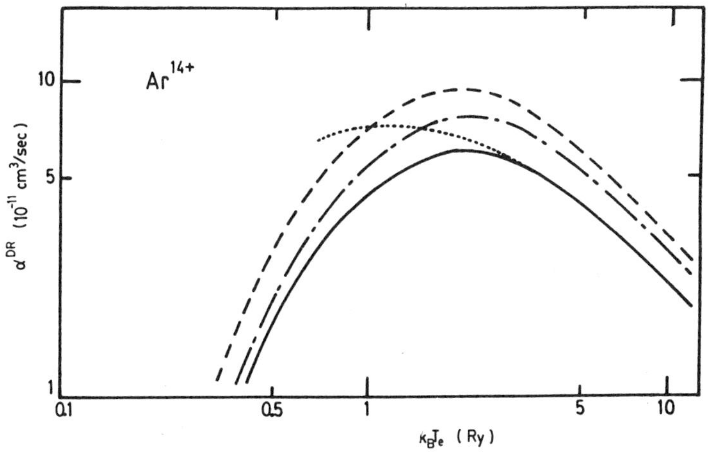

Figure 10. DR rate coefficients α^{DR} vs k_BT_e for the system e + Ar^{14+} are given for the different coupling schemes: active electron couplin, L_{ab} coupling, (dashed curve); core-electron coupling (dash-dot curve); core-electron coupling with configuration interaction (solid curve); core-electron coupling with CI and exchange (dotted curve). The effect of coupling and CI is expected to be more important as the number of electrons in the outermost open shell increases.

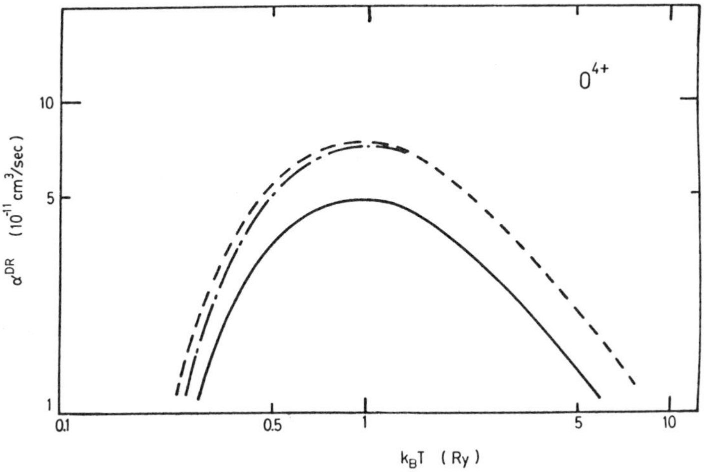

Figure 11. Same as Fig. 10, for the system e + O^{4+}.

Table 2. Effect of configuration mixing. (a) Single configuration calculation for the system e + Fe^{23+} at k_BT_e = 1 keV. (b) Examples of mixing two configurations, with the mixing coefficients $a_{qq'}$. Ref.21.

(a)

Configuration	State	$\omega(d)$	α^{DR}
3s3d	1D	0.0739	2.01-13
$3p^2$	1D	0.0404	4.65-15
$3d^2$	1D	0.0822	1.45-14
3s4p	1P	0.0690	1.70-14
3s4p	3P	0.1353	4.46-14
3s4d	1D	0.1049	6.46-14
3s4d	3D	0.4027	1.11-13
3p4s	1P	0.0833	1.39-14
3p4s	3P	0.1993	3.04-14
3p4d	1P	0.1989	7.32-15
3p4d	3P	0.3247	3.46-16
3d4s	1D	0.1663	9.38-14
3d4s	3D	0.7732	2.36-14
3d4p	1P	0.1455	6.55-15
3d4p	3P	0.4470	2.48-15

(b)

q	Configuration	State	a_{q2}	a_{q1}	$\omega_{CI}(d)$	$\alpha^{DR}_{CI}(d)$
1	3s4p	1P	0.53539	0.84460	0.0422	1.56-14
2	3p4s	1P	0.84460	-0.53539	0.3632	1.33-14
1	3s4p	3P	0.54306	0.83970	0.5654	3.09-16
2	3p4s	3P	0.83970	-0.54306	0.0882	4.15-14
1	3s4p	1P	-0.28875	0.95740	0.1270	7.65-17
2	3p4d	1P	0.95740	0.28875	0.0443	1.31-14
1	3s4p	3P	-0.29584	0.95524	0.3880	8.05-15
2	3p4d	3P	0.95524	0.29584	0.1170	3.83-14
1	3s4p	1P	0.14045	0.99009	0.2275	2.32-14
2	3d4p	1P	0.99009	-0.14045	0.0790	1.66-14
1	3s4p	3P	0.13832	0.99039	0.4260	1.11-14
2	3d4p	3P	0.99039	-0.13832	0.1099	3.44-14
1	3s3d	1D	0.66402	0.74771	0.0486	9.32-14
2	$3p^2$	1D	0.74771	-0.66401	0.0702	7.80-14
1	3s3d	1D	-0.30782	0.95144	0.1060	1.18-15
2	$3d^2$	1D	0.95144	0.30782	0.0629	1.89-13
1	$3d^2$	1D	-0.54203	0.84036	0.0687	2.01-14
2	$3p^2$	1D	0.84036	0.54023	0.1340	9.59-16
1	3s4d	1D	0.60815	0.79382	0.1056	1.17-13
2	3d4s	1D	0.79382	-0.60815	0.6132	3.65-14
1	3s4d	3D	0.60113	0.79915	0.8778	3.99-14
2	3d4s	3D	0.79915	-0.60113	0.4723	1.67-13
1	3s4d	1D	0.50293	0.86433	0.0972	3.11-14
2	3p4p	1D	0.86433	-0.50293	0.1272	4.39-14
1	3s4d	3D	0.46138	0.88720	0.6633	4.81-14
2	3p4d	3D	0.88720	-0.46138	0.4464	1.07-13
1	3p4p	1D	0.56494	0.82514	0.1363	1.01-13
2	3d4s	1D	0.82514	-0.56494	0.2066	3.75-14
1	3p4p	3D	0.65724	0.75368	0.7146	2.33-14
2	3d4s	3D	0.75368	-0.65724	0.7238	2.13-14

Order of Coupling

When more than two electrons are present in a collision system, there are many different ways by which the orbital and spin quantum numbers can be coupled. For example,

$$1s^2 + k_c \ell_c \text{ ------> } 1s2s2p \quad (71)$$
$$\text{(i)} \qquad\qquad\qquad \text{(d)}$$

where the intermediate state (d) can be represented as

$$\{2s2p\ [L_{ab}S_{ab}],\ 1s\ LS\} = \text{coupling 1}$$
$$\{1s2s\ [L_{ta}S_{ta}],\ 2p\ LS\} = \text{coupling 2} \quad (72)$$
$$\{1s2p\ [L_{tb}S_{tb}],\ 2s\ LS\} = \text{coupling 3.}$$

In the absence of any other couplings, such as the spin-orbit interaction V_{so}, the original electronic Hamiltonian $H_e = K_e + V_{ze} + V_{ee}$ commutes with any one of the above three coupling schemes, so that any one of the above is acceptable. However, as shown in Table 3 and Figs 10 and 11, its effect on α is quite drastic. Generally, all three coupling schemes give α^{DR} which are somewhat lower than the angular-momentum-average (AMA) result, sometimes as much as a factor of two. This illustrates the possible danger in calculating the rate coefficient for a single configuration in a particular coupling scheme.

Evidently, there are several ways to rectify the situation: (1) As discussed previously, one can use CI with more than one configurations whose energies lie close to each other. (2) One can resort to intermediate coupling (IC), where V_{so} is introduced to break the symmetry of H_e, as will be discussed next. (3) One can introduce some angle-dependent polarization potential in the core, which again can break the symmetry of H_e; this method is simpler but needs further developments. In practice, it is important that, at least for the dominant subset of intermediate states (d), one of the above methods should be used to estimate the coupling effect.

In addition to the above variations of the LS coupling scheme, there are of course other coupling schemes, such as jj and jK coupling. For configurations involving HRS, the jK coupling scheme seems to be most effective.

Table 3. Effect of different orders of coupling on the Auger and radiative transition probabilities, A_a and A_r, and on the fluorescence yield ω; the intermediate resonance state 1s2s2p of e + Ar^{16+} is examined, where (t, a, b) = (1s,2s,2p). (b) Dominant transitions in the e + Ar^{16+} system for the 1s2s2p intermediate state, calculated in intermediate coupling. The total rate is reflected in the last column of Table 4.

(a)	L_{ab}	S_{ab}	A_a	A_r	ω(d)	ωV_a	α^{DR}(cm^3/sec)
	1	0	8.39(+13)	2.68(+13)	0.242	6.09(+13)	
	1	1	2.55(+13)	8.03(+13)	0.759	5.81(+13)	
							1.23(−13)
	L_{bt}	S_{bt}					
	0	0	8.05(+13)	2.68(+13)	0.250	6.03(+13)	
	0	1	2.92(+13)	8.03(+13)	0.734	6.42(+13)	
							1.29(−13)
	L_{at}	S_{at}					
	1	0	4.03(+13)	1.07(+14)	0.730	8.78(+13)	
	1	1	6.93(+13)	0.0	0.0	0.0	
							9.11(−14)

(b)	Line	A_a	A_r	ω(d)	ωV_a
	1/2−3/2	3.84(+12)	1.02(+14)	0.964	7.33(+12)
	1/2−1/2	1.51(+13)	8.80(+13)	0.853	1.29(+13)
	1/2−3/2	9.80(+13)	6.75(+13)	0.064	1.26(+13)
	1/2−1/2	8.68(+13)	2.09(+13)	0.194	1.68(+13)

α^{DR}(IC) = 5.52(−14) cm^3/sec.

Table 4. Comparison of the DR rate coefficients obtained by different angular momentum coupling procedure, for e + Ar^{16+}, at k_BT_e=68 Ry. The rates are given in units of 10^{-13} cm^3/sec.

State	AMA	$L_{ab}S_{ab}$	I.C.	C.I.
1s2s2p	1.68	1.23	0.55	0.55
1s2s^2	–	–	–	0.02
1s2p^2	4.19	2.38	2.88	2.75
Sum	5.87	3.61	3.43	3.32

Intermediate Coupling

The spin-orbit interaction V_{so} is a relativistic correction, and its effect becomes larger with higher Z. When $V_{so} \ll V_{ee}$, the usual LS coupling is probably more appropriate. However, for Z $\gtrsim$ 10, V_{so} will have a sizable effect. Furthermore, V_{so} is often used to eliminate the ambiguity of different coupling schemes, as already noted. The IC procedure involves a diagonalization of the energy matrix constructed with V_{so} and any one of the LS coupled basis sets, producing the optimum set with the quantum numbers LSJ. A sample calculation is shown in Table 4 and Figs. 10 and 11. In the example of Table 4, the resulting α^{DR} is smaller than those computed in LS coupling. But, in some other cases, IC may increase the rate slightly; again the overall IC effect is generally small. Of course, in addition to V_{so}, other relativistic corrections should also be incorporated. A simple approach formulated by Cowan and Griffin[25] may be useful for ions with Z $\lesssim$ 40. The usual single-component Schroedinger equations are modified to incorporate the mass and Darwin correction terms as well as V_{so}. A noniterative procedure was suggested, and the validity of the approximation tested.

Cascade Effect

The DR rate and cross section formulas given previously contain the sum over the final states f' reached from the intermediate states (d) by radiative emission. Implicit in the definition was the assumption that all the states f' included are stable against further Auger emission. So long as the states (d) cascade down by emitting photons to a stable final state without ever emitting a single Auger electron, the definition of the fluorescence yield given earlier is valid. However, in the course of reaching a stable state, if at least one electron is emitted at some state of the cascade chain, then the fluorescence yield should be modified to take this effect into account[26]. We have

$$\omega(d) = \sum_{f'} A_r(d \rightarrow f') / \Gamma(d) \equiv \sum_{f'} \omega(d \rightarrow f')$$

$$\longrightarrow \sum_{f'} \omega(d \rightarrow f') + \sum_{f'} \sum_{d'} \omega(d \rightarrow d') \, \omega(d' \rightarrow f') + \cdots \tag{73}$$

where d',d", etc. are Auger unstable, while the f' are now stable against Auger emission. The cascade effect obviously always reduces ω and thus reduces the rate, and is often quite large especially for light ions where 's are small. This was pointed out in a previous example. The overall effect on α^{DR} can be as large as 30 per cent.

It is noted that the definition of the rate depends on the choice of the particular rate equations; when the rate equation contains explicitly the states d', d" etc., then the rates which appear in this equation should not be cascade-corrected. Otherwise, the contribution from these states is eliminated twice, i.e. the effect double-counted. This problem does not arise in the simple case in which only the ground state of each ionic species is included in the rate equation.

Overlapping Resonances and Interference

All the existing formulas for σ^{DR} and α^{DR} which are in use are based on the isolated resonance approximation (IRA), i.e. the sum over the intermediate states d in eqn 22 includes only the diagonal terms. In general, this is reasonable since the spacing between the resonances is at least an order of magnitude larger than the corresponding width Γ for low n states. However, there are two cases in which which IRA may break down:

Case 1. Usually for low n, $\Gamma_a \gg \Gamma_r$. But, if Γ_a scales as n^{-3} and Γ_r happens to contain a part which is independent of n, then at some higher n, n_c, where $\Gamma_a \lesssim \Gamma_r$. On the other hand, in most cases, the spacing between the resonances also behaves as n^{-3}. Therefore, eventually at some high n and beyond, Γ_r becomes larger than both the resonance spacing and Γ_a, and thus the problem of overlap occurs. This problem has been studied in some detail[27] using a simple complex potential model[28,9]. Quantum defect theory seems to give a slightly different model[29], but the conclusion reached on the effect of overlapping resonances is the same, i.e. the effect is small. In almost all cases of practical interest, the overlapping region due to a constant Γ_r seems to occur at $n \gg n_c$, where the contribution from this region to the total α^{DR} behaves as n^{-3}, and thus is small. By contrast, the contribution from the $n \lesssim n_c$ region is roughly constant for each n, and thus is very large for large n_c. (For Mg^+, with F=0, $n_c \approx 300$, and for Ca^+, $n_c \approx 500$.)

Case 2. When an external field is imposed in the interaction region where the capture takes place, an additional A_F factor is present in the total width, which broadens the resonance lines even for low n, and, as the field strength increases, eventually distorts the levels into continua. This problem has been considered so far under the assumption that the bound state structure of the levels are perturbed but still intact. Obviously, near the field ionization threshold, the ℓ and n mixing must be so strong that the description of the resonances in terms of n, n_1, m and A_{F}, etc. cannot be realistic. Much more study is needed to properly treat the problem of HRS in external fields. Further discussion is given later. See also Fig. 12. The interference between the two amplitudes T_{fi}^{RR} and T_{fi}^{DR} has not been examined carefully thus far. In the region where the IRA is valid, however, the cross terms should have a negligible effect because of the small widths. On the other hand, in the overlapping resonance region, the interference terms can be sizable, although the overall effect may still be small.

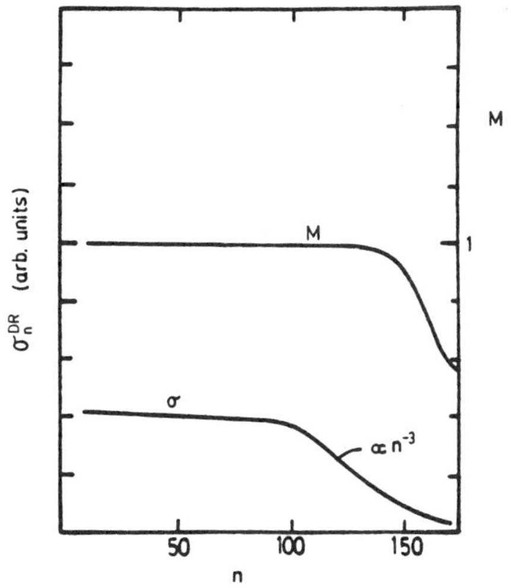

Figure 12. Effect of overlapping resonances on σ^{DR}; $M \equiv (\sigma)_{IRA}/(\sigma)_{MCQDT}$, where IRA denotes the isolated resonance approximation and MCQDT includes the overlapping effect. Note that in the region where M deviates from unity the cross section is already very small.

Fine-structure effect

In many cases in which strong dipole-allowed transitions dominate the DR process, the fs effect is expected to be small. But, there are some exceptions, as was the case for Mg^+ and Ca^+, with $\Delta n_t=0$. Previously, we treated these two cases, assuming that there was only one Auger channel ($3pn\ell$ ---> $3sk_c\ell_c$) for Mg and two Auger channels for Ca, ($4pn\ell$ ---> $3pk_c\ell_c$, $3dk_c'\ell_c'$). However, the spin-orbit coupling in fact splits the 3p and 4p states, with

$$\Delta_{3p}^{fs}(\text{Mg II}) = e_{3p}(P_{3/2}) - e_{3p}(P_{1/2}) = 0.00083 \text{ Ry}$$
$$\Delta_{4p}^{fs}(\text{Ca II}) = e_{4p}(P_{3/2}) - e_{4p}(P_{1/2}) = 0.0020 \text{ Ry} \quad . \tag{74}$$

Therefore, the $p_{1/2}$ channel threshold corresponds to the resonance state $p_{3/2} \cdot n\ell$, with n=23 and n=35 respectively for Ca and Mg. That is, for n higher than 35, the $p_{3/2}$ branch of the Mg resonance manifold will have an additional Auger channel, with $A_a(p_{3/2}n\ell \dashrightarrow p_{1/2}k'_c\ell_c')$. This has the effect of reducing the contribution from the $p_{3/2}$ branch of the resonance manifold. It was found that the total reduction in the cross section is about 15 per cent. (This is indicated in Fig. 19). The cross section in the vicinity of n=35 was smoothed out due to the fact that the $p_{1/2}$ channel involves $n' \lesssim \infty$ and is badly distorted by the electric field F. As a result, the reduction in α^{DR} due to fs is not so abrupt in the vicinity of n=35.

EXTRINSIC ELECTRIC FIELD EFFECTS

The effect of external fields on the DR cross section and rate coefficient has been the subject of intense theoretical and experimental studies during the past two years, stimulated largely by a series of beautiful experiments carried out at JILA by Dunn and his group[30,31] on the Mg^+ system. The experiment is described by Dunn at this workshop. The measured cross section came out to be nearly ten times larger than the theoretical estimate[32]. In addition, the Ca^+ data[33] by Williams is also too high by a factor of 18, as compared with the theory[34]. These discrepancies are partially explained away[35,36] by invoking the field effect, as explained below, but there are still some differences which are yet to be understood. The ORNL data[37] on light ions also showed a strong field mixing effect, and the C^+ data[38] of Mitchell et al seemed to be higher by a factor of 10 as compared with the naive theory[39].

Field Mixing[57,59]

The effect of electric microfields on the DR rate was estimated earlier by Jacobs et al,[40] in the case of Fe^{23+}, $\Delta n_t=0$ excitation, 2s -->2p. Although their estimate was crude and some gross approximations were made, the result suggested that the DR rate could increase by as much as a factor of 3 due to a strong microfield. When a electric field (static) of sufficient strength F is imposed in the interaction region, the HRS to which the projectile electron is captured are badly distorted. For a purely

Coulombic care, the ℓ levels for given n are completely degenerate in the absence of F, and any small F>0 will immediately remove this degeneracy; ℓ is no longer a good quantum number and is replaced by n_1, an "electric field quantum number", for wave functions written in parabolic coordinates; n and m are still good quantum numbers. On the other hand, when the HRS electrons are in a non-Coulombic field of the core, some low-ℓ states may no longer be degenerate and the field mixing of these levels should depend on the relative size of the splitting and the interaction energy $e\,\vec{F}\cdot\vec{r}$. In order to determine more precisely the extent of such mixing, we diagonalize the energy matrix constructed within the manifold of ℓ states of given n

$$\mathbb{D} = \begin{bmatrix} d_{11} & d_{12} & 0 & 0 & 0 & \cdots \\ d_{21} & d_{22} & d_{23} & 0 & 0 & \cdots \\ 0 & d_{32} & d_{33} & d_{34} & 0 & \cdots \\ 0 & 0 & d_{43} & d_{44} & d_{45} & \cdots \\ \cdot & \cdot & \cdot & & & \\ \cdot & \cdot & \cdot & & & \\ \cdot & \cdot & \cdot & & & \end{bmatrix}, \quad \mathbb{A}^{-1}\mathbb{D}\mathbb{A} = \mathbb{D}^{d} = \text{diagonal}, \tag{75}$$

where

$$d_{ii} = \langle n\,\ell_i\,m \mid H_{pol} \mid n\,\ell_i\,m \rangle, \quad \ell_i = |m|, |m|+1, \cdots, n-1$$

$$d_{ij} = \langle n\,\ell_i\,m \mid H_{stk} \mid n\,\ell_j\,m \rangle, \quad i \neq j \quad \text{and} \quad \ell_i = \ell_j \pm 1 \tag{76}$$

with, for example,

$$H_{pol} = H_{HF} + \langle \text{core} \mid V_{ee} \mid \text{core} \rangle_{\text{core}} - \frac{\alpha_{\text{dipole}}}{(r^2 + a^2)^2}$$

$$H_{stk} = e\,F\,r\cos\theta \quad . \tag{77}$$

Typical values of d_{ij} are given in table 5.

Table 5. Matrix elements d_{ij} of the Stark field mixing matrix $\mathbb{D}$ for the $e + Mg^+$ system. (a) Diagonal elements for the first 8 rows and columns of $\mathbb{D}$. They decrease as n^{-3} and as ℓ^{-5} for large ℓ. (b) Off-diagonal elements for the first 9 rows and columns; they are symmetric and nonzero only for $\ell = \ell' \pm 1$, and increase as n^2, and are constant in ℓ for large ℓ. For $\ell \gtrsim 10$, and for each fixed n, all the elements are obtained by the above simple extrapolations. Interpolation in n is smooth, so that only a few n values need to be studied in practice, especially when a large number of n ($\lesssim n_F \approx 50 \sim 100$) is involved.

(a) Diagonal elements $d_{\ell\ell}$

ℓ	n=30	n=40	n=50
0	−6.69(−4)	−2.82(−4)	−1.45(−4)
1	−3.62(−4)	−1.53(−4)	−7.81(−5)
2	−3.28(−5)	−1.38(−5)	−7.09(−6)
3	−9.09(−6)	−3.84(−6)	−1.96(−6)
4	−2.46(−6)	−1.05(−6)	−5.35(−7)
5	−1.99(−7)	−8.45(−8)	−4.33(−8)
6	−8.50(−8)	−3.61(−8)	−1.85(−8)
7	−4.10(−8)	−1.75(−8)	−8.98(−9)
8	−2.17(−8)	−9.27(−9)	−4.77(−9)

(b) Off-diagonal elements $d_{\ell\ell'}$

ℓ	ℓ'	n=30	n=40	n=50
0	1	7.27(−6)	1.29(−5)	2.02(−5)
1	2	6.49(−6)	1.16(−5)	1.18(−4)
2	3	6.36(−6)	1.13(−5)	1.77(−5)
3	4	6.29(−6)	1.12(−5)	1.75(−5)
4	5	6.24(−6)	1.12(−5)	1.75(−5)
5	6	6.20(−6)	1.11(−5)	1.74(−5)
6	7	6.14(−6)	1.11(−5)	1.74(−5)
7	8	6.09(−6)	1.10(−5)	1.73(−5)
8	9	6.02(−6)	1.09(−5)	1.72(−5)

In the limit of d_{ii}=constant, a_{ij} should correspond to full mixing. In this case, the mixed states are given simply by

$$|n\,k\,m\rangle = \sum_{\ell=|m|}^{n-1} a_{n\ell}\,|n\,\ell\,m\rangle$$

$$\longrightarrow \sum_{\ell=|m|}^{n-1} (-)^{n+m-1}\,(2\ell+1)^{1/2} \begin{pmatrix} \lambda & \lambda & \ell \\ m_- & m_+ & m \end{pmatrix} |n\,\ell\,m\rangle, \tag{78}$$

where

$$\lambda = \frac{n-1}{2}, \quad m_{\pm} = \frac{m \pm k}{2}, \quad k = n_1 - n_2$$

$$n_1 = 0,\ 1,\ \cdots,\ n-m-1 \quad \text{and} \quad n_2 = n - n_1 - |m| - 1 \ . \tag{79}$$

Using the a_{ij} thus determined, A_a^F and A_r^F are recalculated and σ^{DR} are evaluated as

$$\bar{\sigma}^F(n) = \left(\frac{4\pi\,Ry}{e_c}\right) \tau_o \frac{g_{core}}{2g_i} \sum_{k,m} \frac{2A_a^F \cdot A_r^F}{\Gamma_a^F + \Gamma_r^F + A_F} \left(\frac{Ry}{\Delta e_c}\right)(\pi a_o^2), \tag{80}$$

which is to be compared with the F=0 result,

$$\sigma^{F=0}(n) = \left(\frac{4\pi Ry}{e_c}\right) \tau_o \frac{g_{core}}{2g_i} \sum_{\ell,m} \frac{2A_a A_r}{\Gamma_a + \Gamma_r} \left(\frac{Ry}{\Delta e_c}\right)(\pi a_o^2) \ . \tag{81}$$

In eqn 80 we used the field-corrected quantities; in the uncoupled representation,

$$A_r^F(n_a \ell_a m_a \longrightarrow n_a \ell_a' m_a') = A_r \quad \text{, independent of F}$$

$$A_a^F(n\,k\,m) = \sum_{\ell} a_{k\ell}^2 A_a(n\,\ell\,m) \tag{82}$$

The field ionization probability A_F is given by Damburg and Kolosov[41] as

$$A_F = (f_1)^{2n_2+m+1} e^{-f_2} / [n^3 n_2! \ (n_2 + m)!] \quad , \tag{83}$$

where

$$f_1 = 4 (-2E_F)^{3/2} / F$$

$$f_2 = f_1 /6 + n^3 f_3 F/4 \tag{84}$$

$$f_3 = n_2 (34n_2 + 34m + 46) + m(7m + 23) + 53/3$$

and

$$E_F = E_0 + E_1 F + E_2 F^2 + \cdots$$

$$E_0 = - 2/(n')^2 \quad , \quad n' = n/2$$

$$E_1 = 3 n' n_{12} \quad , \quad n_{12} = n_1 - n_2 \tag{85}$$

$$E_2 = n'^4 (17n^2 - 3n_{12} - 9m^2 + 19) , \quad \text{etc.}$$

The form (83) is not applicable for very high F; instead of increasing monotonically with F, A_F turns over at some high F and becomes very small for higher F. For $F > F_{max}$ 4/ $6n^3(2n_2+m+1)$, A_F assumes the asymptotic form $A_F \sim (F \ln F)^{2/3}$. The behavior of A_F and A_a^F for different field strengths is shown in Figs. 13-16. The behavior of A_a in ℓ is such that, in eqn 81, the sum over ℓ is usually cut off at $\ell \lesssim 7$ for low n and 4 at high n. On the other hand, the k-sum in eqn 80 can run all the way to n-1. Therefore, in the case of complete mixing, we have, in the limit of $\Gamma_a \gg \Gamma_r$,

$$\bar{\sigma}^{F=0}(n) \sim A_r \cdot \sum_{\ell,m} 1 \sim A_r \cdot \ell_{max}^2$$

$$\bar{\sigma}^{F}(n) \sim A_r \cdot \sum_{k,m} 1 \sim A_r \cdot 2n \ell_{max} \ . \tag{86}$$

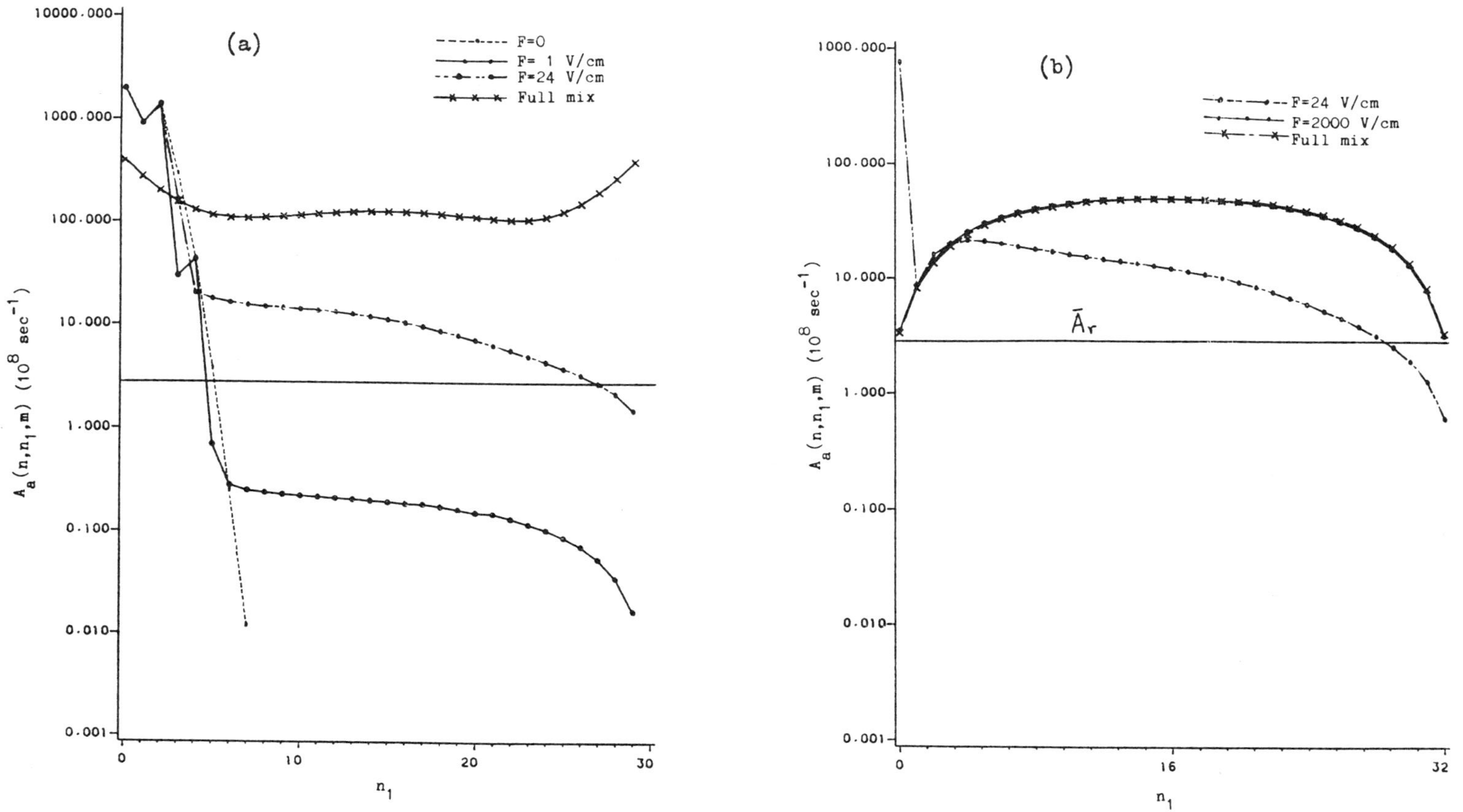

Figure 13. (a) Auger probabilities $A_a^F(n=30, n_1, m=0)$ vs the electric quantum number n_1, for Mg^*, for the $3s \rightarrow 3p$ excitation and different electric field strength F. (b) A_a^F for n=35, m=2. The horizontal line denotes the radiative probability $A_r(3p \rightarrow 3s)$.

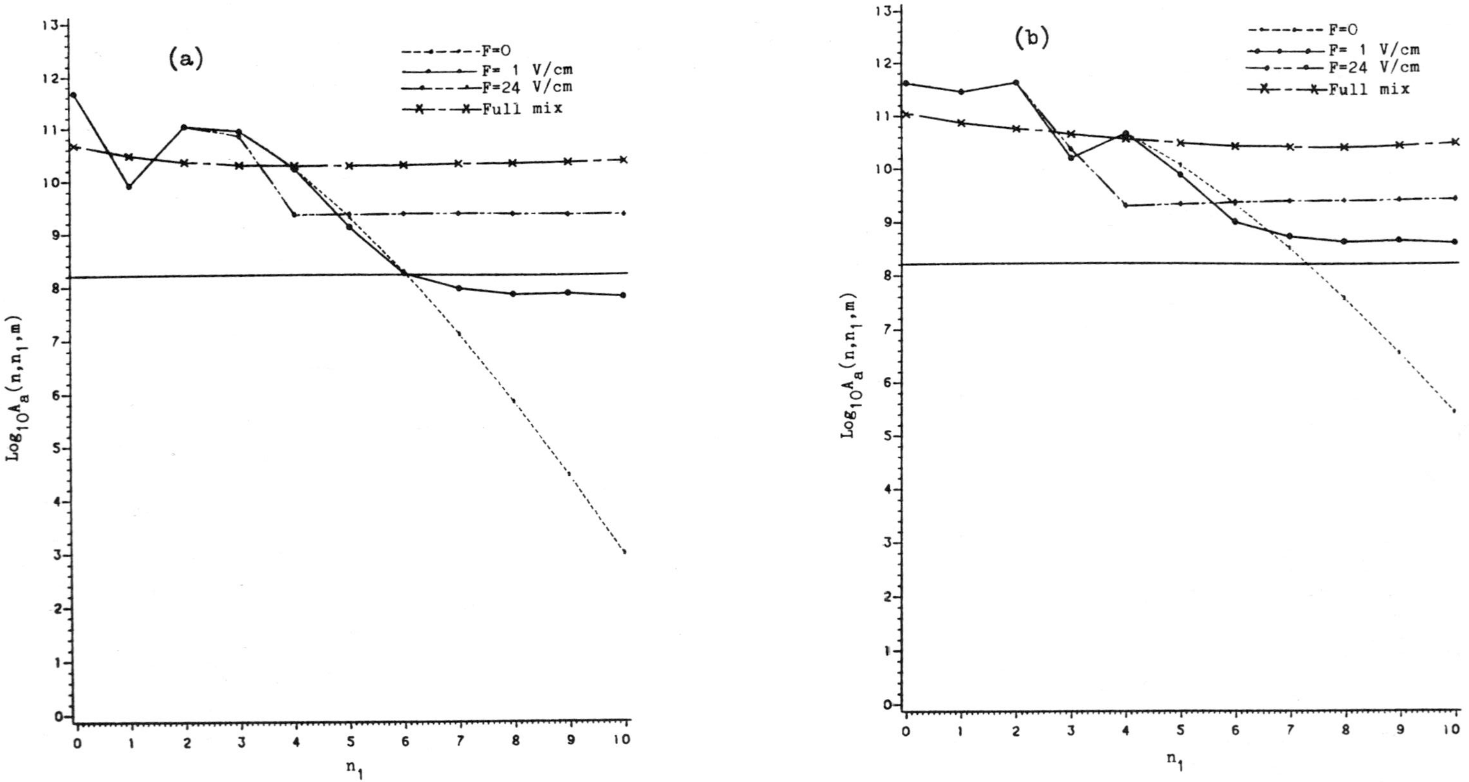

Figure 14. (a) Auger probabilities $A_a^F(n=30, n_1, m=0)$ vs the electric quantum number n_1, for Ca^*, for the 4s→4p excitation and for different electric field strength F. (b) A_a^F for n= 30, m= 0. The horizontal line denotes the radiative decay probability A_r(4p→4s). Here 4p→3d is involved in A_a^F; the effect of field mixing is similar to that in (a).

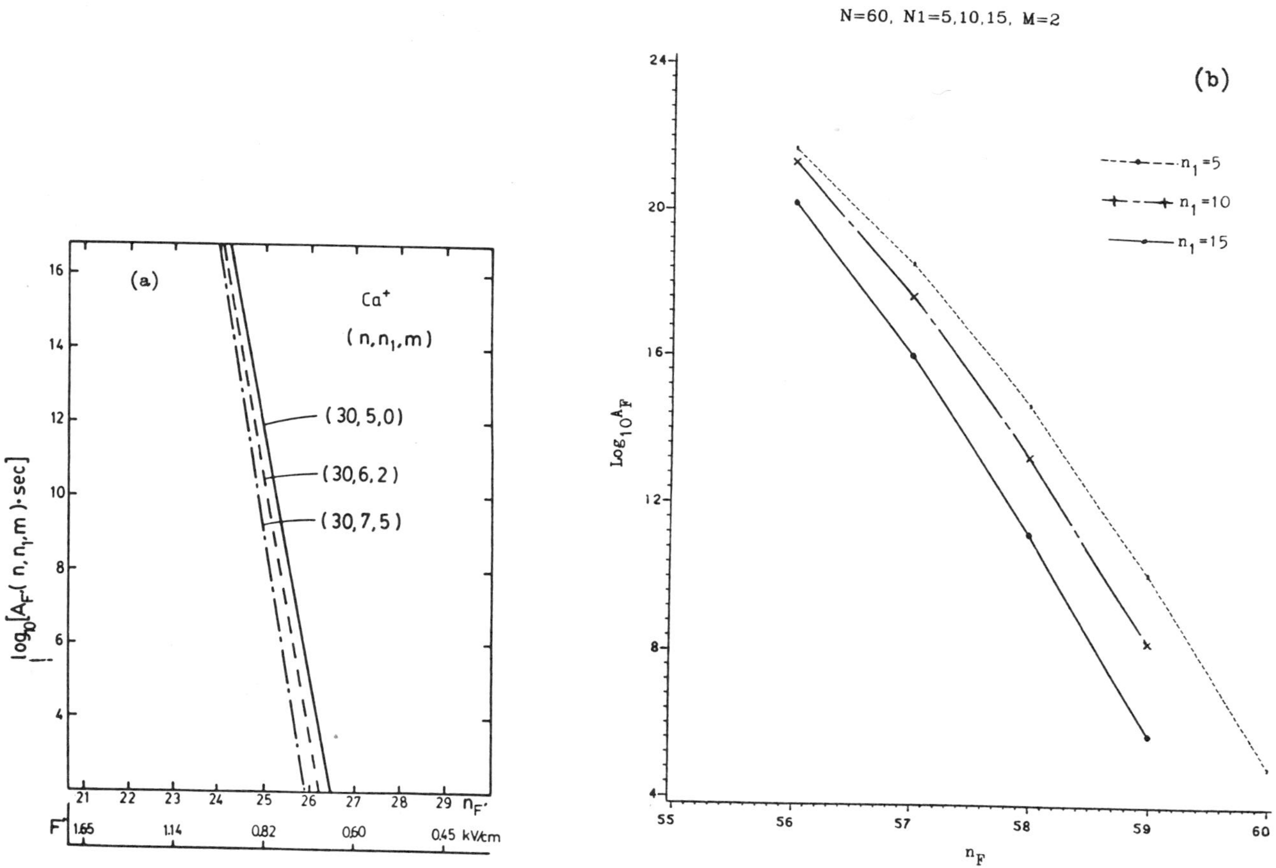

Figure 15. Field ionization probability $A_F(n, n_1, m)$ vs $n_{F'}$ for the system $e + Ca^+$, where $n_{F'}$ is given by $n_{F'} = (6.2 \times 10^8 / F(V/cm))^{1/4}$. The purely Coulombic case is considered here.

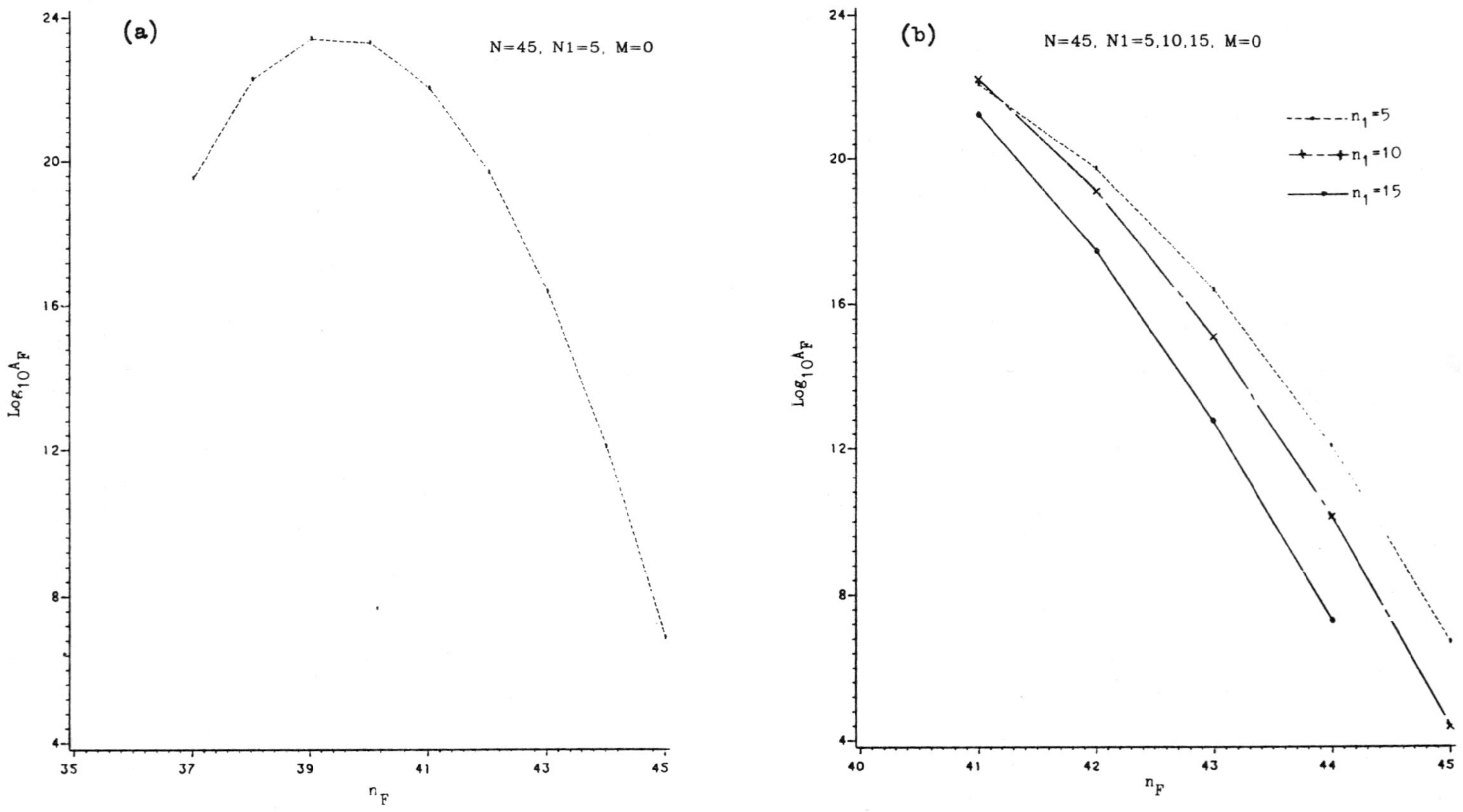

Figure 16. Same as Fig. 15. Note that in (a), an anomalous behavior of A_F is shown for $n_F \lesssim 41$, which corresponds to higher field strength; this is the result of the approximate formula proposed in ref. 41. In practice, this part of A_F has to be modified.

The optimum enhancement factor due to the field mixing is then

$$\eta_{max}(n) = \bar{\sigma}^{F} / \bar{\sigma}^{F=0} \sim 2n/ \ell_{max} \qquad (87)$$

Summing over n, up to n_{max}, we have an estimate of the enhancement factor

$$\eta_{max} = n_{max} / \ell_{max} \qquad (89)$$

In the Mg experiment, we have roughly $n_{max} \approx 65$ and $\ell_{max} \approx 6$, which gives $\eta_{max} \approx 11$, while for the Ca case, $n_{max} \approx 80$ and $\ell_{max} \approx 5$ and thus $\eta_{max} \approx 16$. (The actual situation in the case of Ca was somewhat different from the Mg case in that F in the interaction region was estimated to be less than 0.3 V/cm.) A sizable field was present in the Mg case, but not strong enough to produce complete mixing for all $n \lesssim n_{max}$. In eqn 88, n_{max} was determined from the classical formula $n_{max} = (3.2 \times 10^8 \text{ Volts/cm} \times F_a)^{1/4}$, where F_a is the analyzing field that was used to separate the recombined neutral beam from the charged beam. In both experiments, $F < F_a$. Thus far, the theoretical calculation included only ℓ-mixing within each n, mainly because of computational reasons. However, we know that F is strong enough, especially for those HRS near $n \lesssim n_{max}$, to mix different n states as well. The problem of the field effect near the field ionization threshold is a delicate one, and we are only beginning to understand the complexity of the physics involved. Any time dependence of the field can further complicate the problem. In Fig. 17, we display the results of a calculation for Mg based on the ideas described in this section.

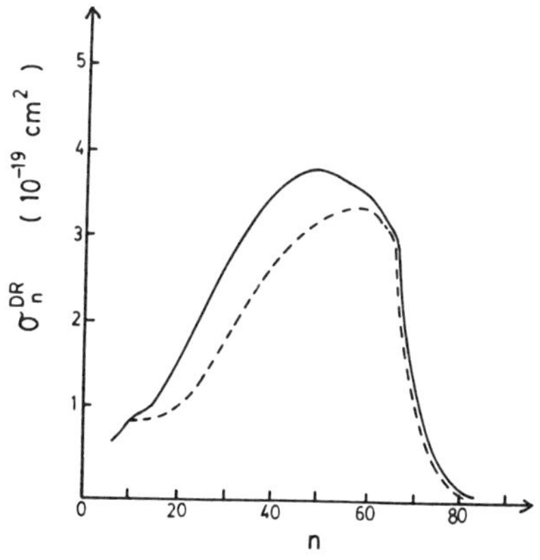

Figure 17. DR cross section for $e + Mg^+$, ($3s \rightarrow 3p$), with the inclusion of the field-mixing ; F = 24 V/cm (solid curve), and F = 8 V/cm (dashed curve). The cutOff field is F' = 36 V/cm.

Field Mapping

In the Mg experiment[31], the neutral beam of $Mg^*(3s, n\ m)$ atoms emerges from the interaction region. Distribution of the Mg^* in different quantum states is predicted by theory, according to the specific field mixing model. To investigate this distribution, a variable-field wedge detector with field F' was placed downstream to field-ionize the neutral beam, step by step. The field ionization probability A_F was seen to be sensitive to the quantum numbers (n, n_1, m). For each fixed (n, n_1, m), A_F stays small for small F', until F' increases to a critical value F_c', beyond which A_F begins to increase very rapidly. We therefore arbitrarily assign the value $A_F^{crit} = 10^6\ sec^{-1}$, and assume that when $A_F \gtrsim A_F^{crit}$ that particular state is 100 per cent ionized. Because of the rapid rise of A_F in the vicinity of F_c', the final result is not sensitive to the particular choice of A_F^{crit}. Of course, the critical field F_c', where $A_F = A_F^{crit}$, is a function of (n, n_1, m), and thus the above procedure carries out a mapping of (n, n_1, m) to F'. Apparently, this is not a 1 to 1 mapping. For convenience, we further set[56]

$$n_{F'} = (3.2 \times 10^8\ \text{Volts/cm} \times F'^{-1})^{1/4} \quad .$$

(Brouillard suggests that the value 6.2 rather than 3.2 is more appropriate.) Therefore, for each $\sigma^{DR}(n, n_1, m)$, we map its value on the $n_{F'}$ variable, as

$$\sigma^{DR}(n_{F'}) \equiv \int Y(n_{F'}, n)\ \sigma^{DR}(n)\ dn \quad . \tag{89}$$

More precisely, we have

$$\begin{aligned} \sigma^{DR}(n_{F'}) &= \sum_{n\, n_1\, m} y(n_{F'};\ n\ n_1\ m)\ \sigma^{DR}(n\ n_1\ m) \\ &= \sum_n \left[\sum_{n_1\, m} y(n_{F'};\ n\ n_1\ m)\ \sigma^{DR}(n\ n_1\ m) \right] \\ &= \sum_n Y(n_{F'}, n)\ \sigma^{DR}(n) \quad , \end{aligned} \tag{90}$$

where

$$\sigma^{DR}(n) = \sum_{n_1\, m} \sigma^{DR}(n\; n_1\; m) \tag{91}$$

and

$$Y(\, n_{F'}\, ,\, n\,) = \sum_{n_1\, m} y(n_{F'};\, n\; n_1\; m)\;\; \sigma^{DR}(n\; n_1\; m)/\,\sigma^{DR}(n)\, . \tag{92}$$

The mapping function $Y(n_{F'}\,,\, n)$ depends specifically on the collision system, as well as on the field-mixing model. It was determined explicitly by a numerical calculation for Mg and also for Ca. Sample Y's are given in Figs. 18-20. The n-dependence of Y is quite smooth, so that the actual calculation was carried out only for a selected set of n's, and then interpolated for other n's. The mapped cross section is given in Fig. 21. As expected, they have a sharp drop near $n_F \approx 54$ for Mg, for example, which corresponds to presence of the analyzing field F_a.

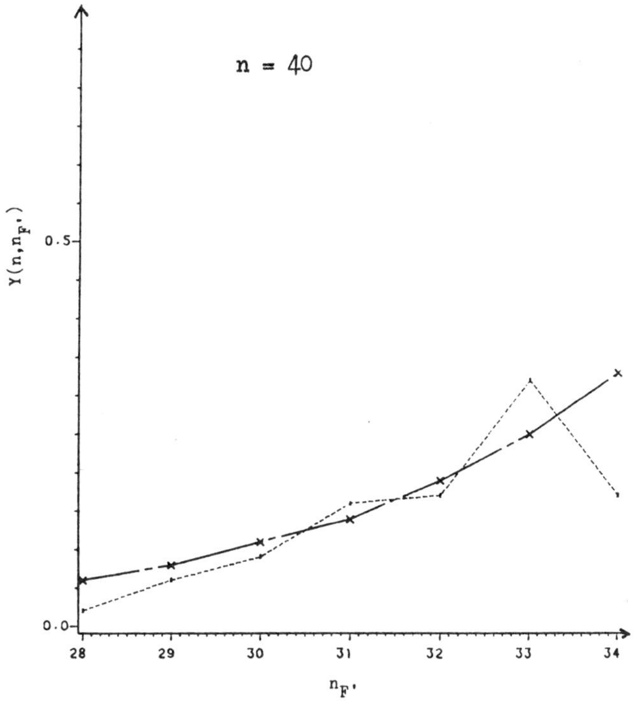

Figure 18. The mapping function $Y(\, n,\, n_{F'})$ at n=40; explicit calculation (dotted curve) is compared with a simple fitted function $\widetilde{Y}$, which is used for all n. This reduces the mapping calculation time by at least a factor of 10.

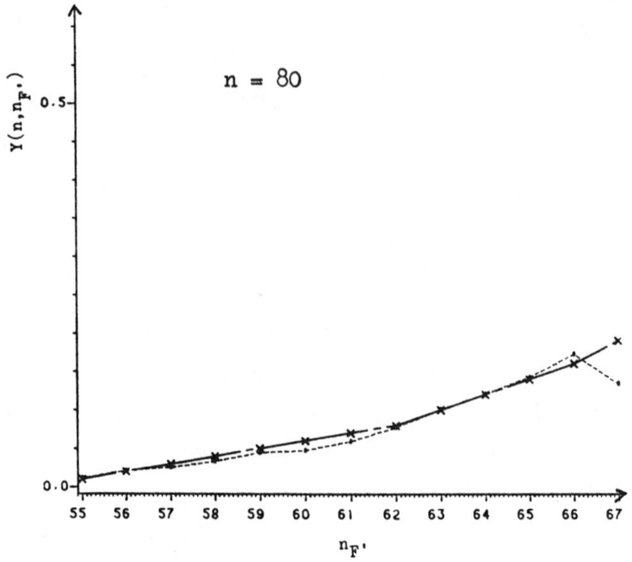

Figure 19. Same as Fig. 18, but for n=80. The fitted function $\tilde{Y}$ is much closer to the actual mapping function Y.

(Figure 20 is given in the next page.)

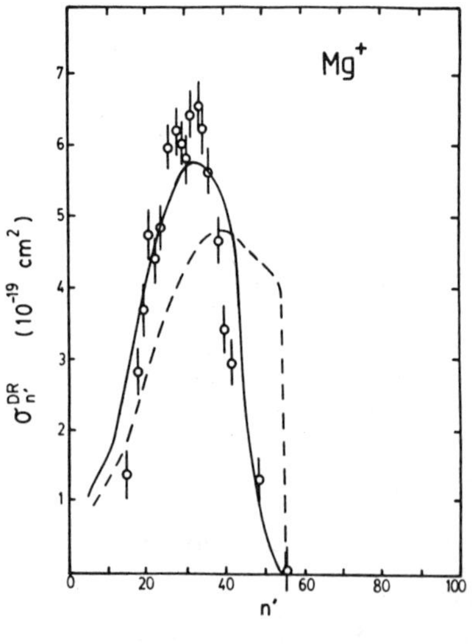

Figure 21. DR cross section is field-mapped in terms of F' and $n_{F'}$ for the Mg^+ + e system; dashed curve. The solid curve assumes a shift in $n_1 \longrightarrow n-1-|m|$ during the collision process. Experimental data are from ref. 31. The mixing field F = 24 V/cm and the cutoff field F_c = 36 v/cm were used.

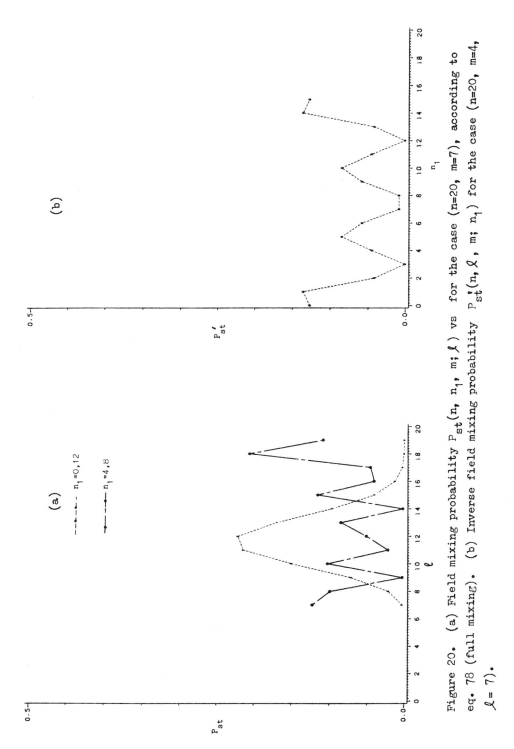

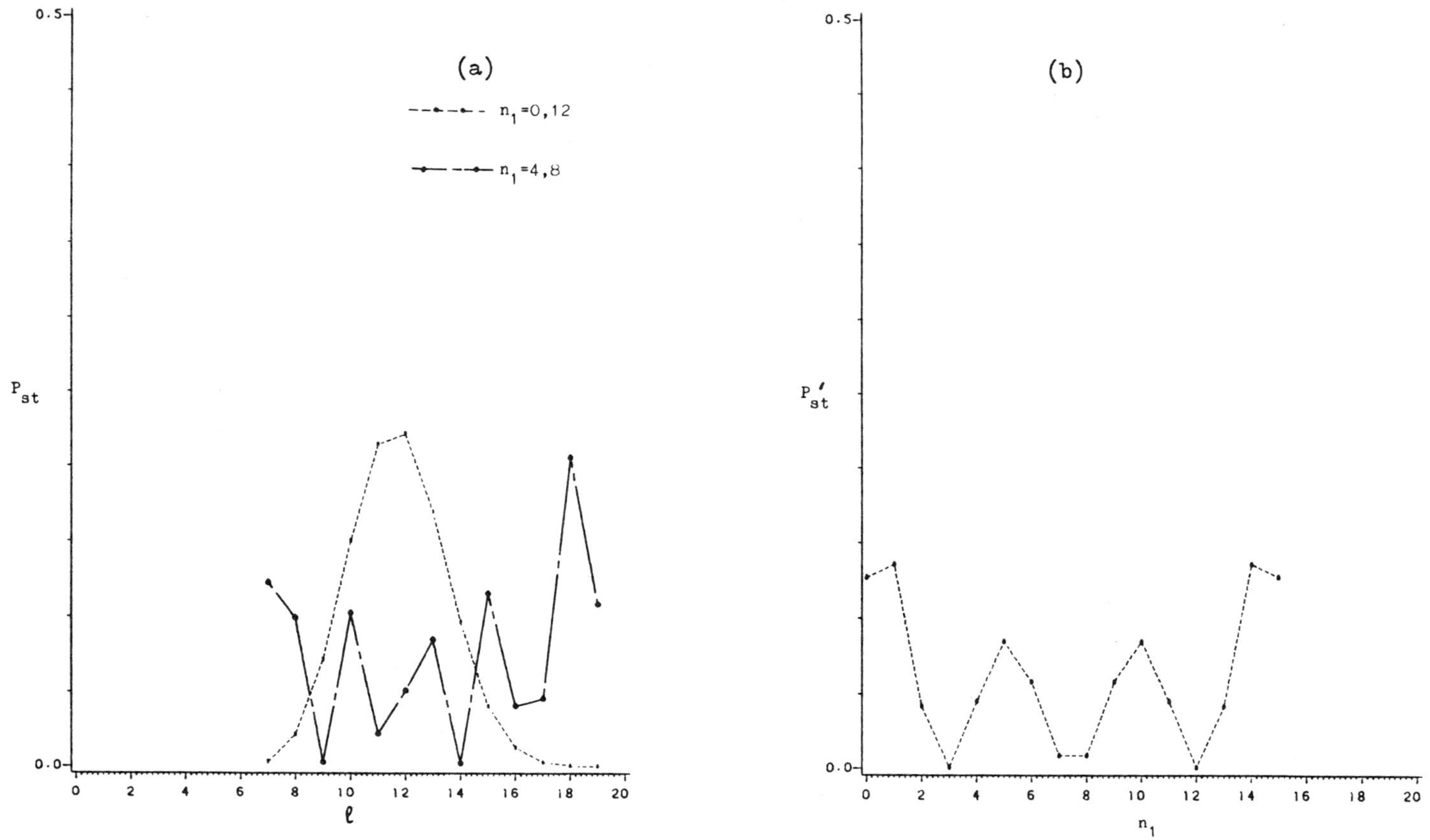

Figure 20. (a) Field mixing probability $P_{st}(n, n_1, m; \ell)$ vs for the case (n=20, m=7), according to eq. 78 (full mixing). (b) Inverse field mixing probability $P'_{st}(n, \ell, m; n_1)$ for the case (n=20, m=4, ℓ= 7).

Field Shift

In a typical experimental setup, different fields can be present in different regions along the beam trajectory. For example, in the Mg experiment, there were at least three field filled regions, each with a different field strength: F in the interaction region, F_a in the analyzer, and F' in the field mapping zone. In the rest frame of Mg*, these fields appear as a time-dependent field $F_{eff}(t)$. Depending on the magnitude of $\partial F_{eff}/\partial t$ and F_{eff} itself, we expect[42] a sizable shift in the quantum numbers (n, n_1, m) for each Mg*. Such field shifts were studied recently both experimentally[42,43] and theoretically[44]. In an effort to partially resolve the discrepancy between field-mapped theoretical cross sections and the most recent experimental data[31], we studied several extreme shift situations. It was found that when $n_1 \dashrightarrow n-1$ (maximum shift), the result seemed to reproduce the experimental shape, as shown in Fig. 21, for both F=24 and 8 Volts/cm. Another model studied simulates the case in which a field of arbitrary strength is suddenly turned off; the result is shown in Fig. 20. Fig. 22 contains the total DR cross section for Mg and Ca. With the inclusion of the field effect, the theoretical cross sections are brought into better agreement with the data, but a lot more remains to be done. (The Mg data points are the old ones; the improved data are about a factor of two higher.)

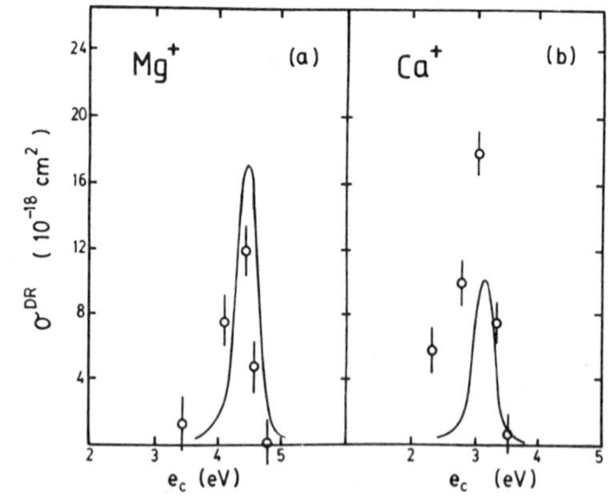

Figure 22. The field-enhanced total DR cross sections for e + Mg^+ and e + Ca^+. Experimental data are from refs. 30, 31, and 33.

***Very recently, improved experimental[31] and theoretical[36] data on the $e + Mg^{+}$ system have become available, in which the energy-integrated total cross section

$$S_{tot}^{DR} = \int \sigma^{DR} \, de_c \quad \text{(in units of cm}^2 \cdot \text{eV)}$$

was compared. We have, for the two different electric field strengths for field mixing

	F = 8 V/cm	F = 24 V/cm
S_{tot}^{DR}-exp	6.2(-18)	9.5(-18)
S_{tot}^{DR}-thy	6.3(-18)	7.4(-18)

The agreement is excellent, especially in view of the extremely difficult and delicate experimental conditions and much uncertainties in the theoretical approximations involved.

RELATED PROCESSES AND SUMMARY

In view of the difficulties in measuring the DR cross section directly, it is of some interest to consider alternate processes which may contain information relevant to DR.

Resonant-Transfer-Excitation

Resonant-transfer-excitation (RTE) is an ion-atom collision process in which an incoming ion is excited as it picks up an electron, thus forming an Auger state. This state may decay subsequently by radiation emission. The target atom thus provides the 'electron beam' in the analogous DR. Tanis et al.[45,46] measured the RTE cross sections for a variety of ions (Si^{11+}, S^{13+}, V^{Z+}, Ca^{Z+}, ...) on many different targets (He, H_2,Ne, Ar, ...) by detecting the 'captured' ions and K x-rays in coincidence. The reaction involves 1s excitation, with $\Delta n_t=0$. The data were analyzed by folding[47] the DR cross section with the Compton profile of the target atom/molecule, as

$$\bar{\sigma}^{RTE} \approx \int \bar{\sigma}^{DR} \, W_B \, d\vec{q} \approx \bar{\sigma}^{DR} \cdot W_B(\bar{q}_{2z})(\Delta e_c/K_i) \quad , \tag{93}$$

$$W_B(\bar{q}_{2z}) \equiv \int d\vec{q}_2 \left| \psi_B(\vec{q}_2) \right|^2_{q_{2z}=q_{2z}} \quad .$$

Fig. 23 contains measured data, and predictions of the simple theory. In most cases, two peaks were observed, corresponding to two distinct groups of intermediate states:

Peak I, for capture to n=2 shell;

peak II, for capture to $n \geqslant 3$ shells.

The peak ratios of these two peaks were examined[48] for different isoelectronic and isonuclear sequences, with excellent agreement between theory and experiment. More recently, the $Ca^{12+} + H_2$ system was studied[49], in which the first peak was not seen; this is entirely consistent with the theoretical prediction. In several of the data, however, the high energy side of the second peak seems to be much larger than the theory, often as much as a factor of two higher. Since this part arises largely from HRS, it is conjectured that the effect of the target core and projectile core field may produce such an enhancement through the initial and final state interactions. (It can be shown that a simple ℓ-mixing of the type discussed earlier cannot account for this increase.)

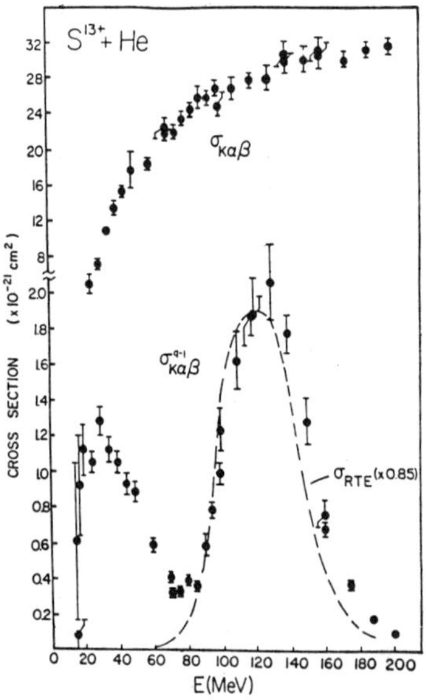

Figure 23. RTE and NTE cross sections for S^{13+} + He, and the total K x-ray production cross section $\sigma_{K\alpha\beta}$, given in ref. 46. The theoretical curve for RTE was obtained[47] using eq. 93 and σ^{DR}. Theory for the NTE peak at lower energies is less certain, especially for lighter projectiles, where the impulse approximation breaks down.

Field-mixed Auger probability A_a^F

Jaffe et al.[50] recently studied the Auger decay rate of Ba(6pnℓ) states in strong electric fields. These doubly excited states were produced by a multistep-laser-excitation technique, and ion current was recorded after the Auger decay occurred. They found that $A_a^F \sim \nu^{-4}$, where ν is the effective principal quantum number. This behavior is expected for the Auger probability with the fully field-mixed Rydberg states (n k m); i.e., since the square of the Clebsch-Gordan coefficients provides an extra ν^{-1} factor multiplied by the usual ν^{-3} behavior. Thus, part of the DR process can be studied, if resonance intermediate states are formed by some means other than electron impact. The experimental data, Fig. 24, are from ref. 50.

In addition to the change in A_a caused by external static electric field, a transient time-dependent field (such as microwave and laser fields) can shift some of the quantum numbers, thus further modifying A_a. Similar situations can also be created for A_r as well, opening up possibly a fertile area of research in which the external environment is experimentally controlled.

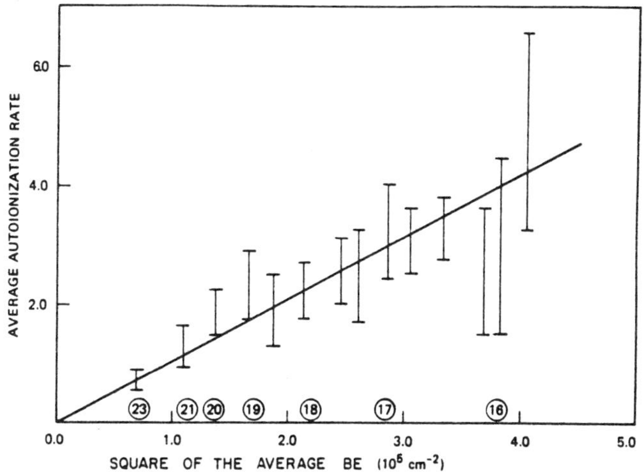

Figure 24. Average autoionization rate A_a vs the square of the average binding energy of the bound stark states of Ba(6pnℓ). The ν^{-4} dependence is consistent with the field-mixed A_a^F, where ν is the effective principal quantum number. Ref. 50.

RE, AI, and PAI

As mentioned earlier and in ref. 7, the structure of the DR theory is such that, in the IRA, resonant excitation, Auger ionization, and photo-Auger ionization are all closely related to DR. Therefore, careful measurements of any of these processes can provide information on the DR cross section. In fact, some theoretical work on RE has already been reported[51-53], while there have been[54] theoretical as well as experimental activities in AI. This topic is being reviewed by Moores and Pindzola at this workshop. Finally, PAI calculation's on PAI (the inverse of DR) and some calculations were carried out[55] earlier. The result was presented in the form of a frequency-weighted cross section, just as in the bremsstrahlung case. The PAI contribution to the total photoionization cross section can be very large, especially when L-and M-shell electron excitations are involved. Additional work is in progress on the Na and Mg sequences.

This work was supported in part by a DOE grant: DE-FG02-85 ER53205. Much of the material presented here was taken from the published work of K. LaGattuta, D. McLaughlin and I. Nasser, especially the theses of DM and IN. I would like to thank Prof. Brouillard and the organizing committee for making possible this very fruitful meeting.

References

1. A. K. Dupree, Adv. Atom. & Molec. Phys. 14, 393 (1978).

2. P. L. Dufton and A. E. Kingston. Adv. Atom. & Molec. Phys. 17, 355 (1981).

3. J. Dubau and S. Volonte', Rep. Prog. Phys. 43, 199 (1980).

4. D. E. Post, R. V. Jensen, C. B. Tartar, W. H. Grasberger and W. A. Lokke, (1977) Atomic Data and Nuclear Data Tables 20. 397.

5. C. deMichelis and M. Mattioli, Nucl. Fusion 21, 677 (1981).

6. H. W. Drawin (1983) Atomic & Molec. Proc. in Controlled Thermonuclear Fusion (S. Flavia, NATO-ASI) p19.

7. Y. Hahn (1983) Comments on Atom & Molec. Phys. 13, 103.

8. Y. Hahn (1985) Adv. Atom & Molec. Phys.21, 123.

9. Y. Hahn and K. LaGattuta, Phys. Rev. A26, 1378 (1982).

10. Y. Hahn and D. W. Rule, J. Phys. B10, 2689 (1977).

11. H. A. Kramer, Phil. Meg. 46, 836 (1923); I. I. Sobelman, Introduction to the theory of Atomic Spectre (Pergamon, NY 1972), p366.

12. C. M. Lee and R. H. Pratt, Phys. Rev. A12, 1825 (1975) and A14, 990 (1976); Y. S. Kim and R. H. Pratt, ibid A27, 2913 (1983).

13. H. A. Bethe and E. E. Salpeter, Quantum Mechanics of one and two electron atoms. (Springer-Verleg, 1974), Secs. 71 & 75.

14. J. Dubau, A. H. Gabriel, M. Loulergue, L. Steanman-Clark and S. Volonti, Mon. Not. Roy. Astro. Soc. 195, 705 (1981).

15. F. Bombarda, F. Bely-Dubau, P. Faucher, M. Cornille, J. Dubau and M. Loulergue and TFR Group. Preprint (1985) EUR-CZA-FC-1261.

16. I. Nasser and Y. Hahn, JQSRT 29, 1 (1983).

17. S. Younger, JQSRT 29, 87 (1983).

18. T. Fujimato, T. Kato and Y. Nakamura, (1982) Inst. Plasma Phys. Nagoya U. IPPJ-AM-23.

19. Y. Hahn, J. N. Gau, R. Luddy and J. A. Retter, JQSRT 23, 65 (1980).

20. L. Roszman, Phys. Rev. A20, 673 (1979).

21. D. J. McLaughlin and Y. Hahn, Phys. Rev. A29, 712 (1984).

22. K. LaGattuta and Y. Hahn, Phys. Rev. A30, 316 (1984).

23. Y. Hahn, J. N. Gau, R. Luddy, M. Dube, and N. Shkolnik, JQSRT 23, 505 (1980) and Y. Hahn, D. McLaughlin and L. LaGattuta (to be published).

24. M. P. Dube, R. Rasoanaivo and Y. Hahn, JQSRT 33, 13 (1985); M. P. Dube and K. LaGattuta (unpublished).

25. R. D. Cowan and D. C. Griffin, J. Opt. Soc. Am. 66, 1010 (1976); R. D. Cowan, The Theory of Atomic Structure and Spectra (U. of Calif. 1981) p200.

26. J. N. Gau and Y. Hahn, JQSRT, 23, 121 (1980).

27. L. LaGattuta and Y. Hahn, Phys. Rev. A31, 1415 (1985).

28. A. P. Hickman, J. Phys. B17, L101 (1984).

29. M. J. Seaton, J. Phys. B17, L531 (1984).

30. D. S. Belic, G. H. Dunn, T. J. Morgan, D. W. Mueller and C. Timmer, Phys. Rev. Lett. 50, 339 (1983).

31. G. H. Dunn, D. S. Belic, N. Djuric and D. W. Mueller, ICAP (1984), and A. Muller et al, Phys. Rev. Lett. 56, 127 (1986).

32. K. LaGattuta and Y. Hahn, J. Phys. B15, 2101 (1982).

33. J. F. Williams, Phys. Rev. A29, 2936 (1984).

34. I. Nasser and Y. Hahn, Phys. Rev. A30, 1558 (1984).

35. K. LaGattuta and Y. Hahn, Phys. Rev. Lett. 51, 558 (1983).

36. K. LaGattuta, I. Nasser and Y. Hahn. Preprint, 1985.

37. D. Griffin, C. Bottcher and M. Pindzola, ORNL preprint 1985.

38. B. A. Mitchell, C. T. Ng, J. L. Forand, D. P. Levac, R. E. Mitchell, A. Sen, D. B. Miko, and J. Wm. McGowan, Phys. Rev. Lett. 50, 335 (1983).

39. L. LaGattuta and Y. Hahn, Phys. Rev. ett. 50, 668 (1983).

40. V. L. Jacobs, J. Davis and P. C. Kepple, Phys. Rev. Lett. 37, 1390 (1976).

41. R. J. Damburg and V. V. Kolosov, J. Phys. B12, 2637 (1979).

42. R. G. Rolfes, D. B. Smith and K. B. MacAdams, J. Phys. B16, L535 (1983).

43. R. G. Hulet and D. Kleppner, Phys. Rev. Lett. 51, 1430 (1983).

44. D. Richards, J. Phys. B17, 1221 (1984).

45. J. A. Tanis, E. M. Bernstein, W. G. Graham, M. Clark, S. M. Shafroth, B. M. Johnson, K. W. Jones and M. Meron, Phys. Rev. Lett. 49, 1325 (1982).

46. J. A. Tanis, S. M. Shafroth, J. E. Willis, M. Clark, J. Swanson, E. N. Strait, Phys. Rev. Lett. 53, 2551 (1984).

47. D. Brandt, Phys. Rev. A27, 1314 (1984).

48. D. J. McLaughlin, I. Nasser and Y. Hahn, Phys. Rev. 31, 1926 (1985).

49. J. A. Tanis et al, Preprint (1985); D.J. McLaughlin and Y. Hahn, Phys. Lett. A112, 389 (1985).

50. S. M. Jaffe, R. Kachru, N. H. Tran, H. B. van Linden van den Heuvell and T. F. Gallagher, Phys. Rev. A30, 1828 (1984).

51. R. D. Cowan, J. Phys. B13, 1471 (1980).

52. P. Faucher and J. Dubau, Phys. Rev. A31, 3672 (1985).

53. M. S. Pindzola, D. C. Griffin and C. Bottcher, Phys. Rev. A32, 822 (1985).

54. D. H. Crandall, 1982. ORNL report ORNL/TM-8453.

55. Y. Hahn, Phys. Lett. A67, 345 (1978); K. LaGattuta and Y. Hahn, Phys. Rev. A25, 411 (1982).

ELECTRON-ION COLLISIONS IN THE AVERAGE-CONFIGURATION DISTORTED-WAVE APPROXIMATION†

Michael S. Pindzola,* Donald C. Griffin,** and Christopher Bottcher

Physics Division, Oak Ridge National Laboratory
Oak Ridge, TN 37831

ABSTRACT

Explicit expressions for the electron-impact excitation, ionization, and resonant-recombination cross sections are derived in the average-configuration distorted-wave approximation. Calculations using these expressions are applied to several types of phenomena in electron-ion scattering where comparison with other theoretical methods and experimental measurements can be made.

I. INTRODUCTION

The theoretical study of electron-ion scattering processes not only extends our understanding of atomic many-body structure and collision dynamics but also has important applications in laboratory and astrophysical plasma research. Even with the advent of high-speed computational machines a detailed knowledge of the time evolution of a many-particle fermion system remains elusive. In the spirit of a NATO Advanced Study a simple introductory approach to the electron-ion scattering problem is presented based on the single-configuration Hartree-Fock model of the atom. In the following sections explicit expressions are derived and numerical results are calculated in the average-configuration distorted-wave approximation for the electron-impact excitation, ionization and resonant-recombination cross sections.

The review is begun in Sect. II by deriving in the uncoupled representation the average-configuration, or array-averaged, radiative transition rate. The same methods are then applied in Sect. III to obtain expressions for the average-configuration autoionization transition

---

†Research supported by the Office of Fusion Energy, U.S. Department of Energy, under Contract No. DE-AC05-84OR21400 with Martin Marietta Energy Systems, Inc.

*Department of Physics, Auburn University, Auburn, AL 36849.

**Department of Physics, Rollins College, Winter Park, FL 32789.

rate. In Sect. IV the average-configuration resonant-recombination cross section is found by applying the principle of detailed balancing to the autoionization rates of Sect. III. Successive substitutions of continuum orbitals for bound orbitals in Sects. V and VI completes the derivation of both the average-configuration excitation and ionization cross sections. The results of calculations using the average-configuration electron-ion scattering cross sections are presented in Sect. VII, where comparisons with both more detailed theoretical methods and experimental measurements are made. Despite its simplicity the average-configuration distorted-wave approximation proves quite useful and sometimes surprisingly accurate in predicting various electron-ion scattering processes.

II. AVERAGE-CONFIGURATION RADIATIVE RATE

Besides forming a simple introduction to the averaging methods, the average-configuration radiative rate proves quite useful in calculating branching ratios for various electron-ion scattering processes. From time-dependent perturbation theory the radiative transition rate from a state i of the initial level to a state f of the final level is given by

$$A_r = \frac{4\omega^3}{3c^3} \left|\langle \gamma_f J_f M_f \right| \sum_{i=1}^{N} \vec{r}_i \left| \gamma_i J_i M_i \rangle\right|^2 , \tag{1}$$

where ω is the transition frequency, c is the speed of light and atomic units are used (1 a.u. = 27.212 eV). The labels J and M are the total angular momentum quantum numbers for the N electron state, while γ represents all other quantum numbers needed to complete the specification. The wavefunction $\langle \vec{r} | \gamma JM\rangle$ is chosen to be a linear combination of antisymmetrized product states where the single-particle spin-orbitals are given by

$$\langle \vec{r} | n\ell m_\ell m_s\rangle = \frac{P_{n\ell}(r)}{r} Y_{\ell m_\ell}(\theta,\phi)\, \chi_{ms}(\sigma) . \tag{2}$$

When the many-particle states $|\gamma JM\rangle$ involved in a transition can be adequately represented by single configurations, it is useful to define an average-configuration radiative transition rate by[1]

$$\bar{A}_r = \frac{4\bar{\omega}^3}{3c^3} \frac{\sum\limits_{\gamma_f J_f M_f} \sum\limits_{\gamma_i J_i M_i} \left|\langle \gamma_f J_f M_f \right| \sum\limits_{i=1}^{N} \vec{r}_i \left| \gamma_i J_i M_i \rangle\right|^2}{\sum\limits_{\gamma_i J_i} (2J_i + 1)} , \tag{3}$$

where $\bar{\omega}$ is the average transition frequency. The average rate $\bar{A}_r$ can be used, for instance, in analyzing optical spectra in those special cases where individual lines are not resolved.

For a given transition between configurations

$$(n_1\ell_1)^{q_1-1} (n_2\ell_2)^{q_2} \rightarrow (n_1\ell_1)^{q_1} (n_2\ell_2)^{q_2-1} , \tag{4}$$

where q is the subshell occupation number, $\bar{A}_r$ may be evaluated using any convenient angular momentum coupling scheme. An example of Eq. (4) is the transition $2s2p^2 \rightarrow 2s^2 2p$, where $q_1 = q_2 = 2$. It is especially instructive and quite simple, however, to evaluate $\bar{A}_r$ in the uncoupled representation. For a radiative transition involving the active orbitals $n_2\ell_2 \rightarrow n_1\ell_1$, the average rate $\bar{A}_r$ of Eq. (3) may also be written as

$$\bar{A}_r = \frac{4\bar{\omega}^3}{3c^3} \frac{N_t(I\rightarrow F)}{G_I} |\langle n_1\ell_1|\vec{r}|n_2\ell_2\rangle|^2_{avg} , \tag{5}$$

where G_I is the total number of states in the initial configuration, $N_t(I\rightarrow F)$ is the total number of single-particle transitions between the initial and final configurations, and $|\langle n_1\ell_1|\vec{r}|n_2\ell_2\rangle|^2_{avg}$ is the average square of the single-particle dipole matrix element. The initial configuration statistical weight is given by

$$G_I = \binom{4\ell_1 + 2}{q_1 - 1}\binom{4\ell_2 + 2}{q_2} , \tag{6}$$

where $\binom{n}{m} = n!/[n-m)!m!]$ is the binomial coefficient. The total number of single-particle transitions is given by

$$N_t(I\rightarrow F) = q_1 \binom{4\ell_1 + 2}{q_1} q_2 \binom{4\ell_2 + 2}{q_2} , \tag{7}$$

An easy way to verify Eq. (7) is to try an example, like $2s\ 2p^2 \rightarrow 2s^2\ 2p$. The active orbitals are $2p\rightarrow 2s$. There are $\binom{6}{2} = 15$ possible uncoupled states

$$(2pm_\ell m_s,\ 2pm'_\ell m'_s)$$

in the active subshell of the initial configuration which can make transitions to the $\binom{2}{2} = 1$ possible uncoupled state

$$(2sm_\ell m_s,\ 2sm'_\ell m'_s)$$

in the active subshell of the final configuration. Since there are 2 single-particle states in each of the uncoupled states in the initial active subshell and 2 single-particle states in the uncoupled state in the final active subshell, the total number of single-particle transitions is (2)(15)(2)(1) = 60.

The average square of the single-particle dipole matrix element is defined by

$$|\langle n_1\ell_1|\vec{r}|n_2\ell_2\rangle|^2_{avg} = \frac{\sum\limits_{m_{\ell_1},m_{\ell_2}} \sum\limits_{m_{s_1},m_{s_2}} \delta_{m_{s_1},m_{s_2}} |\langle n_1\ell_1 m_{\ell_1}|\vec{r}|n_2\ell_2 m_{\ell_2}\rangle|^2}{(4\ell_1 + 2)\ (4\ell_2 + 2)} . \tag{8}$$

If we write $\vec{r}$ in terms of the $\kappa = 1$ spherical harmonic tensor,

$$\vec{r} = r \sum_{\lambda=-1}^{1} C_\lambda^{(1)} , \tag{9}$$

then the dipole matrix element of Eq. (8) is given by

$$\langle n_1\ell_1 m_{\ell_1}|\vec{r}|n_2\ell_2 m_{\ell_2}\rangle$$

$$= D(12) \sum_{\lambda} \langle n_1\ell_1 m_{\ell_1}|C^{(1)}|n_2\ell_2 m_{\ell_2}\rangle$$

$$= D(12) \sum_{\lambda} (-1)^{m_{\ell_2}} \sqrt{(2\ell_1+1)(2\ell_2+1)} \begin{pmatrix} \ell_1 & 1 & \ell_2 \\ 0 & 0 & 0 \end{pmatrix} \begin{pmatrix} \ell_1 & 1 & \ell_2 \\ m_{\ell_1} & \lambda & -m_{\ell_2} \end{pmatrix} , \quad (10)$$

where the radial dipole matrix element is given by

$$D(12) = \int_0^\infty P_{n_1\ell_1}(r)\; r\; P_{n_2\ell_2}(r)\; dr \quad . \quad (11)$$

After squaring the dipole matrix element of Eq. (10), substituting into Eq. (8), and then making use of the properties of sums over 3-j symbols, one obtains

$$|\langle n_1\ell_1|\vec{r}|n_2\ell_2\rangle|^2_{avg} = \frac{2\ell_>\,[D(12)]^2}{(4\ell_1+2)(4\ell_2+2)} \quad . \quad (12)$$

where $\ell_> = \max\{\ell_1,\ell_2\}$.

Substituting G_I of Eq. (6), $N_t(I\to F)$ of Eq. (7), and $|\langle n_1\ell_1|\vec{r}|n_2\ell_2\rangle|^2_{avg}$ of Eq. (12) into Eq. (5) for $\bar{A}_r$, we obtain

$$\bar{A}_r = \frac{8\bar{\omega}^3}{3c^3} \frac{q_1 q_2 \begin{pmatrix} 4\ell_1+2 \\ q_1 \end{pmatrix}\begin{pmatrix} 4\ell_2+2 \\ q_2 \end{pmatrix}}{\begin{pmatrix} 4\ell_1+2 \\ q_1-1 \end{pmatrix}\begin{pmatrix} 4\ell_2+2 \\ q_2 \end{pmatrix}} \frac{\ell_>\,[D(12)]^2}{(4\ell_1+2)(4\ell_2+2)} , \quad (13)$$

or more simply

$$\bar{A}_r = \frac{8\bar{\omega}^3}{3c^3}\, q_2 \frac{(4\ell_2+3-q_1)}{(4\ell_1+2)(4\ell_2+2)}\, \ell_>\,[D(12)]^2 \quad . \quad (14)$$

The configuration energies and bound radial orbitals needed to complete the evaluation of $\bar{A}_r$ for any atomic system may be obtained from the atomic structure code of one's choice, such as Cowan's RCN program[1] or Fischer's MCHF program.[2]

III. AVERAGE-CONFIGURATION AUTOIONIZATION RATE

The autoionization transition rate from a state i of an initial level of an N+1 electron ion to a state f of the final level of an N electron ion is given by

$$A_a = \frac{4}{k_f} |\langle \gamma_f J_f M_f| \sum_{i<j=1}^{N+1} v_{ij} |\gamma_i J_i M_i\rangle|^2 , \quad (15)$$

where $v_{ij} = 1/|\vec{r}_i-\vec{r}_j|$ and k_f is the wavenumber of the continuum final state $|\gamma_f J_f M_f\rangle$, whose asymptotic form is one times a sine function. In the single configuration approximation, an average-configuration autoionization rate is defined by[1]

$$\bar{A}_a = \frac{4}{\bar{k}_f} \frac{\sum_{\gamma_f J_f M_f} \sum_{\gamma_i J_i M_i} |\langle \gamma_f J_f M_f | \sum_{i<j=1}^{N+1} v_{ij} | \gamma_i J_i M_i \rangle|^2}{\sum_{\gamma_i J_f} (2J_i + 1)} , \tag{16}$$

where $\bar{k}_f$ is the average wavenumber. As previously noted,[3] for the first of two types of transitions between configurations given by

$$(n_1\ell_1)^{q_1} (n_2\ell_2)^{q_2} (n_3\ell_3)^{q_3} \rightarrow$$

$$(n_1\ell_1)^{q_1+1} (n_2\ell_2)^{q_2-1} (n_3\ell_3)^{q_3-1} k_4\ell_4 , \tag{17}$$

$\bar{A}_a$ may be evaluated using any convenient angular momentum coupling scheme. An example of Eq. (17) is the transition $2p^53s^23p \rightarrow 2p^63s\ k\ell$, where $q_1 = 5$, $q_2 = 2$ and $q_3 = 1$. For an autoionization transition involving the active orbitals $(n_2\ell_2,\ n_3\ell_3) \rightarrow (n_1\ell_1,\ k_4\ell_4)$, the average rate $\bar{A}_a$ in the uncoupled representation may also be written as

$$\bar{A}_a = \frac{4}{\bar{k}_4} \frac{N_t(I\rightarrow F)}{G_I} |\langle n_1\tilde{\ell_1}k_4\ell_4|\ v\ |n_2\ell_2n_3\ell_3\rangle|^2_{avg} , \tag{18}$$

where G_I is again the total number of states in the initial configuration, $N_t(I\rightarrow F)$ is the total number of transitions between the initial and final configurations, and $|\langle n_1\tilde{\ell_1}k_4\ell_4|\ v\ |n_2\ell_2n_3\ell_3\rangle|^2_{avg}$ is the average square of the two-body Coulomb matrix element. The initial configuration statistical weight is given by

$$G_I = \binom{4\ell_1 + 2}{q_1}\binom{4\ell_2 + 2}{q_2}\binom{4\ell_3 + 2}{q_3} , \tag{19}$$

while the total number of transitions is given by

$$N_t(I\rightarrow F) = (q_1 + 1)\binom{4\ell_1 + 2}{q_1 + 1} q_2 \binom{4\ell_2 + 2}{q_2} q_3 \binom{4\ell_3 + 2}{q_3}\binom{4\ell_4 + 2}{1} , \tag{20}$$

in analogy with Eqs. (6) and (7) of Sect. I.

The average square of the Coulomb matrix element may be separated into direct, exchange and cross terms,

$$|\langle n_1\tilde{\ell_1}k_4\ell_4|\ v\ |n_2\ell_2n_3\ell_3\rangle|^2_{avg}$$

$$= |\langle n_1\ell_1k_4\ell_4|\ v\ |n_2\ell_2n_3\ell_3\rangle|^2_{avg} + |\langle k_4\ell_4n_1\ell_1|\ v\ |n_2\ell_2n_3\ell_3\rangle|^2_{avg}$$

$$- 2\ |\langle n_1\ell_1k_4\ell_4|\ v\ |n_2\ell_2n_3\ell_3\rangle\langle k_4\ell_4n_1\ell_1|\ v\ |n_2\ell_2n_3\ell_3\rangle|_{avg} . \tag{21}$$

The average direct term is defined by

$$|\langle n_1\ell_1k_4\ell_4|\ v\ |n_2\ell_2n_3\ell_3\rangle|^2_{avg}$$

$$= \sum_{\substack{m_{\ell_1},m_{\ell_2}\\ m_{\ell_3},m_{\ell_4}}} \sum_{\substack{m_{s_1},m_{s_2}\\ m_{s_3},m_{s_4}}} \frac{\delta_{m_{s_1},m_{s_2}}\delta_{m_{s_3},m_{s_4}} |\langle n_1\ell_1m_{\ell_1}k_4\ell_4m_{\ell_4}|\ v\ |n_2\ell_2m_{\ell_2}n_3\ell_3m_{\ell_3}\rangle|^2}{(4\ell_1 + 2)\ (4\ell_2 + 2)\ (4\ell_3 + 2)\ (4\ell_4 + 2)} . \tag{22}$$

If we write v in terms of a product of spherical harmonic tensors,

$$v = \sum_{\kappa} \frac{r_<^{\kappa}}{r_>^{\kappa+1}} \sum_{\lambda=-\kappa}^{\kappa} (-1)^{\lambda} C_{-\lambda}^{(\kappa)} C_{\lambda}^{(\kappa)} , \tag{23}$$

then the direct matrix element of Eq. (22) is given by

$$\langle n_1\ell_1 m_{\ell_1} k_4\ell_4 m_{\ell_4}| \, v \, |n_2\ell_2 m_{\ell_2} n_3\ell_3 m_{\ell_3}\rangle$$

$$= \sum_{\kappa} R^{\kappa}(14;23) \sum_{\lambda} (-1)^{\lambda} \langle n_1\ell_1 m_{\ell_1}\, k_4\ell_4 m_{\ell_4}| \, C_{-\lambda}^{(\kappa)} C_{\lambda}^{(\kappa)} \, |n_2\ell_2 m_{\ell_2} n_3\ell_3 m_{\ell_3}\rangle$$

$$= \sum_{\kappa} R^{\kappa}(14;23) \sum_{\lambda} (-1)^{\lambda} (-1)^{m_{\ell_2}+m_{\ell_3}} \sqrt{(2\ell_1+1)(2\ell_2+1)(2\ell_3+1)(2\ell_4+1)}$$

$$\begin{pmatrix} \ell_1 & \kappa & \ell_2 \\ 0 & 0 & 0 \end{pmatrix} \begin{pmatrix} \ell_3 & \kappa & \ell_4 \\ 0 & 0 & 0 \end{pmatrix} \begin{pmatrix} \ell_1 & \kappa & \ell_2 \\ -m_{\ell_1} & -\lambda & m_{\ell_2} \end{pmatrix} \begin{pmatrix} \ell_3 & \kappa & \ell_4 \\ -m_{\ell_3} & \lambda & m_{\ell_4} \end{pmatrix}, \tag{24}$$

where the radial Slater integral is given by

$$R^{\kappa}(14;23) = \int_0^{\infty} \int_0^{\infty} P_{n_1\ell_1}(r) \, P_{k_4\ell_4}(r') \, \frac{r_<^{\kappa}}{r_>^{\kappa+1}} \, P_{n_2\ell_2}(r) \, P_{n_3\ell_3}(r') \, dr \, dr'. \tag{25}$$

After squaring the direct matrix element of Eq. (24), substituting into Eq. (22), and then making use of the properties of sums over 3-j symbols, one obtains

$$|\langle n_1\ell_1 k_4\ell_4| \, v \, |n_2\ell_2 n_3\ell_3\rangle|^2_{avg}$$

$$= \frac{1}{4} \sum_{\kappa} \begin{pmatrix} \ell_1 & \kappa & \ell_2 \\ 0 & 0 & 0 \end{pmatrix}^2 \begin{pmatrix} \ell_3 & \kappa & \ell_4 \\ 0 & 0 & 0 \end{pmatrix}^2 \frac{[R^{\kappa}(14;23)]^2}{(2\kappa + 1)} . \tag{26}$$

The derivation of the average square of the exchange and cross terms of Eq. (21) proceeds in a similar manner to that of the direct term.

Substituting G_I of Eq. (19), $N_t(I \to F)$ of Eq. (20), and the complete expression for $|\langle n_1\ell_1 \tilde{k}_4\ell_4| \, v \, |n_2\ell_2 n_3\ell_3\rangle|^2_{avg}$ into Eq. (18) for $\bar{A}_a$, we obtain

$$\bar{A}_a = \frac{(q_1+1)\, q_2 q_3 \binom{4\ell_1+2}{q_1+1}\binom{4\ell_2+2}{q_2}\binom{4\ell_3+2}{q_3}\binom{4\ell_4+2}{1} M(14;23)}{\bar{k}_4 \binom{4\ell_1+2}{q_1}\binom{4\ell_2+2}{q_2}\binom{4\ell_3+2}{q_3}} , \tag{27}$$

where

$$M(14;23) = \sum_{\kappa} \begin{pmatrix} \ell_1 & \kappa & \ell_2 \\ 0 & 0 & 0 \end{pmatrix}^2 \begin{pmatrix} \ell_3 & \kappa & \ell_4 \\ 0 & 0 & 0 \end{pmatrix}^2 \frac{[R^{\kappa}(14;23)]^2}{(2\kappa+1)}$$

$$+ \sum_{\kappa'} \begin{pmatrix} \ell_1 & \kappa' & \ell_3 \\ 0 & 0 & 0 \end{pmatrix}^2 \begin{pmatrix} \ell_2 & \kappa' & \ell_4 \\ 0 & 0 & 0 \end{pmatrix}^2 \frac{[R^{\kappa'}(41;23)]^2}{(2\kappa'+1)} \qquad (28)$$

$$- \sum_{\kappa} \sum_{\kappa'} (-1)^{\kappa+\kappa'} \begin{Bmatrix} \ell_1 & \ell_2 & \kappa \\ \ell_4 & \ell_3 & \kappa' \end{Bmatrix} \begin{pmatrix} \ell_1 & \kappa & \ell_2 \\ 0 & 0 & 0 \end{pmatrix} \begin{pmatrix} \ell_3 & \kappa & \ell_4 \\ 0 & 0 & 0 \end{pmatrix} \begin{pmatrix} \ell_1 & \kappa' & \ell_3 \\ 0 & 0 & 0 \end{pmatrix} \begin{pmatrix} \ell_2 & \kappa' & \ell_4 \\ 0 & 0 & 0 \end{pmatrix}$$

$$R^{\kappa}(14;23)\; R^{\kappa'}(41;23) ,$$

or more simply

$$\bar{A}_a = q_2 q_3 \frac{(4\ell_1 + 2 - q_1)\,(4\ell_4 + 2)\,M(14;23)}{\bar{k}_4} . \qquad (29)$$

The second of the two types of transitions between configurations is given by

$$(n_1\ell_1)^{q_1}\,(n_2\ell_2)^{q_2} \rightarrow (n_1\ell_1)^{q_1+1}\,(n_2\ell_2)^{q_2-2}\,k_3\ell_3 . \qquad (30)$$

An example of Eq. (30) is the transition $2p^5\,3s^2 \rightarrow 2p^6\,k\ell$; where $q_1 = 5$ and $q_2 = 2$. Following the same steps as outlined above the autoionization rate is given by

$$\bar{A}_a = q_2\,(q_2 - 1) \frac{(4\ell_1 + 2 - q_1)\,(4\ell_2 + 2)\,(4\ell_3 + 2)\,M'(13;22)}{\bar{k}_3\,(4\ell_2 + 1)} , \qquad (31)$$

where

$$M'(13;22) = \sum_{\kappa} \begin{pmatrix} \ell_1 & \kappa & \ell_2 \\ 0 & 0 & 0 \end{pmatrix}^2 \begin{pmatrix} \ell_2 & \kappa & \ell_3 \\ 0 & 0 & 0 \end{pmatrix}^2 \frac{[R^{\kappa}(13;22)]}{(2\kappa+1)} \qquad (32)$$

$$- \frac{1}{2} \sum_{\kappa} \sum_{\kappa'} (-1)^{\kappa+\kappa'} \begin{Bmatrix} \ell_1 & \ell_2 & \kappa \\ \ell_3 & \ell_2 & \kappa' \end{Bmatrix} \begin{pmatrix} \ell_1 & \kappa & \ell_2 \\ 0 & 0 & 0 \end{pmatrix} \begin{pmatrix} \ell_2 & \kappa & \ell_3 \\ 0 & 0 & 0 \end{pmatrix} \begin{pmatrix} \ell_1 & \kappa' & \ell_2 \\ 0 & 0 & 0 \end{pmatrix} \begin{pmatrix} \ell_2 & \kappa' & \ell_3 \\ 0 & 0 & 0 \end{pmatrix}$$

$$R^{\kappa}(13;22)\; R^{\kappa'}(13;22) .$$

The configuration energies and bound radial orbitals needed to evaluate the two types of $\bar{A}_a$ for any atomic system may be obtained from any convenient atomic structure code. Computer subroutines can provide rapid evaluation of the 3-j and 6-j symbols and numerical evaluation of the radial Slater integrals is straightforward. The continuum radial orbitals needed to complete the evaluation of $\bar{A}_a$ may be obtained by solving the radial Schrodinger equation in the distorted-wave approximation. For rapid evaluation of many continuum orbitals a local distorting potential constructed in a semiclassical exchange approximation[4] has proved quite useful. This exchange term simplifies the solution of the differential equation and generally gives results in close agreement with results obtained from a full non-local Hartree-Fock continuum calculation. As will be seen in the next three sections, the expressions for M and M' are the main working equations for the average-configuration method applied to electron scattering processes.

IV. AVERAGE-CONFIGURATION RECOMBINATION CROSS SECTION

The inverse of the autoionizing transition between configurations of Eq. (17) is the recombination process of the type

$$(n_1\ell_1)^{q_1+1}\,(n_2\ell_2)^{q_2-1}\,(n_3\ell_3)^{q_3-1}\,k_i\ell_i \rightarrow (n_1\ell_1)^{q_1}\,(n_2\ell_2)^{q_2}\,(n_3\ell_3)^{q_3}, \quad (33)$$

where we identify the incident electron by replacing $k_4\ell_4 \rightarrow k_i\ell_i$. The recombination process involves the active orbitals $(n_1\ell_1,\ k_i\ell_i) \rightarrow (n_2\ell_2,\ n_3\ell_3)$. From the principle of detailed balancing the average-configuration recombination cross section is given by[1]

$$\bar{\sigma}_{recomb} = \frac{2\pi^2}{k_i^2\ \Delta\varepsilon} \sum_{\ell_i} \frac{G_I}{2G_F}\,\bar{A}_a\ , \quad (34)$$

where G_I is given by Eq. (19), $\bar{A}_a$ is given by Eq. (29),

$$G_F = \binom{4\ell_1+2}{q_1+1}\binom{4\ell_2+2}{q_2-1}\binom{4\ell_3+2}{q_3-1}\ , \quad (35)$$

and $\Delta\varepsilon$ is an energy width larger than the largest resonance width. Substituting the expressions for G_I, G_F, and $\bar{A}_a$ into Eq. (34) gives

$$\bar{\sigma}_{recomb} = \frac{2\pi^2}{k_i^3\ \Delta\varepsilon}\,(q_1+1)\,(4\ell_3+3-q_3)\,(4\ell_2+3-q_2) \sum_{\ell_i} (2\ell_i+1)\ M(23;1i)\ . \quad (36)$$

The inverse of Eq. (30) is the recombination process of the type

$$(n_1\ell_1)^{q_1+1}\,(n_2\ell_2)^{q_2-2}\,k_i\ell_i \rightarrow (n_1\ell_1)^{q_1}\,(n_2\ell_2)^{q_2}\ . \quad (37)$$

The recombination process involves the active orbitals $(n_1\ell_1,\ k_i\ell_i) \rightarrow (n_2\ell_2)^2$. The average-configuration recombination cross section is given by

$$\bar{\sigma}_{recomb} = \frac{2\pi^2}{k_i^3\ \Delta\varepsilon}\,\frac{(q_1+1)\,(4\ell_2+4-q_2)\,(4\ell_2+3-q_2)\,(4\ell_2+2)}{(4\ell_2+1)} \sum_{\ell_i} (2\ell_i+1)\ M'(22;1i). \quad (38)$$

V. AVERAGE-CONFIGURATION EXCITATION CROSS SECTION

A slight modification of the recombination process of Eq. (33) yields the electron excitation scattering process

$$(n_1\ell_1)^{q_1+1}\,(n_2\ell_2)^{q_2-1}\,k_i\ell_i \rightarrow (n_1\ell_1)^{q_1}\,(n_2\ell_2)^{q_2}\,k_f\ell_f\ , \quad (39)$$

where we identify the scattered electron by replacing $n_3\ell_3 \rightarrow k_f\ell_f$ and substituting $q_3 = 1$. The excitation process involves the active electrons $(n_1\ell_1,\ k_i\ell_i) \rightarrow (n_2\ell_2,\ k_f\ell_f)$. The average-configuration excitation cross section is given by

$$\bar{\sigma}_{exc} = \rho_f \ \Delta\varepsilon \sum_{\ell_f} \bar{\sigma}_{recomb} \ , \tag{40}$$

where the density of states ρ_f equals $2/\pi k_f$ for our choice of continuum normalization. Substituting for $\bar{\sigma}_{recomb}$ from Eq. (36) one finds that

$$\bar{\sigma}_{exc} = \frac{8\pi}{\bar{k}_i^3 \ \bar{k}_f} (q_1+1) \ (4\ell_2+3-q_2) \sum_{\ell_i, \ell_f} (2\ell_i+1) \ (2\ell_f+1) \ M(2f;1i) \ . \tag{41}$$

VI. AVERAGE-CONFIGURATION IONIZATION CROSS SECTION

A slight modification of the excitation process of Eq. (39) yields the electron ionization scattering process

$$(n_1\ell_1)^{q_1+1} \ k_i\ell_i \rightarrow (n_1\ell_1)^{q_1} \ k_e\ell_e \ k_f\ell_f \ , \tag{42}$$

where we identify the ejected electron by replacing $n_2\ell_2 \rightarrow k_e\ell_e$ and substituting $q_2 = 1$. The ionization process involves the active electrons $(n_1\ell_1, \ k_i\ell_i) \rightarrow (k_e\ell_e, \ k_f\ell_f)$. The average-configuration differential ionization cross section is given by

$$\frac{\overline{d\sigma_{ion}}}{d\varepsilon} = \rho_e \sum_{\ell_e} \bar{\sigma}_{exc} \ , \tag{43}$$

where $\varepsilon = \bar{k}_e^2/2$ is the ejected electron energy. Substituting for $\bar{\sigma}_{exc}$ from Eq. (41) one finds that

$$\frac{\overline{d\sigma_{ion}}}{d\varepsilon} = \frac{32}{\bar{k}_i^3 \ \bar{k}_e \ \bar{k}_f} (q_1+1) \sum_{\ell_i, \ell_f, \ell_e} (2\ell_i+1) \ (2\ell_e+1) \ (2\ell_f+1) \ M(ef;1i) \ . \tag{44}$$

The total average-configuration ionization cross section is given by

$$\bar{\sigma}_{ion} = \int_0^{E_{max}/2} \frac{\overline{d\sigma_{ion}}}{d\varepsilon} \, d\varepsilon \ , \tag{45}$$

where $E_{max} = (k_e^2 + k_f^2)/2$. Due to the presence of two outgoing continuum orbitals the phase of the interference term in Eq. (28) is arbitrary. The maximum interference approximation of Peterkop[5] takes the negative of the absolute value of the third term on the right hand side of Eq. (28).

VII. CALCULATIONS USING AVERAGE-CONFIGURATION SCATTERING CROSS SECTIONS

The average-configuration scattering cross sections find great utility in the study of electron-impact ionization of atomic ions. Consider the following processes for ionization of an atomic ion labeled A:

$$e^- + A^{n+} \rightarrow A^{(n+1)+} + e^- + e^- \ , \tag{46}$$

$$e^- + A^{n+} \rightarrow (A^{n+})^* + e^- \quad \hookrightarrow A^{(n+1)+} + e^- \ , \tag{47}$$

and

$$e^- + A^{n+} \rightarrow [A^{(n-1)+}]^*$$
$$\hookrightarrow (A^{n+})^* + e^-$$
$$\hookrightarrow A^{(n+1)+} + e^- \quad . \qquad (48)$$

The first process is direct ionization, the second is excitation-autoionization, and the third is resonant-recombination double autoionization. These processes are usually quite independent and for low stages of ionization radiative stabilization of the autoionization stages is generally negligible.

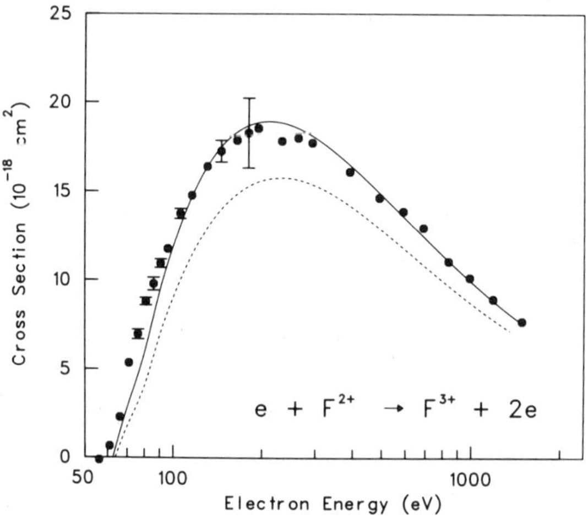

Fig. 1. Electron-impact ionization of F^{2+}. Solid curve: average-configuration distorted-wave calculation for the 2s and 2p direct ionization (Ref. 6); dashed curve: the 3-parameter Lotz formula (Ref. 7); experimental data are from Mueller et al. (Ref. 8).

For electron-impact ionization of F^{2+} the direct process dominates the cross section. In Fig. 1 average-configuration ionization cross section calculations[6] for the transitions

$$1s^2 2s^2 2p^3 k_i \ell_i \rightarrow 1s^2 2s^2 2p^2 k_e \ell_e k_f \ell_f$$
$$\rightarrow 1s^2 2s 2p^3 k_e \ell_e k_f \ell_f \quad , \qquad (49)$$

are compared with the 3-parameter semi-empirical Lotz formula[7] and the experimental crossed-beams measurements of Mueller et al.[8] The break in the theoretical curve at 80 eV is the onset of the 2s ionization. Although the remarkable agreement between theory and experiment in F^{2+} is not typical of other atomic ions, the direct ionization process seems to be described to fairly good accuracy by the average-configuration distorted-wave approximation of Eq. (45).

For electron-impact ionization of S^{4+} both the direct process and the excitation-autoionization process contribute to the total cross section. In Fig. 2 average-configuration excitation and ionization cross section calculations[9] for the transitions

$$2p^6 3s3p k_i \ell_i \rightarrow \begin{cases} 2p^5 3s3p^2 k_f \ell_f \;, \\ 2p^5 3s3p3d k_f \ell_f \;, \\ 2p^5 3s3p4p k_f \ell_f \;, \\ 2p^5 3s3p4d k_f \ell_f \end{cases} \tag{50}$$

and

$$2p^6 3s3p k_i \ell_i \rightarrow \begin{cases} 2p^6 3s k_e \ell_e k_f \ell_f \;, \\ 2p^6 3p k_e \ell_e k_f \ell_f \;, \\ 2p^5 3s3p k_e \ell_e k_f \ell_f \;, \end{cases} \tag{51}$$

are compared with the experimental crossed-beams measurements of Howald et al.[10] The break in the theoretical curve at 170 eV is the onset of the excitation-autoionization processes. Besides the remarkable agreement between theory and experiment, one of the interesting features of the S^{4+} study is confirmation that the metastable $2p^6 3s3p$ configuration dominates the composition of the ion beam, since the agreement between theory and experiment for ionization from the $2p^6 3s^2$ configuration is not as good. Note also that 2p direct ionization would contribute to the formation of S^{6+} except for the fact that a majority of the states of the $2p^5 3s3p$ configuration live long enough to be detected as S^{5+}.

For electron-impact ionization of Ti^{3+} the excitation-autoionization process dominates the cross section. In Fig. 3 average-configuration excitation and ionization cross section calculations[11] for the transitions

$$3p^6 3d\; k_i \ell_i \rightarrow 3p^5 3d^2\; k_f \ell_f \;, \tag{52}$$

and

$$3p^6 3d\; k_i \ell_i \rightarrow 3p^6\; k_e \ell_e\; k_f \ell_f \;, \tag{53}$$

are compared with an intermediate-coupled level to level distorted-wave calculation[12] and the experimental crossed-beams measurements of Falk et al.[13] The $3p \rightarrow 3d$ average-configuration excitation cross section for Ti^{3+} has a total threshold value of 179×10^{-18} cm^2 distributed over 45 different levels. Atomic structure calculations[1] show that only 6 of the 45 levels are autoionizing. The average-configuration curve in Fig. 3 is a statistical distribution of the collision strength to those 6 levels. The more detailed level to level calculation[12] finds instead that the 6 autoionizing levels carry much more collision strength than a

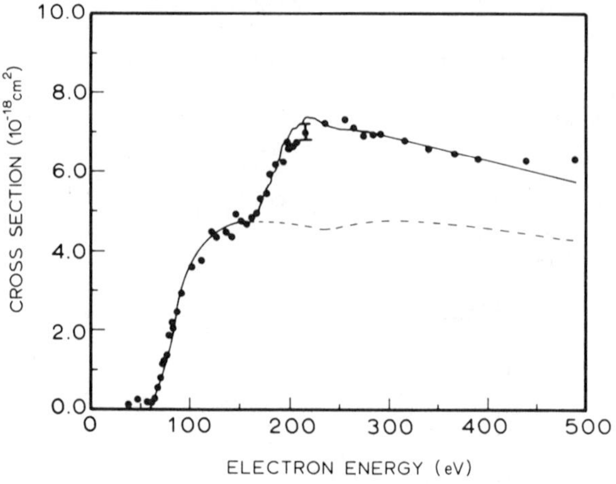

Fig. 2. Electron-impact ionization of S^{4+}. Solid curve: average-configuration distorted-wave calculation for the total ionization including excitation-autoionization of the $2p^63s3p$ configuration (Ref. 9); dashed curve: direct ionization only; experimental data are from Howald et al. (Ref. 10).

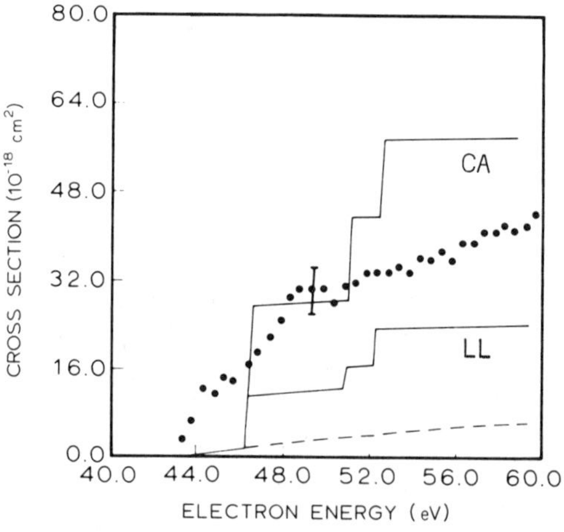

Fig. 3. Near threshold electron-impact ionization of Ti^{3+}. Solid curve marked CA: average-configuration distorted-wave calculation for the total ionization including excitation-autoionization of the $3p^63d$ configuration (Ref. 11); solid curve marked LL: level to level distorted-wave calculation for the total ionization (Ref. 12); dashed curve: direct ionization only; experimental data are from Falk et al. (Ref. 13).

statistical distribution would indicate, although the total value of 179×10^{-18} cm^2 is about the same. The average-configuration distorted-wave method may be in substantial error for those atomic systems in which the levels of the dominant excitation-autoionization configuration straddle the ionization threshold. Note that at higher energies the excitation transitions

$$3p^6 3d\ k_i \ell_i \rightarrow 3p^5 3d 4\ell\ k_f \ell_f \tag{54}$$

will certainly contribute to the total cross section, while on the other hand the 3p direct ionization contributes only to the formation of Ti^{5+}. Recent configuration-interaction level to level distorted-wave calculations[14] and configuration-interaction close-coupling calculations[15,16] have found that theory and experiment are in substantial agreement for Ti^{3+}.

For electron-impact ionization of Sb^{3+} the resonant-recombination double autoionization process contributes to the total ionization cross section. Associated with the dominant excitation-autoionization transition[17]

$$4d^{10} 5s^2\ k_i \ell_i \rightarrow 4d^9 5s^2 4f\ k_f \ell_f\ , \tag{55}$$

are the resonant-recombination transitions

$$4d^{10} 5s^2 k_i \ell_i \rightarrow 4d^9 5s^2 4f n\ell\ . \tag{56}$$

The enhancement of the total ionization cross section below the energy threshold of the $4d \rightarrow 4f$ excitation is given by

$$\bar{\sigma}_{rrda} = \sum_{n\ell} \bar{\sigma}_{recomb}(n\ell)\ \bar{B}_{da}\ (n\ell)\ , \tag{57}$$

where $\bar{B}_{da}(n\ell)$ is the branching ratio for double autoionization. An average-configuration resonant-recombination cross section calculation for the 10d, with an energy bin width of 1 eV, yields 0.13×10^{-18} cm^2.[18] Combined with the many other $n\ell$ configurations the effect of the resonant-recombination process is to substantially enhance the total cross section below the $4d \rightarrow 4f$ threshold. Care must be taken, however, in computing the double autoionization branching ratio. For $4d^9 5s^2$ 4f 10d the dominant decay mode is autoionization to the $4d^9 5s^2$ 5d configuration, which is itself an autoionizing configuration. For many atomic systems the first step decay of the doubly-excited resonant configuration is to a bound configuration, thus $\bar{B}_{da}$ is close to zero and no resonant-recombination enhancement occurs.

The process of the dielectronic recombination of an atomic ion labeled A:

$$\bar{e} + A^{n+} \rightarrow [A^{(n-1)+}]^* \rightarrow A^{(n-1)+} + h\nu\ , \tag{58}$$

can be described using the average-configuration transition rates and scattering cross sections of the previous sections. For dielectronic recombination of Mg^+ the resonant-recombination transitions are

$$2p^6 3s\ k_i \ell_i \rightarrow 2p^6 3p n\ell\ . \tag{59}$$

The average-configuration dielectronic recombination cross section is given by

$$\bar{\sigma}_{dr} = \sum_{n\ell} \bar{\sigma}_{recomb}(n\ell)\ \bar{B}_{rs}(n\ell) \ , \tag{60}$$

where $\bar{B}_{rs}(n\ell)$ is the branching ratio for radiative stabilization. For Mg^+ $\bar{B}_{rs}(n\ell) \to 1$ as $n \to \infty$, thus the total dielectronic recombination cross section peaks at relatively high n values. In Fig. 4 an average-configuration dielectronic recombination cross section calculation[19] is compared to an intermediate-coupled distorted-wave calculation[20] and the experimental crossed-beams measurements of Belic et al.[21] Due to field ionization effects in the experiment only resonances with $n \leqslant 63$ are included in the calculations. The natural Rydberg spectrum of narrow peaks converging to the 3p ionization limit at 4.5 eV has been convoluted with a 0.3 eV Gaussian to simulate the experimental resolution. Although beyond the scope of this review, field mixing effects in the experiment can be incorporated[19] in the average-configuration approximation by changing from a spherical to parabolic coordinate system for the Rydberg electron. The average-configuration results are about 25% high when compared to a more detailed intermediate-coupled matrix-diagonalized distorted-wave calculation.[20]

Although of rather limited utility, the average-configuration scattering cross sections can be used in the study of electron-impact excitation of atomic ions. Consider the following processes for excitation of an atomic ion labeled A:

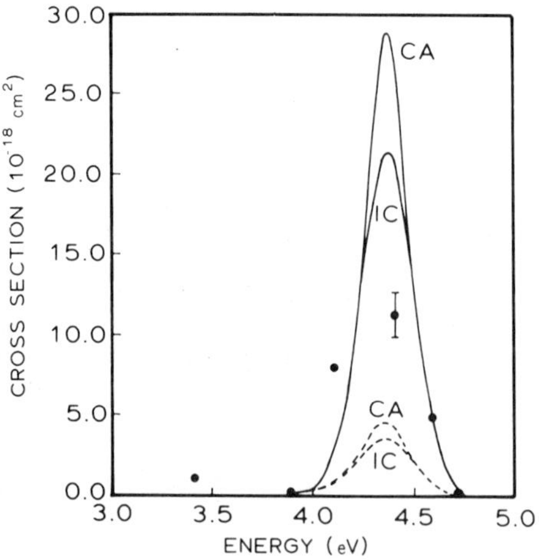

Fig. 4. Dielectronic recombination cross section for Mg^+ including all resonances with $n \leqslant 63$ convoluted with a 0.3 eV Gaussian. Solid curve marked CA: average-configuration distorted-wave calculation including complete field mixing (Ref. 19); solid curve marked IC: intermediate-coupled distorted-wave calculation at a field strength of 24 V/cm (Ref. 20); dashed curve marked CA: average-configuration distorted-wave calculation with no field mixing (Ref. 19); dashed curve marked IC: intermediate-coupled distorted-wave calculation at zero field strength (Ref. 20); experimental data are from Belić et al. (Ref. 21).

$$e^- + A_i^{n+} \rightarrow A_f^{n+} + e^- , \tag{61}$$

and

$$e^- + A_i^{n+} \rightarrow [A^{(n-1)+}]^* \rightarrow A_f^{n+} + e^- . \tag{62}$$

The first process is direct i→f excitation and the second is resonant-recombination autoionization. These processes can sometimes be treated as independent.

For most applications excitation cross sections are needed for level to level or multiplet to multiplet transitions. The average-configuration method, however, can prove quite useful for survey work, analyzing complicated transition arrays in high Z atomic systems, and treating simple alkali-like atomic ions. For the electron-impact 2s→3s excitation of O^{5+} the direct excitation process is substantially enhanced by the resonant-recombination autoionization process. Associated with the excitation transition

$$1s^2 2s k_i \ell_i \rightarrow 1s^2 3s k_f \ell_f , \tag{63}$$

are the resonant-recombination transitions

$$1s^2 2s k_i \ell_i \rightarrow \begin{cases} 1s^2 3p n\ell \\ 1s^2 3d n\ell \\ 1s^2 4\ell' n\ell . \end{cases} \tag{64}$$

The enhancement of the excitation cross section for a single Rydberg series of doubly-excited configurations is given by

$$\bar{\sigma}_{rra} = \sum_{n\ell} \sigma_{recomb}(n\ell)\, \bar{B}_{a,f}(n\ell) , \tag{65}$$

where $\bar{B}_{a,f}(n\ell)$ is the branching ratio for single autoionization leaving the ion in the final configuration, in this case $1s^2 3s$. In Fig. 5 an average-configuration excitation cross section calculation for O^{5+} is presented.[22] The natural spectrum of narrow peaks on top of the smooth background cross section has been convoluted with a 0.3 eV Gaussian to simulate a typical experimental resolution. Care must be taken again in computing the single autoionization branching ratio. Above the $1s^2 3p$ energy threshold at 83 eV, members of $1s^2 3dn\ell$ and $1s^2 4\ell' n\ell$ sequences preferentially decay to the $1s^2 3p$ continuum. Thus large resonance enhancements of the excitation cross sections are confined primarily to the near-threshold energy region. Although not shown, comparison of 2s→3s average-configuration distorted-wave results with other more detailed close-coupling[23,24] calculations is quite good.

VIII. CONCLUSIONS

The average-configuration scattering cross section equations derived in the preceding sections can quite easily be converted to machine code and implemented on any fairly large memory computer. They provide a useful theoretical tool for prediction of electron-impact excitation,

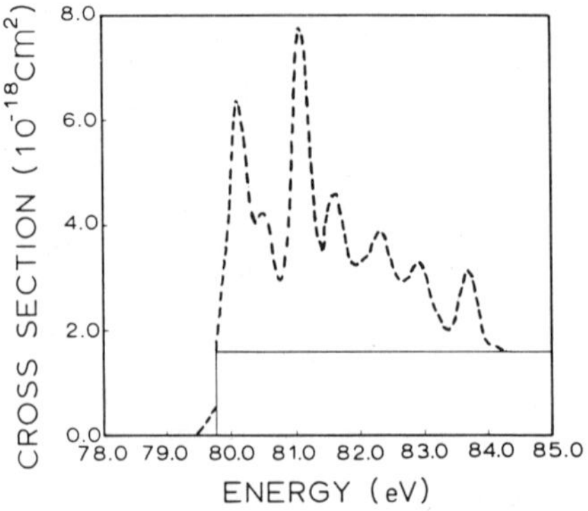

Fig. 5. Electron-impact 2s→3s excitation of O^{5+}. Dashed curve: average-configuration distorted-wave calculation for the total excitation including all resonances convoluted with a 0.3 eV Gaussian (Ref. 22); solid curve: direct excitation only.

ionization, and resonant-recombination processes. If used in conjunction with more detailed theoretical methods and new experimental techniques, they can greatly increase our understanding of the nature of electron-ion collisions.

ACKNOWLEDGMENTS

We wish to thank members of the ORNL-JILA experimental group, A. M. Howald et al. (Ref. 10), for providing us with their experimental data on S^{4+} before publication and for many valuable discussions. We also would like to acknowledge R. D. Cowan for providing us with a copy of his atomic-structure programs and a number of helpful conversations. Finally special thanks goes to D. H. Crandall for his interest and support of our theoretical efforts in electron-ion collision physics.

REFERENCES

1. R. D. Cowan, "The Theory of Atomic Structure and Spectra" (University of California, Berkeley, 1981).
2. C. F. Fischer, "The Hartree-Fock Method for Atoms" (Wiley, New York, 1977).
3. J. N. Gau and Y. Hahn, J. Quant. Spectrosc. Radiat. Transfer 23, 121 (1980).
4. M. E. Riley and D. G. Truhlar, J. Chem. Phys. 63, 2182 (1975).
5. R. K. Peterkop, Zh. Eksp. Teor. Fiz. 41, 1938 (1961) [Sov. Phys. JETP 14, 1377 (1962)].
6. S. M. Younger (private communication); see S. M. Younger, Phys. Rev. A 22, 111 (1980), for a discussion of the numerical method.
7. W. Lotz, Z. Phys. 220, 466 (1969).
8. D. W. Mueller, T. J. Morgan, G. H. Dunn, D. C. Gregory, and D. H. Crandall, Phys. Rev. A 31, 2905 (1985).
9. M. S. Pindzola, D. C. Griffin, and C. Bottcher (to be published).
10. A. M. Howald, D. C. Gregory, F. W. Meyer, R. A. Phaneuf, A. Müller, N. Djuric, and G. H. Dunn (to be published).
11. M. S. Pindzola, D. C. Griffin, C. Bottcher, D. C. Gregory, A. M. Howald, R. A. Phaneuf, D. H. Crandall, G. H. Dunn, D. W. Mueller, and T. J. Morgan, ORNL/TM-9436 (1985).
12. C. Bottcher, D. C. Griffin, and M. S. Pindzola, J. Phys. B 16, L65 (1983).

13. R. A. Falk, G. H. Dunn, D. C. Griffin, C. Bottcher, D. C. Gregory, D. H. Crandall, and M. S. Pindzola, Phys. Rev. Lett. 47, 494 (1981).
14. C. Bottcher, D. C. Griffin, and M. S. Pindzola (to be published).
15. P. G. Burke, W. C. Fon, and A. E. Kingston, J. Phys. B 17, L733 (1984).
16. R.J.W. Henry (private communication).
17. M. S. Pindzola, D. C. Griffin, and C. Bottcher, Phys. Rev. A 27, 2331 (1983).
18. D. C. Griffin, C. Bottcher, M. S. Pindzola, S. M. Younger, D. C. Gregory, and D. H. Crandall, Phys. Rev. A 29, 1729 (1984).
19. D. C. Griffin, M. S. Pindzola, and C. Bottcher, ORNL/TM-9478 (1985).
20. C. Bottcher, D. C. Griffin, and M. S. Pindzola (to be published).
21. D. C. Belic, G. H. Dunn, T. J. Morgan, D. W. Mueller, and C. Timmer, Phys. Rev. Lett. 50, 339 (1983).
22. M. S. Pindzola, D. C. Griffin, and C. Bottcher, Phys. Rev. A 32, 822 (1985).
23. K. Bhadra and R.J.W. Henry, Phys. Rev. A 26, 1848 (1982).
24. R.E.H. Clark, A. L. Merts, J. B. Mann, and L. A. Collins, Phys. Rev. A 27, 1812 (1983).

EXPERIMENTS ON DIELECTRONIC

RECOMBINATION -- A REVIEW

Gordon H. Dunn*

Joint Institute for Laboratory Astrophysics of the National Bureau of Standards and University of Colorado, Boulder Colorado 80309-0440, USA

1. INTRODUCTION

The attention given to the study of dielectronic recombination (DR) during the past three years seems matched only by the challenges to understanding which this intricate and fragile process has afforded. Good progress has been made in meeting these challenges, both experimentally and theoretically, and we focus here on experimental aspects. Since theoretical issues of DR are addressed in portions of the chapters in this volume by Hahn and by Pindzola et al., little attention will be paid here to theory or to complete referencing of such.

Important experimental contributions have come through use of several complementary methods which we will discuss here, but before doing so, we review some of the characteristics of the process. Thus, the reader will be reminded of important experimental variables, the signatures of the process, and the interrelations of these things.

There is a vast literature on DR, too extensive to be cited here. A number of general references[1-15] can be recommended, and although it misses the developments of the past three years, the review by Seaton and Storey[1] is excellent for history, applications, and understanding the general nature of DR. Of course, the early papers by Burgess[12-14] are a must for the reader.

Dielectronic recombination is the resonant capture of an electron by an ion to a doubly excited state of a once-less-charged ion followed by radiative stabilization. It can be represented by the equation

$$X^{n+} + e \underset{k_b}{\overset{k_f}{\rightleftarrows}} (X^{(n-1)+})^{**} \underset{k_r}{\longrightarrow} (X^{(n-1)+})^{*} + h\nu \qquad (1)$$

or by the schematic illustration of Fig 1.

*Staff Member, Quantum Physics Division, National Bureau of Standards.

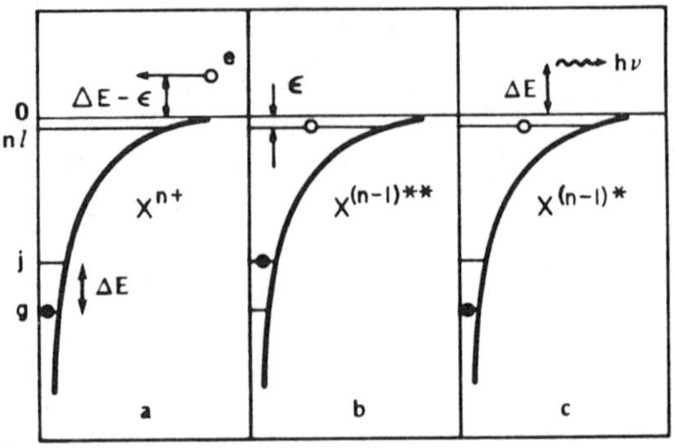

Fig. 1. Cartoon showing sequential "steps" in dielectronic recombination.

In Fig. 1a an electron is incident on an ion with an energy ($\Delta E - \varepsilon$), which is ε less than that (ΔE) needed to excite a bound electron of the ion. The electron gains kinetic energy in the Coulomb field of the ion, so that, in close, it has more than enough kinetic energy to excite the bound electron -- and it does so. Having excited the ionic electron, the incident electron finds itself in Fig. 1b with ε less energy than that needed to escape the ion, and so it is trapped in a Rydberg state of the newly formed doubly excited complex. If the two excited electrons in Fig. 1b "communicate" with each other, the core electron can give its energy back to the captured electron leading to autoionization, with the net effect of a resonance in the elastic (or in some cases, excitation) cross section. Alternatively, the core excited electron (or, less likely, the Rydberg electron) can give up its energy by emitting a photon, and the electron in Fig. 1c is trapped in an excited state -- DR is complete.

From Eq. (1) and the above discussion, one can write the cross section for DR approximately as $\sigma = \sigma_c[A_r/(A_r + A_a)]$, where σ_c is the capture cross section and the quantity in brackets represents the branching ratio between radiative stabilization (rate A_r) and autoionization (rate A_a). The capture cross section and autoionization rate are related by detailed balancing through a volume in phase space and the resonance width, making it possible to write the cross section for DR into a final Rydberg state with quantum numbers $n\ell$ as

$$\sigma_{n\ell} = \sigma_o 2(2\ell + 1)\left(\frac{Aa(n,\ell)\ Ar(n,\ell)}{Aa(n,\ell) + Ar(n,\ell)}\right) \quad . \tag{2}$$

Here σ_o involves statistical weights of the initial state and the electron core after capture, the resonance energy width, and various constants, and $2(2\ell + 1)$ is taken as the statistical weight of the final Rydberg state (n,ℓ).

For DR into a Rydberg state characterized by n, we have $\sigma_n = \sum_{\ell=0}^{n-1} \sigma_{n\ell}$, and the total cross section for DR is $\sigma = \sum_n \sigma_n$. Now, if $A_a \gg A_r$, as is often the case, then the bracketed quantity in Eq. (2) becomes approximtely just equal to A_r. Then σ_n diverges as n^2, since $\sum_{\ell=0}^{n-1} 2(2\ell+1) = 2n^2$. However, $A_a \propto 1/n^3$, i.e., electrons in very distant orbits don't "see" the core excited electron enough to make autoionization likely. Also, A_r is essentially constant with n. Thus, at high enough n, $A_a < A_r$ and $\sigma_n \propto 1/n^3$. Similarly, $A_a(n,\ell)$ decreases rapidly with ℓ [for example, $A_a \propto \exp(-a\ell^2)$, where a is a constant]. For a given (low to moderate) n, and for low ℓ, again $A_a >> A_r$, so that $\sigma_{n\ell} \propto (2\ell+1)\ A_r$, and for high ℓ, $\sigma_{n\ell} \propto (2\ell+1)\exp(-a\ell^2)$, becoming negligible for $\ell > \ell_c) \approx A_r$. The value of ℓ_c may typically lie in the range 7-10. Thus, though the number of resonances which could contribute to DR increases as $2n^2$, only ℓ's for $\ell \lesssim \ell_c$ will contribute substantially, and states with $\ell_g > \ell_c$ form a substantial reservoir of states which, if their autoionization rates were increased, could contribute to DR and enhance the cross section.

Electric fields present in the collision region may mix states of different angular momentum. States in the "reservoir" referred to above with low A_a may become mixed with states of high A_a, thus leading to an effective $A_a > A_r$. Thus, the field is like a "knob" by which the participating numbers of resonances may be tuned in and out. Very large enhancements of the cross section have been demonstrated to occur -- as will be discussed later. This effect was predicted by Jacobs et al.[16] and by Grigoriadi and Fisun[17] for plasma microfields, and has recently been treated by others[18-21] for impressed fields. Dielectronic recombination in fields (DRF) is quite different from that in no fields (DR).

Since $A_r \propto Z^4$ for $\Delta n \neq 0$, then for highly charged ions $A_r < A_a$ only for a few low values of n. The reservoir of "unused" resonances is nonexistent for these low n, and consequently small or no field effects will be expected for ions of high enough charge or for situations where low n are observed for other reasons.

Other environmental effects may also be important in DR experiments: field ionization of Rydberg products, collisional mixing of ℓ states by electrons and ions at high enough density, collisional ionization of Rydberg products, etc. These "secondary" processes affect the observations in an experiment, but not the primary DR process the way mixing by fields does.

The signatures of DR are listed in Table 1. Also shown in the table are important experimental variables which should be measurable or assessable in some way. Finally, the table lists the types of experiments, and within each type makes note of classes of experiment, distinguished by which signatures are observed and/or measured.

The remainder of this paper will discuss the various types and classes of experiment listed in Table 1. Historical order will not be followed for reasons associated with demonstrating early some of the points discussed above.

2. COLLIDING BEAMS

2.1. Coincidence Measurements

In contemplating what would constitute incontrovertible evidence that DR has taken place, one concludes that any one of signatures 1-3, along with signature 5, would be strong evidence, but that an absolutely iron-clad case would be made if one of signatures 1-3 in coincidence with no. 4 were observed along with no. 5.

Belić et al.[22] at JILA and Williams[23] at Perth performed such experiments on Mg^+ and Ca^+ respectively; in both cases product neutrals in

Table 1. Experiments on Dielectronic Recombinaton

Signatures of Dielectronic Recombination

1. Loss of electrons.
2. Loss of ions X^{n+}.
3. Gain of ions $X^{(n-1)+}$.
4. Stabilizing photons $h\nu$.
5. Resonant behavior σ vs. Ee.
6. Ions $X^{(n-1)+}$ in specific Rydberg states.

Experimental Variables on the Reaction

1. Extrinsic electric and magnetic fields.
2. Plasma fields.
3. Secondary colliding partners (electrons, ions, neutrals, photons).
4. Electron energy.
5. Electron energy distribution.
6. State distribution of target ion.

Types of Experiments and Signatures [a, b, c]

		References
1. Colliding Beams		
	[3, 4, 5]	22, 23
	[3, 5, 6]	24
	[3, 5]	27, 28, 29
	[4, (5)]	30, 31
2. Beam-Gas		
	[3, 4, 5]	33-37
3. Beam-Plasma		
	[4, 5]	40
4. Plasma		
	[4]	44-46, 48
	[2, 3]	42, 51, 52, 54-56

coincidence with stabilizing photons were measured, along with the resonant electron energy behavior (signatures 3, 4, and 5). The setups are schematically indicated in Fig. 2. The number N_c of counts in the coincidence peaks after time T were used to calculate DR cross sections from the relation

$$\sigma_{DR} = \frac{(N_c/T)e^2 v_i v_e F}{\xi_\lambda \xi_n i_e i_i (v_i^2 + v_e^2)^{1/2}} \quad . \tag{3}$$

Here v_i and v_e are ion and electron velocities, i_i and i_e are the respective beam currents, F is the beam overlap factor, and ξ_λ and ξ_n are detection efficiencies for photons and neutrals respectively.

The photon detector efficiency ξ_λ could be determined by observing the electron impact excitation of the ion resonance line and using the known cross section for the process to deduce ξ_λ. This also accurately calibrates the energy scale and in addition measures the electron energy distribution (the derivative of the excitation function). The neutral detector efficiency ξ_n could not be directly determined. In both experiments the efficiency of the detector for ions ξ_i was measured. By various routes it was then argued that $\xi_n > \xi_i$ due to Rydberg field ionization and other effects, and ξ_n values of 0.65 and 0.75 were adopted for the Mg^+ and Ca^+ experiments compared to measured ion efficiences of $\xi_i \approx 0.27$. As

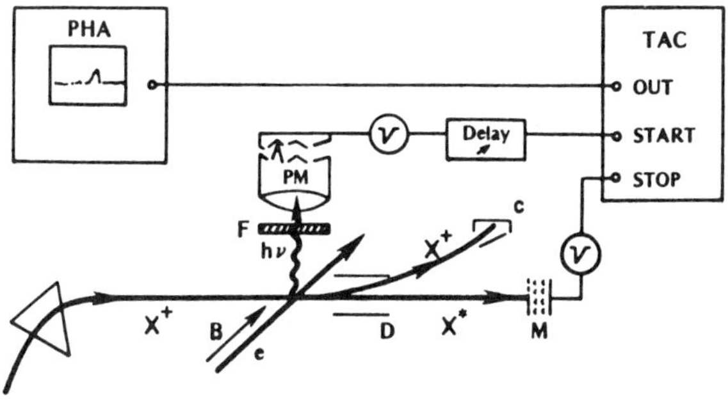

Fig. 2. Schematic of apparatus for coincidence experiments (refs. 22, 23) with signatures 3, 4, 5 (Table 1).

will be seen in Sec. 2.2, there is now evidence that $\xi_n \approx \xi_i$, at least for the Mg^+ case. A very important difference between the two experiments was that there was a magnetic field (~0.02 T) confining the electrons in the Mg^+ experiment. This gave rise to an electric field $\vec{E}_c = \vec{v}_i \times \vec{B}$ (~24 V/ cm) in the collision region. For the Ca^+ measurement, it is believed that only electric fields due to space charge (~0.3 V/cm) were present.

Results of the measurements are shown in Figs. 3 and 4. Also shown are various theoretical curves for DR and for DRF. It is seen that the theoretical DR curves lie substantially below the data. This disparity would be exaggerated further were the figure to incorporate that $\xi_n \approx \xi_i$, as noted above. It was not long after publication of the experimental data, that the hypothesis was made[18] that DRF was being observed; and it is seen that estimates of DRF are in significantly better agreement with experiments than the DR calculations.

2.1. Final Rydberg States -- Field Effects

To more definitively study DRF and unequivocally demonstrate the effects of external fields, experiments were undertaken[24] to measure $\sigma_{n_f}(E_e)$ vs. E_e and $\sigma_{E_e}(n_f)$ vs. n_f for various fields in the collision region, i.e., as the mixing of angular momentum states is varied. Here E_e is electron energy and n_f is an index tied to the principal quantum number which characterizes the final Rydberg state.

Measurements were for the process

$$e + Mg^+(3s) \rightleftarrows Mg^{**}(3p,n_f) \rightarrow Mg^*(3s,n_f) + h\nu$$

using the apparatus schematically illustrated in Fig. 5. Again there was a magnetic field confining the electrons, giving an electric field $\vec{E}_c = \vec{v}_i \times \vec{B}$; and the electric field could be changed by changing $\vec{v}_i$ or $\vec{B}$. Observed signatures in Table 1 are 3, 5, and 6.

Product Rydberg atoms $Mg^*(3s,n_f)$ were allowed to pass into a region with an electric field gradient. Here, atoms in different Rydberg states were field ionized at different spatial locations. Resulting electrons

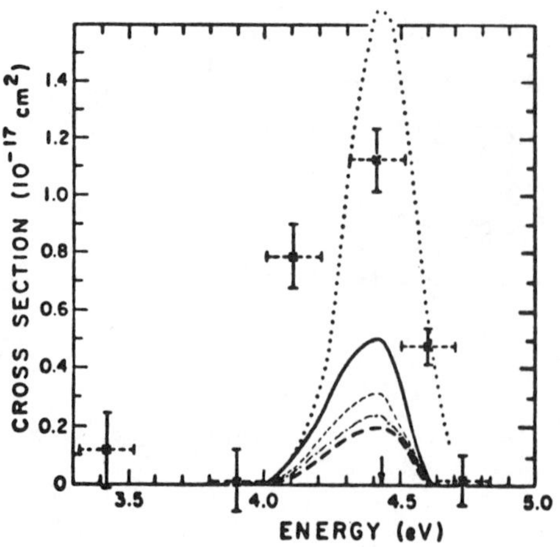

Fig. 3. Points represent cross section results from coincidence experiment (ref. 22) on Mg^+. Lower curves are calculations for DR, upper curve is for DRF. Points reduced with assumption $\xi_n = 0.65$. New evidence (see text, ref. 24) indicates $\xi_n = \xi_i \approx 0.27$, and points should be multiplied by 2.32.

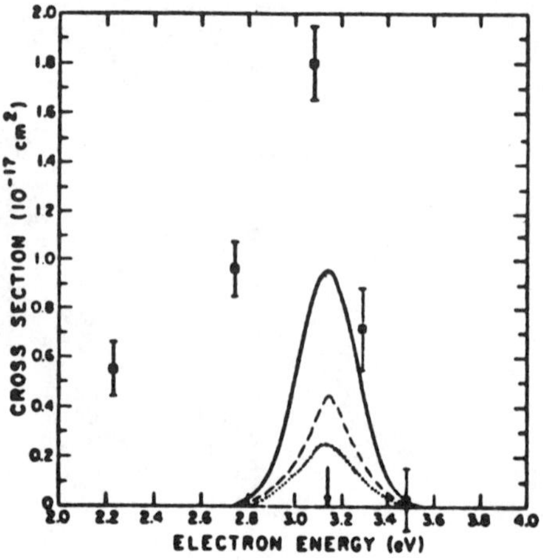

Fig. 4. Coincidence experiment (ref. 23) results for Ca^+. Curves are theoretical for DR. Points reduced assuming $\xi_n = 0.75$. If $\xi_n = \xi_i$, then points should be multiplied by 2.78.

or ions could be detected from a given state, and hence $\sigma_{n_f}(E_e)$ and $\sigma_{E_e}(n_f)$ could be measured.

It is clear from this discussion that n_f is basically a number identifying the electric field value at which a Rydberg atom is field ionized. The classical saddle-point field-ionization relationship is used here, with E_i in volts per centimeter,

$$n_f = (3.2 \times 10^8/E_i)^{1/4} \quad . \tag{4}$$

In a real atom the presence of an electric field leads to Stark splitting, and the field-ionization lifetime $\tau(n_1,n_2,m,n)$ depends on the Stark state quantum numbers n_1,n_2,m,n and on the electric field E_i. Assuming hydrogenic states, and assuming a statistical population of all Stark levels, the detection probability of the detector was calculated as a

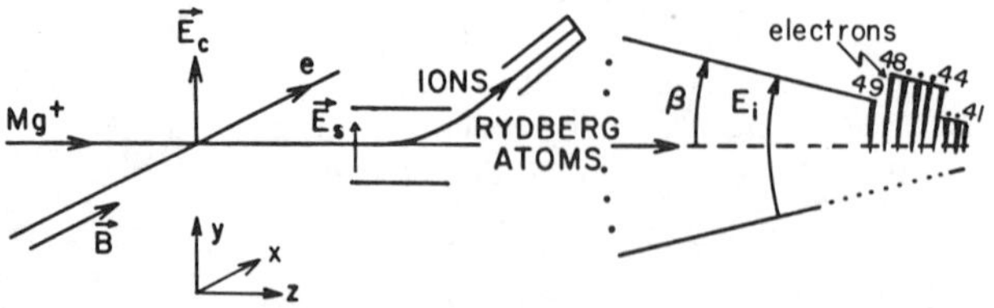

Fig. 5. Apparatus for study of Rydberg state distribution in DRF (ref. 24).

function of principal quantum number n, and it was found that $\bar{n}$ = 1.23 n_f. Thus, an expression $n = (7.3 \times 10^8/E_i)^{1/4}$ analogous to Eq. (4) could have been chosen to index the field (possibly) more directly to n, but the assumptions of hydrogen-like atoms and statistical state populations are special and not necessarily correct. Hence, to emphasize the fact that theoretical results must be "mapped" to compare with experiment, the simple field "index" n_f was chosen.

Figure 6 shows results of measurements for $\sigma(n_f = 33)$ vs. electron energy for two values of extrinsic field in the collision region. Since the electron energy distributions were slightly different for the two measurements, it is appropriate to compare the resonance strengths. The strength of the n_f = 33 resonance represented by

$$S(n_f) = \int \sigma(n_f, E_e)dE_e = \sigma(n_f, 4.43\ \text{eV})\Delta E_e$$

is 3.70×10^{-19} cm^2 eV for E_c = 23.5 V cm^{-1} and is 2.23×10^{-19} cm^2 eV for E_c = 7.24 V cm^{-1}. The ratio is 1.66, and the total uncertainty of these integral values is less than 10%.

Figure 7 shows resonance strenth as defined above vs. n_f for three different fields, clearly demonstrating that extrinsic fields in the collision region affect both cross section magnitude and Rydberg state distribution.

Clearly, if one sums $S(n_f)$ in Fig. 7 for the black points (E_c = 23.5 V cm^{-1}) and makes a reasonable correction for $n_f < 17$, there should result a value equal to that obtained by integrating the cross section curve in Fig. 3. Provided one assumes that $\xi_n = \xi_i$, as discussed in Sec. 2.1, so that the ordinate scale is multiplied by 2.32, there results from Fig. 3, $S = 7.7 \times 10^{-18}$ cm^2 eV. This is to compare with $S = 9.5 \times 10^{-18}$ cm^2 eV from Fig. 7, giving a 23% difference. LaGattuta et al.[25] obtained a theoretical value of 7.4×10^{-18} cm^2 eV and Bottcher et al.[26] obtain about 7.5×10^{-18} cm^2 eV. For E_c = 7.24 V cm^{-1}, Fig. 7 yields

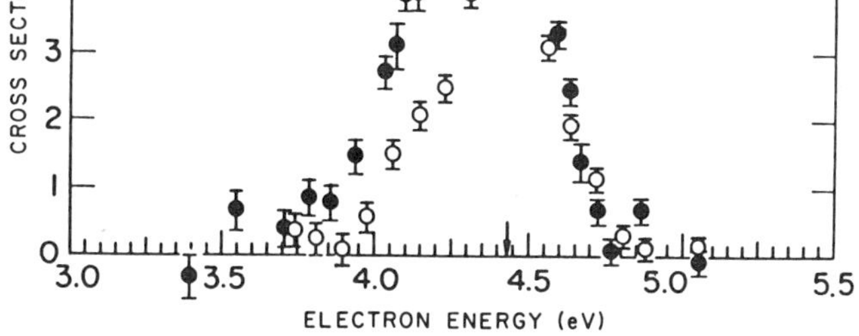

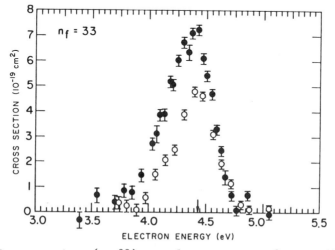

Fig. 6. Cross section $\sigma(n_f=33)$ vs. electron energy for two different extrinsic fields: solid circles, E_c = 23.5 V cm^{-1}; open circles, E_c = 7.24 V cm^{-1}. Uncertainties are one statistical standard deviation. The arrow indicates the excitation threshold energy.

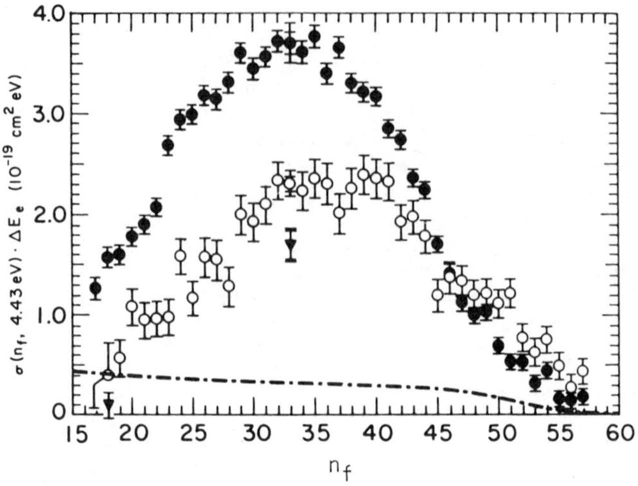

Fig. 7. Resonance strength $S(n_f)$ (see text) vs. field ionization quantum number n_f [see Eq. (4)]. Solid circles, E_c = 23.5 V cm^{-1}; open circles, E_c = 7.24 V cm^{-1}; triangles, E_c = 3.62 V cm^{-1}. The curve is the DR (no fields) calculation of Burgess mapped to n_f. Uncertainties are one statistical standard deviation. For n_f = 33, the relative uncertainties are also indicated. Absolute uncertainties at n_f = 33 are less than 10%.

$S = 6.2 \times 10^{-18}$ cm^2 eV, and this can be compared with the LaGattuta et al.[25] value of 6.3×10^{-18} cm^2 eV and the 6.05×10^{-18} cm^2 eV value of Bottcher et al.[26]

The observed dependences of $S(n_f)$ on n_f seen in Fig. 7 are reasonably consistent with the theoretical predictions both of LaGattuta et al.[25] and of Bottcher et al.[26] Both groups have shown that their theoretical predictions can be mapped to the n_f coordinate to give moderate agreement with the measurements in Fig. 7 provided suitable assumptions are made about the populations within a Stark manifold. Whether these assumptions follow physically from the collision and from field-time histories of the Rydberg particles, is a theoretical mapping issue still not addressed in full detail. Both theoretical groups find better agreement for $E_c \simeq 7$ V cm^{-1}. For $E_c \approx 24$ V cm^{-1} both groups find results substantially below the experimental values.

The experimental results unequivocally demonstrate that DRF is distinctively different from DR, and that when fields are present they must be carefully accounted for. Reasonable agreement with theory demonstrates that the process of DRF is at least moderately well understood.

2.3. Merged Beams

There are two merged beams experiments[27–29] which don't resemble each other nearly as much as the crossed beams coincidence experiments of Sec. 2.1. These are the high electron current, high signal, broad electron energy distribution measurements[28,29] at Oak Ridge National Laboratory (ORNL), and the low electron current, low signal, narrow electron energy distribution measurements[27] at Western Ontario. Because they are so different, they will be discussed separately. Signatures for both experiments are numbers 3 and 5 from Table 1.

2.3.1. The Oak Ridge experiments. The most prolific source of data and the only source of direct cross section measurements for multiply charged ions (however, see Sec. 3) has been the group at ORNL. Quite good signal-to-noise measurements on a range of Li-like,[28] Na-like,[29] and Be-like ions have been made.

Apparently, the only significant drawback to the measurements is the fact that there is a substantial field present which can be neither directly nor accurately assessed. One approach, for example, to determining the fields has been to calculate DRF rates for different fields for a given ion. Then, choosing the results which give best agreement with experiment, a value of the field can be inferred. In effect, the DRF rate for a selected ion in conjunction with theory becomes a field meter. The field so determined can then be used to compare experimental with calculated DRF rates for other ions. Of course, the field so determined is also compared with what one may reasonably estimate the fields to be based on space charge theory, estimated angles of motion in the magnetic field, and other considerations. This method has proven to be moderately -- but not unequivocally -- self consistent.

The experimental apparatus[29] is schematically shown in Fig. 8. Ions of charge q+ from a tandem accelerator enter the interaction region through an axial hole in the cathode of the electron gun. After the interaction region, the ions are charge-state analyzed using an electrostatic deflector (the early experiments[28] of this group employed magnetic analysis). Product ions of charge (q-1)+ from DRF are deflected onto a position sensitive detector (PSD). The dominant source of background is electron pickup by the q+ ions from residual gas or from slit edges.

Ion energies E_i were a few tens (say 30) MeV, electron lab energies E_e were the order of 1 keV, the solenoid field was about 0.02 T, and electron currents were about 30 mA. Effective average fields in the collision region were inferred, as discussed above, to be about 25 V cm^{-1}.

The relative center-of-mass energy of an ion and a collinear electron is given by

$$E_r = \mu[E_e/m_e + E_i/M_i - 2(E_eE_i/m_eM_i)^{1/2}] \qquad (5)$$

where μ is the reduced mass of the electron of mass m_e and ion of mass M_i. In the experimental realization, the beams are not collinear; since the electrons have angular velocities about the axis, there is possible misalignment of the beams, there is angular divergence of the ion beam, etc. However, it is assumed that these effects primarily produce a spread in E_r and that Eq. (5) gives the centroid of E_r.

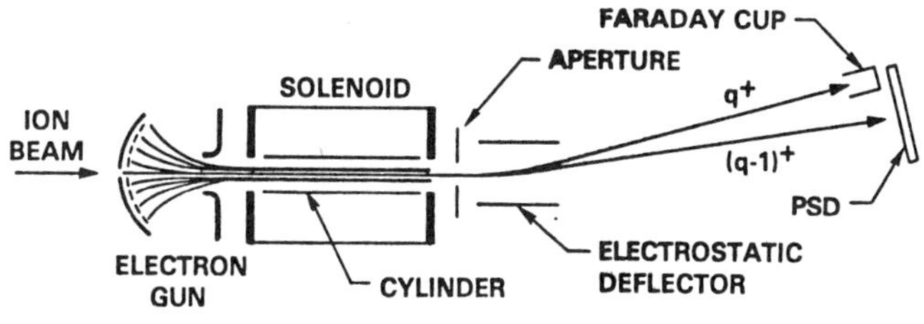

Fig. 8. Merged beams apparatus used at ORNL (ref. 29) for DRF studies with multiply charged ions.

It was found -- as will be seen from data presented below -- that the DRF resonances were quite asymmetric. Empirical electron velocity distributions were deduced from the data. In a similar way to the inference of electric field, the velocity distributions could be deduced from the data for one ion and then used in comparisons with other ions. It was found[29] necessary to propose a two-component distribution given by

$$f_{\parallel} = (\alpha/\sqrt{\pi})\ \exp[-\alpha^2(v_{\parallel}-v_{\parallel 0})^2]$$

and

$$f_{\perp} = 2\beta\ v_{\perp}\ \exp[-\beta^2 v_{\perp}^2]$$

where $v_{\parallel}$ and $v_{\perp}$ are the electron velocity components parallel and transverse to the ion beam, respectively; $v_{\parallel 0}$ is the electron velocity determined by the cathode potential, and α and β are constants.

Though various procedures could be used to change the relative collision energy, the means chosen[29] -- which minimized background and other problems -- was to set the ion energy, set the cathode potential V_c to a particular value, then modulate the potential on the tube surrounding the interaction region so that the interaction energy was alternately at the

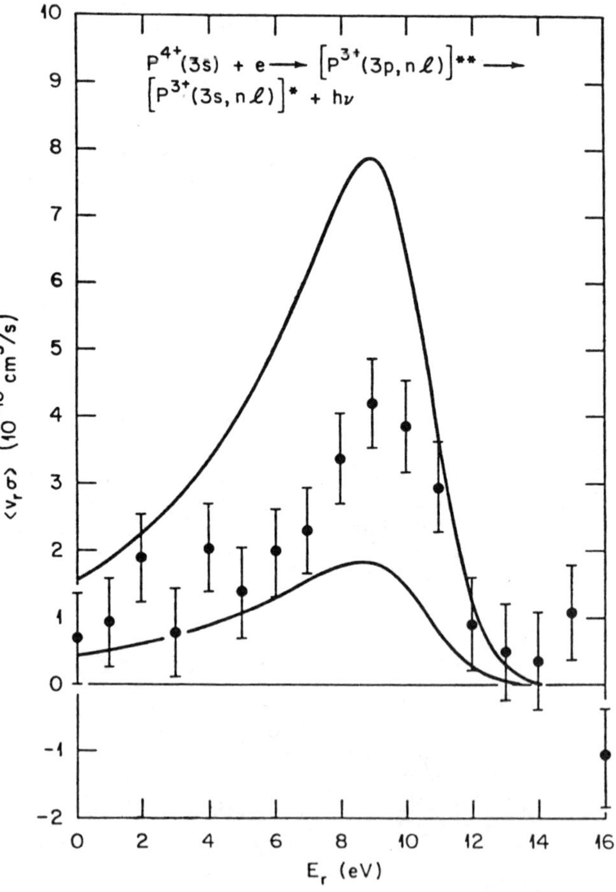

Fig. 9. The DRF rate vs. interaction energy E_r (ref. 29) for P^{4+}. Upper curve is for full field mixing; lower curve is for no mixing (DR).

energy of measurement and an energy known to be well off the DRF resonance. This procedure was not followed in the first work[28] of this group on B^{2+} and C^{3+}, and it was later shown that analysis of the data suffered from background subtraction problems. The measurements have since been re-done for these ions, and results are substantially in disagreement with theory for DR, but in moderate agreement with DRF theory; though experimental results seem to lie slightly above predictions even for full mixing by fields.

In Figs. 9 and 10, experimental results[29] are shown for P^{4+} and S^{5+}. The measured rate $\langle v_r \sigma \rangle$ is shown plotted vs. E_r. The solid curves represent theory[19]: the low curves for no field mixing and the upper curves for complete mixing. The data lie between the curves, and, as was noted earlier, an assumed field of $E_c \approx 25$ V cm^{-1} produces calculated curves in reasonable agreement with the data. The unusual empirical velocity distribution discussed above is seen to fit the data reasonably well. Even for these multiply charged ions, fields continue to play a major role.

2.3.2. Western Ontario experiment. The Western Ontario experiment[27] employs a trochoidal merger and demerger for the electron beam as

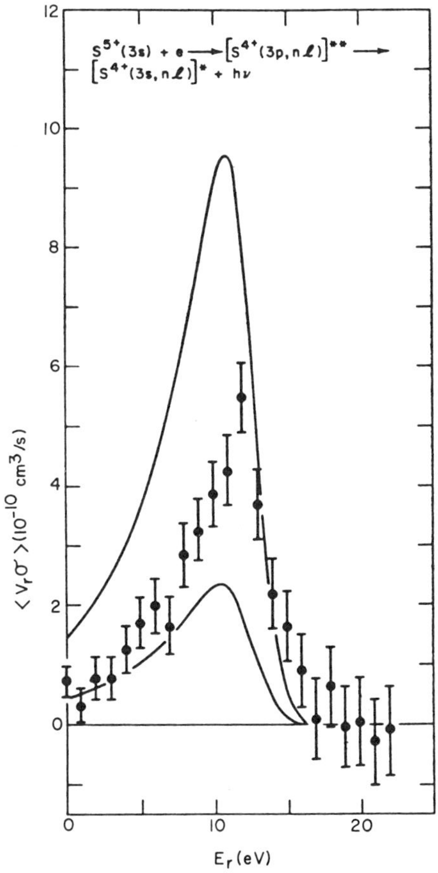

Fig. 10. DRF rates vs. interaction energy E_r for S^{5+} from ref. 29. Curves are as in Fig. 9.

shown in Fig. 11. Low beam currents are used to take advantage of kinematic energy narrowing. It is believed that electron energy resolution of near 0.045 eV was obtained in these experiments -- higher resolution by a factor of 6 than the JILA[22] and Perth[23] experiments and by a factor of more than 100 than the ORNL measurements.[28,29]

However, the good energy resolution extracts its toll in an extremely poor signal-to-noise ratio. Results of measurements on C^+ are shown in Fig. 12. It is impressive to note that the point at 9.08 eV involved

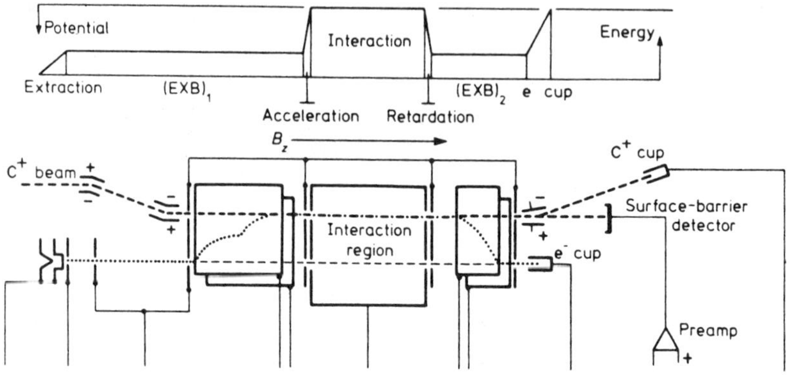

Fig. 11. Merged beams apparatus using torchoidal merger and de-merger and used for DRF experiment on C^+ (ref. 27).

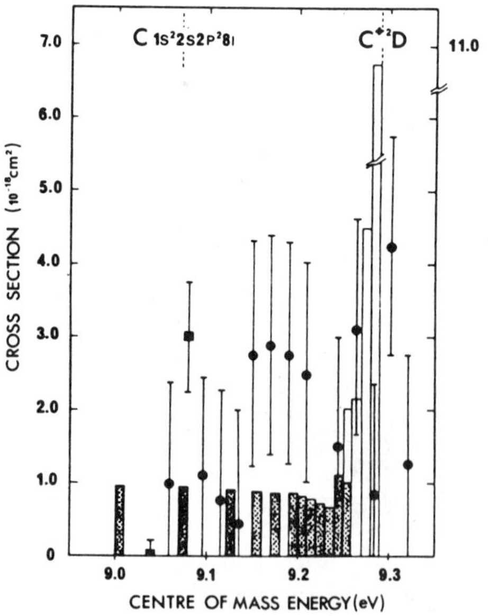

Fig. 12. DRF cross section vs. interaction energy for C^+ (ref. 27). Histogram blocks represent DR theory, and shaded blocks are what are expected to be observed in the experiment.

105 hours of integration! The entire assemblage of points represents over 700 hours of data. The measured cross sections are substantially (the authors say ~×8 in resonance strength) larger than theoretical predictions for DR, also shown in the figure.

The electric field separating C^+ from C^* (3.6 kV cm^{-1}) is large enough that C^* with $n_f > 17$ are field ionized, so the data are for $n_f \lesssim 17$. The only supposed fields in the interaction region should be Lorentz fields $\vec{v}_i \times \vec{B}$ due to small misalignment of the ion beam with the magnetic axis. With an energy resolution of 0.045 eV, one can fix the maximum misalignment and use that to infer a maximum $E_c \approx 1$ V cm^{-1}. It is difficult to see how this small field could be affecting $n_f \leq 17$. Apparently no serious attempts have been made to rationalize these results with DRF theory. It would seem that the disparity between DR theory and experiment must lie elsewhere than in the field effect.

2.4. Crossed Beams -- Observation of Photons Only

Zapesochnyî et al.[30,31] at Uzhgorod State University have collided electrons with He^+ and K^+ ions in a crossed beams configuration and observed photons through a monochromator in the third orthogonal direction. Ions of 13 keV energy and current density of 8×10^{-4} A cm^{-2} were crossed by electrons with current density of 2×10^{-2} A cm^{-2} and stated energy resolution 2.5 eV. Background pressures were $\approx 10^{-7}$ Torr, and signal/background ratios of between 1 and 1/23 were obtained with photon resolution of about 2 nm. They observed peaks in the apparent excitation cross section below the excitation threshold which they attribute to dielectronic recombination -- signatures 4 and 5 from Table 1.

Results of these measurements are shown in Figs. 13 and 14 for He^+ and K^+ respectively. Comparing the resonance strength $S = \int \sigma dE_e$ measured

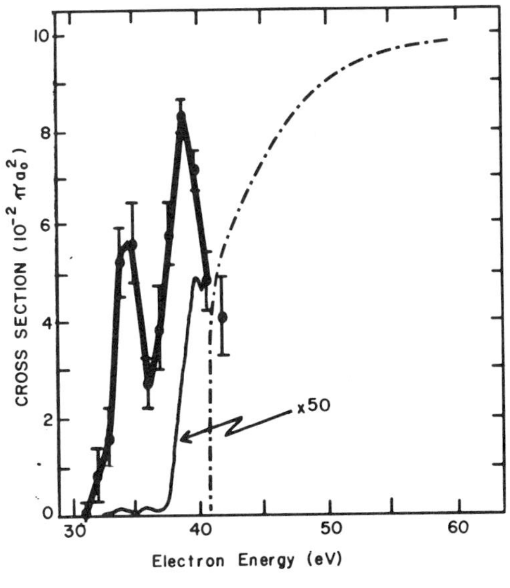

Fig. 13. Cross sections vs. energy from observation (ref. 30) of photons in crossed beams of electrons and He^+. Heavy curve is drawn through experimental points. Light curve is theoretical for DR multiplied by 50. Dot-dashed curve is through experiment for excitation.

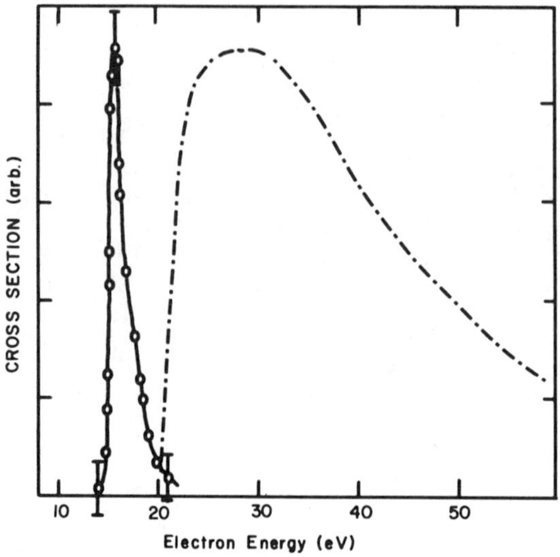

Fig. 14. Cross section in arbitrary units vs. energy from observations (ref. 31) of photons in crossed beams. Points and solid curve are attributed by authors to DR, dot-dashed curve to excitation.

here with that predicted by the Burgess formula yields $S_z/S_B \approx 10^4$! In both cases, the cross section at the resonance peak is comparable to the excitation cross section (also shown). Considering the energy width of the e-beam, this is -- by present understanding -- an impossibility for DR or DRF, and one should probably look for an alternative explanation of the data.

3. BEAM-GAS EXPERIMENTS; RESONANT TRANSFER AND EXCITATION (RTE)

"When the mean kinetic energy ... of electrons ... greatly exceeds the binding energies of electrons to their nuclei, then the electrons behave almost as if they were free," -- Feagin et al.[32]

Starting with the work of Tanis et al.,[33] several groups[34-37] have made use of this to measure a process called Resonant Transfer and Excitation (RTE). The process has been related theoretically to DR by Brandt[38] and by Feagin et al.,[32] provided the above conditions hold. The process is again just Eq. (1), except that the electrons are "loosely bound" on a target atom, and the Compton profile of velocities broadens the observed resonance.

For example, one may write for Li-like sulfur incident on the electrons attached to He gas:

$$S^{13+}(1s^2 2s) + e[He^+] \rightarrow S^{12+}(1s2s^2 2p) +$$
$$+ [He^+] \rightarrow S^{12+}(1s^2 2s^2) + [He^+] = h\nu \quad .$$

Experiments have been done observing signatures 3, 4, and 5 -- i.e. by observing the stabilizing photon in coincidence with a once-less-charged ion and also observing the resonant behavior of σ vs E. The apparatus used by Clark et al.[36] is shown in Fig. 15, and is generically representative of that used by others as well.

Figure 16 shows results[35] for the reaction written above involving Li-like sulfur incident on electrons attached to He. The dashed curve in the figure represents Brandt's adaptation[38] of McLaughlin and Hahn's[39] DR cross section for the process multiplied by 0.85. The agreement is excellent, and is representative of a number of cases which have been studied.[33-37]

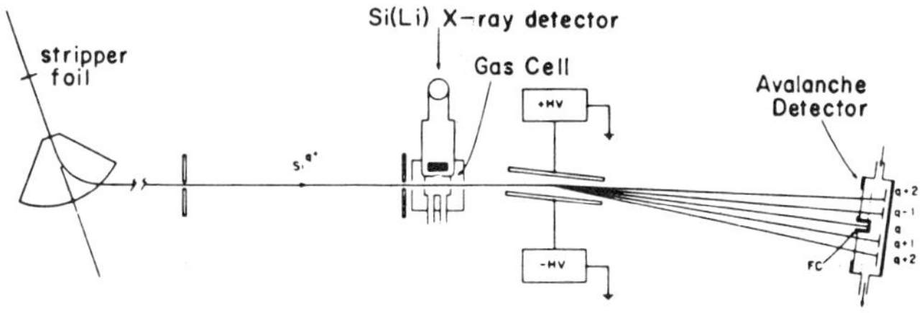

Fig. 15. Apparatus (ref. 36) typical of those used for the DR-related process of RTE.

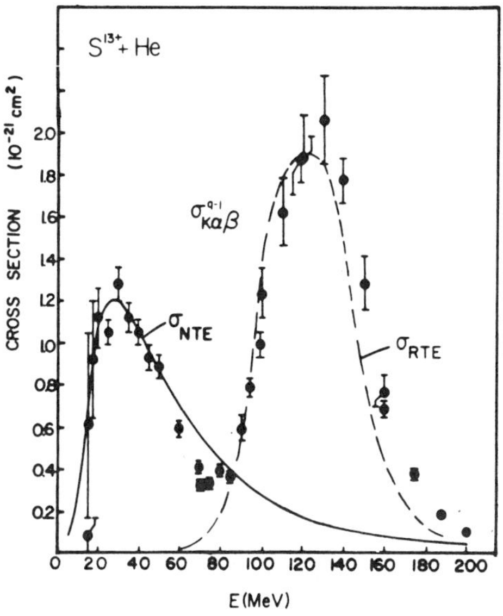

Fig. 16. Cross section vs. ion energy for K X-rays coincident with single electron capture (ref. 35). The maximum near 130 MeV is due to RTE, and the maximum near 30 MeV is attributed to another process. The dashed curve (ref. 38) is calculated using DR cross sections (ref. 39) folded with appropriate energy distributions, and multiplied by 0.85.

In all instances, correlation is to low-lying Rydberg states -- i.e. small n. Thus, fields of the nucleus, etc., don't have a role in mixing. There is no "unused" pool of ℓ states which can be mixed in for these low n. These experiments seem to demonstrate that when such complications are absent, agreement between experiment and simple DR theory is quite good.

4. BEAM-PLASMA EXPERIMENTS

In this method, stabilizing photons and the resonant behavior with energy -- signatures 4 and 5 of Table 1 -- are observed. A target of ions trapped in an electron beam ion source (EBIS) is bombarded by electrons of varying energy, and the stabilizing X-rays from DR are observed through a hole in the cathode as illustrated in Fig. 17.

Thus far, this technique has been used by Briand et al.[40] only for relative observations, though estimates of parameters indicate that experimental observations are of the same order of magnitude as theoretically predicted. To make experiments absolute would require absolute calibration of geometry involved, calibration of the X-ray detector, and a determination of ion density. Electron-beam overlap with the ion target would have to be assessed. Also, excited states of the target ions could be a problem, and fields may be difficult to assess. However, because of the broad range of charge states and targets accessible, it should be considered whether this is worth pursuing.

A modelled[41] intensity spectrum vs electron energy is shown in Fig. 18 for argon ions. The n = 2 DR shows up between 2 keV and 2.4 keV. In this range, the first peak corresponds to He-like recombination, the second and most prominent peak to Li-like recombination, and the third one to B-like recombination. Be-like recombination shows up as a shoulder between the second and third peaks. The n = 3 DR occurs around 2.75 keV, and n = 4 around 2.9 keV. The excitation curve begins around 3.1 keV.

Experimental observations[40] are shown in Fig. 19 and many of the modelled features are observed. It can be concluded that some work remains to make the method into one yielding reliable DR cross section values.

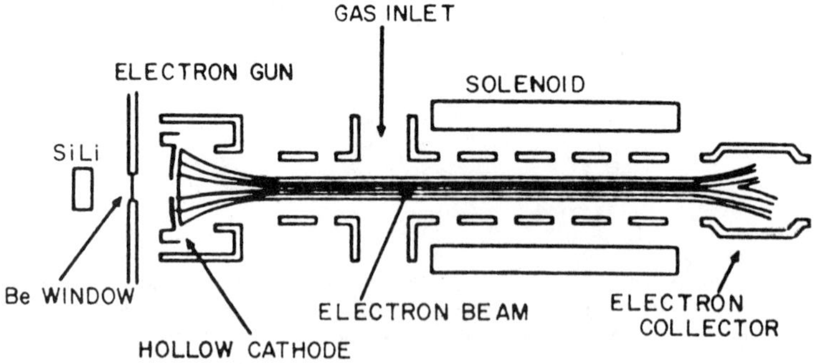

Fig. 17. EBIS ion source used (ref. 40) for study of X-ray production as electron energy is changed.

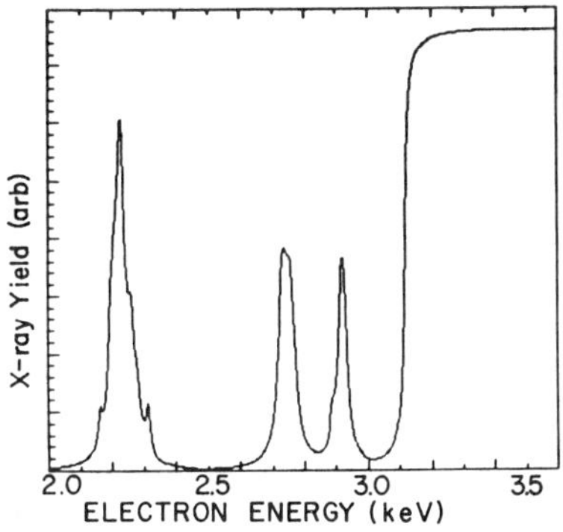

Fig. 18. A modelled intensity spectrum for expected yield vs. energy with argon in EBIS source. See text for explanation of lines (ref. 41).

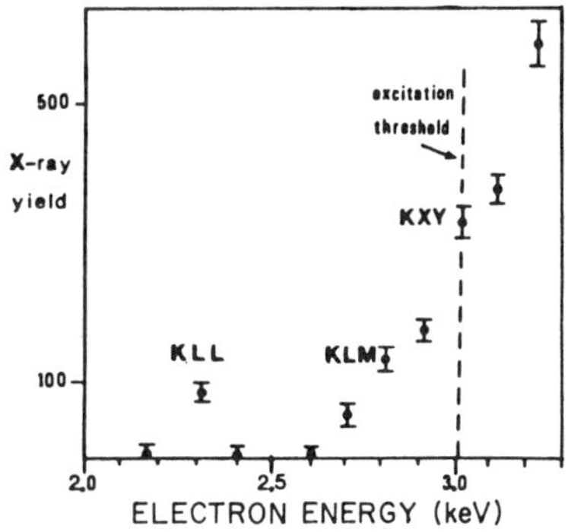

Fig. 19. Measured X-ray yield vs. electron energy in EBIS source (ref. 40). Compare Fig. 18.

5. PLASMA METHODS

The importance of dielectronic recombination is tied intimately to its role in hot plasmas, such as those in stellar coronae, fusion devices, lasers, etc. If knowledge of DR is to help understand plasmas, then it is reasonable to expect that observations on plasmas would help in understanding DR. Indeed, DR satellite lines have long been observed[4] from plasmas, and the first DR rate measurements[42] were made using plasma techniques and a θ-pinch. Two approaches will be discussed separately: the satellite line method and the method based on time behavior modelling.

5.1. Satellite Line Method

For H-like and He-like ions of moderate and high Z, we have the situation that A_r is of the same order of magnitude as A_a, and only low n's will contribute as final Rydberg states. For most of these low n's, the satellite lines can be resolved from the ion resonance line.

It was shown by Bely-Dubau et al.[43] that for these conditions

$$\alpha_d = C_R(T_e) \sum_s I_S/I_R \tag{6}$$

where α_d is the DR rate coefficient, $C_R(T_e)$ is the resonance line excitation rate coefficient, the I_S's are satellite line intensities, and I_R is the resonance line intensity. Given the resolved satellite lines for low n, one can use theory to approximately evaluate contributions from unresolved satellites from higher n. The measurement thus becomes a matter of taking highly resolved spectra under hot plasma conditions where hydrogen-like and helium-like ions are present, and using Eq. (6) (implementing theory for $C_R(T_e)$). Only signature 4 from Table 1 is observed.

Sophisticated spectroscopy has yielded[44-49] beautiful spectra analyzed as above to get α_d. Figure 20 shows a spectrum for hydrogen-like titatium (Ti^{21+}) taken using the Princeton Large Torus (PLT). The data

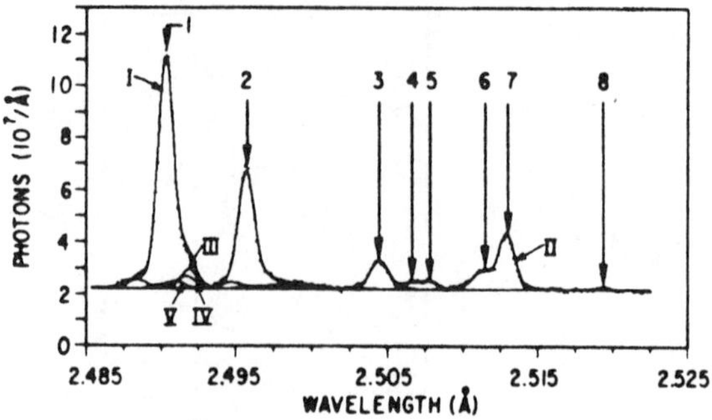

Fig. 20. Experimental (points) and theoretical spectrum for Ti^{21+} taken on PLT (ref. 44).

were recorded by a high-resolution ($\Delta\lambda/\lambda$ = 15,000, at λ = 2.5 Å) curved crystal spectrometer which permitted simultaneous observation of spectrum lines in the wavelength range from 2.485 to 2.525 Å. Also displayed in the figure are various synthetic spectra shown by solid curves, which include dielectronic satellites with n from 2 to 5 (curves II-V). The peaks labelled 3-8 are dielectronic satellites with n = 2. The fit to the observations is excellent after a small shift of 0.0003 Å in the theoretical curve.

The rate coefficient was determined as described above and excellent agreement was found with theory, as shown in Fig. 21. Several theoretical values are shown in the figure and described by the authors.[44] This method has been used to determine the DR rate coefficients for helium-like calcium,[48] titanium[45,48] and iron,[46] and for hydrogen-like titanium.[44] Tokamak plasmas[44-46] as well as laser produced[48] plasmas have been used. In all cases, agreement with theory is similar to that observed in Fig. 21.

The involvement of only low n values insures -- as has been emphasized several times in earlier sections -- that field effects are not important. For these simple structures involving high charge state and low n, the data indicate that the theory works well when no complicating features such as field effects are present. This was also shown by the RTE data in Sec. 3.

5.2. Time Behavior Modelling

Diagnostics of transient plasmas has become a sophisticated and in ways an elegant science in recent years due in large measure to impetus from controlled fusion needs. It is thus possible to monitor as functions of time the temperature and density of electrons, densities of ions and the different stages of ionization, and spatial variations of these quantities. Figure 22 illustrates a setup[50] for determining plasma rate coefficients from such measurements. The DR signatures relevant to these measurements are numbers 2 and 3 in Table 1.

The densities N_z of the ionization stages z are governed by a set of linear coupled rate equations, which is usually taken to be

$$\frac{dN_z}{dt} = N_e(S_{z-1}N_{z-1} - S_zN_z + \alpha_{z+1}N_{z+1} - \alpha_zN_z) \quad , \qquad (7)$$

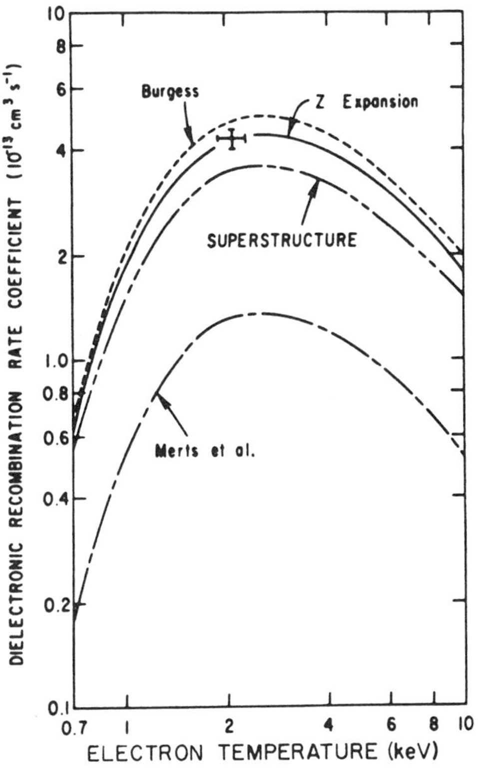

Fig. 21. Measured (ref. 44) rate coefficients for DR of Ti^{21+} obtained by analysis of spectra like Fig. 20. Curves are theoretical.

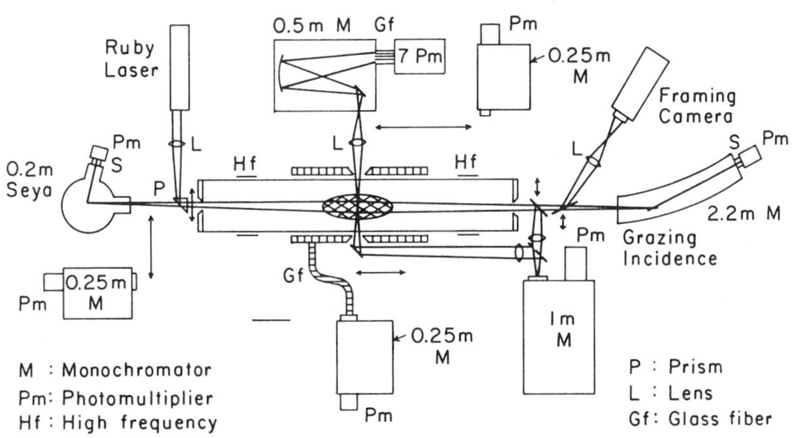

Fig. 22. Schematic overview (ref. 50) of a typical experimental apparatus for plasma rate measurements.

where N_e is the electron density, S_z is the rate coefficient for ionization of ions of charge z, and α_z is the rate coefficient for recombination. These coupled rate equations are solved numerically with the measured electron density and temperature of the plasma as input. One usually assumes values of the rate coefficients at the beginning; e.g., one may start with S_z's calculated from the Lotz formula and α_z's calculated from the Burgess formula. The rate coefficients are then adjusted until a "best fit" is obtained to the measured time histories of the N_z's, and these are declared to be the "measured" values.

Discharge conditions can sometimes be found such that ionization is most important, or conversely adjust it so that recombination is dominant. Such conditions obviously lead to more clean-cut fitting situations. It is usual -- as has been done in Eq. (7) -- to neglect multiple ionization, and this may be a meaningful deficiency.

Recombination rate coefficients deduced in this manner are total effective rate coefficients assumed to be made up of radiative recombination, dielectronic recombination, three-body recombination, and charge transfer recombination. One makes arguments to estimate or neglect the components other than DR, makes the estimated corrections, and arrives at a final DR rate coefficient. Measurements have been made using both θ-pinches and tokamaks.

Brooks et al.[42] were the first to make quantitative determinations of DR rate coefficients. They used the basic approach outlined above to make measurements for Fe^{8+}, Fe^{9+}, Fe^{10+} and Fe^{11+}. The measurements were generally within a factor of two of DR coefficients calculated using simple methods, but were very different when account was taken of the actual density of their plasma. They stated, "We would be amiss not to point out that our effective recombination rates disagree with theoretical values calculated for the actual (high) electron densities in our experiments. Such calculated values are smaller by about an order of magnitude, presumably because cross sections for electron collisions with ions in the contributing high Rydberg levels were overestimated in the calculations." This disparity has not been resolved.

Meng et al.[51,52] made measurements using this method for[52] Ar^{7+} to Ar^{11+} and for[51] C^{4+} and C^{5+}. Generally, their measured values for DR rate coefficients are 2 to 3 times greater than values calculated by Shull and van Steenberg,[53] except that the measured values are more than an order of magnitude larger for the Ne-like Ar^{8+}. They summarize, "At present, no final conclusion is possible on how plasma effects influence the net recombination rate. It is generally accepted that collisions with doubly excited states, which result in ionization, reduce recombination, and collisions leading to angular momentum redistribution enhance it. At higher densities, low-lying levels of the recombining ion are strongly populated resulting in additional recombination channels, and quasistatic ion electric fields as well as Lorentz ($\vec{v} \times \vec{B}$) electric fields induce Stark mixing among the doubly excited states, which again enhances the rate."

Isler et al.[54] made measurements on Fe^{14+}, Fe^{15+}, Fe^{17+}, and Fe^{18+} in the ISX-B tokamak. They found values smaller than given by the Burgess formula by factors ranging from 5 to 1.2. They summarize, "The totality of plasma investigations of dielectronic recombination does not yet present a completely unified picture. In some cases the agreement with theoretical calculations appears quite good, but in others the correspondence is not satisfactory. Both the experiments and theory need more work. In particular, the density dependence of the recombination rates should be examined more extensively."

Breton et al.[55] have examined recombination of highly charged (e.g. Mo^{31+}) molybdenum ions in the recombination dominated part of a sawtooth discharge of a tokamak. Their rate coefficient values are between 0.5 and 1 times values given by the Burgess formula. Most recently, Wang et al.[56] studied Ti^{16+}-Ti^{19+} in the sawtooth oscillations of the TEXT tokamak and determined rate coefficients within 20% (generally lower) of those given by the Burgess formula.

6. SUMMARY

The elegant and fragile process of dielectronic recombination has become much more fully fathomed in the past few years by subjecting the process to scrutiny using several different and complementary techniques. Beams data and plasma data on simple structures at low densities now seem at least moderately well understood. However, the effect of a combination of fields and collisions at higher densities now presents a tantalizing challenge. One may expect that the basic investigations with beams will help clear the way to progress with the density effects problems.

ACKNOWLEDGEMENTS

The author is grateful to many of the authors whose papers are cited here for discussions over the years about the mutually stimulating problems of DR. He is especially grateful to colleagues with whom he has worked closely on DR: D. S. Belić, B. DePaola, N. Djurić, T. Morgan, D. Mueller, A. Müller, and C. Timmer. The author's work in this area has been supported in part by the Office of Fusion Energy, U.S. Department of Energy, and by the National Bureau of Standards.

REFERENCES

1. M. J. Seaton and P. J. Storey, in: "Atomic Processes and Applications," P. G. Burke and B. L. Moiseiwitsch, eds., North Holland, Amsterdam (1977), p. 134.
2. R. H. Bell and M. J. Seaton, J. Phys. B 18:1589 (1985).
3. R. D. Cowan, "The Theory of Atomic Structure and Spectra," University of California Press, Berkeley (1981), p. 549.
4. J. Dubau and S. Volonte, Rep. Prog. Phys. 43:199 (1980).
5. Y. Hahn, in: "Advances in Atomic and Molecular Physics, Vol. 2," D. R. Bates and B. Bederson, eds., Academic Press, New York (1985).
6. See series of papers in "Electronic and Atomic Collisions," J. Eichler, I. V. Hertel, and N. Stolterfoht, eds., North Holland, Amsterdam (1983). S. Datz, p. 795; Y. Hahn, p. 801; G. Dunn et al., p. 809; P. F. Dittner et al., p. 819; J. P. Briand et al., p. 827; A. P. Hickman, p. 833.
7. L. J. Roszman, in: "Physics of Electronic and Atomic Collisions," S. Datz, ed., North Holland, Amsterdam (1982), p. 641.
8. A. L. Merts, R. D. Cowan, and N. H. Magee, Jr., Los Alamos Scientific Laboratory Report No. LA6220-MS (1976) (unpublished).
9. T. Fujimoto, T. Kato, and Y. Nakamura, Dielectronic recombination of hydrogenic ions, Nagoya Institute of Plasma Physics, Report IPPJ-AM-23 (1982).
10. Y. Hahn, see chapter in this volume.
11. M. S. Pindzola, D. C. Griffin, and C. Bottcher, see chapter in this volume.
12. A Burgess, Ap. J. 139:776 (1964).
13. A. Burgess, Ap. J. 141:1588 (1965).

14. A. Burgess, Ann. d'Astrophysique 28:774 (1965).
15. See series of papers in "Atomic Excitation and Recombination in Extended Fields, M. H. Nayfeh and C. W. Clark, eds., Gordon and Breach, New York (1985). Y. Hahn et al., p. 339; V. L. Jacobs, p. 351; D. C. Griffin et al., p. 367; A. R. P. Rau, p. 383; P. F. Dittner et al., p. 393; G. H. Dunn et al., p. 405; J. Kohl et al., p. 421; J. B. A. Mitchell, p. 439; J. A. Tanis et al., p. 453; J. Dubau, p. 469.
16. V. L. Jacobs, J. Davis, and P. C. Kepple, Phys. Rev. Lett. 37:1390 (1976); V. L. Jacobs and J. Davis, Phys. Rev. A 19:776 (1979).
17. A. K. Grigoriadi and O. I. Fisun, Sov. J. Plasma Phys. 8:440 (1982).
18. K. LaGattuta and Y. Hahn, Phys. Rev. Lett. 51:558 (1983).
19. D. C. Griffin, M. S. Pindzola, and C. Bottcher, Oak Ridge National Laboratory Report ORNL/TM-9478 (1985).
20. D. Harmin, private communication (1985).
21. A. P. Hickman, J. Phys. B 17:L101 (1984); also J. Phys. B 18:3219 (1985).
22. D. S. Belić, G. H. Dunn, T. J. Morgan, D. W. Mueller, C. Timmer, Phys. Rev. Lett. 50:339 (1983).
23. J. F. Williams, Phys. Rev. A 29:2936 (1984).
24. A. Müller, D. S. Belić, B. D. DePaola, N. Djurić, G. H. Dunn, D. W. Mueller, and C. Timmer, Phys. Rev. Lett. 56:127 (1986).
25. K. LaGattuta, I. Nasser, Y. Hahn, private communication (1985 manuscript).
26. C. Bottcher, D. C. Griffin, M. S. Pindzola, private communication (1985 manuscripts).
27. J. B. A. Mitchell, C. T. Ng, J. L. Foraud, D. P. Levac, R. E. Mitchell, A. Sen, D. B. Miko, and J. W. McGowan, Phys. Rev. Lett. 50:335 (1983).
28. P. F. Dittner, S. Datz, P. D. Miller, C. D. Moak, P. H. Stelson, C. Bottcher, W. B. Dress, G. D. Alton, and N. Nesković, Phys. Rev. Lett. 51:31 (1983).
29. P. F. Dittner, S. Datz, P. D. MIller, P. L. Pepmiller, and C. M. Fou, Phys. Rev. A 33:124 (1986).
30. I. P. Zapesochnyi, Ya. N. Semenyuk, A. I. Dashchenko, A. I. Imre, and A. I. Zapesochnyi, JETP Lett. 39:142 (1984).
31. I. S. Aleksakhin, A. I. Zapesochnyi, and A. I. Imre, JETP Letters 18:531 (1978).
32. J. M. Feagin, J. S. Briggs, T. M. Reeves, J. Phys. B 17:1057 (1984).
33. J. A. Tanis, E. M. Bernstein, W. G. Graham, M. Clark, S. M. Shafroth, B. M. Johnson, K. W. Jones, and M. Meron, Phys. Rev. Lett. 49:1325 (1982).
34. J. A. Tanis, E. M. Bernstein, W. G. Graham, M. P. Stockli, M. Clark, R. H. McFarland, T. J. Morgan, K. H. Berkner, A. S. Schlachter, and J. W. Stearns, Phys. Rev. Lett. 53:2551 (1984).
35. J. A. Tanis, E. M. Bernstein, M. W. Clark, W. G. Graham, R. H. McFarland, T. J. Morgan, B. M. Johnson, K. W. Jones, and M. Meron, Phys. Rev. A 31:4040 (1985).
36. M. Clark, D. Brandt, J. K. Swenson, and S. M. Shafroth, Phys. Rev. Lett. 54:544 (1985).
37. W. A. Schönfeldt, P. H. Mokler, D. H. H. Hoffmann, and A. Warczak, Zeit. f. Physik (in press).
38. D. Brandt, Phys. Rev. A 27:1314 (1983).
39. D. J. McLaughlin and Y. Hahn, Phys. Lett. 88A:394 (1982).
40. J. P. Briand, P. Charles, J. Arianer, H. Laurent, C. Goldstein, J. Dubau, M. Loulergue, and F. Bely-Dubau, Phys. Rev. Lett. 52:617 (1984).
41. M. Cornille, J. Dubau, M. Loulergue, J. P. Briand, and P. Charles, private communication, to be published (1985).

42. R. L. Brooks, R. U. Datla, and H. R. Griem, Phys. Rev. Lett. 41:107 (1978); R. L. Brooks, R. U. Datla, A. D. Krumbein, and H. R. Griem, Phys. Rev. A 21:1387 (1980).
43. F. Bely-Dubau, A. H. Gabriel, and S. Volonte, Mon. Not. R. Astron. Soc. 189:801 (1979).
44. M. Bitter, S. von Goeler, S. Cohen, K. W. Hil, S. Sesnic, F. Tenney, and J. Timberlake, Phys. Rev. A 29:661 (1984).
45. M. Bitter, K. W. Hill, M. Zarnstorff, S. von Goeler, R. Hulse, L. C. Johnson, N. R. Sauthoff, S. Sesnic, K. M. Young, M. Tavernier, F. Bely-Dubau, P. Faucher, M. Cornille, and J. Dubau, Phys. Rev. A 32:3011 (1985).
46. F. Bely-Dubau, M. Bitter, J. Dubau, P. Faucher, A. H. Gabriel, K. W. Hill, S. von Goeler, N. Sauthoff, and S. Volonte, Phys. Lett. 93A:189 (1983).
47. M. Bitter, S. von Goeler, K. W. Hill, R. Horton, D. Johnson, W. Roney, N. Sauthoff, E. Silver, and W. Stodiek, Phys. Rev. Lett. 47:921 1981).
48. B. N. Chichkov, M. A. Mazing, A. D. Shevelko, and A. M. Urnov, Phys. Lett. 83A:401 (1981).
49. A. Pospieszczyk, Astron. & Astrophys. 39:357 (1975).
50. P. Greve, M. Kato, H.-J. Kunze, and R. S. Hornady, Phys. Rev. A 24, 429 (1981).
51. H. C. Meng, H.-J. Kunze, and T. Schmidt, Phys. Lett. 105A:221 (1984).
52. H. C. Meng, P. Greve, H.-J Kunze, and T. Schmidt, Phys. Rev. A 31, 3276 (1985).
53. J. M. Shull and M. van Steenberg, Ap. J. Suppl. Ser. 48:95 (1982).
54. R. C. Isler, E. C. Crume, D. E. Arnurius, Phys. Rev. A 26:2105 (1982).
55. C. Breton, C. DeMichelis, M. Fingenthal, and M. Mattioli, Phys. Rev. Lett. 41:110 (1978).
56. J. S. Wang, H. R. Griem, R. Hess, W. L. Rowan, and P. T. Kochanski, Report, Plasma Prepring UMLPF #85-003 (July 1985).

EXPERIMENTS ON ELECTRON-IMPACT EXCITATION AND IONIZATION OF IONS

R. A. Phaneuf

Physics Division, Oak Ridge National Laboratory
Oak Ridge, Tennessee 37831 USA

1. INTRODUCTION

The challenge of understanding the characteristics and behavior of the so-called fourth state of matter, the ionized plasma, requires quantitative information on both the structure and collisional properties of ions. The simple glow-discharge tube, the laboratory ion source, the controlled-thermonuclear fusion plasma, the ionosphere, the solar corona — they all share one common attribute — their microscopic and macroscopic properties are controlled to a large degree by inelastic collisions between electrons and ions. The delicate balance produced by the continual interchange of kinetic, radiative and potential energy determines the signature by which such a plasma communicates with its surroundings, namely the light it produces. For even the simplest of plasmas, the interpretation of the spectrum of radiation which is emitted requires detailed knowledge of both the electronic structure of the ionic species present, and of the collision processes by which these ions reach and leave the excited ionic states from which the radiation occurs. At higher particle densities, the transport of radiation within the plasma must also be taken into consideration.

The quantitative study of inelastic collisions between electrons and ions has been motivated for many years by the astrophysical community. The need for accurate collision cross sections has been dramatically intensified during the last decade by the worldwide emphasis on controlled-thermonuclear-fusion research, and its development as a potential energy source. The energetic radiation from highly ionized impurities in such high-temperature plasmas represents a serious mode of energy loss which has thus far prevented such devices from reaching the required ignition temperature ($kT \gtrsim 10$ keV) and confinement condition ($n\tau \gtrsim 10^{14}$ $cm^{-3}\cdot s$). The effect of impurity ions on plasma confinement is also poorly understood, and may perhaps even be beneficial under certain circumstances [1]. Accurate models for plasmas require reliable cross section data for electron-impact excitation, ionization and recombination of these impurity ions.

Since the pioneering intersecting-beams experiments of Dolder, Harrison and Thonemann [2] in 1961 on electron-impact ionization, and of Dance, Harrison and Smith [3] on electron-impact excitation in 1966, considerable progress has been made in electron-ion collisions, particularly in the case of ionization. This careful early work set a high standard for research in the field, and it is interesting to note that despite significant refinements in such areas as detectors, optical calibration standards, data acquisition, computer control and ion sources, the basic methodology of these experiments remains unaltered. Useful data have been produced by other techniques, such as monitoring spectral-line emission from magnetically confined plasmas [4], or by storing ions in a trap [5], but it is generally accepted that the colliding-beams method is capable of producing the most definitive basic data, since all the important parameters may be directly measured, and diagnostic checks performed. Such techniques form the subject of lectures to be presented at this Advanced Study Institute by P. Defrance and will therefore not be addressed here, except for purposes of categorization of experiments. Similarly, the theory of excitation and ionization of ions will be discussed in detail in the lectures of M. S. Pindzola and D. L. Moores and will not be addressed here per se.

The subjects of electron-impact excitation and ionization of ions have been comprehensively reviewed by D. H. Crandall [6] and E. Salzborn [7], respectively, at the preceding NATO Advanced Study Institute on this same topic, held at Baddeck, Canada, in 1981. Useful reviews have also been given at two other NATO Advanced Study Institutes on related themes by Dolder [8] on excitation and ionization of positive ions, and by Crandall [9] on collisions of electrons with multiply charged ions.

The present report will review some of the fundamental aspects and then focus on highlighting some of the important experimental results which have been produced during the interim period.

2. ELECTRON-IMPACT EXCITATION

2.1 Fundamentals

2.1.1 Threshold Behavior

The cross section for excitation of an ion by a free electron has the distinct characteristic of being finite, and often having its maximum value, at the threshold electron energy for the process. This threshold corresponds to the difference in energy between the initial and final ionic states. This is in contrast to that for excitation of a neutral atom, which rises from zero at threshold. The characteristic for ion excitation can only be accounted for quantum-mechanically and is a consequence of the Coulomb attraction between the electron and the ion. In this near-threshold region, resonances often play an important role in the excitation process and must be taken into account in accurate theoretical predictions of the cross section. At asymptotically high energies, perturbation methods generally give an accurate account of the cross section. Thus, ion excitation by electrons is most important in the energy region where theory is most difficult and least reliable. Accurate experimental data are needed near threshold to benchmark the calculations.

2.1.2 High-Energy Behavior

At high collision energies (i.e., above roughly ten times the excitation threshold), the energy-dependence of the cross section is well-defined and depends on the specific type of excitation process.

For dipole-allowed transitions ($\Delta\ell=1$, $\Delta s=0$), the cross section dependence on electron energy E follows the well-known Bethe form:

$$\sigma = \frac{A}{E}\ln(BE) \quad \text{(A,B are constants)} \qquad 2.1$$

For dipole-forbidden, spin-allowed transitions ($\Delta\ell\neq1$, $\Delta s=0$), the cross section has the dependence:

$$\sigma = \frac{C}{E} \quad \text{(C is a constant)} \qquad 2.2$$

For dipole- and spin-forbidden transitions ($\Delta\ell\neq1$, $\Delta s\neq0$), the cross section falls off even more rapidly:

$$\sigma = \frac{D}{E^3} \quad \text{(D is a constant)} \qquad 2.3$$

2.1.3 Excitation and Emission Cross Sections

The signature of the excitation event is generally the emission of a photon as the excited level radiatively decays. If the excited level can only be populated directly by electron impact, a so-called level excitation cross section may be obtained from measurements of the photon emission as the ion radiatively decays. The branching ratio for emission of that wavelength and the polarization of the radiation must also in general be taken into account. If the branching ratio is unknown, or if the excited level may be populated by cascading from higher levels excited by electron impact, a photon emission cross section may be determined. It is possible for a given measured cross section to represent an excitation cross section at energies up to the threshold for the first cascading level and an emission cross section thereafter. The general form of the energy dependence of a photon emission cross section may therefore be some combination of those given above:

$$\sigma = \frac{A}{E}\ln(BE) + \frac{D}{E^3} \qquad 2.4$$

where A, B, and D are constants. Note that the C/E term (Eq. 2.2) need not be included in the general case, since the term (A/E)ln(BE) may be expressed as the sum of two terms, one proportional to (lnE)/E and one proportional to 1/E.

2.1.4 Polarization of the Radiation

In general, the emitted radiation will be polarized, and this must be considered in determining cross sections from radiated photon intensities. Measurement of the polarization itself is independent of the absolute calibration of the photodetector and provides a stringent test of theory, since it describes ratios of individual sublevel excitation cross sections. For dipole-allowed transitions, it has been shown that the constant "B" in the Bethe formula (2.1) can be determined from the energy E_0 at which the polarization passes through zero [10]. The constant B in Eq. 2.1 is given by

$$B = e^3/E_0 \,. \qquad 2.5$$

The constant "A" in Eq. 2.1 can be calculated from the optical oscillator strength "f" for the transition, if it is known, and from the transition energy ΔE:

$$A = \frac{4\pi a_0^2 f}{\Delta E} \tag{2.6}$$

where a_0 is the Bohr radius. Thus the absolute magnitude of the cross section at high energies may be determined in the Bethe approximation by measuring only the polarization of the emitted radiation. This provides an important internal consistency check for both experiment and theory, as well as an independent verification of the absolute calibration of the experiment.

2.1.5 Collision Strengths and Predictor Formulae

Theoretical calculations on electron-impact excitation are most commonly presented in terms of the so-called "collision strength," defined as follows:

$$\Omega = \omega \, \sigma \, E \tag{2.7}$$

where ω is the statistical weight of the initial or lower level, equal to (2S+1)(2L+1), or (2J+1) if fine-structure is taken into account; E is the collision energy in Rydbergs; and σ is the cross section in units of πa_0^2. This is the collisional analog of the oscillator strength for radiative transitions. For dipole-allowed transitions, a "Bethe plot" of the collision strength versus ln(E) becomes linear at high energies, the slope being proportional to the optical oscillator strength (see Eqs. 2.1 and 2.6).

Both astrophysicists and plasma physicists frequently use the well-known "g-bar" or "Gaunt-factor" formula of Seaton [11] and van Regemorter [12] to obtain estimates of excitation cross sections:

$$\sigma = \frac{8\pi}{\sqrt{3}} \cdot \frac{f}{E} \cdot \frac{\bar{g}}{\Delta E} \cdot \pi a_0^2 \tag{2.8}$$

where E is the incident electron energy (in Rydbergs), ΔE is the transition energy (in Rydbergs), f is the optical oscillator strength for the transition, a_0 is the Bohr radius, and $\bar{g}$ is the "effective Gaunt factor". For singly charged ions, a value of $\bar{g}=0.2$ has been found to reproduce the available body of experimental results reasonably near threshold, to well within the advertised factor of two. Crandall [6] and Younger and Wiese [13] have performed critical analyses of the values which have been recommended and used for $\bar{g}$ for multiply charged ions. For dipole-allowed $\Delta n=0$ transitions, a value near unity seems optimal. For $\Delta n \neq 0$ and dipole-forbidden transitions, the value ranges between 0.05 and 0.7, depending on the transition. Clearly, no simple formula will describe the excitation process to a high degree of accuracy for all ions, but it can be expected that Eq. 2.8 will probably be reliable to within a factor of two or so near threshold, provided that $\bar{g}$ is chosen as prescribed.

It should be noted that when the Gaunt-factor formula is used with a constant value for $\bar{g}$, it does not have the correct asymptotic (Bethe) energy dependence for dipole-allowed transitions (Eq. 2.1). In the original formulation, an energy-dependent prescription for $\bar{g}$ was given to take this into account [12]. At higher energies, the recommendation is:

$$\bar{g} = \frac{\sqrt{3}}{2\pi} \ln(E/\Delta E) \qquad 2.9$$

The usual procedure is to use a constant value of $\bar{g}$ from threshold to the energy where Eq. 2.9 produces that value, and to use Eq. 2.9 for higher energies.

2.1.6 Detailed Theoretical Methods

A spectrum of theoretical formulations and approximations has been applied to electron-impact excitation of ions. Highly recommended for the experimentalist or nonspecialist is the introduction to theory presented by Crandall [6]. Further details may be found in reviews by Henry [14] and Robb [15], and in other papers and reviews cited therein. Since reference will be made to a number of theoretical methods in comparing to experiments, a few comments are in order. In approximately increasing order of sophistication, some of the methods which have been applied are:

Semiclassical Binary-Encounter Approximation
Plane-Wave Born Approximation
Glauber Approximation
Coulomb-Born Approximation
Coulomb-Distorted-Wave Approximation
Close-Coupling or R-Matrix Method.

The first three cannot account for the finite cross section at threshold for ion excitation by electron impact but often give reasonable results at high energies. The last four have in specific cases each reproduced experiment very well but have proven inadequate in others. The close-coupling or R-matrix method is in principle the most accurate but may produce inferior results to the distorted-wave method, unless all important levels and couplings are explicitly taken into account. Allowance for other effects such as electron exchange, electron-electron correlations in the target wave functions, recombination resonances and unitarization can each be important in specific cases and can rearrange the general hierarchy of approximations from that given above.

2.2 Colliding-Beams Experiments

Although theoretical calculations of ion excitation by electrons are numerous, the experiments are extremely sparse. All total cross section and rate-coefficient measurements reported to date have been based on the detection of photons radiated as the excited level decays. Photon dispersion, finite detection solid angle and detector efficiency all reduce signal count rates from intersecting-beams experiments to extremely low levels. Even with these constraints, highly accurate cross section measurements have been made for a number of ions. In fact, absolute-cross-section measurements for ions are generally more accurate than those for atoms, since the charge on the target species allows its density to be much more accurately quantified. Absolute optical calibration becomes more difficult at the shorter wavelengths which are characteristic of the multiply-charged ions which are important in high-temperature plasmas.

To date, accurate crossed-beams data exist for at least three multiply charged ions. Useful data on excitation of ionic inner-shell electrons has also been obtained indirectly in a number of crossed-beams experiments for electron-impact ionization, via the excitation-autoionization process. This excitation process often produces a distinctive signature in the total ionization cross section beginning at the threshold energy for the inner-shell excitation.

2.2.1 Hydrogen-Like Ions

The first ion excitation to be studied by the crossed-beams method was the 1s-2s transition in the simplest one-electron ion, He^+. Dance et al. [3] and Dolder and Peart [16] crossed beams of electrons with a He^+ ion beam and detected the 30.4 nm Ly-a radiation emitted when the product 2s levels were Stark-mixed with 2p levels in an electric field. The independent experiments are consistent with each other, as shown in Fig. 1, when normalized to theory at high energy. However, when normalized in this way, the measured cross sections near the 41 eV threshold are only about half the best close-coupling estimates [17,18]. This is especially confusing, since structure in the measured cross sections just above the threshold has been clearly resolved [16,19] and is reproduced quite well by theory [18].

Dashchenko and coworkers [20] have measured the relative cross section for 1s-2p excitation of He^+ from threshold to 300 eV. For this dipole-allowed excitation, they find excellent agreement with the theoretical energy dependence over the entire energy range.

Crandall [6] has suggested that theory may be overestimating the effect of the monopole ($\Delta\ell=0$) contribution in the 1s-2s excitation. Dolder [8] has recently re-examined the experiments and concluded that it is not impossible that the electron beam could have produced molecular ions from background gas, and that these ions could have been trapped by the space charge of the electrons, reacting with the He^+ in some mysterious way. Considering the care with which the experiment was performed, this would seem far-fetched, were it not for the fact that a similar unexplained effect was observed at high electron currents in an experiment by Gregory et al. [21] on 2s-2p excitation of N^{4+}, right near 50 eV! The 1s-2s excitation of He^+ remains an outstanding issue and will likely only be resolved by further experimental work.

2.2.2 Helium-Like Ions

The only He-like ion to be studied by the beams method thus far is Li^+. Rogers and coworkers [22] made absolute measurements of the 548.5 nm radiation emitted following the spin-forbidden 1^1S-2^3P excitation. Their results are compared in Fig. 2 to a number of theoretical calculations [23-27]. Near threshold the cross section is comparable in magnitude to that for 1s-2s excitation of He^+ and shows well-defined structure due to the effect of resonances in the threshold region. The data also exhibit the expected E^{-3} dependence at high energies. The qualitative agreement with 5-state close-coupling calculations is good, but quantitative agreement with distorted-wave theory is better in this case.

2.2.3 Lithium-Like Ions

The most extensive experimental studies to date have been for the lithium isoelectronic sequence. Ions of this electronic structure have strong resonance lines, which are very prominent in the radiation emitted by plasmas. Absolute cross sections have thus far been measured for 2s-2p excitation of Be^+, C^{3+} and N^{4+}. The Be^+ case, represented in Fig. 3, is often cited as a benchmark for both theory [28] and experiment [29]. Figure 3b shows a comparison of the experiment with a number of theoretical calculations, the latter converging towards the measurements, as the

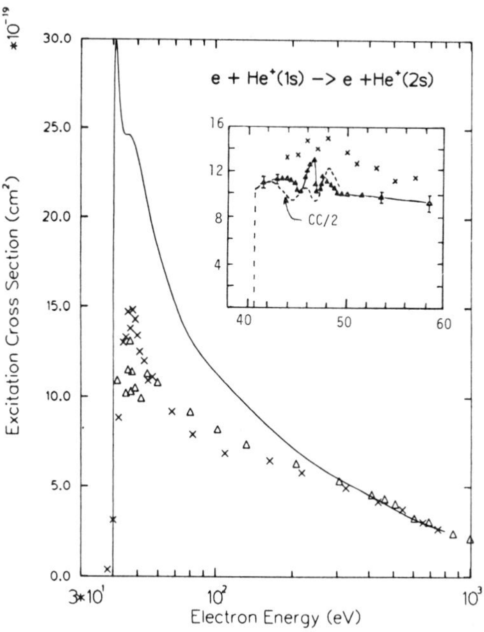

Fig. 1. Comparison of experimental and theoretical cross sections for 1s-2s excitation of He^+ by electron impact. Triangles are measurements of Dolder and Peart [16] and crosses are those of Dance et al. [3]. The solid curve represents a close-coupling calculation by Burke et al. [17], which has been joined smoothly to the Born approximation at high energy. The inset compares the structure in the measured cross sections near threshold to close-coupling calculations [18], which have been divided by 2 (dashed curve).

degree of sophistication increases. A disconcerting discrepancy of about 15% still remains with the best close-coupling calculations, however, over the entire energy range, well outside the range of absolute accuracy of the experiment (9%). Inclusion of additional basis states in the calculations fails to improve the agreement further, suggesting convergence of the close-coupling calculations. In this case, the polarization of the emitted radiation was also measured in the experiment, and the internal

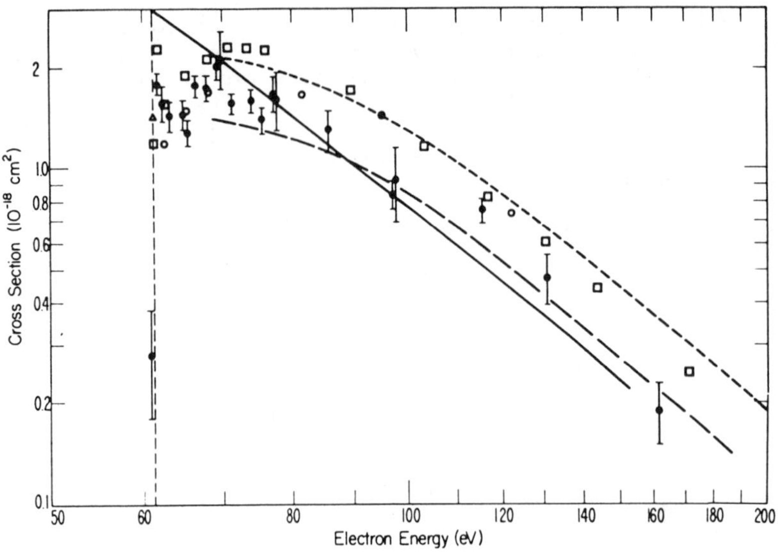

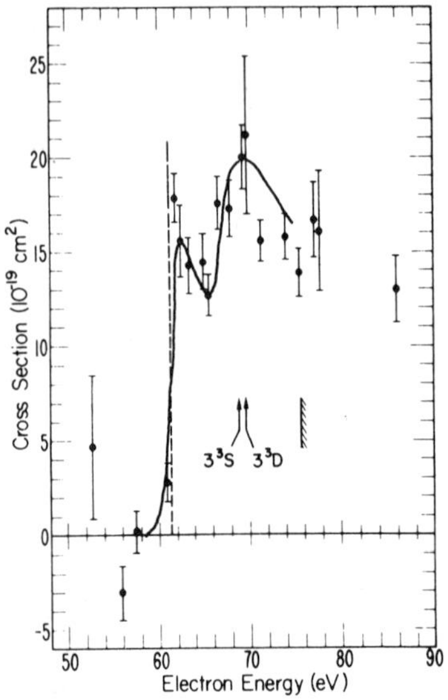

Fig. 2. Absolute experimental cross sections for 584.5 nm emission from $1^1S \rightarrow n^3\ell$ excitation of Li^+ by electron impact, from Rogers et al. [19]. The measurements (circles with flags) are compared to a number of theoretical predictions above: solid curve, Coulomb-Oppenheimer (Ref. 23); long-dashed curve, distorted-wave (Ref. 24); short-dashed curve, distorted-wave (Ref. 25); squares, 5-state close-coupling (Ref. 26). On the left is a blow-up of the threshold region and shows definite structure due to resonances. The solid curve at the left is a recent 5-state close-coupling calculation of Christensen and Norcross [27] convoluted with the 1 eV FWHM experimental energy spread.

consistency check at high energies noted in Section 2.1.4 was applied. The results, shown in Figs. 3c and 3d, demonstrate the internal consistency of the experiment and verify its absolute photometric calibration. This suggests that all the physics necessary to define the problem may not be included in even the most sophisticated theories. For example, it has been suggested that the role of electron-electron correlations in the target wave functions may be more important than originally thought for this seemingly simple case of a single valence electron outside a closed shell [30].

For 2s-2p excitation of multiply charged Li-like ions, the situation is much improved. Absolute experiments for C^{3+} [31] and N^{4+} [21] are in excellent agreement with both close-coupling and Coulomb-distorted-wave calculations over the whole energy range of the measurements. The comparison with 2-state close-coupling calculations of van Wyngaarden and

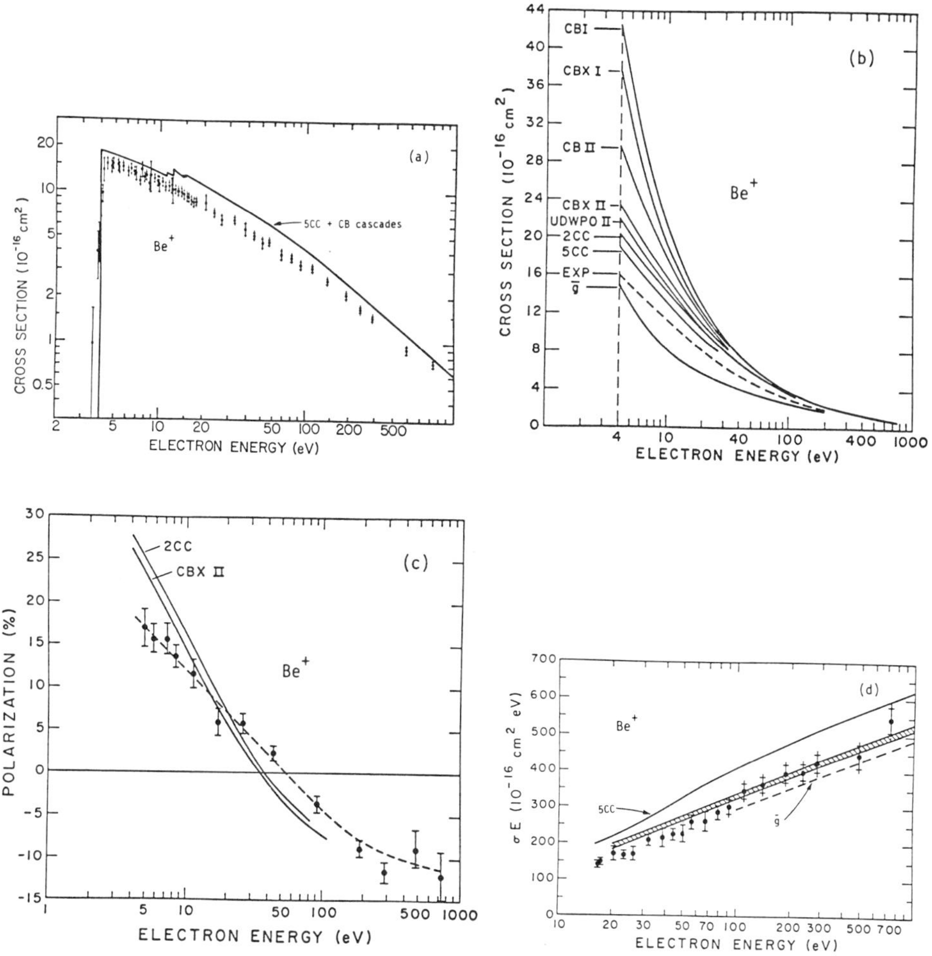

Fig. 3. Comparison of experimental and theoretical results for 313-nm emission resulting from 2s-2p excitation of Li-like Be^+, from Taylor et al. [29]. The references to and a discussion of the various theoretical methods are given by Hayes et al. [28]. (a) Comparison of 5 cc calculation of Hayes et al. (with CB cascade contributions) to experimental emission cross section. (b) Comparison of a number of theoretical approximations to experiment. (c) Comparison of experimental and theoretical polarization fractions for the unresolved 313-nm doublet. (d) Bethe plot comparing the actual experimental collision strengths and theoretical values to those deduced from the polarization measurements alone (hatched band), as discussed in Section 2.1.4.

Henry [32] for N^{4+} is shown in Fig. 4. As was noted by Gregory and coworkers [21], the increasing dominance of the Coulomb field causes the 2s and 2p levels to be relatively further separated from those of higher n, evidently reducing the importance of correlation effects for these ions. If this trend is indeed general, it has important ramifications for the physics of high-temperature plasmas, where, at least for the present, theory must be relied upon for the more highly charged ions. Further measurements on multiply charged ions will be needed to either prove or disprove this hypothesis.

Some experimental information is also available for 1s-2p inner-shell excitation of Li-like ions via the excitation-autoionization process. In this case the excitation of an inner-shell electron is manifest in the measured total electron-impact ionization cross section as a sharp jump which onsets at the threshold for the excitation process. This mechanism was first observed experimentally by Peart and Dolder [33] in cross sections for electron-impact ionization of Ba^{+}. Data for 1s-nℓ excitation now exist for six members of the lithium sequence: Be^{+} [34], B^{2+} [35], C^{3+} and N^{4+} [36], O^{5+} [35-37], and Ne^{7+} [38]. The implementation of electron-cyclotron-resonance (ECR) multicharged ion sources in crossed-beams experiments at Louvain-la-Neuve in Belgium and Oak Ridge in the United States during the last year or so has made possible more accurate measurements for O^{5+} [35,37] and new data for Ne^{7+} [38]. The earlier-reported O^{5+} data from Oak Ridge [36] were limited by low ion-beam intensity and had large uncertainties. They suggested a 1s-2p excitation cross section that was roughly three times that predicted theoretically. This cast doubts on the ability of theory to predict scalings of cross sections

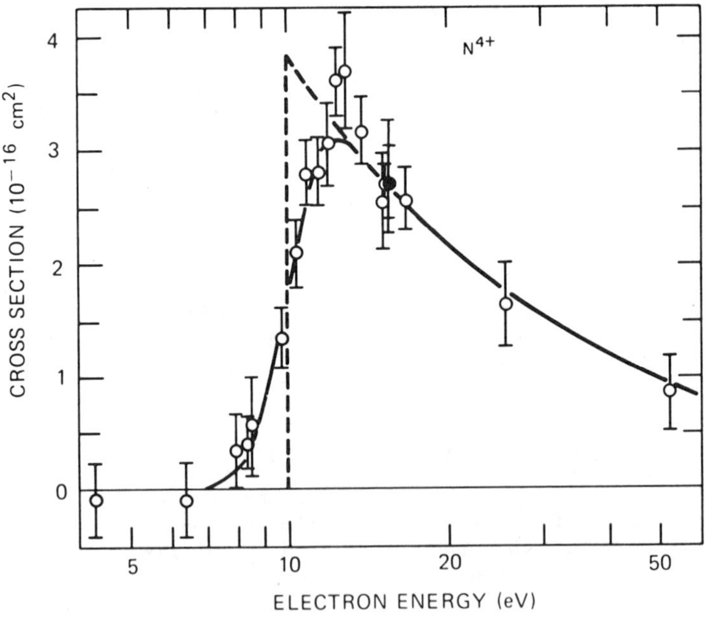

Fig. 4. Absolute experimental cross sections for 2s-2p excitation of N^{4+} as a function of electron energy, from Gregory et al. [21]. The solid curve represents a convolution of two-state close-coupling theory [31] with the experimental electron energy distribution.

to higher ionic charges. Figure 5 compares the earlier and recent data for O^{5+} from ORNL. The newer, more precise O^{5+} data from both laboratories are in near-perfect agreement and remove the discrepancy for the higher charge states. Table 1 is adapted from a recent paper by Crandall et al. [35], in which the experimental data for the Li-sequence have been carefully re-analyzed. The comparison between experiment and the various theoretical estimates for the 1s-2p excitation cross section just above its threshold is seen to be quite favorable for the multicharged members of this sequence.

In estimating excitation cross sections from ionization data in this way, questions such as branching ratios for radiative decay versus autoionization of the core-excited levels, and interference between the direct and indirect channels must also be considered. These issues and other examples of excitation-autoionization will be addressed later in the discussion of electron-impact ionization in Section 3.3.1.

2.2.4 Sodium-Like Ions

The only other isoelectronic sequence where more than one member has been studied in crossed-beams experiments is that of sodium, where measurements have been made for Mg^+ [39-42] and Al^{2+} [43]. The measurements for both ions are for 3s-3p excitation of the unresolved fine-structure doublet and have been made absolute by radiometric calibration.

The independently-absolute data for Mg^+ of the JILA and Uzhgorod groups are in agreement within their relative total uncertainties and are also within 15% of close-coupling and distorted-wave calculations near threshold. Figure 6 shows non-absolute measurements by Zapesochnyi and coworkers [41] with an improved energy resolution of 0.3 eV, which show well-resolved structure in the near-threshold region due to recombination resonances. Angular differential cross sections have also been measured for this ion [42] and will be discussed in the next section.

In the case of Al^{2+}, the absolute experiment of Belić et al. [43] and distorted-wave-with-exchange theory are in near-perfect agreement over the entire energy range of the measurements, suggesting again that theory may become more reliable for multiply charged Na-like ions, as was the case for the Li-sequence.

2.2.5 Angular Differential Measurements

Probably the most significant innovation in the study of ion excitation in the last few years has been the development at Jet Propulsion Laboratory by Chutjian's group [44] of an apparatus to measure inelastic angular scattering of electrons from ions. The measurements of the scattered electrons are differential in both inelastic energy loss and angle. Thus far, Cd^+ [44], Zn^+ [45] and Mg^+ [42] have been studied using this technique. In Fig. 7, angular scattering measurements for 3s-3p excitation of Mg^+ at an electron energy of 35 eV are compared to the 5-state close coupling calculation of Msezane and Henry [45]. The measurements have been normalized to theory at 12°, but the agreement in shape is encouraging. Similar results shown in Fig. 8 for 4s-4p excitation of Zn^+ are in comparable agreement with close-coupling calculations. A plot of the experimental differential cross section versus electron energy at a fixed scattering angle of 14° shows that the agreement with theory is much worse at lower electron energies, where the inelastic energy loss becomes

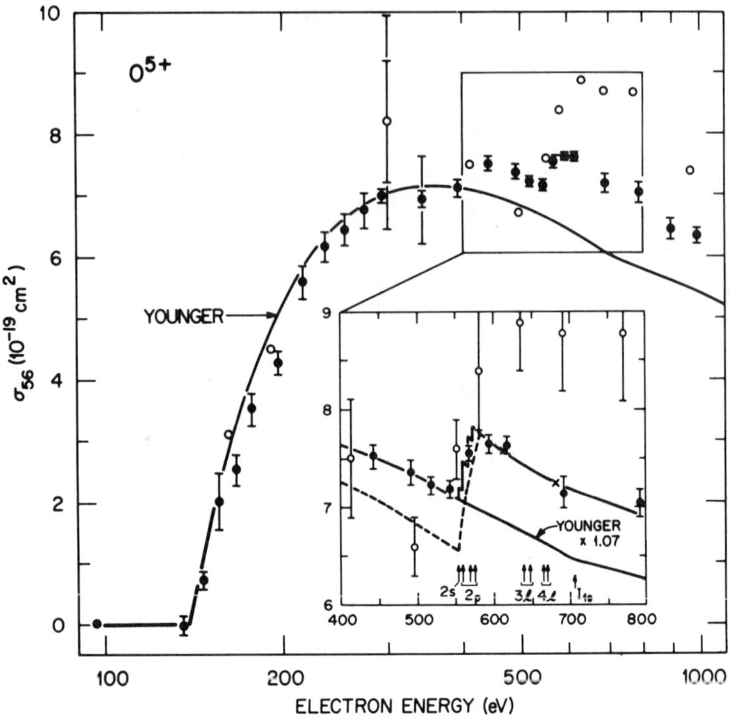

Fig. 5. Experimental cross sections for electron-impact ionization of Li-like O^{5+}, showing contributions due to $1s \to n\ell$ excitation-autoionization, from Crandall et al. [35]. The open circles are earlier data from Ref. 36, and the solid points are recent, more accurate measurements. The solid curve is a scaled distorted-wave direct ionization calculation by Younger [60]. The inset compares the measurements with close-coupling calculations for the $1s-n\ell$ excitations (see Table 1).

Table 1. $1s^22s \to 1s2s2\ell$ Excitation cross sections in 10^{-19} cm^2, adapted from Ref. 35

Ion	Energy (eV)	Previously Deduced Exp.	Six-State Close Coupling	Presently Deduced Exp.	Ratio Exp/6cc Theory
Be^+	125	17	9.3	20 ± 8	2.1 ± 0.9
B^{2+}	208		4.1	4.0 ± 1.0	1.0 ± 0.2
C^{3+}	325	3.2	2.24	2.3 ± 0.7	1.0 ± 0.3
N^{4+}	460	1.8	1.27	1.6 ± 0.4	1.3 ± 0.3
O^{5+}	612	2.8	0.74	0.9 ± 0.3	1.2 ± 0.4

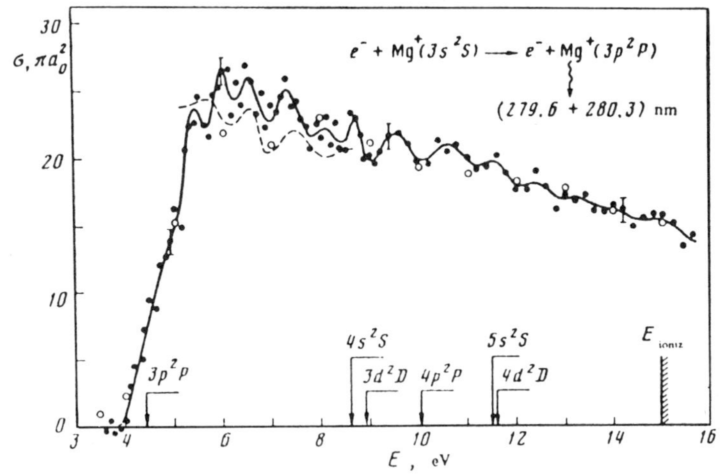

Fig. 6. Cross sections near threshold for 280 nm emission from 3s-3p excitation of Mg^+, from Zapesochnyi et al. [41]. The open circles are earlier, absolute measurements [39] and the solid points are normalized measurements with improved energy resolution. The solid curve is generated from the measurements by "digital filtering," and the dashed curve is a theoretical calculation.

an appreciable fraction of the incident electron energy. Whether this discrepancy indicates a problem with the theory, or the experiment, or both, remains unresolved at the present time.

These differential scattering measurements from ions are technically difficult and represent a significant advancement. Clearly, such experiments will have more to say in the next few years and will provide new motivation to the theorists.

2.2.6 Other Ions

Some of the most reliable experimental absolute cross section data are for singly-charged, heavier ions, where the cross sections are generally larger, and the wavelengths of the transitions are in or close to the visible region of the optical spectrum, where radiometric calibration is more tractable. As pointed out by Crandall [6], 4s-4p excitation of Ca^+ by Taylor and Dunn [46] must rank as one of the most meticulous absolute experiments to be performed in atomic collision physics.

A list of excitation experiments for positive atomic ions from intersecting-beams experiments up to 1982 have been compiled and tabulated by Dolder [8]. Apart from the work already mentioned, this author is aware of only two other ions for which measurements have been made since that review.

The JILA group of Stefani and coworkers [47] measured absolute cross sections for emission of 141.4 nm radiation resulting from 4^1S - 4^1P excitation of Ga^+. Their data, shown in Fig. 9, are about 60% below the Gaunt-factor prediction near threshold, and 20% above it at high energies. No other theoretical predictions were available.

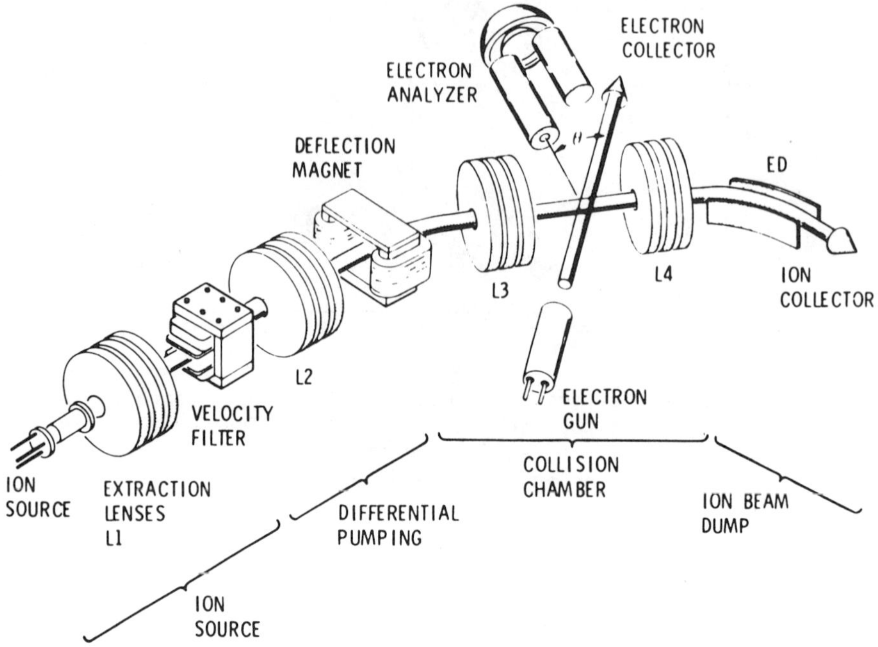

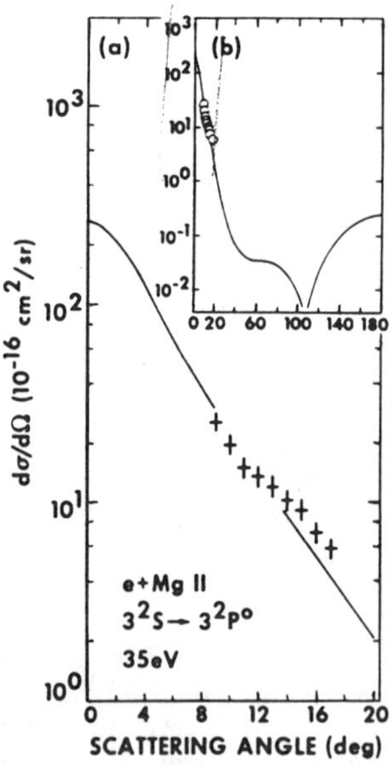

Fig. 7. Angular scattering apparatus of Chutjian [44] and differential cross section measured as a function of scattering angle for excitation of Cd^+ at an electron energy of 35 eV, from Ref. 42. Solid curve is a close-coupling calculation of Msezane and Henry [42], to which the measurements have been normalized.

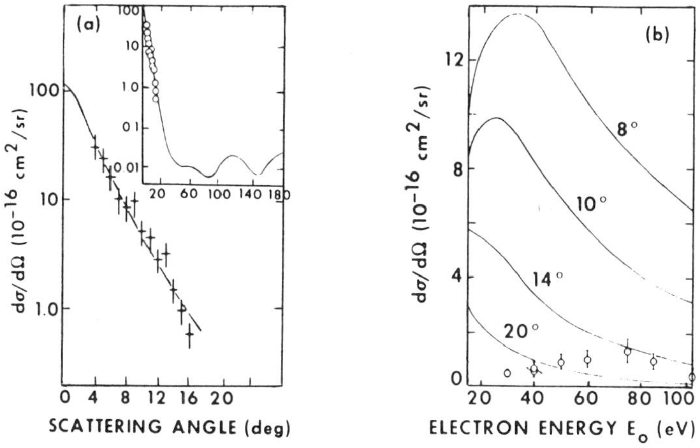

Fig. 8. Measured and calculated angular differential scattering cross sections from Chutjian et al. [45] for 4s-4p excitation of Zn^+. The data are presented as a function of scattering angle for fixed electron energy of 75 eV in (a), and as functions of the electron energy for fixed scattering angles in (b). The experimental data in (b) correspond to an angle of 14°.

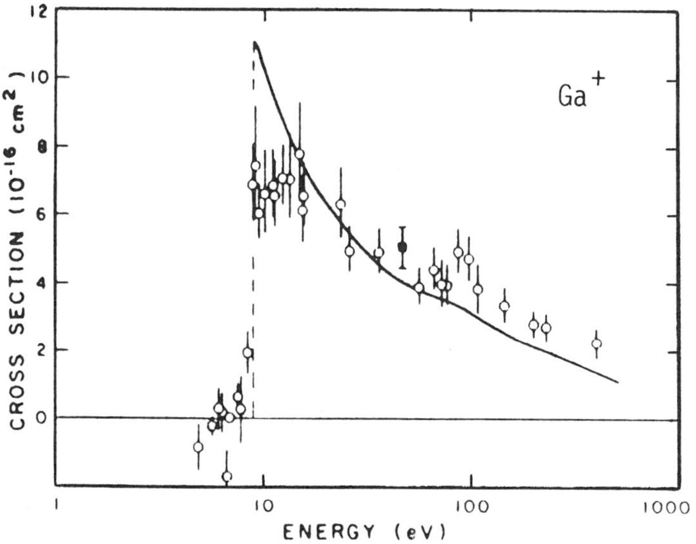

Fig. 9. Absolute cross section measurements for 141.4 nm emission resulting from 4^1S - 4^1P excitation of Ga^+, from Stefani et al. [47]. Solid curve is the Gaunt-factor or $\bar{g}$-formula.

Hane and coworkers at Nagoya University have recently developed a crossed beams apparatus to study electron-impact excitation of ions and have reported absolute cross sections for excitation of eight different transitions in Cd^+ which are relevant to the the formation of the 441.6 nm laser line [48,49]. Their results for $5^2S \rightarrow 5^2P^0$ excitation are presented in Fig. 10.

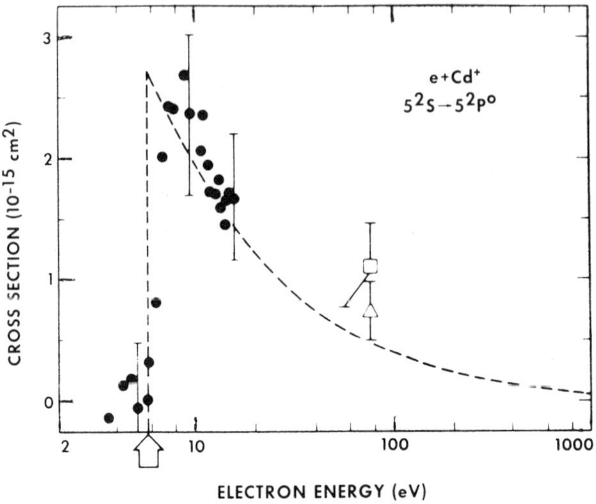

Fig. 10. Absolute cross sections for 5^2S–$5^2P°$ excitation of Cd^+, compared to the $\bar{g}$-formula prediction. The solid points are the absolute data of Hane et al. [49] for the $^2P°_{3/2}$ level. The square is the integrated differential cross section measured by Chutjian [44] for the unresolved $^2P°_{3/2}$, $^2P°_{1/2}$ doublet. The triangle is the estimate of the latter for the $^2P°_{3/2}$ state alone, assuming cross section ratios proportional to statistical weights. The figure is taken from Ref. 44.

2.3 Excitation Rate Coefficients from Plasmas

Collisional excitation rate coefficients have been determined experimentally for a number of ions by detecting absolutely the spectral emission from well-behaved and well-diagnosed plasmas. The reviews of Kunze [4] and Gabriel and Jordan [50] are standard references for such experiments. Work on Li-like ions by Kunze and Johnson [51] and by Boland and coworkers [52] are representative examples of some of the careful work which has been done. Results for 2s-2p excitation of C^{3+} and N^{4+} are consistent with those obtained from crossed-beams measurements.

The collisional excitation rate coefficient α(T) is defined as follows:

$$\alpha(T) = \int_{E_0}^{\infty} v\ \sigma(E)\ f(E)\ dE \qquad 2.10$$

where v is the electron velocity, E the electron energy, E_0 is the threshold energy for the process, and f(E) is the electron energy distribution function, which is usually a Maxwellian with some electron temperature, T. The rate coefficient is expressed in units of cm^3s^{-1}.

Magnetically confined plasmas such as pinches have been the most widely used for determinations of excitation rates. The method is based on a number of assumptions about the behavior of the plasma, the most important of which form part of the basis for the so-called coronal-equilibrium model. These are that the state which is being excited to produce the relevant radiation is populated only by excitation from the initial level in collisions with electrons, and that the excited level is depopulated only by radiative decay. No other processes should play a role in the dynamics of the states involved. The extent to which these conditions apply must be considered in each case, and an allowance made for any departures.

To determine the rate coefficient, the densities of the electrons and ions in the initial state must be determined and also the electron temperature. All of these must be homogeneous over the radiating volume. In addition, the photon detection system must be absolutely calibrated, as in the intersecting beams method. Thus excitation rate coefficients determined in this way are subject to large uncertainties, and the averaging over a Maxwellian energy distribution causes any structure in the cross section, and the physical information contained therein, to be lost. Nevertheless, the method is important in that it has provided the only available information on ions in high ionization stages, and improvements have been made in plasma diagnostic techniques to characterize the plasma.

Experimental rate coefficients have been reported recently for Li-like Ne^{7+} by the Ruhr University group of Chang and coworkers [53]. Their rate coefficients at 75 eV for 2s-3ℓ and 2s-4ℓ excitation, and those of Kunze and Johnson [54] and Hadad and McWhirter [55] are presented in Fig. 11. The measurements agree with close-coupling [56] and Coulomb-Born [57] calculations to well within the experimental uncertainties of typically 50%.

The Maryland group of Wang and coworkers report excitation rate measurements for a number of strong transitions in Fe^{9+} [58], and in Fe^{7+}, Fe^{8+} and Fe^{10+} [59] from a theta-pinch plasma of hydrogen seeded with 1% Fe. The absolute uncertainties of their rate coefficients are estimated to be 30-40%, and their rates generally agree within these limits with distorted-wave theory where it is available.

2.4 Other Methods

It has been clear for some time that because of low photon intensities and the difficulties of calibration in the extreme ultraviolet and soft X-ray regions, new experimental approaches are needed to benchmark theory for higher ionization stages. Such a course is being pursued by the JILA group led by G. H. Dunn. They are constructing a merged electron-ion beams apparatus in which the excitation process will be detected by electron energy-loss spectroscopy. The method is limited to the near-threshold region, but that is where the cross section is usually largest, the most important in plasmas, and where theory is least reliable. Collection of the scattered electrons will be facilitated by immersion of the interaction and analysis regions in an axial magnetic field. The inherent advantage of the merging-beams technique of improved

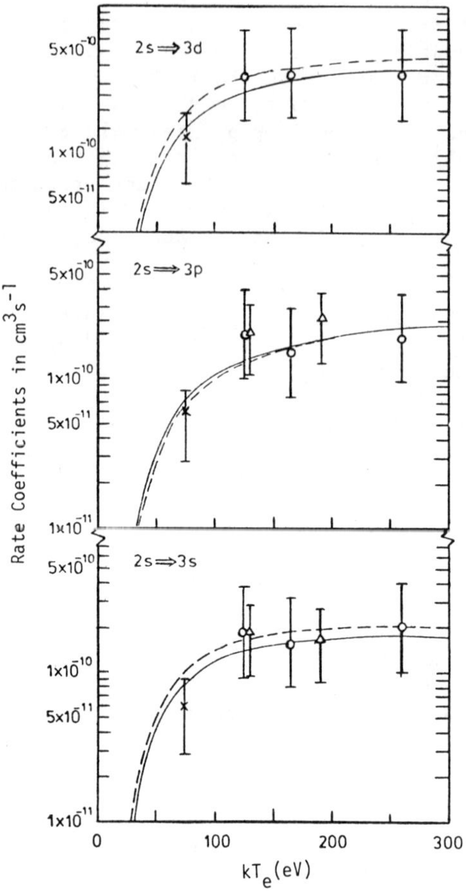

Fig. 11. Plasma rate coefficient measurements for Li-like Ne^{7+} as a function of electron temperature, from Chang et al. [53]. Crosses are their measurements, circles are data of Kunze and Johnson [54], and triangles are from Haddad and McWhirter [55]. Solid curves are 5-state close-coupling calculations [56], and dashed curves are Coulomb-Born results [57].

energy resolution promises important experimental information about the effects of resonances, which thus far have been barely discernible in crossed-beams experiments.

3. ELECTRON-IMPACT IONIZATION

3.1 Fundamentals

3.1.1 Ionization Mechanisms

Ionization is a complex process, since a large number of mechanisms can cause the ejection of electrons from bound states of atomic systems into the continuum. The most important ionization mechanisms which have been identified for ions are summarized below, and examples of each are given.

a. Direct Ionization:

$$e + X^{+q} \rightarrow e + X^{+q+n} + ne \qquad 3.1$$

In direct or "knock-out" ionization, the collisional transfer of energy to the target electron exceeds its binding energy, allowing it to escape the ion. The ejection of more than a single electron by this mechanism ($n>1$) is a higher-order collisional interaction and is relatively improbable. In the case of He^+, only direct ionization is possible, and the results of the first crossed-beams experiment to measure electron-impact ionization of an ion of Dolder and Peart [2], shown in Fig. 12, serve as an example of the behavior of the direct ionization cross section, which is represented well by distorted-wave theory [60].

b. Ionization-Autoionization:

$$e + X^{+q} \rightarrow e + (X^{+q+1})^* + e$$
$$\qquad\qquad\qquad \hookrightarrow X^{+q+2} + e \qquad 3.2$$

The ejection of a more tightly bound inner-shell electron leaves an ion in a core-excited state which can subsequently autoionize, resulting in the ejection of one or more additional electrons, depending on the excitation energy of the "hole" state, and the branching ratio for radiative versus Auger decay (fluorescence yield). In this example, the excited state $(X^{+q+1})^*$ decays by single Auger emission, yielding a net double ionization process. For deeper inner-shell vacancies, successive Auger emissions, or simultaneous multiple Auger emissions may also be possible, resulting in net multiple ionization. The ionization-autoionization process is the dominant mechanism for multiple ionization of ions. A comparison in Fig. 13 of experimental cross sections [61,62] for electron-impact double ionization of Ar^{2+}, Ar^{3+} and Ar^{4+} from a recent paper by Pindzola et al. [61] shows the dominance of 2p ionization-autoionization for these ions.

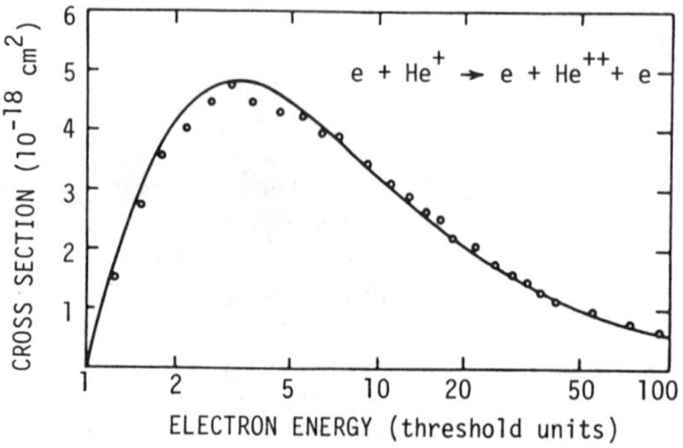

Fig. 12. Cross section for electron-impact ionization of He^+ as a function of electron energy in threshold units. Points are crossed-beams measurements of Dolder et al. [2], and curve is distorted-wave exchange calculation of Younger [60].

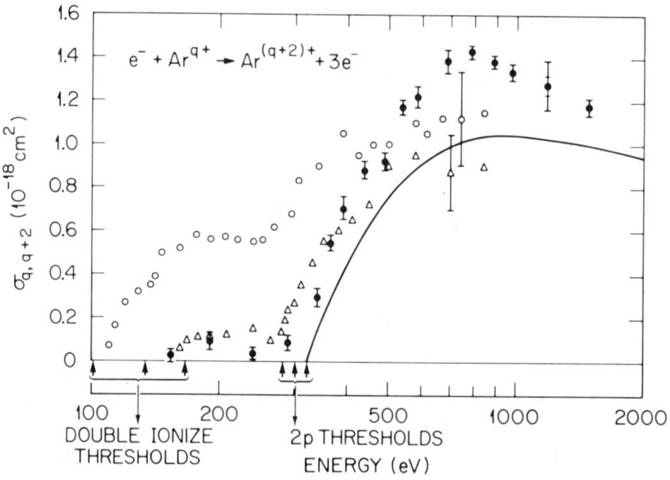

Fig. 13. Experimental cross sections for double ionization of Ar^{2+} (open circles, Ref. 62), Ar^{3+} (triangles, Ref. 62), and Ar^{4+} (solid circles, Ref. 61), showing dominance of 2p-ionization-autoionization. The solid curve is a distorted-wave calculation for direct 2p ionization.

c. Excitation-Autoionization:

$$e + X^{+q} \rightarrow e + (X^{+q})^* \quad \hookrightarrow X^{+q+1} + e \qquad 3.3$$

The electron-impact excitation of an ionic inner-shell electron can also produce a core-excited ion which can decay by the Auger process, leading to net ionization. As noted in Section 2.2.1 the cross section for excitation of an ion by electron impact is finite and most often largest at threshold. Thus the excitation-autoionization process for a strong, well-isolated transition often produces a distinctive jump in the measured total ionization cross section at the threshold energy for the excitation process. As noted earlier in Section 2.2.3, this mechanism can provide valuable information about the excitation process. Figure 14 presents the results of the first experimental demonstration of this process by Peart and Dolder [33,63] for singly charged alkaline-earth ions. Measurements by Falk and Dunn [64] for Be^+ are also presented. This sequence remains one of the more dramatic demonstrations of the potentially dominant role of excitation-autoionization in ionization.

d. Resonant-Recombination Double Autoionization:

$$e + X^{+q} \rightarrow (X^{+q-1})^{**} \quad \hookrightarrow (X^{+q})^* + e \quad \hookrightarrow X^{+q+1} + e \qquad 3.4$$

In this case the incident electron resonantly recombines with the target ion to form a doubly-excited intermediate state, which subsequently decays by two-step Auger emission. The excitation process here is analogous to dielectronic recombination, except that it is associated with the excitation of an inner-shell electron. This ionization mechanism was first postulated by LaGattuta and Hahn [65] in 1981, and is predicted to make a non-negligible contribution to the electron-impact ionization of Fe^{15+}, as shown in Fig. 15. A large contribution due to the excitation-autoionization process is also predicted for this Na-like system. Henry and Msezane [66] have suggested a variation on this resonant process whereby the doubly-excited intermediate state may in special cases decay by simultaneous double Auger emission, which they have labelled "auto-double ionization".

3.1.2 Threshold Behavior

The threshold ionization of atoms by electron impact has received considerable attention, both experimentally and theoretically, and is often considered a subfield in itself. According to the Wannier theory [67,68], the threshold behavior is primarily determined by the Coulomb interaction of the two outgoing electrons, a classical treatment of which leads to a threshold cross section:

$$\sigma_{ion} \propto (E-I)^n \quad , \quad n = \frac{1}{4}\left[\left(\frac{100Q-9}{4Q-1}\right)^{1/2} - 1\right] \qquad 3.5$$

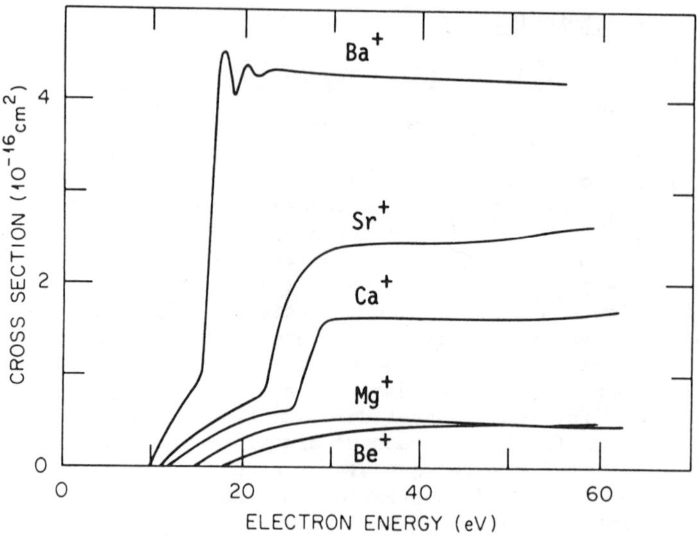

Fig. 14. Role of inner-shell excitation-autoionization in electron-impact ionization of the alkaline-earth ions. The curves are experimental data of Peart and Dolder [33,63], except for Be^+, which is from Falk and Dunn [64]. The sharp changes in slope indicate the onset of the excitation-autoionization process.

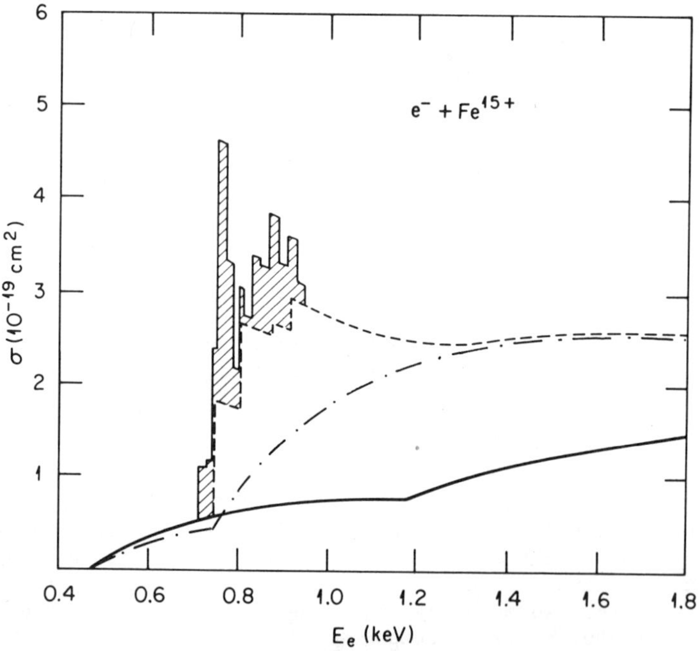

Fig. 15. Theoretical cross sections for electron-impact ionization of Fe^{15+}, from Ref. 65. Solid curve represents the Lotz semiempirical formula for direct ionization, with the dashed curve added to represent excitation-autoionization contributions. The hatched area is added to represent the resonant-recombination double-autoionization contribution. The chain-dashed curve is the prediction of the Burgess-Chidichimo semi-empirical formula. The predictor formulae are discussed in Section 3.1.4.

where E is the incident electron energy, I is the ionization potential and Q is the charge on the residual ion. For neutral atoms, Q=1 and n=1.127, a value which has been verified experimentally for He to a high level of precision [69]. For single ionization of ions of initial charge q, the prediction for n is given in the table below.

q	n
1	1.056
2	1.036
3	1.026
10	1.009
100	1.001

The predicted departures from a linear ionization onset with energy for ions are very small and become even less significant with increasing ionic charge. While nearly all existing experiments for ions are consistent with these predictions, none is of sufficient precision to quantify such small departures from linearity.

3.1.3 High-Energy Behavior

At asymptotically high energies, as was the case for dipole-allowed excitation, the energy dependence of the ionization cross section is given by the well-known Bethe form:

$$\sigma = \frac{A}{E} \ln (BE) \qquad \text{(A,B are constants)} \tag{3.6}$$

This asymptotic dependence is expected even when indirect ionization mechanisms such as excitation-autoionization make appreciable contributions to the ionization cross section, since these must either fall off as ln(E)/E or faster, depending on the type of excitation (see Section 2.1.2).

3.1.4 Predictor Formulae

A number of formulae have been proposed to predict cross sections for electron-impact ionization of ions, beginning with the classical theory of J. J. Thomson [70] in 1912. As noted by Salzborn [7], this theory has serious inadequacies but has provided useful information about approximate scalings of cross sections.

a. The Lotz Formula

The most successful and extensively used predictor formula for modelling of plasmas is that of Lotz [71]:

$$\sigma(E) = \sum_{j=1}^{3} a_j \, r_j \, \frac{\ln(E/I_j)}{E \, I_j} \{1 - b_j \exp[- c_j(E/I_j - 1)]\} \tag{3.7}$$

where

E = incident electron energy in eV

I_j = ionization energy of subshell j in eV

r_j = number of electrons in subshell j

$\sigma(E)$ = ionization cross section in cm^2

a_j, b_j, and c_j are constants determined by experiment, theory, or educated guesswork. Values for these have been tabulated for a number of ions by W. Lotz [72]. For ions of charge 4 or greater, it is customary to set $b_j = c_j = 0$ and $a_j = 4.5 \times 10^{-14}$, which gives the usual one-parameter Lotz formula.

The Lotz semiempirical formula is based in part on Coulomb-Born calculations for hydrogenic ions, and has the correct Bethe energy dependence at high energies. It also approximates very closely the expected linear onset in the near-threshold region, as discussed in the previous section. As will be seen in examples later, the Lotz prediction is generally reliable to within 25% or so in predicting the direct ionization cross section.

The Lotz formula includes only direct ionization processes. As we have seen, indirect ionization mechanisms such as excitation-autoionization can make important contributions to the total ionization cross section, and even dominate in some cases.

b. The Burgess-Chidichimo Formula

Burgess and Chidichimo [73] have recently proposed a general empirical formula similar to that of Lotz which allows for inner-shell excitation-autoionization. Their formula is

$$\sigma(E) = C \sum_j a_j (I_H/I_j)^2 (I_j/E) \ln(E/I_j) W(E/I_j) \pi a_0^2 \qquad 3.8$$

where

I_H = 13.6058 eV

$a_0 = 5.2917 \times 10^{-9}$ cm

$$W(E/I_j) = \begin{cases} 0 & E \leqslant I_j \\ [\ln(E/I_j)]^{n} I_j/E & E > I_j \end{cases}$$

n is given by Eq. 3.5 ($n \approx 1$)

C = 2.30 is an empirical constant

I_j = ionization (or excitation) energy of the jth shell (or subshell)

a_j = occupation number of jth shell (or subshell)

The function $W(E/I_j)$ assures the Wannier threshold behavior discussed in the previous section but approaches unity at very high energies, thus preserving the asymptotic Bethe energy dependence. They determined the constant C empirically by comparison to all the available experimental cross section data. A value of C=2.30 was observed to fit the data with a 23% rms deviation.

Excitation-autoionization is included in the same manner as direct inner-shell ionization, except that the lowest excitation threshold energy for the subshell is used for E_i. Thus the finite cross section for ion excitation at threshold cannot be accounted for. Nevertheless, their approach does make some allowance for excitation-autoionization, and in cases where there are a number of excitations from a subshell occurring over a range of threshold energies, it can give reasonable predictions for the cross section. As is the case for inner-shell ionization, the lowest excitation threshold energy must be known.

The Burgess-Chidichimo formula must be considered an improvement over the Lotz formula, since the empirical constant is chosen from a much broader data base, and since the contributions of direct and indirect ionization have been separated. A comparison of their empirical prediction with the Lotz formula and detailed theoretical calculations is given in Fig. 15 for Na-like Fe^{15+}, where 3s-nℓ excitation-autoionization is expected to make a substantial contribution.

3.1.5 Theoretical Approaches

Since ionization theory will be addressed specifically by the lectures of Pindzola and Moores, only a list of the methods which have been applied to ionization of ions will be given here. The introduction to theoretical approaches given by Salzborn [7] is especially recommended to the experimentalist or non-specialist. In roughly increasing order of sophistication, some of the theoretical methods which have been applied to electron-impact ionization are:

Classical Binary Encounter
Impact Parameter Method
Plane-Wave Born Approximation
Exchange-Classical Impact-Parameter Method
Coulomb-Born Approximation
Coulomb-Distorted-Wave Approximation
Close-Coupling/Coulomb-Born Approximation.

As in excitation, inclusion of effects such as exchange and electron-electron correlation within a given calculation can change the hierarchy. Contributions due to excitation-autoionization have generally been estimated by calculating separately the inner-shell excitation cross section, scaling it by the appropriate branching ratio for autoionization versus radiative decay, and simply adding this to the ionization cross section. The Close-Coupling/Coulomb-Born method [74] is an attempt to include, in a unified treatment, both direct and indirect ionization. Thus, implicitly accounted for are possible effects due to interference between direct and indirect ionization channels, as well as resonances.

3.2 Experimental Methods

Again, experimental techniques for electron-impact ionization of ions will be addressed specifically by P. Defrance, and only a few comments are in order here, in order to place the results to be presented into a proper perspective. A recent review by Dunn [75] gives an insightful and critical discussion of these methods and of the history and current status of experimental research in this field.

3.2.1 Intersecting-Beams Method

As was noted in the case of excitation, the most accurate and detailed data on the ionization process has come from the intersecting-beams method. A beam of ions is crossed with an electron beam, and the fraction of ions whose charge has increased is measured as a function of electron-beam energy. In the case of ionization, all of the fast product ions may in principle be collected and detected with unit efficiency, giving an advantage in signal of three orders of magnitude or more over excitation experiments. Thus both the sensitivity and accuracy of the method is improved. Differences of more than 10% between absolute cross sections measured by different groups are uncommon. The beams method also offers the highest energy resolution and thus far has provided the only information about the onset and contributions of indirect mechanisms or the role of metastable states of the parent ion.

3.2.2 Plasma Spectroscopy Method

In this technique, ionization rate coefficients (see Eq. 2.10) are determined from measurements of the time evolution of the intensity of spectral-line emission from a well-diagnosed plasma. The same requirements and constraints apply that were discussed for excitation rates in Section 2.3, with the exception that absolute calibration of the photon detectors is not required to deduce ionization rates. These rate coefficients are adjustable parameters in a set of coupled rate equations describing the time histories of properly chosen spectral lines representing each of the ionization stages of the ion of interest. Although such measurements are generally considered to be less accurate than the beams data and give little or no information about specific ionization mechanisms, the only published data on the higher ionization stages ($q>7$) have been obtained using either this or the related trapped ion method.

3.2.3 Trapped-Ion Method

Absolute cross sections for electron-impact ionization have also been obtained by trapping ions in some combination of electric and/or magnetic fields and bombarding them with energetic electrons. One of the most definitive applications of this concept is the work of Donets and Ovsyannikov [76], who analyzed the time histories of the charge states of trapped ions produced in an electron-beam ion source (EBIS). The charge-state distribution is determined by extracting the ions and performing a time-of-flight analysis. The modeling and fitting procedure for deducing the rate coefficients is similar to the plasma spectroscopy method outlined above, but the method offers the advantage of a monoenergetic, well-defined electron beam, and thus yields cross sections rather than Maxwellian-averaged rate coefficients. The accuracy of the data obtained are intermediate between the intersecting-beams and plasma-spectroscopic methods. For the heavier ions, multiple ionization has been shown by the beams experiments to be non-negligible, and thus the sequential single-ionization model which has been applied may be inadequate for determining accurate cross sections.

3.3 Experimental Results for Ionization

Unlike the situation for electron-impact excitation, there has been vigorous experimental activity and substantial progress made in electron-impact ionization during the last several years. It will be impossible to give a comprehensive review of this work during the time alotted, so specific examples will be presented which are particularly representative of various aspects of the ionization process. Emphasis will be given to work reported since the last meeting of this series.

3.3.1 Single Ionization

The hydrogen-like ion is the simplest for which electron-impact ionization can occur, and such collisions may be regarded as a true three-body systems. Since only direct ionization is possible, measurements for H-like ions provide an important benchmark for theory. The Z^{-4} scaling of the cross section as a function of electron energy in threshold units, predicted by the Coulomb-Born and distorted-wave approximations, is compared to the available data in Fig. 16. The experimental results for He^+ are crossed-beams data of Dolder et al. [2], and the data for the higher-q ions are from the trapped-ion measurements of Donets and Osvyannikov [76].

The role of 1s-nℓ excitation-autoionization (EA) in electron-impact ionization of Li-like ions was discussed in Section 2.2.3, with regard to the information it can provide about the excitation process. We consider here its relative importance as an ionization mechanism. The crossed-beams data of Falk and Dunn [34] for Be^+, shown in Fig. 17, indicate a relatively small EA contribution in this case. They find a good fit to the data by adding the g-formula for the excitation cross section to the

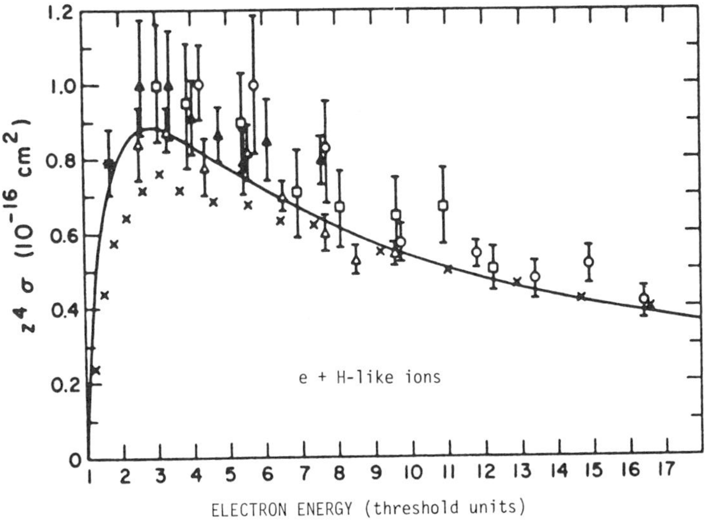

Fig. 16. Comparison of scaled electron-impact ionization cross sections plotted in threshold energy units with distorted-wave-exchange calculations [60] for Z=128. X - He^+ [2]; ○ - C^{5+}, □ - N^{6+}, △ - O^{7+}, ▲ - Ne^{9+}, * - Ar^{17+} [76]. Figure is adapted from Ref. 75.

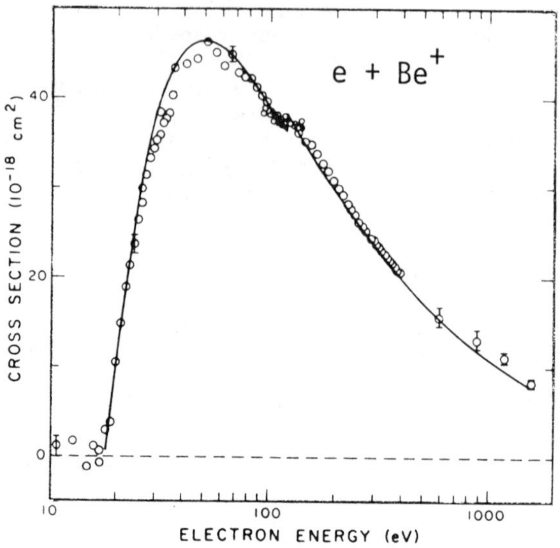

Fig. 17. Experimental cross sections of Falk and Dunn [34] for electron-impact ionization of Li-like Be^+, compared to the sum of the Lotz and $\bar{g}$ formulae for direct ionization and 1s-nℓ excitation autoionization, respectively.

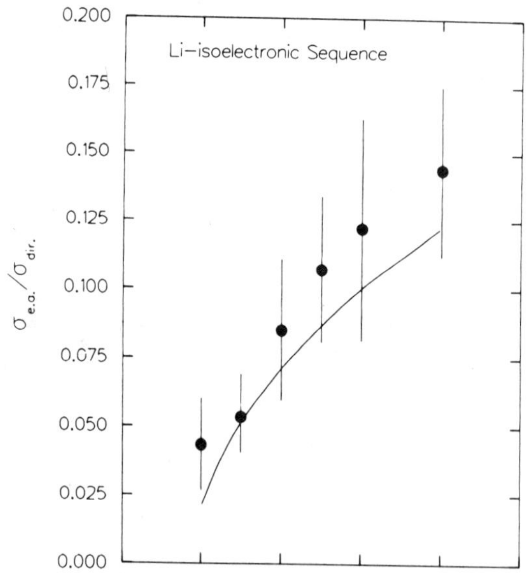

Fig. 18. Comparison for the Li-isoelectronic sequence of experimental data (points) and theoretical predictions (curve) for the ratios of the maximum 1s-nℓ EA contribution to the peak direct ionization cross section.

Lotz-formula scaled by 0.95. Scaled Coulomb-Born calculations do not compare favorably with the measurements in this case, even at high energies, where they underestimate them by about 30%. Similar results for Li-like O^{5+} were presented earlier in Fig. 5 and show a relatively larger contribution due to 1s-nℓ excitation-autoionization.

Since accurate experimental data now exist for six members of the Li sequence, it is instructive to reconsider the relative importance of the 1s-nℓ excitation-autoionization mechanism for these ions. The graph shows the ratio of the EA contribution just above its threshold to the peak direct ionization cross section, as a function of nuclear charge Z. The points are experimental [34,35,38], and the solid curve results from taking the ratio of close-coupling excitation cross sections [32] to the measured peak direct ionization cross section. The predicted trend is followed by the measurements, although the latter are systematically slightly higher.

For higher Z, it is no longer appropriate to ignore the radiative decay of the 1s2s2ℓ core-excited states. This occurs because the radiative decay rates scale along an isoelectronic sequence roughly as Z^4, while the autoionizing rates are almost independent of Z. The branching ratio for autoionization versus radiative decay must always be considered when estimating contributions of inner-shell excitations to ionization.

Systematic studies along ionic sequences can be particularly useful in identifying trends and assessing the importance of indirect ionization mechanisms. Griffin and coworkers [77] have completed a combined experimental-theoretical study of single ionization in the Xe-isonuclear sequence, from Xe^{2+} to Xe^{6+}. A comparison in Fig. 19 of their experimental cross sections with Lotz predictions for direct ionization shows large contributions due to indirect ionization, which increase in relative importance with increasing ionic charge. Distorted-wave calculations were made for excitations of the type 4d-nℓ (where nℓ=4f, 5d and 5f), which were found to account for most of the indirect contributions. Some of the important energy levels which have been calculated for the Xe-isonuclear sequence are shown in Fig. 20 for Xe^{+}, Xe^{3+} and Xe^{5+}. The Xe^{6+} ion has the simplest electronic structure ($4d^{10}\,5s^2$), and the agreement of the calculations with experiment is seen in Fig. 21 to be excellent in this case.

All these Xe-isonuclear data show onsets of the EA process at electron energies which are several eV less than those predicted by theory and are suggestive of the role of the resonant-recombination double-autoionization process. Griffin and coworkers made calculations of this process for the Xe^{6+} case but found that although the resonant recombination cross section was appreciable, the calculated branching to double autoionization was too small (0.05) to account for the observed features in the single ionization cross section in this case.

3.3.2 Multiple Ionization

Cross sections for the ejection of more than a single electron from an ion in an electron-ion collision were first reported in 1980 by Müller and Frodl [78] for Ar^{q+} (q=1,2,3). They concluded that ejection of a single L-shell electron, followed by autoionization, was responsible for almost all of the observed multiple ionization. Some of this data were presented in Fig. 13. For complex ions, this ionization-autoionization mechanism becomes relatively more important for the higher ionization stages, since the cross sections for inner-shell ionization depend only weakly on charge state, while those for the ejection of outer-shell

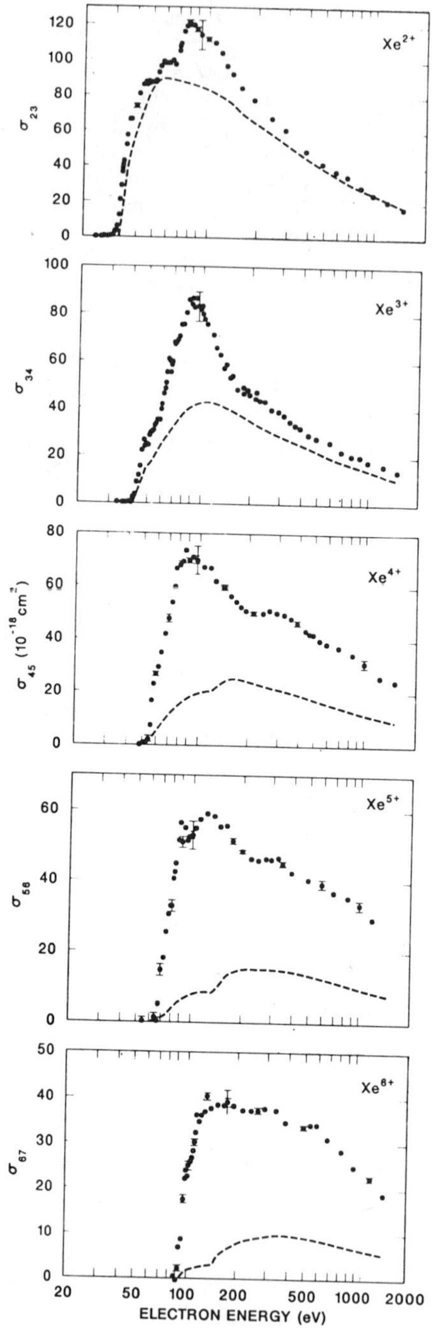

Fig. 19. Comparison of experimental cross sections [77] for electron-impact ionization of Xe^{q+} ions (q=2-6), showing increasing dominance of excitation-autoionization with increasing ionic charge.

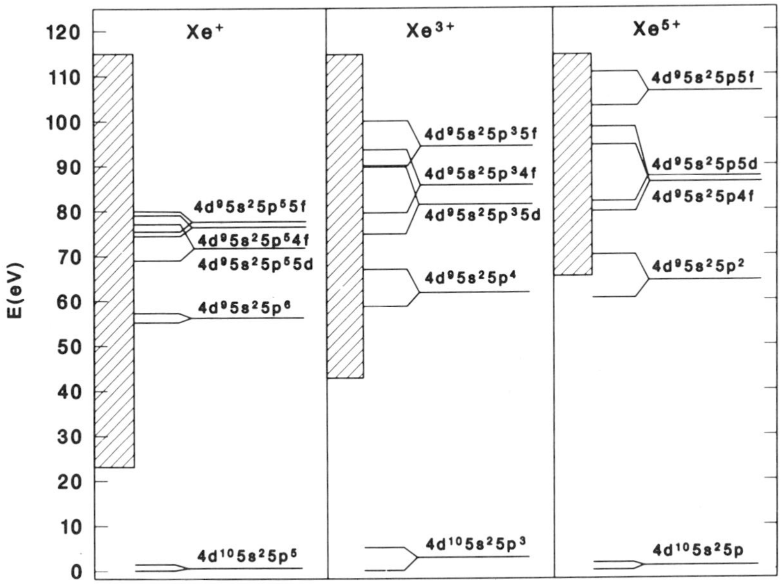

Fig. 20. Calculated core-excited energy levels important in 4d excitation-autoionization of Xe^{+}, Xe^{3+} and Xe^{5+}, from Ref. 77. The hatched region indicates the ionization continuum.

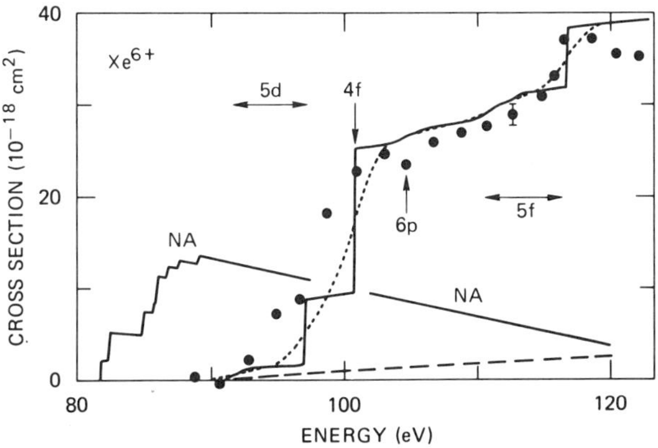

Fig. 21. Comparison to experiment of distorted-wave 4d-nℓ excitation cross sections added to the Lotz direct ionization cross section (long-dashed curve) for Xe^{6+}. The curve marked NA represents the sum of the 4d excitations which are non-autoionizing. The short-dashed curve is a convolution of the theory with the experimental electron energy distribution. Both theory and experiment are from Griffin et al. [77].

electrons fall off roughly as the inverse-square of the ionization potential. The direct ejection of more than one outer-shell electron in a single collision is not a first-order process, and quantum-mechanical methods have not yet been developed to treat this problem. The only theoretical method which has been applied to such collisions is the classical binary-encounter approximation [79].

The Giessen group have recently remeasured the cross sections for double ionization of Ar^{+} and Ar^{4+} with higher precision [80]. Their results are presented in Fig. 22. For Ar^{4+} they measure a small double-ionization cross section at energies below the 320 eV threshold for 2p ionization. To account for this, they apply the binary-encounter model for ejection of two outer-shell electrons. This small cross section added to the 2s and 2p direct ionization cross sections predicted by the Lotz formula gives a reasonable account of the experimental data, although the classical model overestimates the small low-energy contribution.

Pindzola and coworkers [61] have recently published a combined experimental-theoretical study of double ionization of quadruply-charged noble gas ions and also have examined the double-ionization data from the Giessen group for Xe^{q+} ions [81]. On the basis of distorted-wave calculations, they conclude that double ionization of rare-gas ions is dominated by the inner-shell ionization-autoionization mechanism. Their calculations and the Lotz formula for the sum of 3d and 3p ionization are compared to the experimental data for Kr^{4+} in Fig. 23. A blowup of the threshold region shows an almost negligible contribution from direct double ionization, which has a threshold of 141 eV. The distorted wave theory is in reasonable agreement with experiment up to about 300 eV but overestimates the peak by about 30%.

A similar comparison with experimental data from the Oak Ridge and Giessen groups for double ionization of Xe^{4+} shows the dominant contribution of 4d and 4p ionization-autoionization in this case. In the configuration-averaged calculation, the fraction of 4d vacancy levels which autoionize is difficult to accurately assess. The upper curve represents the sum of cross sections for ionization of all 4p and 4d levels, while the lower results from multiplication of the total by the fraction of 4d levels which autoionize (1/3). This assumes that the total 4d ionization cross section may be partitioned according to the statistical weights of the individual 4d levels, which may not be the case. Complicated level-by-level calculations would be required to resolve this issue. In any case, the experimental data fall comfortably between these two extremes, and it may be safely assumed that 4d and 4p ionization-autoionization dominate the double ionization cross section in this case also. Their distorted-wave calculations for the Xe^{q+} ions are strongly influenced by term-dependence in the ejected-electron continuum and by ground-state correlations.

Another interesting manifestation of the clear signature of the ionization-autoionization process in multiple ionization was reported by Achenbach and coworkers [81], who observed pronounced resonance-like "humps" in the experimental double ionization cross sections for Xe^{+} and I^{+}. Figure 24 shows their results, and demonstrates the striking correspondence between these structures and measured 4d photoionization cross sections for Xe atoms [82]. Why this correspondence should exist is not yet fully understood, but it is likely that the answers lie in the detailed atomic structures of these complex ions. Electron-impact excitations of the type $4d\text{-}n\ell$, followed by double autoionization are probably making important contributions to the double ionization of these ions.

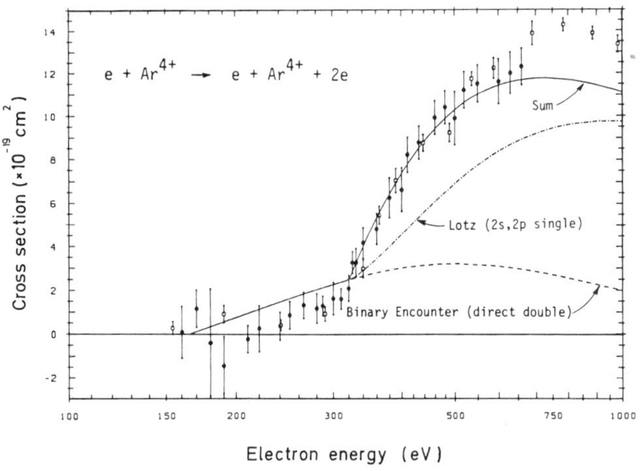

Fig. 22. Experimental cross sections for electron-impact double ionization of Ar^{4+}, from Müller et al. [80]. Solid and open circles are data from the Giessen and Oak Ridge groups, respectively. Solid curve is the sum of binary-encounter direct-double ionization (dashed) and Lotz 2s and 2p direct ionization cross sections. Chain-dashed curve is the binary-encounter estimate for 2ℓ direct ionization.

3.4 General Remarks about Electron-Impact Ionization of Ions

Analysis of the steadily increasing body of experimental data on the ionization of positive atomic ions by electron impact leads one to some general conclusions about the behavior of the cross sections and the relative importance of the various ionization mechanisms. A few such observations are listed below.

1. Direct single ionization is quite well represented by distorted-wave theory and predicted by semiempirical formulae with reasonable accuracy.

2. Direct multiple ionization is negligible in most cases.

3. Inner-shell excitation-autoionization can be an important or even dominant ionization mechanism in some cases and negligible in others.

4. It is probable that resonant-recombination multiple-autoionization or related resonant processes are non-negligible for complex ions whenever the excitation-autoionization process is dominant.

5. Indirect ionization mechanisms become relatively more important as the ionic charge increases along an isoelectronic or isonuclear sequence.

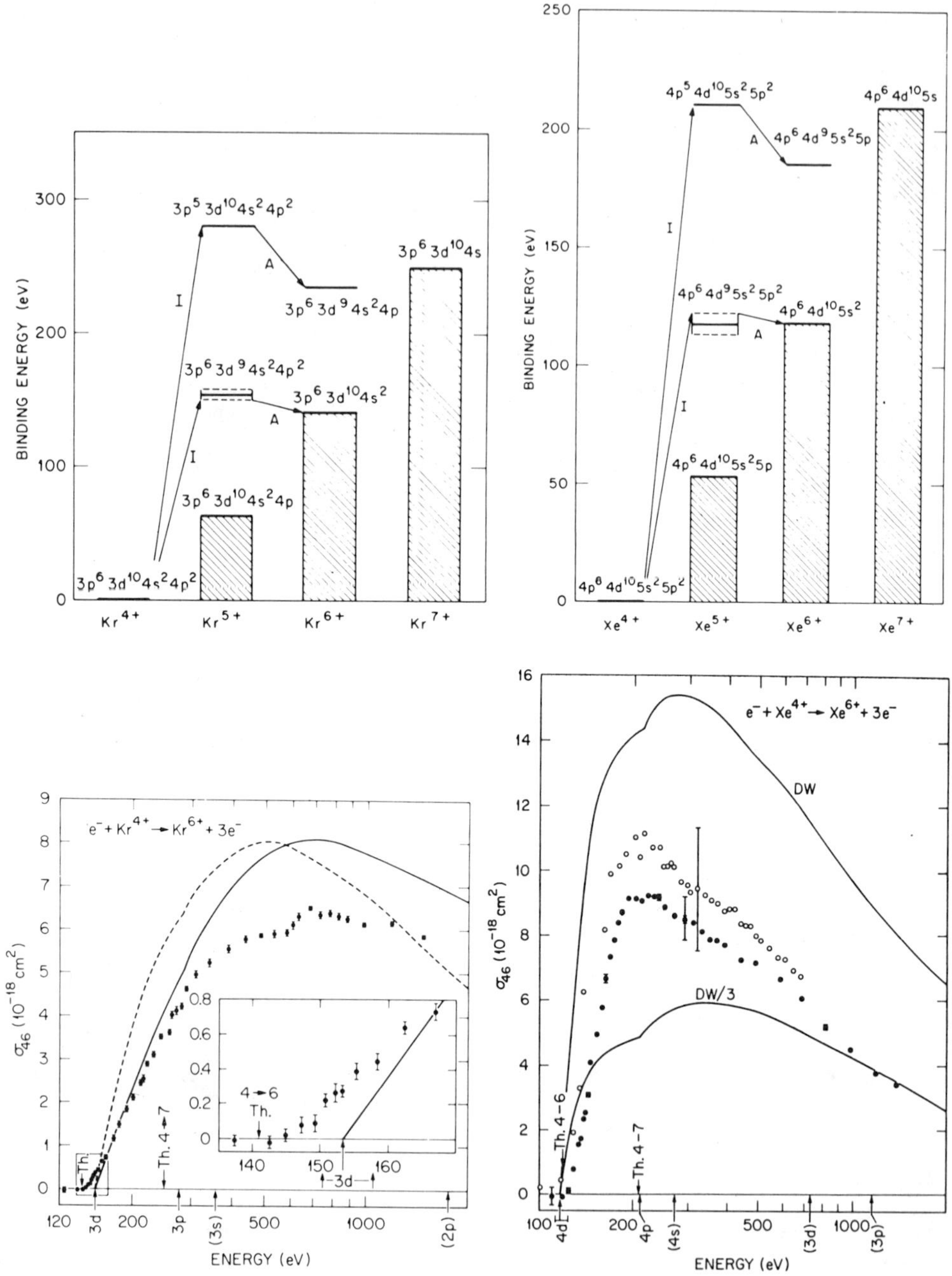

Fig. 23. Relevant energy levels for ionization-autoionization of Kr^{4+} and Xe^{6+}. Comparison of experiment and theory for double ionization of Kr^{4+} and Xe^{4+}. Solid points are Oak Ridge data [61] and open points are data from the Giessen group [81].

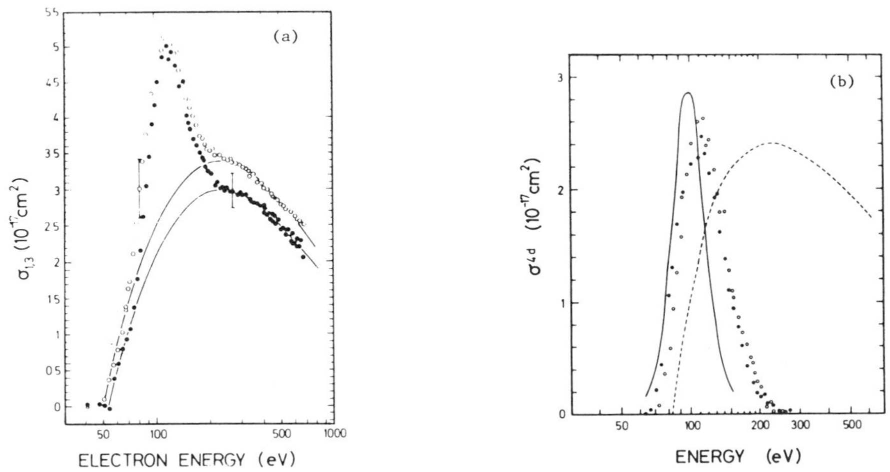

Fig. 24. Comparison of experimental cross sections for electron-impact double ionization of Xe^+ (solid circles) and I^+ (open circles), from Achenbach et al. [81]. Curves in (a) represent a fit of the Lotz-formula to the low and high-energy data. Data in (b) represent the residual "hump" obtained by subtracting the solid curves in (a) from the total. Solid curve in (b) is the measured 4d-photoionization cross section [82], and dashed curve is the 4d ionization cross section predicted by the Lotz formula.

6. Contributions to ionization from indirect mechanisms are generally more significant for ions with complex electronic structure than for simple ions.

7. Inner-shell ionization-autoionization is the dominant multiple ionization mechanism for most multicharged ions.

8. Multiple ionization of partially-stripped heavy ions is not always negligible compared to single ionization, and should be considered in accurate plasma models, or determination of excitation and ionization rate coefficients.

Because of the many possible pathways, ionization is a complex process. These observations are intended to characterize the ionization of ions in a general way, and should not be misinterpreted as hard and fast rules. Specific exceptions may be found for many of them, and undoubtedly more exceptions and more characterizations will be discovered as our data base for ionization expands. New advanced ion sources and unprecedented computational facilities have opened the door wider to the systematic experimental and theoretical investigations which will advance our understanding of both ionization and excitation phenomena.

ACKNOWLEDGMENTS

The author is indebted to a number of colleagues who are largely responsible for any insights and much of the information contained in this report. Particular recognition is due those with whom I have worked most closely on such experiments: D. H. Crandall, G. H. Dunn, D. C. Gregory, and P. O. Taylor. C. Bottcher, D. C. Griffin, and M. S. Pindzola have contributed countless discussions and patient explanations to whatever understanding I may have of the theory of electron-impact phenomena. In addition, my associations with A. Müller, E. Salzborn, and P. Defrance have been most stimulating and fruitful. Special thanks are due to colleagues whose research is quoted in this work, and also to F. M. Ownby and H. T. Hunter for their assistance in preparing the manuscript.

This work was supported by the Office of Fusion Energy of the United States Department of Energy, under Contract No. DE-AC05-84OR21400 with Martin Marietta Energy Systems, Inc.

REFERENCES

1. M.F.A. Harrison, "The Plasma Boundary region and the Role of Atomic and Molecular Processes," in Atomic and Molecular Physics of Controlled Thermonuclear Fusion, ed. C. J. Joachain and D. E. Post, NATO Advanced Study Institute Series B: Physics, Vol. 101 (Plenum, New York and London, 1983), pp. 473-475.

2. K. Dolder, M.F.A. Harrison and P. C. Thoneman, Proc. Roy. Soc. A 264, 367 (1961).

3. D. F. Dance, M.F.A. Harrison and A.C.H. Smith, Proc. Roy. Soc. A 290, 73 (1966).

4. H.-J. Kunze, Space Sci. Rev. 13, 565 (1972).

5. F. A. Baker and J. B. Hasted, Phil. Trans. Roy. Soc. A 261, 33 (1966).

6. D. H. Crandall, "Electron-Impact Excitation of Ions," in Physics of Ion-Ion and Electron-Ion Collisions, ed. F. Brouillard and J. W. McGowan, NATO Advanced Study Institute Series B: Physics, Vol. 83 (Plenum, New York and London, 1983), pp. 201-238.

7. E. Salzborn, "Electron-Impact Ionization of Ions," ibid., pp. 239-277.

8. K. T. Dolder, "Experimental Aspects of Electron-Impact Ionization and Excitation of Positive Ions," in Atomic and Molecular Physics of Controlled Thermonuclear Fusion, ed. C. J. Joachain and D. E. Post, NATO Advanced Study Institute Series B: Physics, Vol. 101 (Plenum, New York and London, 1983), pp. 213-243.

9. D. H. Crandall, "Electron-Ion Collisions," in Atomic Physics of Highly Ionized Atoms, ed. R. Marrus, NATO Advanced Study Institute Series B: Physics, Vol. 96 (Plenum, New York and London, 1983), pp. 399-453.

10. D.W.O. Heddle, J. Phys. B 12, 489 (1979).

11. M. J. Seaton, "The Theory of Excitation and Ionization by Electron Impact," in Atomic and Molecular Processes, ed. D. R. Bates (Academic, New York, 1962), p. 374.

12. H. Van Regemorter, Astrophys. J. 136, 906 (1962).

13. S. M. Younger and W. L. Wiese, J. Quant. Spectrosc. Rad. Transfer 22, 161 (1979).

14. R.J.W. Henry, "Excitation of Positive Ions by Electron Impact," Phys. Reports 68, 1 (1981).

15. W. D. Robb, "Theoretical Studies of Electron Impact Excitation of Positive Ions," in Atomic and Molecular Processes in Controlled Thermonuclear Fusion, ed. M.R.C. McDowell and A. M. Ferendeci, NATO Advanced Study Institute Series B: Physics, Vol. B53 (Plenum, New York and London, 1980), pp. 245-265.

16. K. T. Dolder and B. Peart, J. Phys. B 6, 2415 (1973).

17. P. G. Burke and A. J. Taylor, J. Phys. B 2, 44 (1969).

18. S. Ormonde, W. Whitaker, and L. Lipsky, Phys. Rev. Lett. 19, 1161 (1967).

19. N. R. Daly and R. E. Powell, Phys. Rev. Lett. 19, 1165 (1967).

20. A. I. Dashchenko, I. P. Zapesochnyi, A. I. Imre, V. S. Bukstich, F. F. Danch and V. A. Kel'man, Sov. Phys. JETP 40, 249 (1975).

21. D. C. Gregory, G. H. Dunn, R. A. Phaneuf and D. H. Crandall, Phys. Rev. A 20, 410 (1979).

22. W. T. Rogers, J. Olsen and G. H. Dunn, Phys. Rev. A 18, 1353 (1978).

23. J. A. Tully, from Ref. 22.

24. D. W. Norcross and D. G. Hummer, from Ref. 22.

25. A. K. Bhatia and A. Temkin, J. Phys. B 10, 2893 (1977).

26. W. D. Robb, from Ref. 22.

27. R. B. Christensen and D. W. Norcross, Phys. Rev. A 31, 142 (1985).

28. M. A. Hayes, D. W. Norcross, J. B. Mann and W. D. Robb, J. Phys. B 10, L429 (1977).

29. P. O. Taylor, R. A. Phaneuf and G. H. Dunn, Phys. Rev. A 22, 435 (1980).

30. D. W. Norcross, private communication (1984).

31. P. O. Taylor, D. C. Gregory, G. H. Dunn, R. A. Phaneuf and D. H. Crandall, Phys. Rev. Lett. 39, 1256 (1977).

32. W. L. van Wyngaarden and R.J.W. Henry, J. Phys. B 9, 146 (1976).

33. B. Peart and K. Dolder, J. Phys. B 1, 872 (1968).

34. R. A. Falk and G. H. Dunn, Phys. Rev. A 27, 754 (1983).

35. D. H. Crandall, R. A Phaneuf, D. C. Gregory, A. M. Howald. D. W. Mueller, T. J. Morgan, G. H. Dunn and R. J. W. Henry, "Electron-Impact Excitation of B^{2+} and O^{5+}: Excitation-Autoionization of Li-like Ions," Phys. Rev. A (1986, in press).

36. D. H. Crandall, R. A. Phaneuf, B. E. Hasselquist and D. C. Gregory, J. Phys. B 12, L249 (1979).

37. P. Defrance, S. Chantrenne, S. Rachafi, D. Belic, D. Gregory and F. Brouillard, Proc. 2nd Eur. Conf. Atom. Mol. Phys. (ECAMP), Amsterdam (1985).

38. S. Chantrenne, P. Defrance, S. Rachafi, D. Belic and F. Brouillard, XIV ICPEAC, Abstracts of Contributed Papers, Palo Alto, CA (1985), p. 299.

39. I. P. Zapesochnyi, V. A. Kel'man, A. I. Imre, A. I. Dashchenko and F. F. Danch, Sov. Phys. JETP 42, 989 (1976).

40. D. S. Belic, G. H. Dunn, T. J. Morgan, D. W. Mueller and C. Timmer, Phys. Rev. Lett. 50, 339 (1983), and unpublished data.

41. I. P. Zapesochnyi, A. I. Dashchenko, V. I. Frontov, A. I. Imre, A. N. Gomonai, V. I. Len'del, V. T. Navrotskii and E. P. Sabad, JETP Lett. 39, 51 (1984).

42. I. D. Williams, A. Chutjian, A. Z. Msezane and R. J. W. Henry, XIV ICPEAC, Abstracts of Contributed Papers, Palo Alto, CA (1985), p. 289.

43. D. S. Belic, R. A. Falk, G. H. Dunn, D. Gregory and C. Cisneros, Bull. Am. Phys. Soc. 26, 1315 (1981), unpublished data.

44. A. Chutjian, Phys. Rev. A 29, 64 (1984).

45. A. Chutjian, A. Z. Msezane and R.J.W. Henry, Phys. Rev. Lett. 50, 1357 (1983).

46. P. O. Taylor and G. H. Dunn, Phys. Rev. A 8, 2304 (1973).

47. G. Stefani, R. Camilloni, G. H. Dunn and W. T. Rogers, Phys. Rev. A 25, 2996 (1982).

48. K. Hane, T. Goto and S. Hattori, Phys. Rev. A 27, 124 (1983).

49. K. Hane, T. Goto and S. Hattori, J. Phys. B 16, 629 (1983).

50. A. H. Gabriel and C. Jordan, "Interpretation of Spectral Intensities from Laboratory and Astrophysical Plasmas," in Case Studies in Atomic Collision Physics II, ed. E. W. McDaniel and M. R. C. McDowell, North Holland, Amsterdam (1972), pp 209-291.

51. H.-J. Kunze and W. D. Johnson III, Phys. Rev. A 3, 1384 (1981).

52. B. C. Boland, F. C. Jahoda, T. J. L. Jones and R. W. P. McWhirter, J. Phys. B 3, 1134 (1970).

53. C. C. Chang, P. Greve, K.-H. Kolk and H.-J. Kunze, Physica Scripta 29, 132 (1984).

54. H.-J. Kunze and W. D. Johnson, III, Phys. Rev. A 27, 2249 (1983).

55. G. N. Haddad and R.W.P. McWhirter, J. Phys. B 6, 715 (1973).

56. W. L. van Wyngaarden and R.J.W. Henry, Can. J. Phys. 54, 2019 (1976).

57. O. Bely and D. Petrini, Astron. Astrophys. 6, 318 (1970).

58. J. S. Wang, R. U. Datla and H. R. Griem, Phys. Rev. A 29, 1558 (1984).

59. J. S. Wang and H. R. Griem, Phys. Rev. A 27, 2249 (1983).

60. S. M. Younger, Phys. Rev. A 21, 111 (1980).

61. M. S. Pindzola, D. C. Griffin, C. Bottcher, D. H. Crandall and D. C. Gregory, Phys. Rev. A 29, 1749 (1984).

62. A. Müller and R. Frodl, Phys. Rev. Lett. 44, 29 (1980).

63. B. Peart and K. T. Dolder, J. Phys. B 8, 56 (1975).

64. R. A. Falk and G. H. Dunn, Phys. Rev. A 27, 754 (1983).

65. K. J. LaGattuta and Y. Hahn, Phys. Rev. A 24, 2273 (1981).

66. R.J.W. Henry and A. Z. Msezane, Phys. Rev. A 26, 2545 (1982).

67. G. Wannier, Phys. Rev. 90, 817 (1953).

68. F. H. Read, J. Phys. B 17, 3965 (1984).

69. S. Cvejanovic and F. H. Read, J. Phys. B 7, 1841 (1974).

70. J. J. Thomson, Phil. Mag. 23, 449 (1912).

71. W. Lotz, Z. Physik 206, 205 (1968).

72. W. Lotz, Z. Physik 216, 241 (1968); ibid. 220, 466 (1969).

73. A. Burgess and M. C. Chidichimo, Mon. Not. Roy. Astr. Soc. 203, 1269 (1983).

74. B. Jakubowicz and D. L. Moores, J. Phys. B 14, 3733 (1981).

75. G. H. Dunn, "Electron-Ion Ionization" in Electron-Impact Ionization, ed. T. D. Märk and G. H. Dunn, Springer-Verlag, Wien and New York (1985), pp. 277-316.

76. E. D. Donets and V. P. Ovsyannikov, Sov. Phys. JETP 53, 466 (1981).

77. D. C. Griffin, C. Bottcher, M. S. Pindzola, S. M. Younger, D. C. Gregory, and D. H. Crandall, Phys. Rev. A 29, 1729 (1984).

78. A. Muller and R. Frodl, Phys. Rev. Lett. 44, 29 (1980).

79. M. Gryzinski, Phys. Rev. 138, A336 (1965).

80. A. Müller, K. Tinschert, C. Achenbach, and E. Salzborn, J. Phys. B 18, 3011 (1985).

81. C. Achenbach, A. Müller, E. Salzborn, and R. Becker, Phys. Rev. Lett. 50, 2070 (1983).

82. R. Haensel, G. Keitel, P. Schreiber, and C. Kunz, Phys. Rev. 188, 1375 (1969).

ELECTRON IMPACT EXCITATION AND IONIZATION OF IONS

EXPERIMENTAL METHODS

Pierre Defrance

Université Catholique de Louvain, Institut de Physique
Chemin du cyclotron, 2 - B1348 Louvain-la-Neuve
Belgium

INTRODUCTION

The experimental studies of electron impact ionization and excitation of ions are in constant progress since the first crossed electron-ion beam experiment has been performed in 1961.

Various approaches have been used for those studies. One is based on a quantitative analysis of the time evolution of plasma emitted radiation. This method gives information on ionization and excitation rates and requires additional measurements of electron temperature and density and the knowledge of the other processes taking place in the plasma. This makes the detailed information very difficult to obtain from those observations. This approach is covered by other lecturers, M. Harrison[1] and J. Dubau[2], and shall not be considered here.

A second approach makes use of ion traps in which a well defined ion species can be stored in various configurations, electron beams, electric and magnetic fields. Some of those devices are also used as ion sources and the time evolution of the charge spectra can produce quantitative information on the ionization process.

The third approach involves two beams interacting in a restricted volume. In those experiments the particles are in a well defined state when they interact. The actual relative velocity is also well-known so that detailed information can be obtained.

In these lectures, fundamental relations are established for a study of the electron impact ionisation and excitation processes in a restricted case of ion trap that is the electron beam ion source and in the interacting beam approach.

A detailed analysis of the most important encountered problems is also given, while the discussion of results is to be found in the lecture of Ron Phaneuf[3].

BASIC RELATIONS

Collision experiments involve two particles supposed to be discernable, particles 1 and 2. In the general case, a number of particles 1 collide at a velocity v_1 with a target composed of particles 2. That target is supposed to be at rest in the laboratory. Due to their big mass and small velocity, ions are naturally considered as the target components and subsequently, electrons are the projectiles. The subscripts e and i shall be used for the notation of all the parameters characterizing the electrons and ions respectively.

The importance of a defined process is related to the probability P that it can occur during the experiment. That probability is the ratio of the number of events dN_T to the number of trials N_T : $P = dN_T/N_T$.

In the classical picture of a collision, this ratio is the probability that a projectile interacts with the target, or it can be considered as a measure of the opacity of the target. The opacity of a target of density n_i, surface dS and thickness dx is :

$$\frac{dN_T}{N_T} = \frac{\sigma n_i dS dx}{dS} \tag{1}$$

where σ is the area associated with a single target particle. The number of trials during a short time interval dt is related to the flux of particles j_e :

$$dN_T = j_e \, dS \, dt \tag{2}$$

By introducing (2) in (1) :

$$\frac{dN_T}{j_e dS dt} = \frac{\sigma n_i dS dx}{dS}$$

or

$$\frac{dN_T}{dt} = \sigma \, n_i \, j_e \, dV \tag{3}$$

with $dV = dS\,dx$ is the volume occupied by the target.
The parameter σ is the cross-section or "total cross-section" for the process under investigation and characterize the absorption of the total incident flux without any description of the products distribution in the phase-space.

The flux j_e of the incident particles of velocity v_e is :

$$j_e = n_e \, v_e$$

where n_e is the electron density.
An alternative form of (3) is :

$$\frac{dN_T}{dt} = \sigma \, n_i \, n_e \, v_e \, dV \tag{4}$$

THE ELECTRON BEAM ION SOURCE

The basic formulae (3) or (4) have been widely used to produce a lot of data concerning atomic or molecular targets. In the case of ions, the difficulty to create a stationary target of well known density has limited their use. Hasted[4] has given a precise description of those ion traps.

The conditions for this use are :

(a) ions must be created and trapped in a restricted volume in order to form a stationary target. This volume must be at a very good vacuum in order to reduce competing processes, especially charge exchange reactions.

(b) an electron beam of well defined velocity and flux must cross the ion trap.

(c) events must be observable !

In the case of ionization cross-section measurements, equation (3) is written :

$$\frac{dN_q}{d(j_e t)} = - \sigma_{q,q+1} N_q + \sigma_{q-1,q} N_{q-1} \tag{5}$$

where $N_q = \int_{volume} n_q dV$ is the total number of trapped ions of charge q, and with the assumption that j_e is uniform over the volume. In (5) only single ionization is taken into account.

Equation (5) shows that by measuring the charge state distribution and its derivative with respect to the ionization factor (jt), a set of equations of the form (5) can be written for a fixed electron energy. The solution of this set of equations is a set of cross-sections for the single ionization of the observed charge states.

This method has been successfully applied by Donets[5] and his collaborators for the design of a multiply charged ion source and for ionization cross-section measurements.

His apparatus, the Electron Beam Ion Source[6] (EBIS) is shown schematically in Fig. 1.

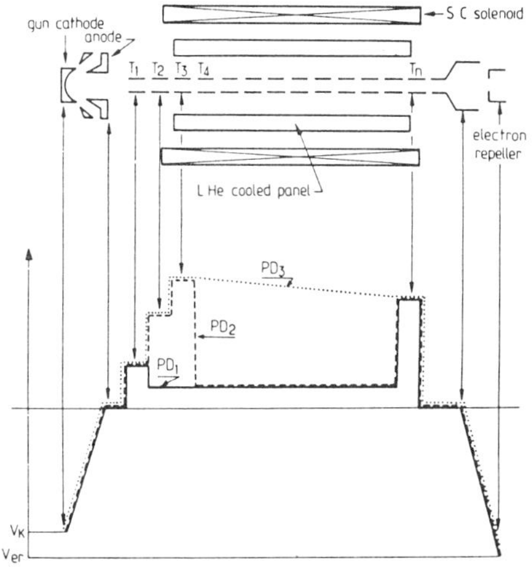

Fig. 1. Schematic drawing of an EBIS with the different potential distribution along the source.

It consists of a superconducting solenoïd of 1.4 meter long producing a magnetic field of order of a few tesla ($\lesssim$ 5T). A high current electron gun is located outside of the magnetic field. The electron beam moves along the magnetic field axis. Due to magnetic compression, very high current densities can be achieved : of order of 2 x 10^3 A/cm^2 for a total current of 0.5 A at an electron energy limited to 60 keV. A set of isolated cylinders are surrounding the electron beam. Various potential distributions can be defined, in order to introduce PD1, to trap PD2 or to extract the ions PD3. In PD1, gas is introduced at the entrance of the source, low charge state ions are created and diffuse to the central cylinders. When the potential distribution appears, ions are trapped during a confinement time τ, up to 10 sec. During this period ions are successively ionized and they are trapped in the potential well created by the electron beam space-charge.
After the confinment time τ, the potential distribution PD3 gives the ions the opportunity to be extracted out of the source. A time of flight spectrometer is used to analyse the charge state distribution.

Spectra are taken after different confinment time τ, showing the time evolution to the highest charges for Argon (Fig. 2).

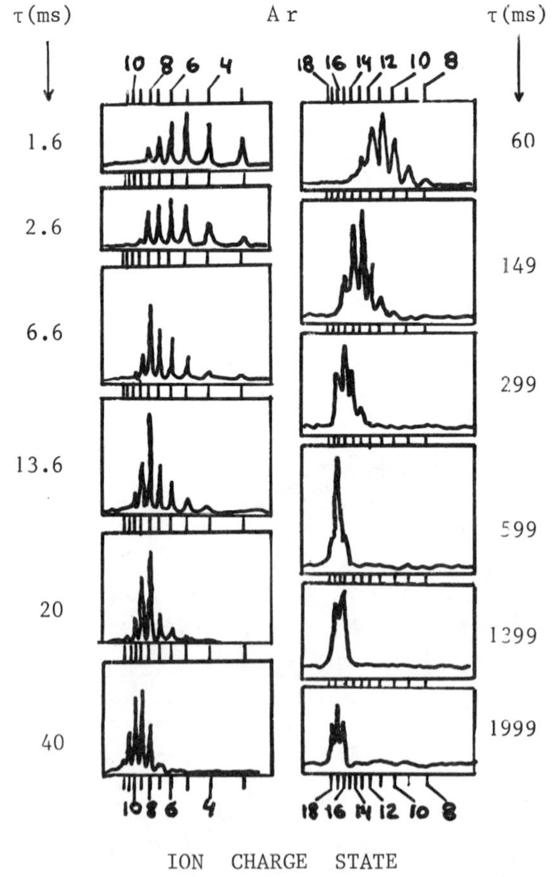

Fig. 2. Time evolution of the Argon ion charge state spectrum[7].

From the spectra, the normalised charge states distributions are estimated at each time (Fig. 3). The time derivatives can be deduced and the set of equations is written and solved.

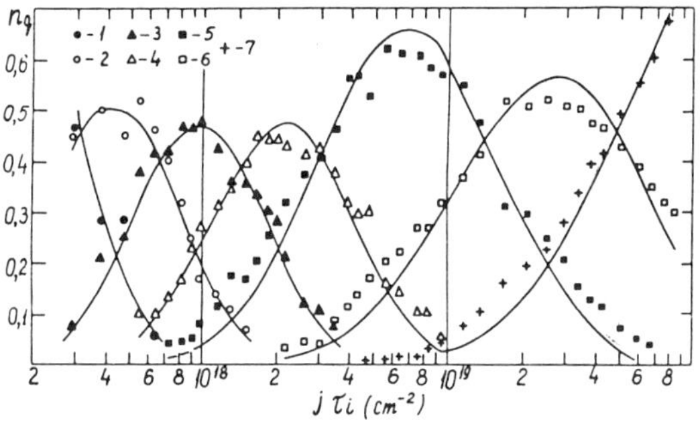

Fig. 3. Charge state distribution evolution of nitrogen ions[5].

The relative precision of the data obtained by means of that method is clearly shown on the histograms of successive values of $\sigma_{q,q+1}$ for Nitrogen Ions (Fig. 4).

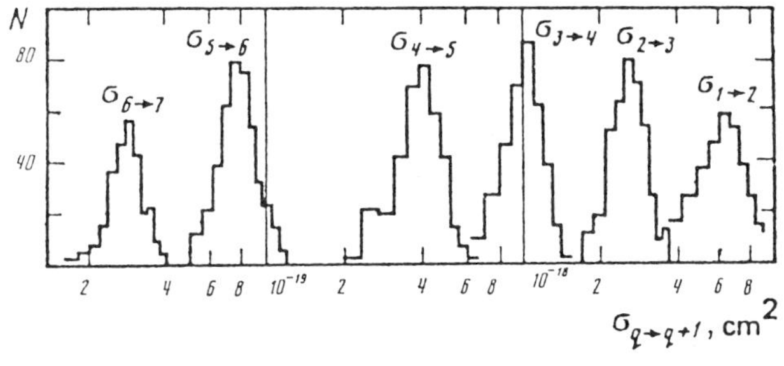

Fig. 4. Histograms for the Nitrogen Ions (E_e=5.45 keV)[5].

Final data are those yielding the smallest deviation from the measured charge state distribution evolution.
No absolute estimation of the error is given by the authors.

This method gives the general shape of the cross-sections, but is not able to give detailed information especially when secondary processes occur : due to the lack of precision, sharp enhancements of the cross-sections cannot be observed. These experiments where also limited by the charge resolution of the analysing system which does not separate charge states higher than 15 so that only mean values for charges group are given. Let us mention that the energy dependence of individual σ are obtained for Xenon ions up to 51+.

BEAM-BEAM INTERACTION

Some of the problems of ion traps disappear when ions are extracted from a source and are given some kinetic energy so that standard analysing systems can isolate ions of a well defined species.
On the other hand, the major disadvantage of this "moving" target, is the subsequent lowering of the ion density :

$$n_i = \frac{j_i}{v_i} = \frac{j_i}{\sqrt{\frac{2qe}{m_i} V_i}} \tag{6}$$

By increasing the accelerating voltage V_i from 1 V to 10 kV, the target density decreases by two orders of magnitude. A second disadvantage is the interaction of the ion beam with the residual gas atoms or molecules. This interaction becomes relatively more important.

Crossed beams

Both beams are supposed to be monokinetic and the volume in which they interact is shown on the Figure 5 together with the system of axis. The angle between the velocities is θ. The densities of both particle beams are supposed to be independent of the position along their own axis, y and y' for electrons and ions respectively.

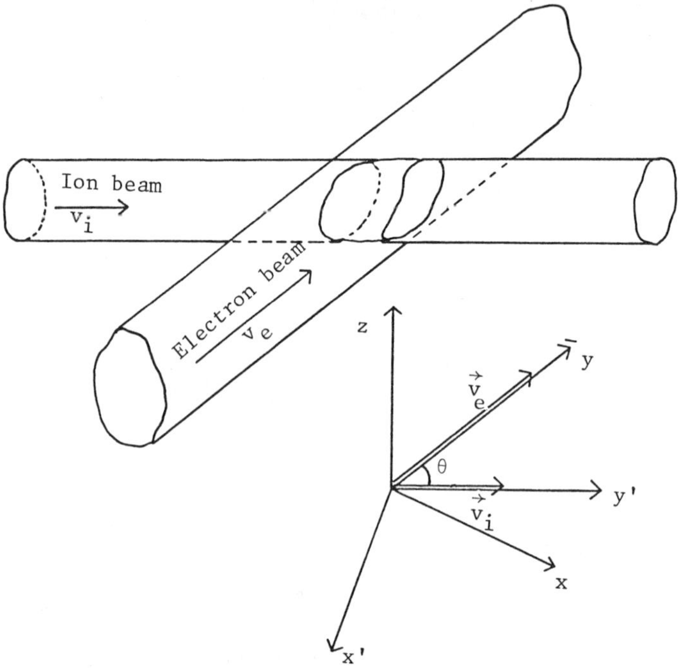

Fig. 5.

The count rate N is estimated by integrating (4) over the interaction volume in the following way :

(a) particles density are related to the flux :

$$n_e(x,z) = \frac{j_e(x,z)}{v_e}$$

$$n_i(x',z) = \frac{j_i'(x',z)}{v_i} \tag{7}$$

(b) The relative velocity v between two colliding particles is :

$$v = |\vec{v}_e - \vec{v}_i| = v_e^2 + v_i^2 - 2v_e v_i \cos\theta \tag{8}$$

(c) The volume element in the beam coordinate system is :

$$dV = dy\,dy'\sin\theta dz = \left(\frac{dx}{\sin\theta}\right)\left(\frac{dx'}{\sin\theta}\right) dz \sin\theta \qquad (9)$$

With (7), (8) and (9), the formula (4) becomes :

$$N = \sigma \int_{\text{Volume}} \frac{j_e(x,z)(j_i(x',z)\ \sqrt{v_e^2+v_i^2-2v_e v_i \cos\theta}\ dx\,dx'\,dz}{v_e \quad v_i \quad \sin\theta}$$

In this expression, integration over x and x' are separated in the form :

$$J_e(z) = \int_{w_e} j_e(x,z)dx \quad \text{and} \quad J_i(z) = \int_{w_i} j_i(x',z)dx' \qquad (10)$$

J_e and J_i are the flux of particles crossing a slice of width w_e and w_i respectively and height dz at the position z.
The count rate is written :

$$N = \frac{\sigma v}{v_e v_i \sin\theta} \int_z J_e(z)\ J_i(z)\ dz \qquad (11)$$

To measure the total cross-section for a given process in a beam-beam experiment, the velocities of both beams and their angle must be well defined. The count rate and the overlap integral of vertical density profiles must be measured.
Formula (11) is usually written in a different form, by defining the "form-factor" F :

$$F = \frac{\int J_e(z)dz \int J_i(z)dz}{\int J_e(z)J_i(z)dz} \qquad (12)$$

$$N = \frac{\sigma v I_e I_i}{v_e v_i F \sin\theta} \qquad (13)$$

In (12), the total particles current are :

$$I_e = \int J_e(z)dz$$

and

$$I_i = \int J_i(z)dz \qquad (14)$$

The form-factor has the dimension of a length and its meaning is easily understood with the aid of the example of an homogeneous electron beam and having a rectangular cross-section.
In this case (Fig. 6)

$$J_e(z) = I_e/h_e.$$

The overlap integral is obtained for $|z| \leq h_i/2$, and F is :

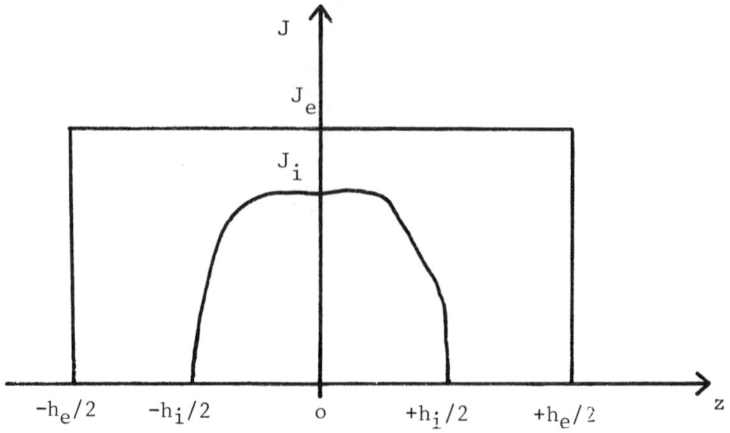

Fig. 6.

$$F = \frac{I_e I_i}{\int_{-h_i/2}^{+h_i/2} \frac{J_e}{h_e} J_i(z)dz} = h_e \tag{15}$$

where h_e and h_i are the heights of the electrons and ion beams respectively with $h_e > h_i$. The form factor F in this particular case is the size of the largest beam, here the electron beam.

Merging beams

If both beams are parallel to their common axis, the y axis (θ=o) they are superimposed and called "merging beams". Formula (4) is transformed by means of the relations : $v = |v_e - v_i|$
$dV = dx\,dy\,dz$.

The count rate N is :

$$N = \frac{\sigma v}{v_e\, v_i} \int_{\text{volume}} j_e(x,y,z)\, j_i(x,y,z)\, dx\, dy\, dz$$

If the interaction length L is small enough so that density profiles can be considered as uniform along the y axis, the formula becomes :

$$N = \frac{\sigma v L}{v_e\, v_i} \int_{\substack{\text{beam}\\ \text{cross-section}}} j_e(x,z)\, j_i(x,z)\, dx\, dz \tag{16}$$

That relation is usually written in the same form as (13) by defining the total particles currents I_e, I_i and the form-factor F :

$$I_e = \int j_e(x,z)\, dx\, dz$$

$$I_i = \int j_i(x,y)\, dx\, dz$$

$$F = \frac{\int j_e(x,z)\,dx\,dz \int j_i(x,z)\,dx\,dz}{\int j_e(x,z)\,j_i(x,z)\,dx\,dz} \qquad (17)$$

Finally the count rate N is given by :

$$N = \frac{\sigma v L \; I_e I_i}{v_e \; v_i \; F} \qquad (18)$$

The actual form factor has now dimension of an area.
Again, in the case where the biggest beam -usually the electron beam- has a uniform current distribution, the current density is given by I_e/S where S is the area of the electron beam cross-section.

In order to obtain absolute values of σ, it is necessary to measure the density profiles of both beams along the z direction for crossed beams and also along the x direction for merging beams.
Different methods have been developped to measure the density profiles.

In the most popular one, two slits of height Δz are moved through the beams and the transmitted currents ΔI_e and ΔI_i are simultaneously recorded at different positions. A numerical integration of the data gives the F value. An other method makes use of a plate moving at constant velocity through the beams. The derivative of the transmitted current I_t is the one dimensional density profile.
In the merging beam case, two one-dimensional profiles are required and if the interaction length is long, those profiles must be taken at different positions along the y axis.

The difficulty of those measurements is mainly due to the fact that they must be taken before or after the experiment.
As a consequence, fluctuations of the density profiles can occur giving a possible source of error. Moreover, the profiles are not taken at the crossing of the beams and errors can result from this procedure.

Finally, in the case of crossed or merged beam experiments with storage rings, the measurement of density profiles is completely impossible due to the destructive character of the method.

<u>The animated beam method</u>

The fundamental limit of the described method is due to the presence of the form factor which takes into account the inhomogeneities of both beams: one slice of a beam always crosses the same slice of the other beam.

But it is immediately seen that if one beam is homogeneous, the overlap integral in formula (11) becomes :

$$J_e \int J_i(z)\,dz = J_e \; I_i \qquad (19)$$

An homogeneous beam is not convenient to handle, because it is produced by strongly collimating the beam, so that the intensity is considerably reduced.

The difficulty can be passed round by alternative method to produce an homogeneous beam.

Let us suppose that a thin slice of thickness dz' of the electron beam is moved through the ion beam at a constant speed u over a height H.
In this case (Fig. 7), the electron density is independent of the position of the slice :

$$dJ'_e = J_e(z')\,\frac{dz'}{H}$$

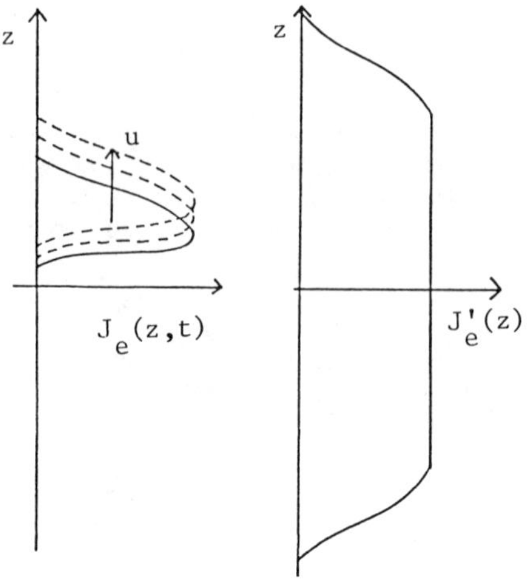

Fig. 7.

The z' axis is parallel to the z axis, with an origin moving at velocity u so that z' = z - ut.
The spatial density produced by the sweeping electron beam is :

$$J'_e = \int dJ'_e = \int \frac{J_e(z')dz'}{H} = \frac{I_e}{H} \qquad (20)$$

This relation holds if all the slices have the same velocity u and if the electron beam axis has always the same direction. This also assumes that the intensity is constant over the period of the displacement. The mean counting rate is then, from (11), (19) and (20), assuming that at the beams no longer overlap at the position z = ± H/2 :

$$N = \frac{\sigma v}{v_e\, v_i\, \sin\theta}\,\frac{I_e I_i}{H}\ .$$

The actual form factor is the amplitude of the electron beam displacement. This means that the counting rate is the same as in the case where a uniform electron beam of current I and height H crosses the ion beam.
In spite of its simplicity, the preceding formula is not used in this form but in a slightly modified one, by estimating the total numbers of events K produced during the time T of one complete movement :

$$K = \int_{\text{period}} N dt = \frac{\sigma v}{v_e\, v_i\, \sin\theta}\,\frac{I_e I_i T}{H}$$

Let us remind that H/T = u.
The final formula is :

$$K = \frac{\sigma \sqrt{v_e^2+v_i^2-2v_e v_i \cos\theta}}{v_e\, v_i \sin\theta}\,\frac{I_e I_i}{u} \qquad (21)$$

The parameters of (21) do not include any density profile so that their measurement is not needed any more.
The relation with initial method is found by comparison of (13) and (21):

$$\boxed{NF = Ku} \tag{22}$$

This method has been successfully applied in Louvain-la-Neuve by P. Defrance et al[8] for the electron impact ionization of light ions or atoms as well as multiply charged ions. The ribbon electron beam is swept electrostatically, by applying a ramp voltage on a deflector. The velocity u is estimated by measuring the time Δt that the electron beam spent between two thin wires located on both sides of the ion beam.

This method has also been recently applied in Giessen by A. Müller et al[9] in a somewhat different manner : the electron beam is swept mechanically through the ion beam at a low velocity while ion current and velocity are recorded as a function of the position of the electron gun. This method allows the estimation of the signal by taking into account the possible fluctuations of the velocity and the ion beam intensity.

The animated beam method can also be applied to merging beam experiments, especially in experiments involving ion storage rings, where direct form factor measurements are impossible.
In that case, a uniform density electron beam is produced by moving the electron beam in both x and y directions, like a TV spot, but by keeping the beam axis parallel to the ion beam axis.
In that case the density J'_e is :

$$J'_e = \frac{I_e}{S} = \frac{I_e}{WH} \tag{23}$$

S being the area of the "screen" occupied by the electron beam during the time T for one complete scan.
In a similar way as in the crossed beam, the total numbers of counts K produced in one scan is derived from (16) :

$$K = \frac{\sigma v L I_e I_i}{v_e v_i} \left(\frac{T}{S} \right) \tag{24}$$

Let us suppose that n lines, each of period t are present in a scan and that the frequency of the scan along the x axis is ν. It follows that:

$$\frac{T}{S} = \frac{T}{WH} = \frac{1}{t\left(\frac{W}{t} \right)\left(\frac{H}{T} \right)} = \frac{\nu}{u_x u_y}$$

The final formula is :

$$K = \frac{\sigma v L I_e I_i}{v_e v_i} \frac{\nu}{u_x u_y} \tag{25}$$

Again, no density profile measurement is required. The difficulty is now to assume a constant velocity displacement of the electron beam and being sure that the two beams are kept parallel.
The connection between the formulas (18) and (25) is :

$$\frac{K u_x u_y}{\nu} = NF \tag{26}$$

The choice of the angle

In the design of an experiment, the first parameter to be defined is the angle of the beams. This choice is essentially made on the basis of the relative velocity needed to observe the phenomena. The relative velocity is connected to the angle of the particles (formula (8)) :

$$v = |\vec{v}_e - \vec{v}_i| = \sqrt{v_e^2 + v_i^2 - 2v_e v_i \cos\theta} \qquad (8)$$

For fixed ion and electron beam energies the relative velocity varies with the θ angle as shown on Figure 8.

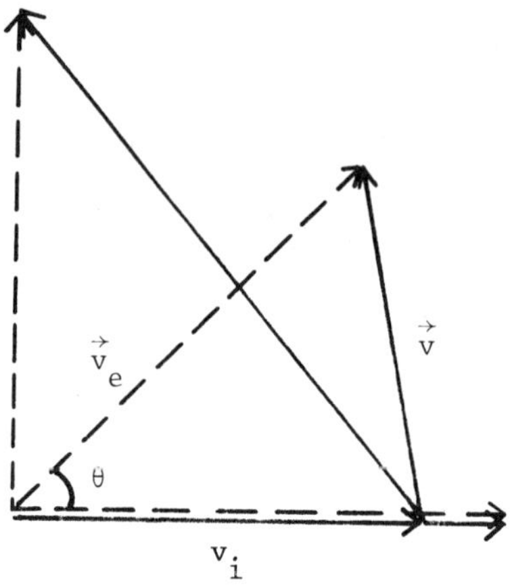

Fig. 8.

It is clearly seen that for small θ(inclined beams) or better for θ=o (merging beams) very low relative velocities are obtained. The zero velocity is found in the particular case of θ=0 and v_e=v_i. This arrangment give the opportunity to observe phenomena appearing at very low C.M. energy, like dissociative and dielectronic recombination.
Details of experimental parameters are found in the lecture of B. Mitchell.

The worldwide used arrangment for electron impact ionization and excitation experiments involves beams crossing at right angle. Inclined beams have been used for detachment experiments only.
At right angle, the relative energy E_r calculated in the center of mass is :

$$E_r = \left(\frac{1}{1+\frac{m_e}{m_i}}\right)\left(V_e + \frac{m_e}{m_i}\, q\, V_i\right) \quad (\text{eV}) \qquad (27)$$

where V_e and qV_i are the electrons and ions energies (in eV) respectively. For ions, m_e/m_i is negligeable, and the relative energy is approximately :

$$E_r = V_e + \frac{m_e}{m_i} qV_i \qquad (28)$$

The first term of (28) is the actual electron energy in the collision region. In most cases the second term is small with respect to the first one, usually less than 1 eV.

The error on the relative energy ΔE_r is :

$$\Delta E_r = \Delta V_e + \frac{m_e}{m_i} qV_i = 2 \sqrt{q \frac{V_e V_i m_e}{m_i}} \Delta\theta \qquad (28)$$

The third term is derived from the angular term in (8), taking into account the possible error on the angle $\Delta\theta$.
In (28), the first term is dominant in usual working conditions.

Two effects reduce the actual electrons kinetic energy V_e : the contact potential V_c at the cathode and the negative potential created by space charge effect in the electron beam itself. The energy expressed in eV is :

$$V_e = V_p - V_c + \int_{center}^{plate} E(z)dz \qquad (29)$$

Those corrections can be estimated by numerical calculation. The last term is usually obtained by assuming that the electron density distribution is uniform in the collision region. This procedure leads to corrections of order of a few eV.

Another method consists on the direct measurement of a well known cross-section threshold, excitation or ionization. The experimentally derived threshold is compared to the known value, from the theory, or from other sources. The difference gives the correction needed.

The electron energy distribution must also be estimated especially in the cases where sharp effects can be expected, like in excitation experiments. The electron energy spread is due to two effects. The first one is the temperature of the cathode which gives the electrons a maxwellian distribution with a width depending on the temperature :

$$\Delta V = 0.172 \times 10^{-3} \, T \qquad (30)$$

The second one is the electrons space-charge which creates an electric field varying from 0 on the axis to a maximum at the edge. The subsequent energy distribution can be approximated by a parabola in the case of a ribbon-electron beam, and the width of the energy distribution is generally of order of one eV.

IONISATION CROSS-SECTION MEASUREMENTS

The absolute determination of the electron impact ionization cross-sections requires the measurement of the total fluxes of the parent ion beam and of the products :

$$e + X^{q+} \rightarrow X^{(q+k)+} + (k+1)e^- \qquad (31)$$

The schematic arrangement of such an experimental set-up is shown in Figure 9. It consists of three parts :
(a) The parent ion beam production : ion source, charge selection and defining slits.
(b) The electron beam production and the collision region, described in the preceding section.
(c) The analysis region : magnetic or electrostatic charge separator, Faraday cup for the ion beam and single particle detector for the products of reaction (31).

APPARATUS (schematic)

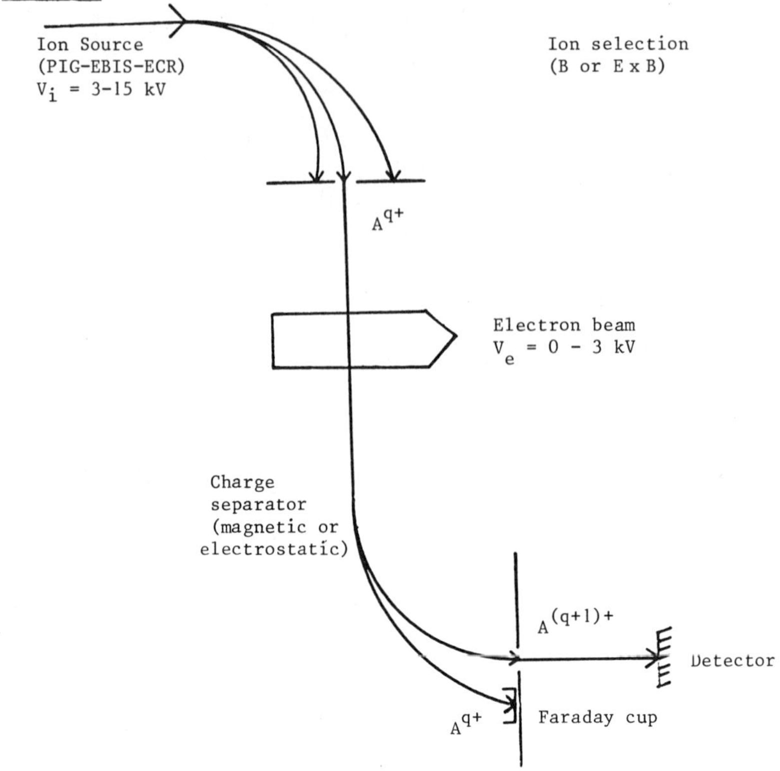

Fig. 9.

The parent beam production

The ion beam production has received a fundamental improvement in the last years through the development of two new types of ion sources : the Electron Beam Ion Source (EBIS) and the Electron Cyclotron Resonance (ECRIS) Ion Source. Those have been reviewed by H. Winter[10] together with more conventional ones. The remarkable feature of the two mentioned sources is the enlarged charge state distribution : values higher than 20 are available for Atomic Collision Physics. The EBIS has been described in a preceding section of this chapter.

The ECRIS design is based on the microwave electron heating in the presence of a magnetic bottle field configuration

A multipolar (hexapolar or octupolar) magnetic field is superimposed to the mirror fields in order to trap the plasma and in order to eliminate plasma instabilities. The mirror fields are adjusted to that electron cyclotron rotation takes place at the microwave frequency ω_c on a closed surface inside the source volume :

$$\omega_c = \frac{e\,B_s}{m_e} \tag{32}$$

In those conditions, the electron temperature increases mainly by resonant heating : in the simple case of circularly polarized microwaves in a uniform magnetic field, the phase angle between the wave electric field E_0 and the electron velocity is constant, so that the subsequent electron acceleration is also constant and the time dependence of the electron energy is given by :

$$\frac{1}{2} m_e v_e^2 = \frac{1}{2} m_e v_o^2 + \frac{1}{2m_e} (E_o \; et)^2 + e v_o E_o t \cos\phi \qquad (33)$$

In the source, the electron kinetic energy is limited by losses due to collisions with ions or atoms. The energy gain is also limited because the magnetic field is inhomogeneous : electrons leave the resonance region and oscillate between the mirrors. During the oscillations, the phase angle can be modified by an undefined quantity. This effect reduces the efficiency of the resonant heating. It has been seen that electrons may also gain energy by stochastic heating when they cross several times the resonant surface.
By those processes, electrons may reach temperature of a few tens keV and they are able to produce high charge states by successive collisions with ions trapped in the source even in the case of metallic ions.

According to the high electron temperature in the plasma and the feeding gas pressure, the ion charge state distribution results from the equilibrium between the ionization reactions and electron capture through charge exchange with gas molecules or atoms.

That capture is usually made in highly excited states that may decay radiatively to the ground state and to the metastable states. As a consequence, some long lived excited states may reach the collision region and interact with the electron beam.

For example 2^3S heliumlike N^{5+} ions have been observed in a percentage of some 4%. In a recent experiment on the electron impact ionization of Fe^{9+} by Gregory et al[12], an important signal has been observed below the first ionization threshold. This feature has been fully explained by Pindzola et al[13] through the contribution of the dominant metastable state population.

Another type of metastable ions has been observed in the sodiumlike serie : the long lived autoionizing states. Ions in those states have been seen[14] to produce a very high background in electron impact ionization experiments for the lowest members of the sequence Mg^+Al^{2+}, Si^{3+} and Ar^{7+}.

The analysis region

The product ions do not change their velocity in the reaction and the radius of curvature of parent ions (R') and the products (R) in the charge separator are related to their charge :

$$R'q = R(q+k) \qquad (34)$$

where k is the number of ejected electrons. The ion current is collected on a Faraday cup that is quite simple to handle, with the precaution that secondary emission of electrons or ions do not affect the measurement and that a 100% collection is achieved.
The detectors are of the secondary electron emission type : channel electron multiplier, channel plates or others.
The absolute determination of the product flux requires to solve some fundamental problems :

(a) The signal is superimposed to a background which must be determined.
(b) The interaction of two charged particle beams can reduce the transmission of the products to the detector.
(c) Tests must show that spurious contributions are not included in the count rate.
(d) The detector efficiency must be precisely estimated.

The background determination

In a crossed electron ion beam ionization experiment, the product ions are detected together with the background ions which are mainly due to stripping on residual gas in the vacuum chamber :

$$X^{q+} + y \rightarrow X^{(q+k)+} + \ldots \tag{35a}$$

The measured count rate N_m, is the sum of the true signal N and the products of the reaction (35) N_S :

$$N_m = N + N_S \tag{36}$$

This indicates that all of the experiments must have means of descriminating true events from this background. The energy loss in a stripping reaction is small with respect to the ions kinetic energy, so that conventional energy analysis cannot separate the ions formed in reactions (29) and (35).

The background countrate N_S can be estimated if the electron beam is turned off. In order to take into account the ion beam fluctuations and the possible background modulations, the electron beam is turned on and off repeatedly at a frequency of order of 1 kHz. The counting rate is introduced in two scalers alternatively, one when the electron beam is "on", the second one when the electron beam is "off". The counting sequence is illustrated in Figure 10 (a). The true signal N is calculated by substracting the second number from the first one.

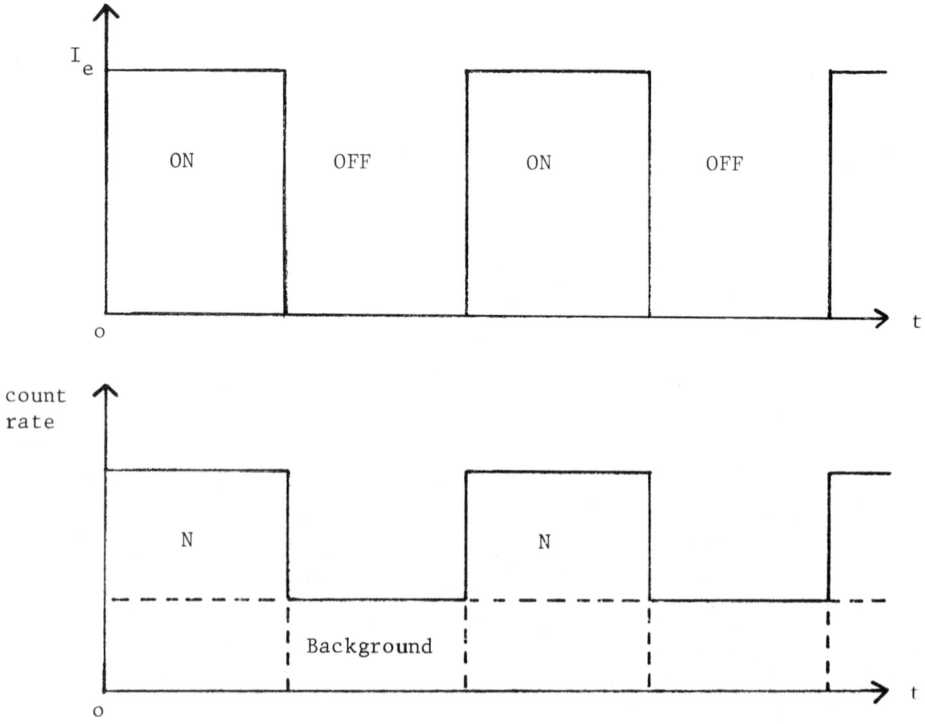

Fig. 10 (a)

In the animated beam method, the background appears naturally when the beams do not cross any more, so that the electron current modulation is not required. The signal delivered by the detector is recorded by a multichannel analyser working in the multiscaling mode and synchronised with the electron beam displacement.
The true signal appears as a peak superimposed on a flat background, and the number of events K is obtained from integration of the peak. This is illustrated in Figure 10 (b) together with the electron beam sweeping voltage.

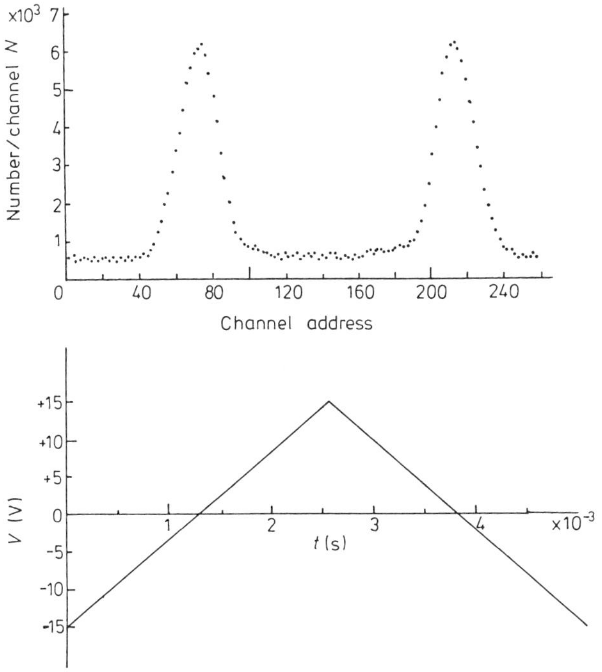

Fig. 10 (b).

An additional advantage of this method is the absence of possible background modulation due to pressure modulation when the electron beam is turned ON and OFF.

Beam-beam interaction

The space-charge density of the electron beam creates an electric field that modifies the ion trajectories and various effects result from this situation :

(i) An electron beam is an efficient electrostatic ion lens trajectories are bend toward the ion beam axis so that some ions can miss the Faraday cup or the ion detector. The size of those and the beam optics must be designed in order to achieve a 100% collection when the electron beam is on and when it is off.

(ii) In addition in the animated beam method, the ions are deflected by the electron beam electric field and the same conclusion as in (i) can be drawn. A detailed analysis of those effects is given by Harrison[15] by describing the electron beam as an infinitely wide ribbon.

Secondary ion trapping

The electron beam ionises residual gas atoms or molecules : those ions are formed at very low energies, so that they are easily trapped in the electron beam voltage depression. As a consequence, they reduce the space-charge repulsion in the electron beam and they also constitute a target that interacts with the ion beam[16] :

$$X^{q+} + Y^{+} \rightarrow X^{(q+1)+} + \ldots \qquad (36)$$

That reaction takes place in phase with the electron beam so that its contribution to the signal cannot be directly substracted.
This contribution is estimated in the following way :

Let us suppose that the density of trapped secondary ions, n^+ is related to the electron density n_e :

$$n^{+} = a\, n_e = \frac{a\, j_e}{v_e} \qquad (37)$$

From (3) the reaction rate N^+ due to the reaction (36) is :

$$N^{+} = \sigma^{+} \int_{\text{volume}} j_i(y,z)\, n^{+}(x,y,z)\, dx\,dy\,dz$$

where σ^+ is the cross-section for the process (36).
Integration of the expression gives :

$$N^{+} = \frac{\sigma^{+} a\, I_e I_i}{v_e F} \qquad (38)$$

and, in the animated beam method :

$$K^{+} = \frac{\sigma^{+} a\, I_e I_i}{v_e u} \qquad (38')$$

The corresponding measured total counting rates N_m and K_m are

$$N_m = N + N^{+} \quad \text{and} \quad K_m = K + K^{+}$$

so that the apparent cross-section σ_m is related to the parameters in the form :

$$\sigma_m = \sigma + \sigma^{+} a \frac{v_i}{v} \qquad (39)$$

In order to eliminate this spurious contribution, slow ions must be extracted out of the trap by applying a small electric field close to the interaction volume. The field modulation gives the opportunity to observe the importance of reaction (36) in the actual experiment.
An example is the study of the double ionization H^- for which a factor of ten between the true cross-section and the apparent one has been observed[17]. The Figure 11 shows the apparent cross-section as a function of the extracting voltage : for negative voltages, slow ions are extracted and for positive one, they are trapped.

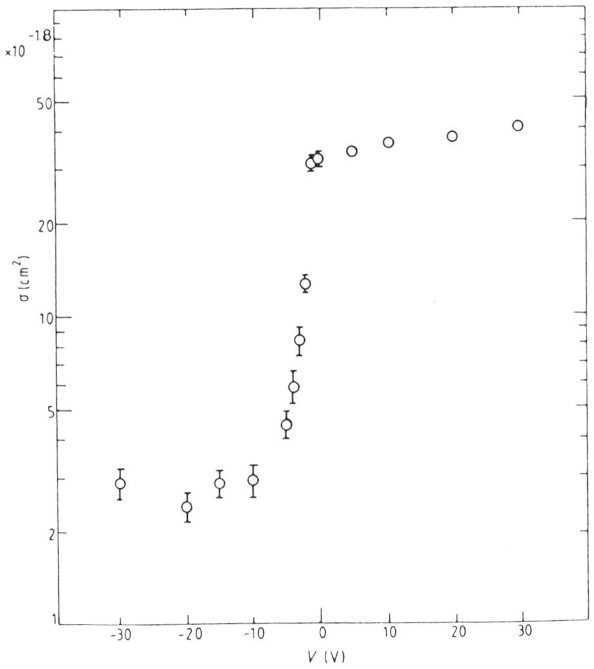

Fig. 11.

This effect can be attributed to the molecular ions H_2^+ because the vacuum chamber was pumped by turbomolecular pumps and the lightest molecules H_2 where probably the component of residual gas.
The cross-section for the process :

$$H^- + H_2^+ \rightarrow H^+ + H_2 + e^- \tag{40}$$

has been measured by Szücs et al[18] in a merging beam apparatus. The value for this process at the corresponding center of mass energy is 4.2×10^{-16} cm^2 $\pm$ 1.4×10^{-16} cm^2.

Values of a are estimated by (39) : they have been found equal to 0.74 ± 0.22 and 1.10 ± 0.33 at 300 eV and 330 eV electron energy respectively. Those figures are very close to unity and indicates that the full space charge neutralisation is achieved.

Finally some tests must be made in order to detect the possible problems or spurious contributions.

For example, the true signal is linearly dependent on the particle currents and on nonlinear observed dependance indicates the presence of spurious reactions.

It is also important to check the threshold of the reaction under study; a significant signal below this threshold can be due to transmission problems associated with the beam-beam interaction or due to the presence of a metastable component in the primary ion beam.

Detector efficiency measurement

Some ions do not produce detectable pulses when they strikes the detector. This number must be estimated, in order to obtain the true signal N :

$$N_m = \gamma N \tag{41}$$

The detection efficiency γ is defined as the ratio of the measured count rate (N_m) and the actual number of ions (N).
The simplest method to measure γ is to compare the electrical current I_i produced by a number of ions N reaching the detector. This current is measured by means of a Faraday cup. The detection efficiency is simply :

$$\gamma = \frac{N_m q e}{I_i} \tag{42}$$

This kind of measurement is limited by the maximum counting rate of standard detectors, of order of 10^4 ions/sec so that measured currents are of order of 10^{-15} A and they require high sensitivity electrometers. An alternative method has been used in Giessen : the ion beam is collimated, so that the transmitted beam is supposed to be more or less homogeneous[18]. A small part of this collimated beam passes through a hole and reaches the detector. This fraction is :

$$i_t = I_i \times A_2 / A_1 \tag{43}$$

where A_1 and A_2 are the area of the first and the second hole respectively.

The detector efficiency is estimated in two steps :
(i) The current density distribution is measured by moving the second hole together with the detector across the collimated ion beam to that the absolute detection efficiency is determined for a small area.
(ii) The second hole is fixed and the detector done is moved in order to scan its whole area so that relative counting efficiencies are estimated.

Another method has been proposed by F. Brouillard[20]. In order to measure higher currents and to transmit to the detector a small fraction of this one only, the beam is swept at a high velocity through a slit located in front of the detector.

The transmitted current i_t is :

$$i_t = \nu T I_i \tag{44}$$

where T is the mean time that the beam spend at the slit and ν is the frequency of the crossings. At high frequencies and with a current I_i of order of 10^{-9} A, the transmitted current is easily measurable by means of a Faraday cup. Values of T of order of 1 nsec are obtained.

At low frequencies, the Faraday cup is removed and the count rate N_m is related to the transmitted current :

$$\gamma = \frac{N_m q_e}{\nu T I_i} \tag{45}$$

That expression does not take into account the probability that two or more ions arriving during the same crossing cannot be separated by the detection system. This estimation is made in the following way. Let P(n) be the probability that n ions pass through the slit during one crossing. P(n) is given by the Poisson law :

$$P(n) = \frac{1}{n!} \left(\frac{T}{\tau} \right)^n e^{-T/\tau}$$

with $\tau = qe/I_i$ is the average time separating two ions in the beam. The probability p(n) that one pulse is delivered when n ions reach the detector in a time smaller than the response of the detector : $p(n) = 1 - (1-\gamma)^n$. The probability of one pulse over one crossing is :

$$\Sigma p(n)P(n) = 1 - e^{-T/\tau}$$

If the sweep is repeated at frequency ν, the counting rate is :

$$N_m = \nu(1 - e^{-\gamma T I_i/qe}) \tag{46}$$

The experimental set-up is shown in Figure 12. The slit is located at the bottom of the first Faraday cup.

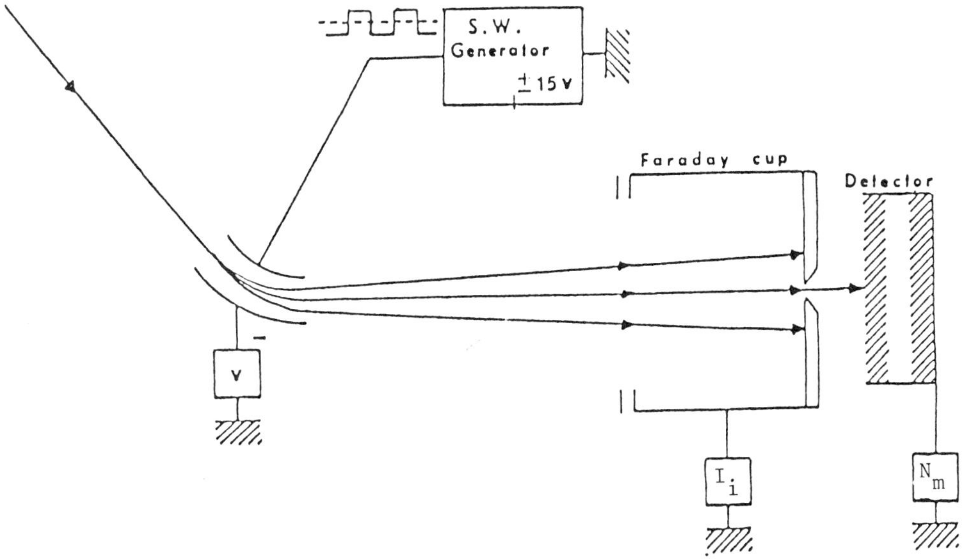

Fig. 12.

In Figure 13, are shown the linear dependence of the transmitted current (a), measured by the Faraday cup (2) and the linear dependence of the counting rate (b). From the slope of the transmitted current (44), T is estimated. From the slope of the counting rate, the detection efficiency is estimated by (46). This method has been applied for various types of detectors and for various types of particles, atoms, negative ions and multiply charged ions.

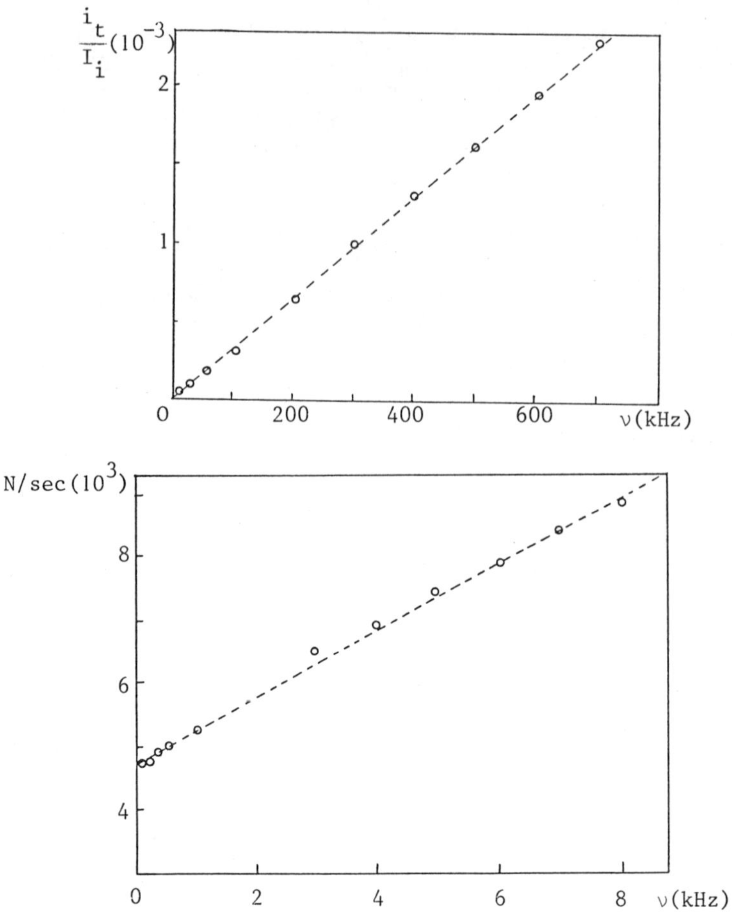

Fig. 13

EXCITATION CROSS-SECTION MEASUREMENTS

Excitation cross-sections are measured in beam-beam experiments in three different ways :

(a) optical detection method,
(b) electron loss analysis method,
(c) excitation autoionization method.

Those methods are not overlapping : the first one, (a) is only of use for excitation of dipole transitions, so that de-excitation of states takes place in the collision region. It is a total cross-section measurement. The second (b) is a differential cross-section measurement by means of the analysis of the energy lost by incident electrons. It can be used for all type of transitions. The third (c) is a total cross-section measurement of autoionizing states excitation.

Optical detection method

This method is applied in a 90° crossed electron ion beam experiment by P.O. Taylor and G.H. Dunn[21], (Fig. 14).
The detection system is composed of lenses, a filter, a polarizer and a photon detector. This system is placed at a fixed angle. The count rate in this device is only a small part of the total count rate, due to different factors :

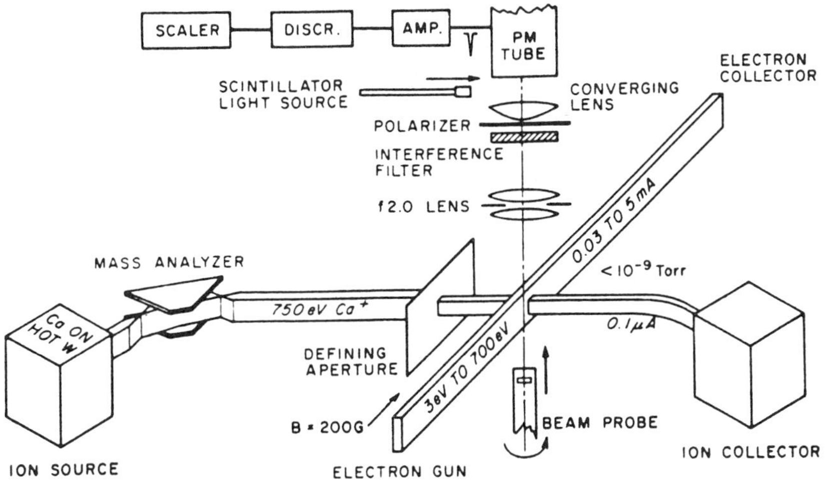

Fig. 14.

(a) the radiation is emitted anisotropically over the space, as a function of the angle θ between the detection system axis and the electron beam axis. From the total count rate, a fraction

$$\alpha = \frac{1 - P\cos^2\theta}{1 - P/3} \tag{47}$$

is collected at the angle θ, where P is the polarisation of the light

$$P = N_{//} - N_{\perp} \ / \ N_{//} + N_{\perp} \tag{48}$$

where $N_{//}$ and N are respectively the measured count rate with the polarizer axis parallel and perpendicular to the electron beam. Let us note that in the case where the θ angle is choozen, so that $\cos^2\theta = 1/3$, that is for $\theta = 54.7°$, the polarisation is not to be measured, because the result is independent of P.

(b) The count rate is reduced by the absolute photon detection efficiency for a specific wavelength D_λ due to the detector and the finite solid angle for photons emitted at the position (0,0,0) of the collision region.

(c) The form factor F must now include the probability that the photons emitted by excited ions are detected when they are emitted from each point of the collision region :

$$F = \frac{\int J_e(z)dz \int J_i(z)dz}{\int J_e(z)J_i(z)\eta(z,\lambda)dz} \tag{49}$$

$\eta(z,\lambda)$ is the detection efficiency relative to D_λ. The function $\eta(z,\lambda)$ includes three terms :

$$\eta(z,\lambda) = D_R(z,\lambda) - I_1 + I_2 \tag{50}$$

The term I_1 is introduced because some of the excited ions radiate outside of the electron beam, while I_2 takes into account the ions radiating outside of the electron beam but from which photons are still detected. I_1 and I_2 takes into account the life time τ of the transition.

$$I_1(z,\lambda) = \frac{\int_0^{w_e} e^{-x/v_i\tau} D_R(x,z,\lambda)dx}{\int_0^{w_e} D_R(x,o,\lambda)dx} \quad (51)$$

$$I_2(z,\lambda) = (e^{w_e/v_i\tau} - 1) \frac{\int_{w_e}^{\infty} e^{-x/v_i\tau} D_R(x,z,\lambda)}{\int_0^{w_e} D_R(x,o,\lambda)dx} \quad (52)$$

A precise cross-section estimation implies the knowledge of the parameters P, D_λ,$D(x,z,\lambda)$ and of the density profiles $J_e(z)$ and $J_i(z)$.

(a) The measurement of P implies the use of polarizer. In order to avoid this determination, theoretical values have been used by Gregory et al[22]. That correction is of order of a few percent only.
(b) D_λ is measured by using a uniform, nonpolarized monochromatic light source of the same size as the collision region. The source is calibrated by comparison with blackbody emission.
(c) $D_R(z,\lambda)$ which is the quantity giving the most important contribution to the corrected form factor is obtained by means of a movable line light source having approximately the width of the electron beam. That source is moved along the ion beam axis and the z direction. An important background in those experiments is produced by the electron beam and by the ion beam. The true signal must be determined by a double pulsing sequence. As a consequence measurements need often more than one full day of signal acquisition.

Electron energy loss

In 1982, A. Chutjan and W.R. Newell[23] reported the first measurements of differential cross-sections for inelastic scattering of electrons from an ion. Angular differential cross-sections constitute a more sensitive test for theory. They also provide a means to study the dipole forbidden transitions :

$$e(E,\theta=o)+X^{q+}(nl) \rightarrow X^{q+}(n'l')+e(E-\Delta E,\theta)$$

For this purpose, a conventional 90° crossed-beam set up is designed including an hemispherical electrostatic analyser located at a small angle θ(6° to 17°) with respect to the electron beam axis. The analyze energy resolution is 0.55 eV (full width at half maximum). The counting rate N is related to the differential cross-section $d\sigma/d\Omega$:

$$N = \frac{\frac{d\sigma(\theta)}{d\Omega} I_e I_i \phi\Delta\Omega}{v_e v_i F} \quad (53)$$

In this expression ϕ is the product of the analyser transmission and the detector efficiency; $\Delta\Omega$ is the solid angle observed by the analyser.

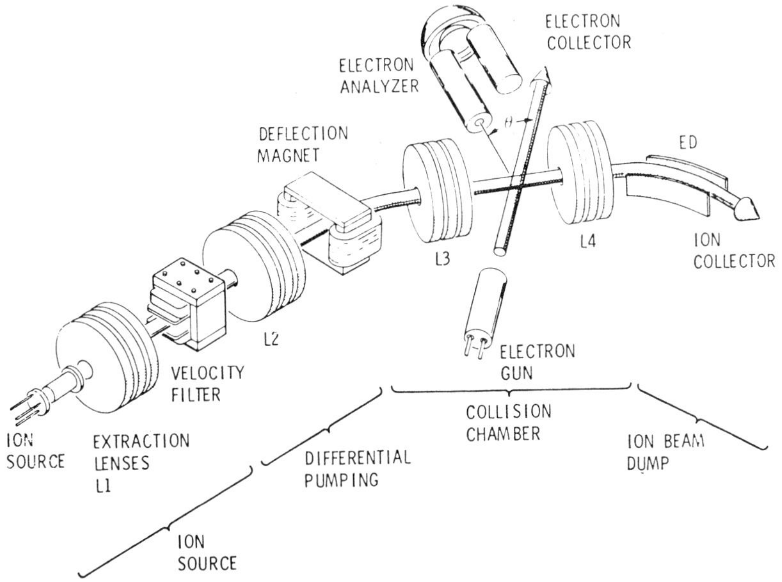

Fig. 15.

An example of the signal as a function of the energy loss (Fig. 16) shows the peaks corresponding to the transitions under study[24].

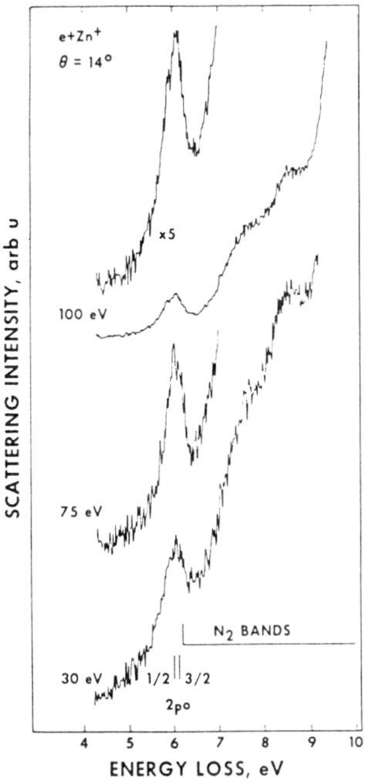

Fig. 16.

Relative differential cross-sections are immediately obtained at each energy by measuring the ratios of peak area.

In this experiment, some of the parameters, ϕ, $\Delta\Omega$ and F, where not measured, but only estimated by numerical calculation. This procedure introduce systematic errors in the measurements.

Three singly charged ions have been considered in these experiments, Zn^+, Cd^+ and Mg^+. The method seems to be promising and its application to the electron impact excitation of multiply charged ions can certainly produce detailed information.

Excitation-autoionization

The excitation to autoionizing states is particular interest : it contributes to the total ionization cross-section and can be studied by means of the same methods as the direct ionization.

The contribution of this process appear as an excitation process superimposed to the direct process, so that absolute excitation cross-sections are obtained in principle from ionization measurements. Various features reduce the field of these investigations :

(a) the true signal cannot be separated from the direct ionization one so that it is not really possible to estimate precisely the cross-section except at the threshold,

(b) the electron energy distribution has usually a width of order of 1 eV, so that when multiple excitation-autoionization processes appear in a short energy range, they can not be resolved.

ACKNOWLEDGMENTS

This work has been supported by the "Institut Interuniversitaire pour les Sciences Nucléaires".
I am indebted to the mentioned co-workers for their participation to this work.
I acknowledge fruitful conversations with Ron Phaneuf, Don Gregory and Alfred Müller during the preparation of these lectures.

REFERENCES

1. M. Harrison, This Institute.
2. J. Dubau, This Institute.
3. R. Phaneuf, This Institute.
4. J. Hasted, "Confinment of Ions for Collision Studies", in Physics of Ion-Ion and Electron-Ion Collisions, ed. F. Brouillard and J.W. McGowan, NATO-ASI Series B, Vol. 83 (Plenum, New-York and London, 1983).
5. E.D. Donets and P. Ovsyannikov, Sov. Phys. JETP 53 (3):466 (1981).
6. J. Arianer, Nucl. Inst. and Methods, B9:516 (1985).
7. E.D. Donets, Physica Scripta, Vol. T3:11 (1983).
8. P. Defrance, F. Brouillard, W. Claeys and G. Van Wassenhove, J. Phys. B : At. Mol. Phys. 14:103 (1981).
9. A. Müller, K. Huber, K. Tinschert, R. Becker and E. Salzborn, J. Phys. B : At. Mol. Phys. 14:2993 (1985).
10. H. Winter, "Atomic Physics of Highly Ionized Atoms, Ed. R. Marrus (Plenum, New-York and London, 1983).
11. Lichtenberg : Phase-Space Dynamics of Particles, J. Wiley and Sons, New-York (1969).

12. D.C. Gregory, F.W. Meyer, A. Müller and P. Defrance (to be published) (1985).
13. M.S. Pindzola, D.C. Griffin and C. Bottcher (to be published) (1985).
14. A.M. Howald, D.C. Gregory, F.W. Meyer, R. Phaneuf, A. Müller, N. Djuric and G.H. Dunn, Phys. Rev. A (1985), (submitted for publication).
15. M. Harrison, Methods of experimental Physics, Vol. 7A, Academic Press New-York pp 95-115 (1968), Ed. B. Bederson and W.L. Fite.
16. A.J. Dixon, M.F.A. Harrison and A.C.H. Smith, J. Phys. B : At. Mol. Phys. 9, 2617 (1976).
17. P. Defrance, W. Claeys and F. Brouillard, J. Phys. B 15:3509 (1982).
18. S. Szücs, S. Rachafi and M. Terao (unpublished) (1983).
19. K. Rinn, A. Müller, H. Eichenauer and E. Salzborn, R. Sc. Instr. 53(6):829 (1982).
20. F. Brouillard, S. Chantrenne, W. Claeys, A. Cornet and P. Defrance, XIII ICPEAC, Berlin, Book of abstracts p. 713, Ed. J. Eichler, W. Fritsch, I.V. Hertel, N. Stolterfoht and U. Wille (1983).
21. P.O. Taylor and G.H. Dunn, Phys. Rev. A 8:3204 (1973).
22. D. Gregory, G.H. Dunn, R. Phaneuf and D. Crandall, Phys. Rev. A 20:410 (1979).
23. A. Chutjan and W.R. Newell, Phys. Rev. A 26:2271 (1982).
24. A. Chutjan, High Temperature Science 17, 135 (1984).

DISSOCIATIVE RECOMBINATION OF MOLECULAR IONS

J.B.A. Mitchell

Dept. of Physics and Centre for Chemical Physics

The University of Western Ontario

London, Ontario, Canada. N6A 3K7

INTRODUCTION

The dissociative recombination of molecular ions with electrons plays a central role in the physics and chemistry of ionized systems. Although simple in concept it displays an underlying complexity which has frustrated many attempts to obtain a clear understanding of its nature. In fact almost without exception it is impossible to say with confidence that the recombination of any molecular species is well understood. The problem lies partially within the inherent complexity of molecular systems with their infinite series of excited rydberg states which play an important role in the mechanism for recombination. An additional concern is the fact that as yet no absolute demonstration of the predicted mechanisms for recombination has been performed. Many experimental measurements relating to recombination have been made. All however have either inherent uncertainties concerning the identity of initial and final excitation states or have been performed on systems where recombination can proceed through a variety of channels so that no individual one can be singled out.

A number of reviews on molecular ion recombination have been published. Bardsley and Biondi (1970), Bardsley (1979), Berry and Leach (1981), Maruyama et al (1981), Eletskii and Smirnov (1982), Mitchell and McGowan (1983), McGowan and Mitchell (1984). This current review will concentrate upon a comparison of theoretical and experimental studies and in particular developments since the last NATO ASI on the Physics of Electron-Ion and Ion-Ion Collisions will be highlighted.

The recombination of small molecular ions generally proceeds via a mechanism in which the resulting electron-ion compound state dissociates into neutral fragments. This is called dissociative recombination. Energy has to be imparted to a molecule to ionize it. During recombination at least part of this energy must be removed and for dissociative recombination excess energy is carried away in the forms of kinetic energy and internal energy of the dissociation products. In this way the recombination process is stabilized so that the system does not become reionized. The initial electron-ion compound state formed when the electron is first captured, has a potential energy greater than the ionization limit of the molecule so it can autoionize within a typical time of 10^{-13} secs. For the stabilization of the recombination to be efficient it must have a characteristic time of the same order or smaller than this. The time for the electron-ion state to dissociate into neutral fragments is typically comparable to the lifetime against auto-ionization so dissociative recombination can be a very efficient process. For atomic ions, stabilization occurs via radiative emission of photons with a characteristic time of $\sim 10^{-9}$ secs. This process does not compete very effectively with auto-ionization and so atomic ions have very small probabilities for recombination.

Batès and Massey (1947) first proposed the mechanism for dissociative recombination in order to explain the diurnal variation of the electron density in the earth's ionosphere. Earlier proposals based upon atomic ion recombination proved to have too long a timescale since predicted recombination rates were small. The much more efficient dissociative recombination mechanism allows molecular ions, formed via photoionization during the daytime, to act as an efficient sink for electrons at night.

Bates and Massey's mechanism for dissociative recombination can be understood with reference to figure 1. This shows the ground electronic state of a diatomic molecular ion AB^+. The latter is intersected by a repulsive doubly excited state of the neutral, AB_1^{**} and this dissociates to the limit $A^* + B$. Consider an electron with kinetic energy Ee approaching the ion AB^+. The total energy of the electron-ion system is then equal to the potential energy of AB^+ plus Ee. It can be seen from figure 1 that this energy is degenerate with the energy of AB_1^{**} for any value of Ee provided that AB_1^{**} intersects AB^+ in the vicinity of its minimum.

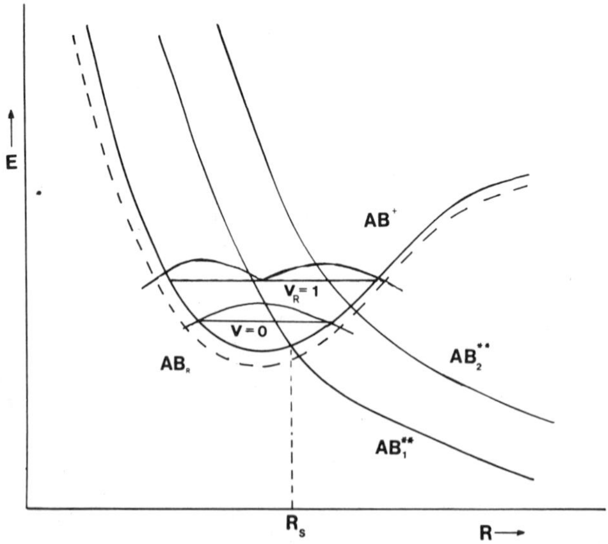

Fig. 1. Schematic representation of the states involved in the dissociative recombination of molecular ions.

It is possible then for the system to make a transition

$e + AB^+ \rightarrow AB_1^{**}$

i.e. a compound state can be formed. Since however this state lies above the ionization limit, it can decay back to AB^+ and a free electron. This is called autoionization. AB_1^{**} is repulsive however and so A^* and B will move rapidly apart. As they do their kinetic energy increases and so the potential energy of the system decreases. When their separation is greater than Rs, then AB_1^{**} can no longer autoionize and the recombination is stabilized. AB_1^{**} then continues towards dissociation

$AB_1^{**} \rightarrow A^* + B$

The above model presupposes however, that a state such as AB_1^{**} exists and that there is a good overlap between the nuclear wavefunctions of the AB^+ and AB_1^{**}.

It can be seen from figure 1 that if the ion is vibrationally excited then the overlap between the nuclear wavefunctions of AB^+ and AB_1^{**} is small. In this case there is a small probability of the initial electron capture occurring except for high energies. Note that the electron capture proceeds via a vertical, Franck-Condon transition between AB^+ and AB_1^{**}.

On the other hand, if the only available dissociating state is AB_2^{**} then the reverse is true. Ground state ions will not recombine with low energy electrons via this mechanism whereas vibrationally excited states will.

<u>Experimental techniques</u>

The probability of electron-ion recombination is commonly quantified in terms of a rate coefficient $\alpha(cm^3s^{-1})$ defined by:

$$\frac{dNo}{dt} = \alpha N_i N_e \qquad (1)$$

where N_e, N_i and N_o are the number of electrons, ions and recombined neutrals in a given system. The term α is related to the collision cross section $\sigma(cm^2)$ by

$$\alpha = \int v_e \sigma(v_e) f(v_e) dv_e \qquad (2)$$

where v_e is the electron velocity of the reaction products and $f(v_e)$ is the velocity distribution. For most ionized media $f(v_e)$ is a Maxwellian distribution.

A wide variety of experimental techniques have been used to study molecular ion recombination. Essentially they fall into two categories, those which measure the disappearance of ions or electrons due to recombination and those where the neutrals thus formed are detected directly. Intersecting beam techniques such as Merged (Auerbach et al 1977), Inclined (Peart and Dolder 1974), crossed beams (Phaneuf et al 1975, Vogler and Dunn 1975) and laser induced fluorescence (Zipf 1980a,b) belong to the latter category. Microwave afterglow (Mehr and Biondi, 1969, Shiu et al 1977, Oskam and Mittelstadt 1963), Shock tube (Cunningham & Hobson 1969), Pulse radiolysis (Rebbert & Ausloos 1972) Maier & Fessenden (1975), Flame studies (Hayhurst and Telford 1971, Burdett and Telford 1979), Flowing afterglow langmuir probe (Alge, Adams and Smith 1983), Trapped ion (Walls and Dunn 1974, Mathur et al 1978) belong to the former.

Details and characteristics of these techniques have been discussed elsewhere (Mitchell and McGowan, 1983, McGowan and Mitchell 1984) and only the three techniques which are currently most active will be discussed here.

Microwave Afterglow

The afterglow technique was used to provide the first measurements of dissociative recombination more than thirty years ago and is still actively pursued. In the experiments, fig. (2) a glow discharge plasma is formed in a reaction vessel and when the exciting mechanism is turned off the decay of the plasma electrons is observed as a function of time by studying the reflection of low energy microwaves. The change in the electron density is given by:

$$\frac{\partial N_e}{\partial t} = \Sigma P_i - \Sigma L_i - \nabla . \Gamma_e \qquad (3)$$

where P_i = formation rate

L_i = Loss rate due to atomic processes

$\nabla . \Gamma_e$ = Loss rate due to diffusion.

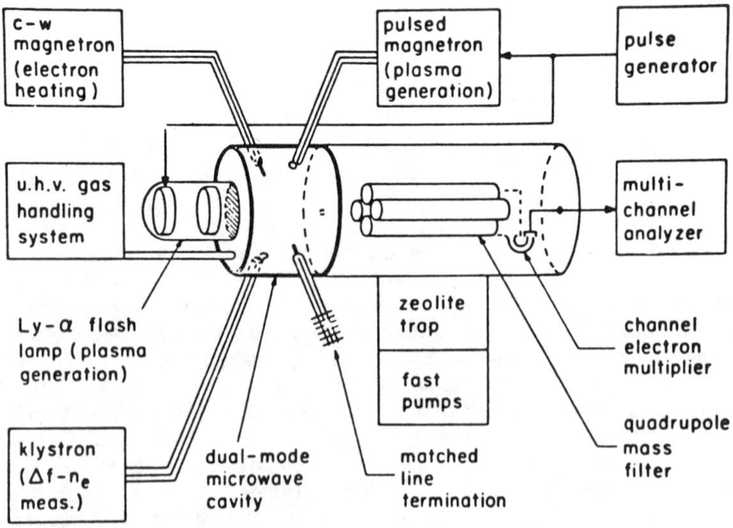

Fig. 2. Schematic diagram of Biondi's microwave afterglow apparatus. (Mehr and Biondi, 1969).

Care has to be taken to make sure that all contributions to each of the three terms on the RHS of equation (3) are taken account of. For example, metastable collisions can give rise to associative ionization thus increasing P_i; if the pressure is too high then three body effects contribute to L_i. One must also consider the importance of electron heating processes such as superelastic collision of electrons with excited atoms and molecules.

Ambipolar diffusion losses are usually minimized by using an appropriate buffer gas. Under ideal conditions:

$$\Sigma P_i = 0$$

$$\nabla . \Gamma_e = 0$$

$$L_i = \alpha N_i N_e$$

$$N_i = N_e$$

Then

$$\frac{dN_e}{dt} = - \alpha N_e^2 \tag{4}$$

and so

$$\frac{1}{N_e}(r,t) = \frac{1}{N_e}(r,o) + \alpha t \tag{5}$$

Hence the rate coefficient α can be determined from the slope of the $1/N_e$ vs. t plot. In practice, however, ambipolar diffusion giving rise to non-uniform electron distributions cannot be ignored requiring computer generated solutions of equation (3). Under normal conditions, $T_e = T_i = T_{gas}$, that is, the system has time to thermalize. It is possible however, to raise T_e, the electron temperature by microwave heating and so study the temperature dependance of the dissociative recombination rate coefficient. In this case T_i remains equal to T_{gas}.

The major problems associated with this technique are concerned with the identification of the ion under study and its excitation state. Clustering processes occur in the plasmas under certain conditions and these give rise to species such as $H_3O^+.(H_2O)_n$, $(N_4)^+$, etc. which compete for electrons in the plasma with the primary ion under study in the recombination process. The use of a mass spectrometer for the monitoring of plasma conditions is mandatory for the proper measurement of individual recombination rates of the various ion species.

A more complete description of a microwave afterglow apparatus employing mass spectrometric sampling may be found in Mehr and Biondi (1969). Determination of the state of excitation of the ion under study, is a major problem in recombination measurements. The problem lies in the fact that molecular ions can be electronically, vibrationally and rotationally excited. These states are often long lived compared to the lifetime of the ion in the system. In the past most experimenters have inferred the presence or absence of excitation by studying the rate coefficient for the recombination as the experimental conditions are varied. For example inert buffer gases are used in the plasma afterglow technique and the pressure of these gases can determine the rate of vibrational relaxation of the ion under study.

Recently Zipf (1980 a,b) however has introduced the use of laser induced photofluorescence for the determination of initial ion excitation in afterglow measurements. In this technique, a 1 MW tuneable dye laser is used to excite a particular transition from a given initial vibrational state (e.g. v = 0) and the fluorescence arising from the relaxation of the upper state is measured. Since the absolute laser flux entering the cavity is known then the population of the state under study can be determined directly.

Both Zipf (1980a,b) and Shiu et al (1977a,b,78) have further extended the afterglow technique to the examination of the excitation states of the dissociation products by measuring the light emitted when these states relax. Measurements such as these are vital to our eventual understanding of the mechanisms for dissociative recombination. Unfortunately, however, plasma techniques are capable of measuring rate coefficients only, because the observations are averaged over the electron velocity distribution of the plasma. They are therefore not as sensitive as the intersecting beam methods capable of measuring cross sections.

INTERSECTING BEAM EXPERIMENTS

These are experiments in which a beam of electrons is made to collide with a beam of ions and the products of the recombination are subsequently measured. Three variations have been used, namely crossed beams, inclined beams and merged beams.

Comprehensive reviews of intersecting beam techniques have been given by Harrison (1966), Dolder (1969), Brouillard and Claeys (1983) and Auerbach et al (1977).

Considering the case of a monenergetic ion beam with energy E_i intersecting a monoenergetic electron beam of energy E_e at some angle θ, then the collision energy in the centre of mass frame for this system is given by:

$$E_{cm}=\tfrac{1}{2}\mu(\bar{v}_i - \bar{v}_e)^2=\mu[E_i/m_i+E_e/m_e-2(E_iE_e/m_im_e)^{\frac{1}{2}}\cos\theta] \qquad (6)$$

where m_i, m_e, are the ion and electron masses.

$$\mu = m_em_i/(m_e + m_1) = m_e$$

μ = the reduced mass of the system.

v_r = relative velocity of the ion and the electron.

We can define a quantity called the reduced ion energy E_+:

$$E_+ = \frac{m_e}{m_i} . E_i \qquad (7)$$

so that equation (6) becomes simplified to:

$$E_{cm} = E_+ + E_e - 2(E_+E_e)^{\frac{1}{2}} \cos\theta. \qquad (8)$$

When θ is small, this reduces to

$$E_{cm} \simeq (E_+^{\frac{1}{2}} - E_e^{\frac{1}{2}})^2 + (E_+E_e)^{\frac{1}{2}}\theta^2 \qquad (9)$$

Hence when $E_+ = E_e$, the minimum achievable centre of mass energy is limited by the value of θ.

By differentiating equation (6) with respect to E_i, E_e and θ one can obtain the following expression for the energy resolution ΔE_{cm}. (Assuming Gaussian distributions for ΔE_i, ΔE_e, and $\Delta\theta$).

$$\Delta E_{cm} = \left[\{[1-(E_+/E_e)^{\frac{1}{2}}]\Delta E_e\}^2 + \{[1-(E_e/E_+)^{\frac{1}{2}}]\Delta E_+\}^2 + [2(E_e E_+)^{\frac{1}{2}}\theta\Delta\theta]^2\right]^{\frac{1}{2}} \qquad (10)$$

When $E_+ \simeq E_e$ the contributions due to ΔE_e and ΔE_i become negligible and the energy resolution is dominated by the angular term. This enables one to achieve very high resolution in the merged beam case provided θ can be made small.

Three groups have used intersecting beams for the study of dissociative recombination. Dunn, at JILA, used a crossed beam experiment for the study of the formation of D(2p) and D(n=4) atoms during e + D_2^+ recombination in the energy range from 0.6 - 7eV. Dolder, at Newcastle upon Tyne extended the energy range down to 0.3eV by using the inclined beam technique and measured total cross sections for e + D_2^+, H_2^+ and H_3^+ collisions. The most extensive beam studies of dissociative recombination have been performed at the University of Western Ontario, using a merged beam apparatus. To date more than 30 different species have been studied. McGowan and Mitchell (1984).

In Dunn's experiments, the recombination was measured by detecting the photons emitted during the decay of the excited dissociation products. This is technically difficult to do but detailed information concerning the decay channels of the process can be obtained in this way.

The other experiments both depend on measuring the number of neutrals formed as a result of the electron ion recombination. Because space does not permit detailed description of each of the apparatuses, only a brief outline of the MEIBE I (Merged Electron Ion Beam Experiment) at the University of Western Ontario will be given here.

Ions are produced in an rf ion source mounted in the terminal of a 400 Kev Van de Graaff accelerator. After focussing and mass analysis, the ion beam enters the ultra high vacuum experimental chamber where after being offset electrostatically to remove neutrals it passes through the interaction region. After collision with the electrons, the ion beam is analysed electrostatically to remove neutrals formed as a result of both electron-ion and background collisions and the primary ions are collected in a Faraday cup. The neutrals are allowed to strike a surface barrier detector and are subsequently counted.

The electron beam is formed in a Pierce type electron gun and is subsequently merged with the ion beam using a trochoidal analyser. This device operates by having a magnetic field axial to the electron beam and an electric field perpendicular to it. The electrons undergo a precise spiralling motion in the analyzer and when they emerge, if the correct conditions are present, the input and output vectors of the beam will be identical but the axis of the beam will be shifted to a new axis offset from the original. This new axis is made to coincide with the axis of the ion beam so that merging occurs.

After the interaction with the ion beam, the electrons are then "demerged" using a second trochoidal analyzer before being collected in a Faraday cup. A schematic diagram of the apparatus is given in Fig. (3).

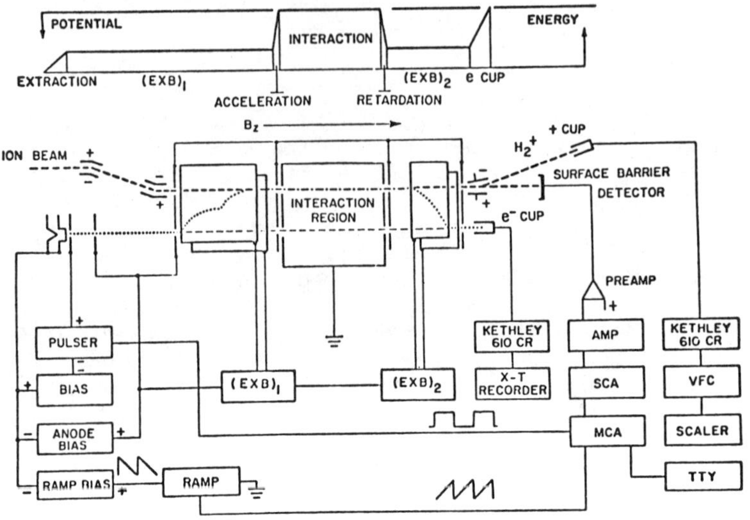

Fig. 3. Schematic diagram of the Merged Electron-Ion Beam Experiment (MEIBE I) at the University of Western Ontario (Auerbach et al 1977).

The recombination cross section may be obtained using the following expression:

$$\sigma(E_{cm}) = \frac{C_n e^2 F}{I_e I_i L}\left|\frac{\bar{v}_i \bar{v}_e}{\bar{v}_i - \bar{v}_e}\right| \simeq \frac{C_n e^2 F}{I_e I_i L}\left|\frac{v_i v_e}{[(v_i - v_e)^2 + v_i v_e \theta^2]^{\frac{1}{2}}}\right| \qquad \text{--- (11)}$$

where C_n = neutral count rate

I_e, I_i = Electron and Ion beam currents

v_e, v_i = Electron and Ion beam velocities

θ = intersection angle

L = intersection length

$$F = \left[\iint i_e(x,y)\,dxdy \iint i_i(x,y)dxdy\right]\left[\iint i_e(x,y).i_i(x,y)dxdy\right]^{-1} \qquad \text{--- (12)}$$

The form factor F is determined by measuring the beam density distributions and overlap at three places along the intersection length (Keyser et al 1979) and so absolute cross sections may be obtained using this apparatus.

A major reduction in background noise levels can be achieved by exploiting the energy resolving properties of the surface barrier detector. Neutral atoms arising from dissociative collisions of the primary beam with the background gas giving ion-atom pairs, arrive with only a fraction of the beam energy and so can be distinguished from other processes such as dissociative recombination and dissociative charge exchange where the resulting neutrals carry the total beam energy. Separation of these latter two processes is then accomplished by modulating the electron beam and counting the neutrals in and out of phase with the modulation using gated scalers.

The merged beam technique represents a very significant advance over previous methods for studying dissociative recombination. Some of its advantages and disadvantages are outlined in table (I).

TABLE I

ADVANTAGES

1. Absolute cross sections can be measured over a very wide energy range, a few meV to many eV.

2. Collision energy is accurately known since nearly monoenergetic beams are used, the only uncertainty coming from the uncertainty in θ. Furthermore, energy de-amplication during conversion from the laboratory to the centre of mass frame of reference makes very high resolution measurements possible (Few meV). Lower limit is determined primarily by intersection angle.

3. Signal to background ratio higher than for inclined beam experiments.

4. Use of nuclear counting techniques allows the neutral products to be studied. Measurement of final state excitation and branching ratios are possible by improving the energy resolution of the counting system and exploiting the energy amplification upon transforming from the centre of mass to the laboratory frame of reference.

FLOWING AFTERGLOW LANGMUIR PROBE

This technique has been used recently to measure a number of dissociative recombination rate coefficients. (Alge et al 1983, Adams et al 1984). It has been described in detail in the review by Smith and Adams (1983) at the previous NATO ASI on the Physics of Ion-Ion and Electron-Ion Collisions.

Briefly a helium carrier gas is introduced into a stainless steel flow tube and is made to pass through the tube under the action of a Roots pump. The gas is excited upstream using a microwave cavity and He^*, He^+ and electrons are generated Fig. (4). Typical carrier gas pressures in the range 0.6-1.0 Torr are used and electron densities up to $7 \times 10^{10} cm^{-3}$ are produced.

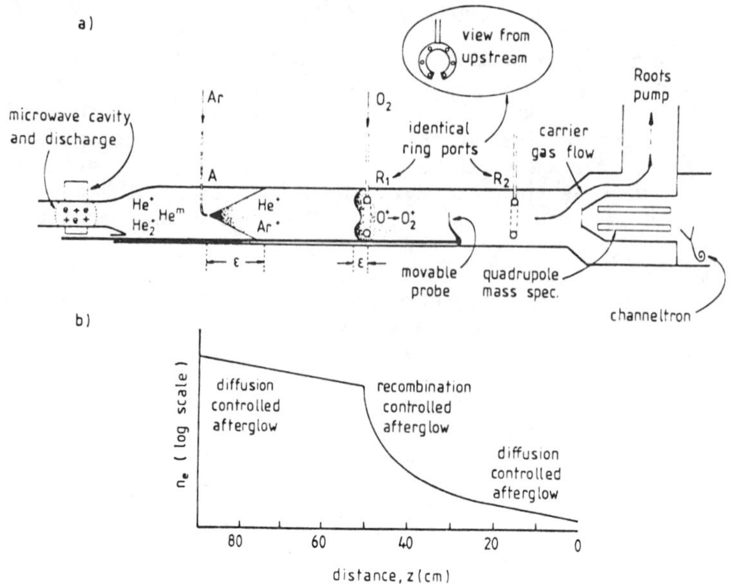

Fig. 4. The FALP experiment showing details of the gas inlet ports, diagnostic instruments, etc... Also shown is a typical electron density profile along the flow tube during recombination studies (Alge et al, 1983).

A moveable langmuir probe is used to measure absolute ion and electron densities. Argon is introduced at port A and this serves to remove the metastable Helium atoms through the reaction:

$$He^* + Ar \rightarrow He + Ar^+ + e$$

Reactant gas is introduced via ports R_1 or R_2 and this is ionized via charge exchange collisions with He^+ ions. The electron density is measured as a function of distance along the tube (fig. (4b) and the rate coefficient for molecular ion recombination is determined from the equation

$$\frac{\partial Ne}{\partial t} = D_a \nabla^2 N_e - \alpha Ne \cdot N_i$$

where D_a is the ambipolar diffusion coefficient. By operating at high helium pressures, the diffusion term becomes negligible so that recombination dominates. A quadrupole mass spectrometer is used to identify the reactant ions.

The FALP apparatus can be operated at a range of temperatures from 80-600K by cooling or heating the tube. At lower temperatures, the formation of cluster ions becomes a problem.

The ions are believed to be de-excited to their vibrational and rotational ground states because of thermalizing collisions with both the reactant and carrier gases.

To date the ions O_2^+, NO^+, NH_4^+, HCO^+, N_2H^+, CH_5^+ have been studied using this apparatus but the most significant study has been of H_3^+ recombination. The FALP study has indicated that this process has a very small rate coefficient. More will be said of the significance of this result later.

THEORETICAL STUDIES

(i) Potential Energy Curves

Calculation of the cross section for dissociative recombination begins with the determination of the relevant potential energy curves for the system. The normal potential curves for a diatomic molecule represent stationary states which are "adiabatic" in form. That is, they represent the energy that the molecule would have in a given electronic configuration for a range of inter-nuclear separations if the atoms were at rest with respect to each other. These permanent states can be observed directly by means of optical spectroscopy and so their energies can be measured experimentally with great accuracy. The molecule can exist not only in its ground state but also in a host of excited states representing configurations where one or both of the electrons is excited.

The energies of these states may be calculated from the Schrodinger equation:

$$H\Psi = E\Psi \qquad \text{--- (1)}$$

where H is the Hamiltonian, Ψ the wavefunction and E the total energy of the complete system, i.e. nuclei and electrons. H can be written in the form:

$$H = H_e + T_R \qquad \text{--- (2)}$$

where T_R is the nuclear kinetic energy operator.

$$T_R = -\left[\hbar^2/2M\right]\nabla_R^2 \qquad \text{--- (3)}$$

M being the reduced mass of the two nuclei $[M = m_A m_B/m_A+m_B]$ and R is the internuclear separation. H_e is called the electronic Hamiltonian and represents the kinetic and potential energy of the electrons plus the potential energy of the nuclei due to their electrostatic repulsion.

$$H_o = \sum_{i=1}^{N} \left[-(\hbar^2/2m)\nabla_i^2 - \frac{Z_A e^2}{r_{iA}} - \frac{Z_B e^2}{r_{iB}} + \sum_{j=i+1}^{N} \frac{e^2}{r}\right] + \frac{Z_A Z e_B^2}{R} \qquad \text{--- (4)}$$

where the sum is over N electrons, m is the electron mass, r_{iA} and r_{iB} are the distances between the electron i and the nuclei A and B and r_{ij} is the distance between electron i and another electron j. Z is the nuclear charge. The total wavefunction for the system can be expanded in terms of electron wave functions $\phi_i(r,R)$ and nuclear wavefunctions

$$\Psi = \sum_i \phi_i(r,R)\,\chi_i(R) \qquad \text{--- (5)}$$

Multiplying equation (1) on the left by ϕ_j^* and integrating over all electronic co-ordinates allows a set of coupled equations for the nuclear wavefunction to be written down. Upon rearranging terms:

$$\left[T_R + T''_{ii} + V_{ii}(R) - E\right] \chi_i(R) = \sum_{j\neq i}^{N} \left[V_{ij} + T'_{ij} + T''_{ij}\right] \chi_j(R) \qquad \text{--- (6)}$$

where $V_{ab} = \langle\phi_a|H_e|\phi_b\rangle$

T' and T" result from the term $\nabla^2\phi\,\chi$

Since $\nabla^2\phi\,\chi = \phi\nabla^2\chi + 2\nabla\phi.\nabla\chi + \chi\nabla^2\phi$

then $T'_{ij} = -2(\hbar^2/2M)\ \langle\phi_i|\nabla_R|\phi_j\rangle.\nabla_R$

$T''_{ij} = -(\hbar^2/2M)\ \langle\phi_i|\nabla_R^2|\phi_i\rangle$

Because of the huge difference in mass between the nuclei and the electrons, the nuclei can essentially be assumed to be at rest when compared with the electronic motion. The Born-Oppenheimer approximation involves neglecting the T' and T" terms which are much smaller than the other terms in equation (6) which then reduces to

$$\left[T_R + V_{ii}(R) - E\right] \chi_i(R) = - \sum_{j\neq i}^{N} V_{ij}\chi_j \qquad \text{--- (7)}$$

Note that in this equation, the V_{ab} terms contain the potential energies of the nuclear motion together with electronic coupling terms for given ϕ's. The problem then is to find suitable ϕ functions so that V may be evaluated. It should be remembered that for stationary solutions the Ψ functions must diagonalize the total Hamiltonian H.

If we take ϕ_i to be the stationary eigenvalue of H_e, i.e. the solution to:

$$H_e\phi_i = V_{ii}\phi_i \qquad \text{--- (8)}$$

then equation (7) is considerably simplified

since $\langle\phi_i|H_e|\phi_j\rangle = V_{ii}\delta_{ij}$

where $\delta_{ii} = 1 \quad \delta_{ij} = 0$

These ϕ_i states are referred to as adiabatic states.

Equation (7) becomes

$$\left[T_R + V_{ii} - E\right]\chi_i = 0 \qquad \text{--- (9)}$$

The essential feature of this equation is that the states χ_i are uncoupled, i.e. there are no coupling terms between different states so that they are permanent states. These are the normal molecular states observed by optical spectroscopy.

The problem with these states from the point of view of describing dissociative recombination is that the non-crossing rule of Von Neumann and Wigner (1929) states that two potential energy curves V_{ii} and V_{jj} may not cross if they have the same symmetry (spin, parity and angular momentum).

This would seem to be a major stumbling block. However the fact is that this rule does not apply to states ϕ_i which do not diagonalize H_e. It is possible to choose wavefunctions ϕ_i which do not diagonalize the electronic Hamiltonian H_e but which when combined with suitable nuclear wavefunctions χ_i do diagonalize the total Hamiltonian. This is of course necessary for the states to have definite energies. Such states are referred to as diabatic states and are discussed very lucidly by O'Malley (1971) from which most of the preceding discussion has been taken. The main point to notice is that for these states, the right hand side of equation (7) is no longer zero and so there is coupling between different states which allows non-optical transitions between different states to occur.

For the case of dissociative recombination or attachment, the most suitable set of wavefunctions ϕ_i is one containing potential scattering terms and terms which describe the resonant electron-ion compound state.

Calculation of the energies of these diabatic states is beyond the scope of this lecture and the interested reader is referred to the reviews of Guberman 1983, Michels 1981, and O'Malley 1969. The results and consequences of such calculations for individual systems will be discussed later.

(ii) Calculation of the Cross Section

Once the intermediate AB^{**} state has been identified then calculations of the cross section for dissociative recombination can proceed. The overall problem is to determine the probability of a transition from an electron-ion scattering continuum to an atom-atom dissociation continuum. In practice this involves calculating the probability of the capture of a free electron into a doubly excited resonance state AB^{**} followed by the determination of the probability of dissociation of this state. Thus the dissociative recombination cross section is often represented:

$$\sigma_{DR} = \sigma_{CAP} \times S_F$$

where S_F, the survival factor has a value of unity or less and represents the dissociation probability. The coupling between the initial scattering state, the intermediate resonance and the final dissociation continuum is complicated by the presence of an infinite number of Rydberg states of the neutral molecule whose potential energy curves lie beneath that of the molecular ion ground state. Initially we shall ignore these and consider only direct dissociative recombination which only involves the three states already alluded to.

This problem has been treated by O'Malley 1966, 1971, 1981, Bardsley 1968a,b, Bottcher 1976.

The dissociative recombination cross section is given by:

$$\sigma = \left\{\frac{4\pi^3}{k^2}\right\} g\,|T|^2 \qquad \text{--- (10)}$$

where $k = p/\hbar$ is the electron wave number (p is the momentum), g is the spin-weighting factor, the ratio of the statistical weights between the two states $\hbar = h/2\pi$ where h is Planck's constant and T is given by

$$T = \exp(i\rho) \langle \chi_i | V_a | \chi_d \rangle$$

Here χ_i is the vibrational wave function of the initial ion AB^+

χ_d is the vibrational wavefunction of the resonance state and V_a is the electronic matrix element

$$V_a = \langle \Phi_i | H_e | \Phi_r \rangle$$

The resonance width Γ_a is given by

$$\Gamma_a = 2\pi |V_a|^2$$

where the lifetime of the resonance against autoionization is $\hbar/\Gamma a$.
ρ is the complex phaseshift for the resonance AB^{**}. The square of the $e^{i\rho}$ term is the survival factor.
Equation (10) can be written in an alternative form.

$$\sigma = \text{const.} \frac{\Gamma_a}{E} \times FC \times SF$$

where Γ_a is the autoionization width of the resonance

E is the energy of the incoming electron

FC, the Frank-Condon factor is the overlap integral between the vibrational wavefunctions of the initial and resonant states and SF is the survival factor.

It is interesting to note that if there is a good overlap, then the cross section should vary as E^{-1}.

The case of e-H_2^+ recombination is of particular importance since it may be approached by ab initio techniques and a number of studies of direct dissociative recombination have been performed for this ion. Bauer and Wu (1956), Wilkins (1966), Bottcher (1976), Zhdanov & Chibisov (1978) Rai Dastidar & Rai Dastidar (1979), Derkits et al (1979), Zhdanov (1980), Schneider et al (1985).

Generally, it is accepted that the lowest repulsive state, the $^1\Sigma_g^+$ $1\sigma_u^2$ state dominates the electron capture and accurate calculations of the energy and wavefunctions of this state have been performed. The results of Guberman (1983) are shown in fig. (5) and they indicate a curve crossing at the v=1 level. This generally agrees with calculations of Hazi et al (1983). Although the earlier studies of Bottcher & Docken (1974) predicted a curve crossing in the vicinity of the v=2 level.

The most recent calculation by Schneider et al (1985) actually predicts a curve crossing in the vicinity of v=0 although the authors

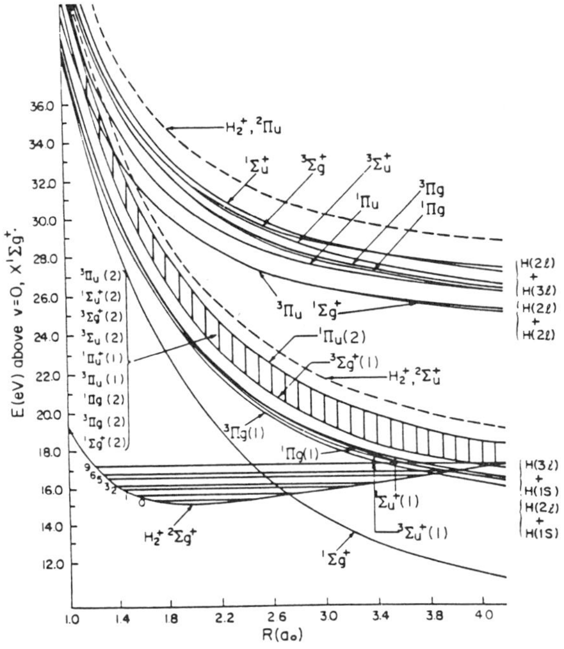

Fig. 5. Potential energy curves for H_2^+ and H_2 as calculated by Guberman, 1983.

point out that they treat the state as a rydberg state which it is not. The validity of this calculation is therefore suspect.

Other calculations, Zhdanov and Chibisov (1978), Derkits et al (1979), Zhdanov (1980) and Schneider et al (1985) have considered contributions due to higher repulsive states of the form $1\sigma_u\ nl$ which are rydberg states of the $^2\Sigma_u^+\ 1\sigma_u$ state of H_2^+. These states probably play an important role for higher electron energies and for vibrationally excited ions.

The results of Derkits et al (1979) for the recombination cross section through the $1\sigma_u 3s\ ^1\Sigma_u^+$ and $^3\Sigma_u^+$ states of H_2 for H_2^+ ions in various vibrational states are shown in fig. (6). It can be seen that these cross sections clearly reflect the Franck-Condon overlap between the repulsive state wavefunction and that of the various vibrational states of the H_2^+ ion.

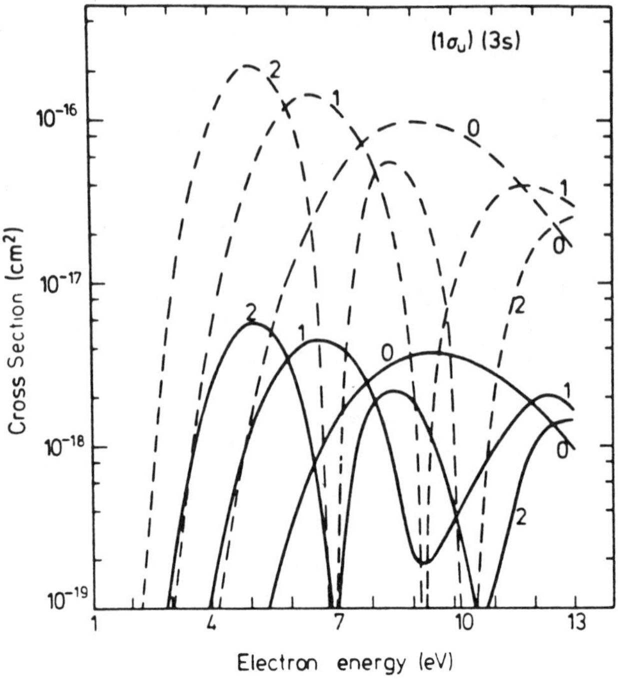

Fig. 6. Cross sections for recombination of electrons with H_2^+ in various initial vibrational states. ——, calculated with allowance for autoionization; ---, calculated without allowance for autoionization (Derkits et al, 1979).

The Influence of Rydberg States

As mentioned earlier the transition from the electron-ion continuum to the dissociation continuum is complicated by the presence of Rydberg states of the neutral molecule which lie just below the H_2^+ ground state or potential curve See fig. (1). These states affect both the initial electron capture process and the subsequent dissociation.

Vibrationally excited levels of high lying rydberg states can lie above the ionization limit of the molecule and can therefore autoionize. The influence of such states upon the ionization thresholds of simple molecules has been amply demonstrated by the high resolution spectroscopic studies of Dehmer and Chupka (1976).

It is conceivable therefore that the reverse of autoionization, namely electron capture can occur via these states. In other words the electron is first captured into a vibrationally excited Rydberg state which can subsequently decay either via autoionization or via predissociation through the diabatic AB^{**} state discussed previously. Since the electron can only be captured when its energy coincides with the energy difference between the vibrational levels of the Rydberg and ion states this process is essentially resonant.

Our understanding of the influence of these Rydberg states on the recombination process has matured considerably since the indirect mechanism was first proposed. The early studies of Bardsley (1968a,b) treated the direct and indirect processes separately but recent work by Giusti (1980), O'Malley (1981) and Giusti et al (1983) have used a combined formalism allowing the interference between the two processes to be studied.

Giusti-Suzor et al (1983) have studied the recombination of H_2^+ in the ground and excited vibrational states using both configuration interaction and MQDT approaches.

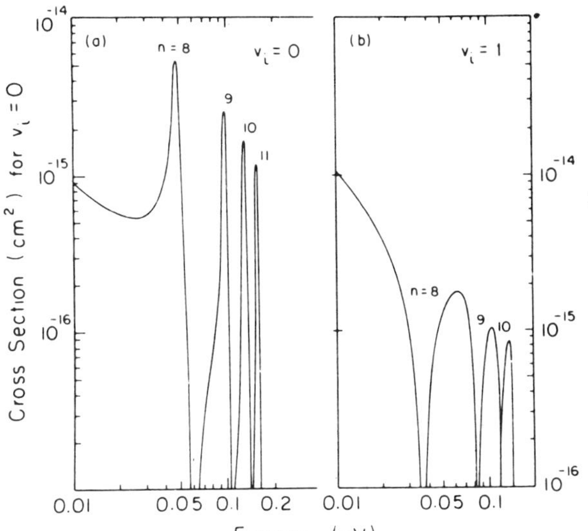

Fig. 7. Cross section for dissociative recombination of H_2^+ computed using the configuration interaction method and a hypothetical resonance curve. (a) for ions with v=0. (b) for v=1. Dips in (b) arise from Rydberg states with v=2. Each series should extend to n=∞ at the vibrational excitation threshold (close to 0.27 eV).

Fig. (7) shows their results for H_2^+ in v=0 and v=1 levels recombining through a hypothetical state which crosses the H_2^+ state at R=2.8A, i.e. well away from the Franck-Condon region of the v=0 level. It can be seen that for H_2^+(v=0), indirect recombination gives rise to peaks while for H_2^+(v=1) it leads to dips in the total cross section. The predicted resonances are typically, a few meV in width. This work is discussed in more detail by Giusti in this volume.

O'Malley (1981) has used an alternative approach in his analysis of the effects of Rydberg states on dissociative recombination. He neglected electron capture via Rydberg states which gives rise to narrow resonance structure but treated the coupling between the dissociating AB^{**} state and low n value Rydberg states AB_R. This coupling is very important as it determines the final states of the neutral recombination products. Fig. (8) shows how transitions will occur between the states AB^{**} and AB_R.

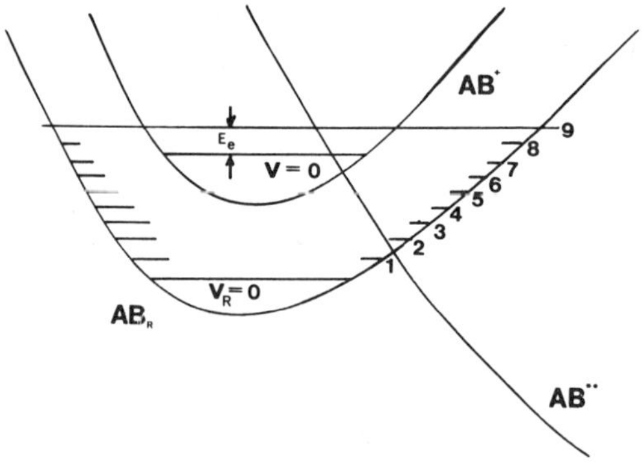

Fig. 8. Schematic illustrating the resonant aspects of the crossing of the dissociating curve responsible for the initial electron capture with low lying Rydberg states.

Now since energy is conserved this means that a transition will satisfy the condition:

$$(V_{AB^{**}} + KE_{AB^{**}}) = (V_{AB_R} + KE_{AB_R})$$

Thus a transition is most likely for the condition that AB^{**} and AB_R both start out with the same potential energy. This can only occur for discrete values of VAB_R corresponding to vibrational levels. Since these vibrational levels can themselves decay via autoionization or dissociation they can be treated as resonances and so resonance formalism (Lippmann and O'Malley 1970) can be used to describe the coupling between AB^{**} and AB_R.

O'Malley has shown that introduction of this coupling serves to modulate the zeroth order cross section (neglecting coupling) close to the resonance energy so that

$$\sigma = \sigma_o\left[\cos \eta_r + e^{-\rho}\sin \eta_r\right]^2$$

where $\eta_r = \tan^{-1}\left[\tfrac{1}{2}\Gamma_r/(E_r - E_e)\right]$

Γ_r is the resonance width given by:

$$\Gamma_r = 2\pi|V_r|^2 \quad \text{and} \quad V_r = \langle \chi_{AB_r} | V_{dr} | \chi_{AB}^{**} \rangle$$

where V_{dr} represents the coupling between the two states. E_r is the energy of the resonance (above AB^+) and E_e is the electron energy. The term $e^{-\rho}$ is the survival factor which is unity for the case of the resonance being completely stable against autoionization and equal to zero if the resonance is fully autoionizing. The ratio of σ/σ_0 is plotted against η_r, i.e. against E_e in the vicinity of the resonance, in Fig. (9). It can be seen that for the case of $e^{-\rho} = 0$, the coupling between the states leads to a dip in the cross section the width of which is estimated by O'Malley to be 30-50meV. If the resonance does not autoionize then it displays a Fano-Beutler profile (Fano 1961), i.e. giving rise to both a local increase and decrease of σ, due to interference between the directly and indirectly dissociating channels.

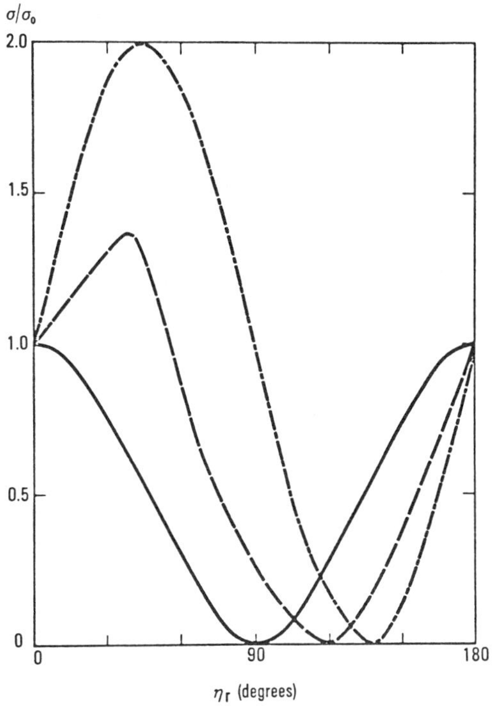

Fig. 9. The ratio of the cross section for dissociative recombination including Tydberg state interaction to that without plotted against η_r.

(——) $e^{-\rho} = 0$. (- -) $e^{-\rho} = 0.5$. (- · -) $e^{-\rho} = 1.0$.

H_2^+ Studies

Much of the impetus for this work came from the work of Auerbach et al (1977) who discovered a host of resonant structures in the cross sections for e + H_2^+ and e + H_3^+ recombination. Subsequent studies however, (D'Angelo 1979, Ng, 1982, Sen 1985) intended to reproduce these resonances have revealed only minimal structure. Following the measurements of Auerbach et al modifications were made to the MEIBE apparatus which although successful in improving the signal to background ratio, may possibly have deteriorated the energy resolution. In addition recent studies of vibrational population control, Sen (1985), have shown that the rf source used in the original measurements is less than suitable because of the short residence time of the ions within the source.

Clearly before the theoretical calculations can be properly compared with actual measurements a much improved experiment incorporating well defined vibrational populations and improved energy resolution will have to be performed. Planning for this is underway.

An aspect of dissociative recombination that has received more attention experimentally that theoretically is the identity of the final products. In particular Phaneuf et al (1975) and Vogler and Dunn (1975) have measured the partial cross section for the recombination of D_2^+ leading to $D^*(n=4)$ and $D^*(2p)$ respectively. In each case they found that the measured partial cross section accounted for only about 10% of the total cross section for e + D_2^+ recombination (Peart & Dolder 1973).

The measurements were however made for energies greater than 0.6eV and no attempt was made to alter the vibrational population of the D_2^+ ions. A variety of exit channels leading to higher n states was therefore available.

For low energy collisions involving vibrationally cold ions only the dissociational limit giving n=2 atoms is energetically available. Peart and Dolder (1975) have studied the cross section for ion pair formation following dissociative recombination of H_2^+, i.e.

$$e + H_2^+ \rightarrow H^+ + H^-$$

They found that the cross section was small, less than 0.1% of the total cross section (Peart and Dolder, 1974). This is interesting since the dissociating state involved in the electron capture actually tends to a limit of $H^+ + H^-$ at large R. (O'Malley, 1969).

The fact that neutrals are the dominant product of dissociative recombination indicates that curve crossings between the resonant dissociating state and the low n Rydberg states play a vital role in the recombination process. (O'Malley 1981).

Recent developments at UWO have opened up new opportunities for the study of H_2^+. As mentioned earlier, the H_2^+ formed by electron impact ionization of molecular hydrogen has all 19 available vibrational states populated. (Von Busch & Dunn 1972). It is in principle possible to selectively remove many of the excited vibrational levels by suitable manipulation of ion source chemistry.

Chupka et al (1968) have shown that the reactions:

$$H_2^+(v) + He \rightarrow HeH^+ + H$$

$$H_2^+(v) + Ne \rightarrow NeH^+ + H$$

are endothermic. Vibrational excitation of the H_2^+ ions corresponding to v>2 and v>1 respectively allows the reactions to proceed rapidly. By mixing helium or neon with the hydrogen gas in the ion source, it is in principle possible to react out the excited levels. Therefore the H_2^+ beams extracted under these conditions should have only the v=0,1,2 or v=0, 1 levels populated. This was tried by Auerbach et al (1977) and it was observed that the measured cross section changed by about a factor of 2 for a hydrogen/helium mixture and by a further factor of 1.6 for a neon/hydrogen mixture. Recent modelling calculations by Sen (1985) however suggest that the reasonably short residence time for ions in the rf source used by Auerbach et al precluded the complete de-excitation. Indeed Sen's analysis indicates that states up to and including v=6 remain populated under normal source conditions.

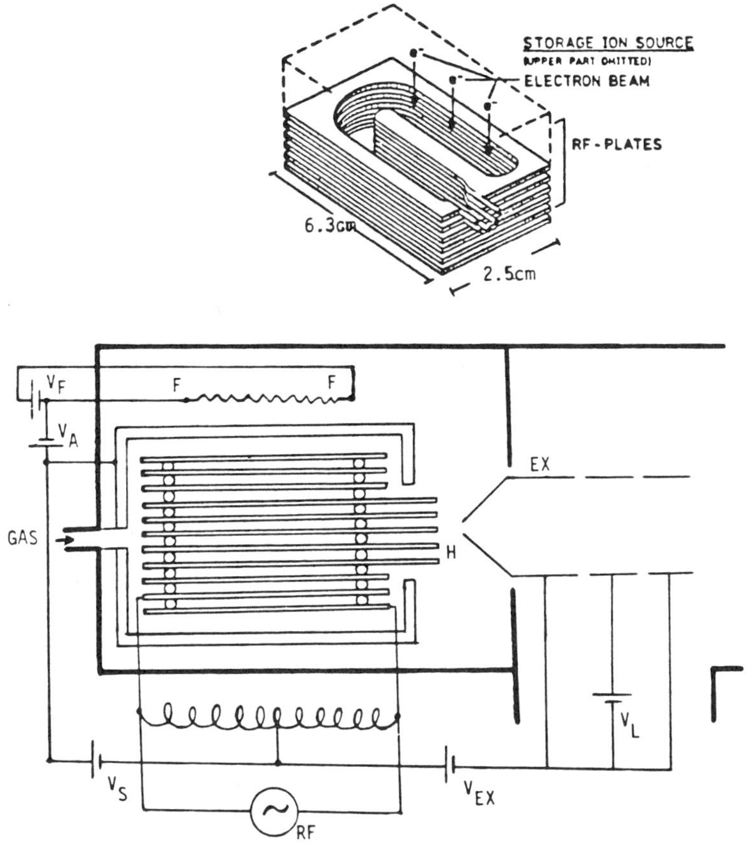

Fig. 10. The Teloy storage ion source showing the arrangement of the plates with the U-holes and the octupolar exit geometry. FF : Filamen; H : Octupolar exit hole; EX : Extractor; RF : Radiofrequency source; V_F : Filament voltage; V_A : Electron accelerating voltage; V_S : Suppressor voltage; V_{EX} : Extraction voltage; V_L : Lens voltage (Sen, 1985).

A radio frequency storage ion source has been constructed at UWO following a design by Teloy & Gehrlich (1974). In this source, ions are formed by electron impact and are confined in a potential well formed by an oscillating electric field applied to a stack of electrodes Fig. (10). Ions diffuse along the U-shaped channel before eventually being extracted through the hexapolar extraction lens. Residence times of several milliseconds can be achieved in this source.

Sen et al (1985) have examined the excitation state of ions extracted from this source by using a low energy crossed beams apparatus to study the collisional dissociation of molecular ions in collision with a beam of helium atoms. Fig. (11) shows the relative cross section for the formation of H^+ from H_2^+ + He collisions for ions formed under different source conditions. It can be seen that for low pressures and pure hydrogen source gas, the process exhibits a low reaction threshold indicating the presence of vibrationally excited H_2^+ ions. When neon is added and the source is operated at elevated pressures, then the excited states are quenched leaving only v=0 and 1 as predicted. Similar results were found for helium except that the observed threshold indicated the presence of v=2 ions in the beam. Fig. (12).

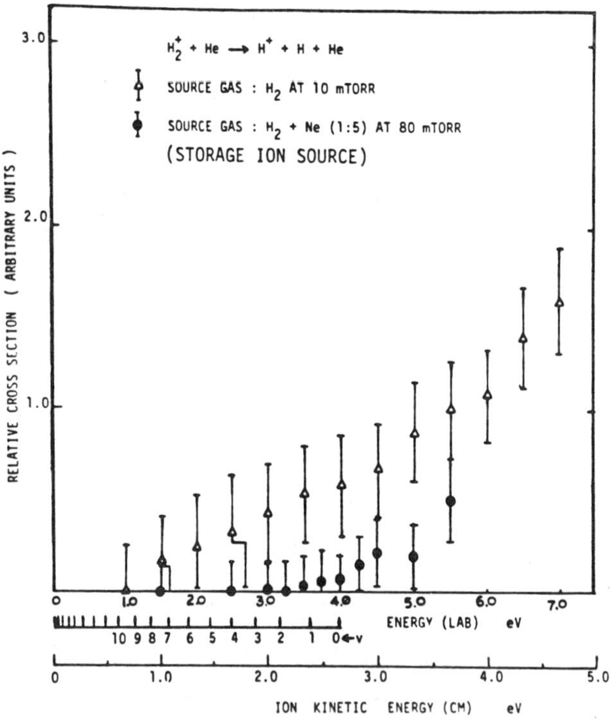

Fig. 11. Relative cross sections for the collisional induced dissociation of H_2^+ ions formed in the storage ion source with pure hydrogen gas and with a mixture of hydrogen and neon, (1:5). Threshold energies for the different vibrational levels of H_2^+ are also shown (Sen et al, 1985).

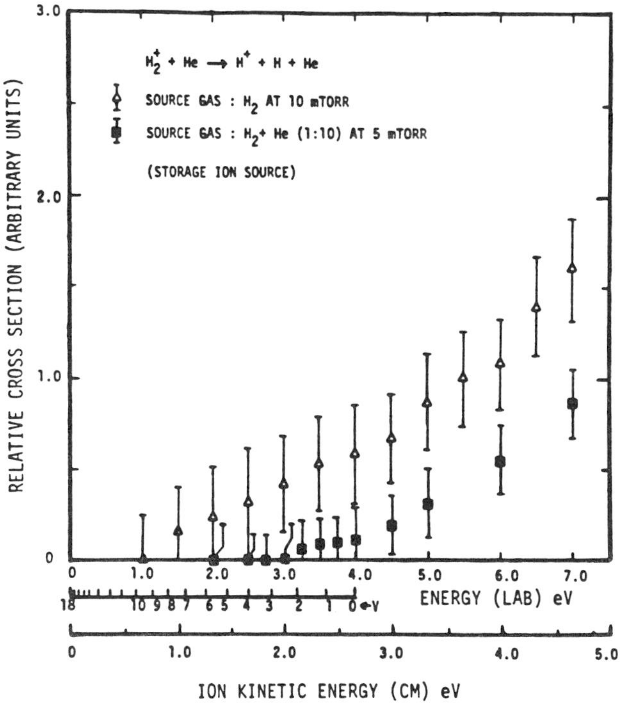

Fig. 12. Relative cross sections for the collisional induced dissociation of H_2^+ ions formed in a storage ion source with pure hydrogen gas and with a mixture of hydrogen and helium, (1:10). Threshold energies for the different vibrational levels of H_2^+ are also shown (Sen et al, 1985).

It is intended to install this source in the MEIBE apparatus and to use it to investigate $H_2^+(v=0,1)$ and $H_2^+(v=0,1,2)$ recombination. In particular it will be interesting to see if the Franck-Condon signature on the collision cross section will be observed, i.e. a deviation from a simple E^{-1} dependance.

Another experiment which is especially important to our understanding of the relative roles of direct and indirect recombination is the search for the predicted resonances in the cross section. Auerbach et al (1977) apparently observed a rich structure in the cross section. Subsequent experiments however, D'Angelo 1979, Ng 1982, Sen 1985, have failed to reproduce the earlier results although in fact some structure is visible in their data.

Currently the energy resolution of the MEIBE apparatus is estimated to be ~40 meV and this is due mainly to electrons leaving the cathode with transverse thermal velocities. Since they are travelling in an axial magnetic field, they perform a spiralling motion with a diameter given by $(D = 67.4V_{\perp}^{\frac{1}{2}}/B)$mm and a pitch of $(212V_{\parallel}^{\frac{1}{2}}/B)$mm (B in Gauss, $V_{\perp}$ and $V_{\parallel}$ are kinetic energies associated with perpendicular and parallel velocities, in eV) Taylor et al (1974). Given the operating conditions of MEIBE I this translates to ~2 mm and ~8cms respectively.

Because of the spiralling it is possible for electrons with large transverse velocities to wriggle through thin, small apertures and so remain in the beam. Plans are underway to build a new electron gun with thick diaphragms with small diameter/length ratios which will trap spiralling electrons allowing axially moving electrons to pass unhindered. Such guns have been designed and used in H. Brongersma's laboratory (Private Communication 1985).

This should greatly improve the resolution of the MEIBE apparatus. With good resolution and well defined ion beams the search for the resonances stands a much better chance for success.

H_3^+ Studies

The simplest polyatomic ion is of course H_3^+ which is a major component of hydrogen plasmas. In fact when molecular hydrogen gas is ionized to form H_2^+, the reaction

$$H_2^+ + H_2 \rightarrow H_3^+ + H$$

rapidly displaces the H_2^+ ion.

Several experimental measurements of the dissociative recombination of H_3^+ have been performed. Leu et al (1973), Peart & Dolder (1974b), Auerbach et al 1977, MacDonald et al (1984) Mathur et al (1978) and Mitchell et al (1984). There is fairly good agreement about the general nature of the cross section as measured in these experiments. The most recent study however Smith et al (1984), Adams et al (1984) performed using the FALP technique has indicated that in fact H_3^+ has a cross section much smaller than previously measured. This finding agrees with theoretical predictions concerning H_3^+. In particular Michels & Hobbs (1984) have found that not only does H_3^+ not have a suitable curve crossing for the direct dissociative recombination of ground state ions, it does not have a large probability for indirect recombination either (Fig. 13). This apparent agreement between theory and experiment has caused a considerable stir particularly because of the great importance of H_3^+ recombination.

Plans at U.W.O.

Sen and Mitchell (1985) have demonstrated that H_3^+ ions with very small internal energies can be produced using the rf storage ion source. This source will again be used with the MEIBE I apparatus to re-examine the $e+H_3^+$ cross section. The wide energy range available with the merged beam technique should allow a definitive measurement to be made. The

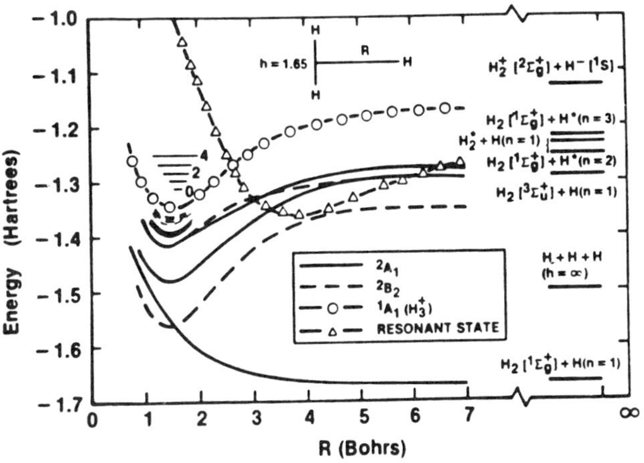

Fig. 13. Potential energy curves fur H_3^+ and H_3 showing the intersection with the dissociative curve responsible for e-H_3^+ recombination (Michels and Hobbs, 1984).

cross section may indeed be small at low energies but for E_{cm}>1.0eV it should be observed to rise to much larger values as the crossing states become accessible. This experiment should be performed within the coming year.

BRANCHING RATIOS FOR DECAY CHANNELS

Much of the modelling of interstellar chemistry relies on having a clear understanding of the branching ratios for the various decay channels following the dissociative recombination of polyatomic molecular ions. Until recently however very little information has been available on this subject. Modelling of processes occurring during the pulse radiolysis of methane and acetylene have allowed estimates of the main decay channels for the dissociative recombination of CH_4^+ and $C_2H_2^+$ to be obtained. (Rebbert et al. 1972, 1973a,b). On the theoretical side, Herbst (1978) has used a statistical phase space theory to predict branching ratios for $HCNH^+$, H_3O^+, CH_3^+ and NH_4^+ while Kulander and Guest (1979) have studied the decay of H_3^+ following recombination.

In 1983 the first direct measurements of the branching ratio for a polyatomic ion recombination were made using the MEIBE I apparatus at the University of Western Ontario, (Mitchell et al 1983). The process studied was:

$$e + H_3^+ \rightarrow H + H + H \quad \text{--- I}$$

$$\rightarrow H_2 + H \quad \text{--- II}$$

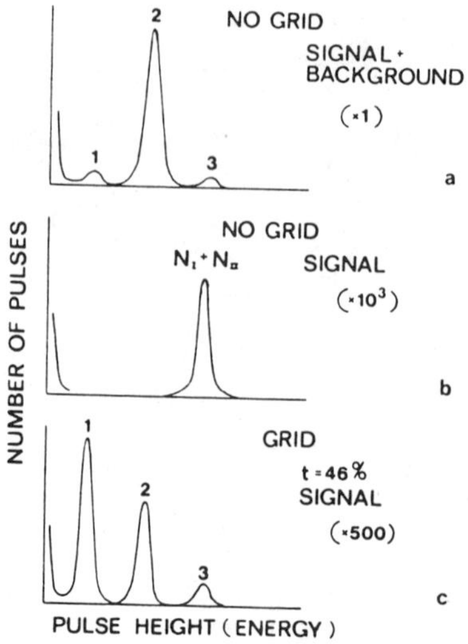

Fig. 14. Pulse height distributions of detector pulses for (a) H_3^+ beam, no grid, signal plus background; (b) no grid, signal only; (c) 46% transmission grid, signal only.

To understand the measurement it is necessary to understand the response of a surface barrier detector to dissociation products. When H_3^+ dissociates, its kinetic energy is shared among the dissociation products according to their mass. Therefore a 300 KeV H_3^+ ion will give rise to three hydrogen atoms each with 100 KeV or a 200 KeV H_2 molecule and a 100 KeV H atom. Since all the products arrive at the detector essentially simultaneously, they yield the same pulse height as a single 300 KeV particle. Hence under normal circumstances channels I and II are indistinguishable. Fig. (14) shows the output from the surface barrier detector for H_3^+ in the MEIBE apparatus. The peaks coresponding to 100 and 200 and 300 KeV particles are populated by neutrals arising from interactions with the background gas:

$$H_3^+ + X \rightarrow H + H_2^+ + X$$

$$H_3^+ + X \rightarrow (H + H) + H^+ + X$$

$$H_3^+ + X \rightarrow (H + H) + H + X^+$$

The 300 KeV peak at low centre of mass energy will also contain a contribution due to dissociative recombination. At higher centre of mass energies, the 100 and 200 KeV peaks will contain contributions from dissociative excitation. (See later). Background signals are separated from electron-ion signals by modulating the electron beam and counting neutrals in and out of phase with the modulation.

If a metallic grid of known transmission probability t is placed in front of the detector, then the probability of two particles traversing the grid is proportional to t^2 and of three particles to t^3. Taking account of the probability for a particle not to pass through the grid being 1-t then it is easily shown that the number of counts in each of the channels is given by

$$N_{100} = 3t(1-t)^2 N_I + t(1-t)N_{II}$$

$$N_{200} = 3t^2(1-t)N_I + t(1-t)N_{II}$$

$$N_{300} = t^3 N_I + t^2 N_{II}$$

where N_I and N_{II} are the number of recombination events decaying to channel I and II respectively. By using single channel analysers, N_{100}, N_{200} and N_{300} can be individually determined and so N_I and N_{II} can be found by solving these equations. Hence the partial cross sections for recombination to channel I and II can be calculated separately. Fig. (15) shows the results that were obtained using H_3^+ ions that were produced in an rf ion source. It has been estimated using a model by Blakely et al (1977) that only 35% of these ions were in the ground state. It can be seen that over the measured energy range, channel I dominates channel II by about a factor of 2:1. Kulander and Guest (1978) predicted that for ground vibrational state H_3^+ and low electron energies only channel I is available for decay. For H_3^+ ions with more than 1 eV

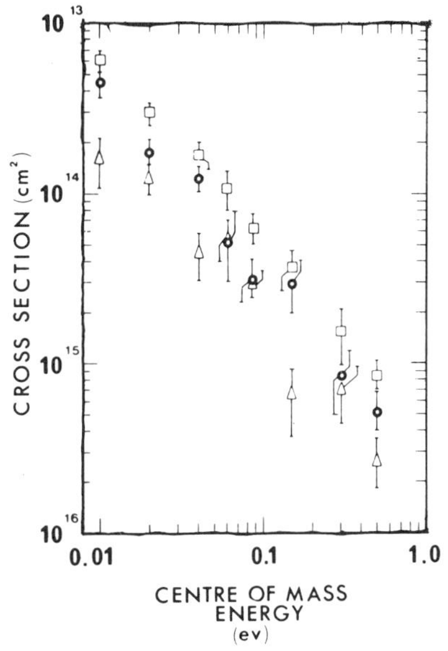

Fig. 15. Cross sections for the dissociative recombination of H_3^+ + e leading to H+H+H (circles) and to H_2+H (triangles). Total cross section also shown (squares). (Mitchell et al, 1983).

of internal energy then channel II should dominate. Thus given that the ions used in this study had a range of internal energies, there is qualitative agreement between the experimental and theoretical studies.

It will be particularly interesting to repeat these studies using ground state H_3^+ ions from the Teloy source.

Further details concerning this technique together with the analysis for more complex triatomic ions such as H_2D^+ or HD_2^+ has been published recently. (Forand et al 1985).

DISSOCIATIVE EXCITATION

Peart and Dolder (1974c, 1975b) have studied the electron impact dissociation or dissociative excitation of H_3^+ and by comparing the experimentally measured threshold for this reaction with the theoretical predicted threshold, they concluded that their H_3^+ ions were de-excited. Their results are shown in fig. (17). This is a very useful measurement as it employs exactly the same experimental configuration as that used for the dissociative recombination measurement. Peart and Dolder examined the reaction

$$e + H_3^+ \rightarrow H^+ + (H + H) + e$$

They did not have sufficient energy resolution in their detector to allow the other dissociation process:

$$e + H_3^+ \rightarrow H + H_2^+ + e$$

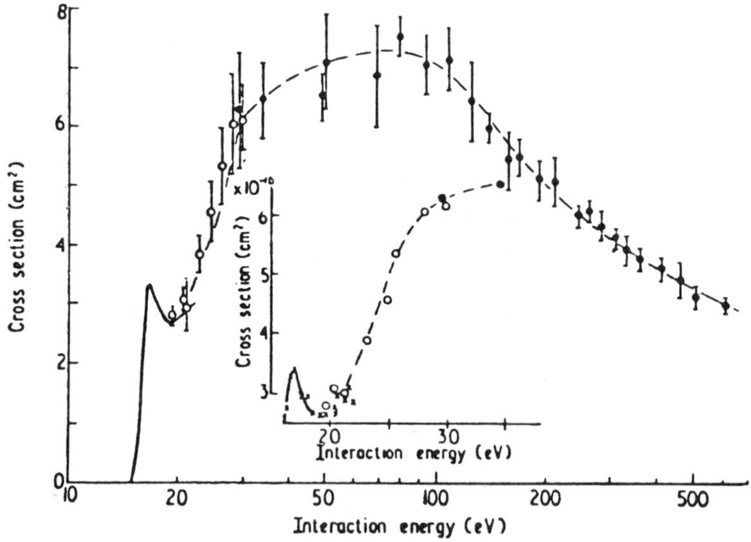

Fig. 16. Measured cross sections for the production of protons by collisions between electrons and H_3^+ ions (Peart and Dolder, 1974c, 1975b).

to be examined. The measurement was performed by counting the H_2 molecules or pairs of H atoms resulting from the dissociation. This is done by using a single channel analyser to select out the $H_2/2H$ channel from the pulse height distribution [see Fig. (14)].

The main source of error in the interpretation of the results lies in the accuracy of the theoretical prediction for the threshold.

Dissociative excitation proceeds via an electronic transition from a bound electronic state of the ion to an excited repulsive state. Usually the initial state is the ground electronic state which may be vibrationally or rotationally excited.

For H_3^+ this state represents a configuration where in equilibrium the three hydrogen atoms are located at the vertices of an equilateral triangle of side 1.65 - 1.66 Bohr. The total energy of this state, below the threshold for dissociation to $H^+ + H^+ + H^+$, is estimated to be -1.34 to -1.35 Hartrees (-36.46 to -36.72 eV) (Kawaoka and Borkman 1971). The vibrational levels of this state have been calculated by Carney and Porter (1976), Carney (1980).

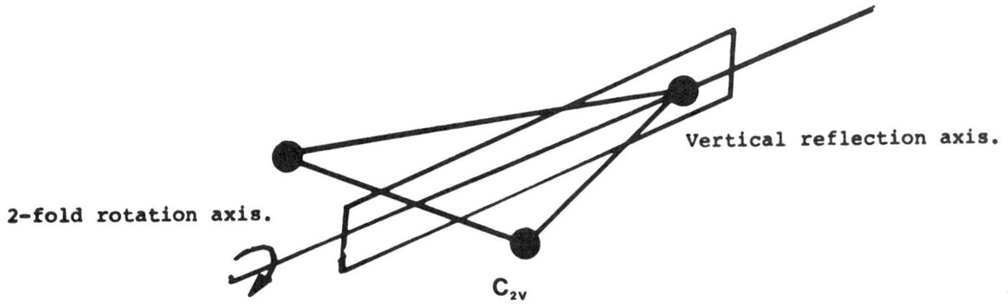

Fig. 17a. C_{2v} symmetry group.

H_3^+ can also exist in excited triangular or linear configurations. We shall neglect the latter for the purposes of the following discussion. The molecule can vibrate either symmetrically or assymmetrically and in these modes it will remain invariant under a number of symmetry operations. (Reflection about axis, rotation etc.) For a full discussion see for example Herzberg (1966), Bright Wilson et al (1955). Group theory can be used to describe molecules in particular conditions which display sets of vibrational modes characteristic of those conditions.

Thus for H_3^+ vibrating in an asymmetric (or isoceles) mode, it can be described via the C_{2v} symmetry group. See Fig. (17). H_3^+ vibrating in a symmetrical (equilateral) mode is described by D_{3h} symmetry. Now it

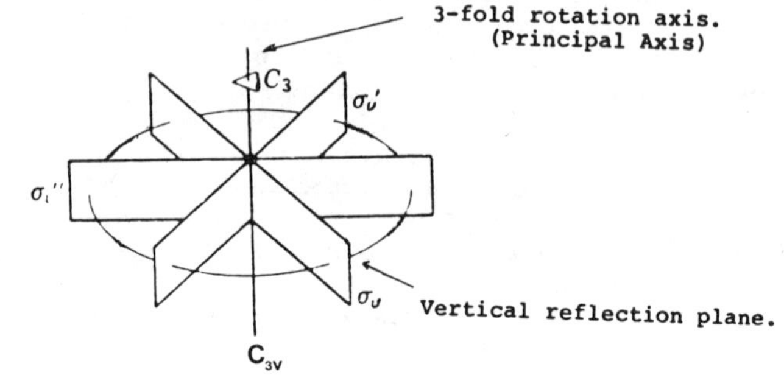

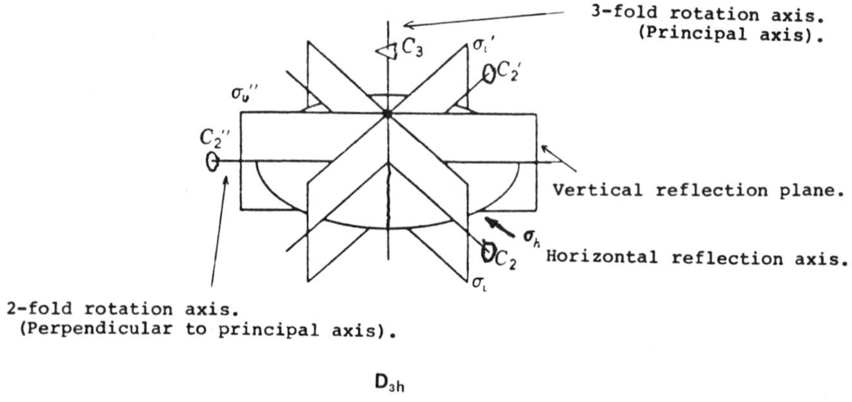

Fig. 17b. C_{3v} and D_{3h} symmetry groups.

should be noticed that these two symmetry groups are not isolated from each other, they do overlap. They are said to be degenerate. A molecule which is vibrating in an asymmetric fashion does pass through a configuration in which it is equilateral [see Fig. (18)].

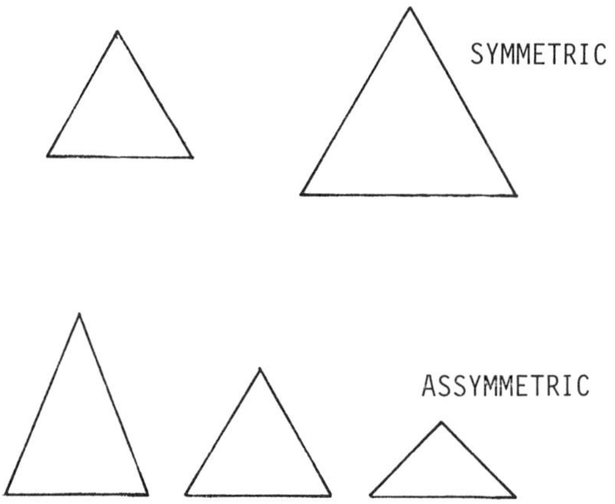

Fig. 18. Vibrational modes for a triangular molecule.

H_3^+ in a D_{3h} configuration has electronic states designated as follows:

$$A_1', A_1'', A_2', A_2'', E', E''$$

which can be either singlet or triplet. E states are degenerate. Possible electronic states in the more restricted C_{2v} symmetry are designated:

$$A_1, A_2, B_1, B_2$$

again singlet or triplet.

Because D_{3h} and C_{2v} symmetries are degenerate under some configurations then there are equivalences between these states. These are shown in Table (II).

D_{3h}	C_{2v}
A_1'	A_1
A_2'	B_2
E'	A_1, B_2
A_1''	A_2
A_2''	B_1
E''	A_2, B_1

The final and perhaps most important point to make here is that a symmetrical vibrational mode will go to an asymptotic dissociation limit of:

$$H + H + H^+$$

while an asymmetric vibration will yield

$$H + H_2^+$$

$$\text{or} \quad H_2 + H^+$$

A number of calculations of the potential energy surfaces for H_3^+ have been performed. Fig. (19) shows the results of Kawaoka and Borkman (1971) who calculated the energies of the ground $^1A_1$ and excited states in D_{3h} symmetry. R is the internuclear separation which for an equilateral configuration is independant of which atoms are examined. Of course it is also possible for the atoms to move asymmetrically and so the energies of these states should also be calculated in C_{2v} symmetry as well. This has been done by Conroy (1969) and by Bauschlicher et al 1973.

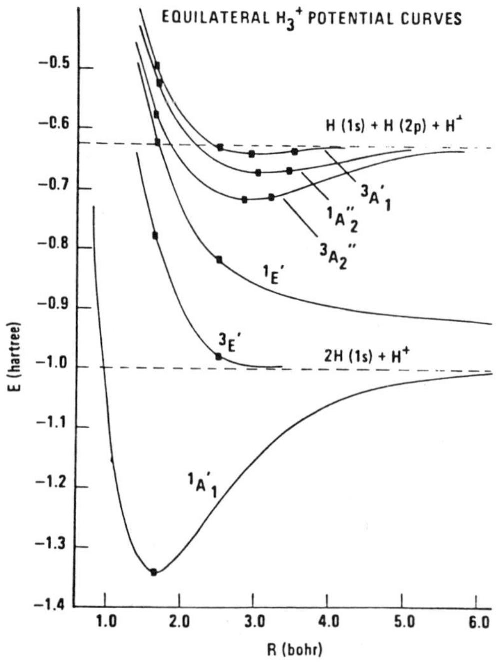

Fig. 19. Potential energy curves for the ground and excited states of H_3^+ in D_{3h} symmetry. (Kawaoka and Borkman, 1971).

For example the lowest $^1E'$ state of Kawaoka and Borkman (1971) translates to $^1A_1$ and $^1B_2$ states in C_{2v} symmetry. The $^1A_1$ state is shown in fig. (21). Note the difference in the meanings of r and R which are explained in the diagram. It can be seen that when R is large, the $^1A_1$ state is bound with respect to dissociation of the H_A, H_B pair. In this case the third hydrogen atom Hc is essentially separate so that the molecule is really a complex of $H_2^+ + H$. As H_C approaches the other two however, the potential energy of the minimum rises until in the equilateral configuration the state is unbound. At this point the $^1A_1$ state in C_{2v} and the $^1E'$ state in D_{3h} are essentially the same state. The situation is perhaps more clearly seen in Fig. (21) which shows three dimensional plots of the ground and excited $^1A_1$ states. (Bauschlicher et al 1973).

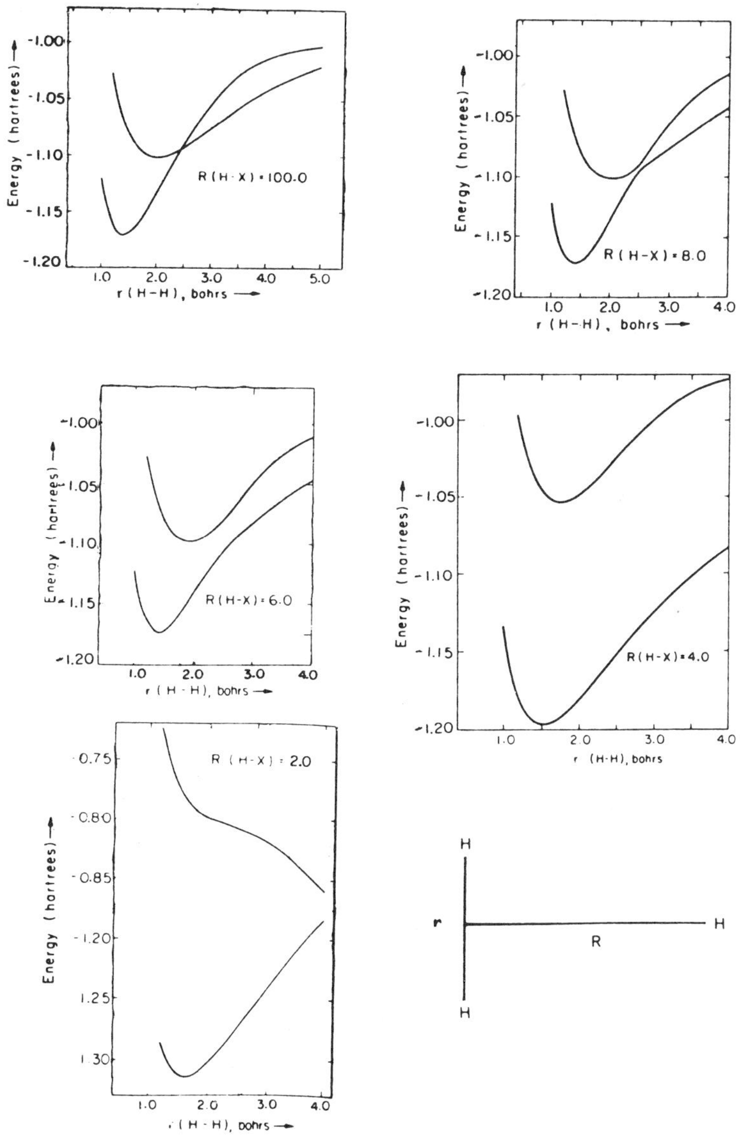

Fig. 20. Potential energy curves for the ground and excited $^1A_1$ states of H_3^+ (Bauschlicher et al, 1973).

For the $^1B_2$ component, Conroy (1969) has shown that this is weakly stable with respect to dissociation to $H_2^+(^2\Sigma_g)$ + H(2p) and less so to $H_2(^1\Sigma_u)$ + H^+. Fig. (22). The H_2 ($^1\Sigma_u$) state is however repulsive so it in turn would dissociate to H(1S) + H(1S). The $^1B_2$ component is unstable with respect to dissociation to $H_2^+(^2\Sigma_u)$ + H(1S) but again the H_2^+ state is unstable and will further dissociate to H(1S) + H^+. Hence all transitions leading to the $^1B_2$ component are likely to give rise to dissociation via the H + H + H^+ channel.

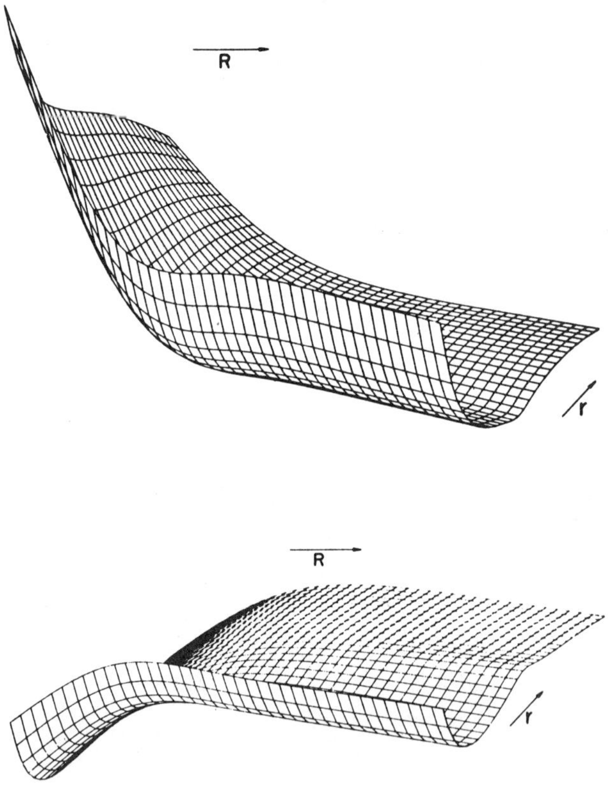

Fig. 21. Perspective plots of the lowest and first excited $^1A_1$ states of H_3^+ (Bauschlicher et al, 1973).

Now let us consider the case of electron impact on H_3^+. The case for photon impact has been examined by Kulander and Bottcher (1978). Electronic transitions between these two states occur in a time very much less than typical vibrational periods so that such transitions occur vertically i.e. they are Franck-Condon transitions. Thus from Fig. (19) it can be seen that an energy of ~19 eV is required for an H_3^+ ion in its electronic and vibrational ground state (symmetrical) to be excited to the $^1E'$ state. Note this energy takes account of the energy of the ground vibrational level above the minimum of the $^1A_1{}^1$ ground state (0.54 eV - Carney and Porter 1976).

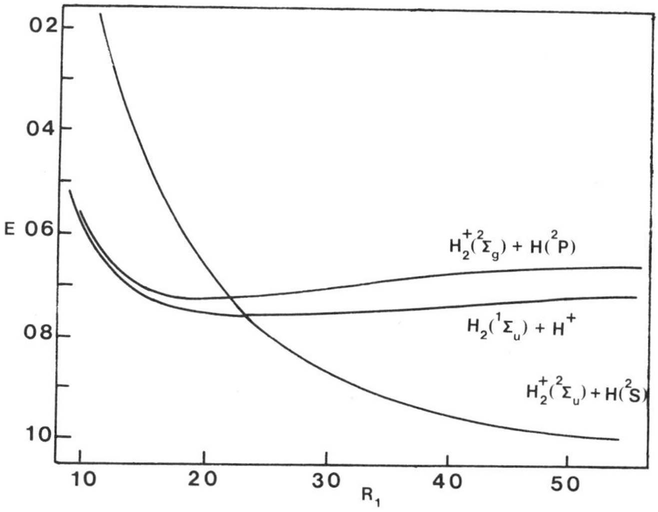

Fig. 22. Potential energy curves for the $^1B_2$ states of H_3^+ (Conroy,1969).

Once excited then the molecule can either dissociate into H + H + H^+, i.e. R and r → ∞ or into H_2^+ + H, R → ∞, r remains finite. Kulander and Bottcher (1978) point out that there is a ridge on the $^1E^1$ ($^1A_1$) surface which tends to make the H + H + H^+ channel more favoured provided the initial H_3^+ ion is not too excited. Their calculation treated the states adiabatically so that an avoided crossing occurs leading to the distortion of the potential surface. A subsequent calculation, Kulander and Heller (1978) used diabatic states so that no avoided crossing occurred. This means that dissociation of H_2^+ + H might then be favoured due to the steeper gradient. Which of these two scenarios is more accurate is not yet known due to lack of experimental evidence.

Dipole allowed transitions involve transitions between states with similar spins and so a transition to the $^3E'$ state Fig. (19) is forbidden for photon impact. For electron impact however, exchange of the projectile and target electrons will permit such a transition to occur. In C_{2v} symmetry, this state translates to $^3A_1$ and $^3B_2$ states. These states dissociate to $H_2^+(^2\Sigma_g^+)$ + H(1S) and to $H_2^+(^2\Sigma_u^+)$ + H(1S) respectively, the latter dissociating further to H^+ + H(1s) + H(1s). In its symmetric $^3E'$ form it dissociates to H(1s) + H(1s) + H^+.

For transitions from the ground vibrational state of H_3^+ the energy required to reach the $^3E'$ state is 14.76eV. [Kawaoka & Borkman (1971), Carney & Porter (1976)].

An interesting feature of the dissociative excitation of H_3^+ is the fact that exchange processes are only effective over a narrow energy range and so the transition to the $^3E'$ state appears as a narrow peak in the cross section at threshold. See Fig. (16). The transition to the $^1E'$ state however is allowed and so it rises to broad maximum far above the transition threshold.

Acknowledgements

I should like to acknowledge the many useful discussions which I have had with Bill McGowan, Norman Bardsley, Tom O'Malley, Chris Bottcher, Anik Giusti-Suzor, Ken Kulander and Harvey Michels. The assistance of co-workers listed in the references has been highly valued. Special thanks are also due to the Canadian National Science and Engineering Research Council, the U.S. Department of Energy and the U.S. Air Force Office of Scientific Research for their support of the merged beam program at U.W.O.

References

1. N.G. Adams, D. Smith and E. Alge. J. Chem. Phys. 81, 1778, 1984.
2. E. Alge, N.G. Adams and D. Smith. J. Phys. B, 16, 1433, 1983.
3. D. Auerbach, R. Cacak, R. Caudano, T.D. Gaily, C.J. Keyser, J. Wm. McGowan, J.B.A. Mitchell, S.F.J. Wilk. J. Phys. B. 10, 3797, 1977.
4. J.N. Bardsley and M.A. Biondi. Adv. At. Mol. Phys. (eds. D.R. Bates and I. Estermann). Vol. 6. Academic Press, NY, 1970, p.1.
5. J.N. Bardsley. Symp. Electron-Molecule Collisions. Invited Papers (ed. I. Shimamura and M. Matsuzawa) (U. of Tokyo, 1979), p.121.
6. J.N. Bardsley, J. Phys. B, 1, 349, 1968.
7. J.N. Bardsley, J. Phys. B, 1 365, 1968.
8. D.R. Bates and H.S.W. Massey, Proc. Roy. Soc. A192, 1, 1947.
9. E. Bauer and T.Y. Wu, Can. J. Phys. 34, 1436, 1956.
10. C.W. Bauschlicher, S.W. O'Neill, R.K. Preston, H.F. Schaefer III and C.F. Bender. J. Chem. Phys. 59, 1286, 1973.
11. R.S. Berry and S. Leach. Adv. Electron. Electron Phys. 57 1, 1981.
12. C.R. Blakeley, M.L. Vestal and J.H. Futrell, J. Chem. Phys. 66, 2392, 1977.
13. C. Bottcher and K. Docken, J. Phys. B. 7, L5, 1974.
14. C. Bottcher. J. Phys. B, 9, 2899, 1976.
15. E. Bright-Wilson, J.C. Derius and P.C. Cross. Molecular Vibrations Dover Pub. Inc., New York, 1955.

16. N.A. Burdett and A.N. Hayhurst. Comb. Flame 34, 119, 1979.
17. F. Brouillard and W. Claeys in Physics of Ion-Ion and Electron Ion Collisions (eds. F. Brouillard and J. Wm. McGowan), (NATO ASI, Baddeck, Nova Scotia), Plenum, NY. p.415, 1983.
18. G.D. Carney and R.H. Porter. J. Chem. Phys. 65, 3547, 1976.
19. G.D. Carney. Mol. Phys. 39, 923, 1980.
20. A. Carrington and R.A. Kennedy. J. Chem. Phys. 81, 91, 1984.
21. H. Conroy. J. Chem. Phys. 51, 3979, 1969.
22. W.A. Chupka and M.E. Russell. J. Chem. Phys. 49, 5426, 1968.
23. A.J. Cunningham and R.M. Hobson. J. Res. NBS A76, 329, 1972.
24. V. D'Angelo. MSc Thesis, University of Western Ontario, 1979.
25. C. Derkits, J.N. Bardsley and J.M. Wadehra. J. Phys. B, 12, 1529, 1979.
26. P.M. Dehmer and W.A. Chupka. J. Chem. Phys. 65, 2243, 1976.
27. K.T. Dolder in Case Studies in Atomic Collision Physics, Vol. 1 (ed. E.W. McDaniel and M.R.C. McDowell) (North-Holland, Amsterdam) p.249, 1969.
28. A.V. Eletskii and B.M. Smirnov. Sov. Phys. Usp. 25, 13, 1982.
29. U. Fano. Phys. Rev. 124, 1866, 1961.
30. L. Forand, J.B.A. Mitchell and J. Wm. McGowan. J. Phys. E, 18, 623, 1985.
31. S.L. Guberman. J. Chem Phys. 78, 1404, 1983.
32. S.L. Guberman, in Physics of Ion-Ion and Electron-Ion Collisions (eds. F. Brouillard and J. Wm. McGowan) (NATO ASI Baddeck, Nova Scotia), Plenum New York, p.167, 1981.
33. A. Giusti. J. Phys. B, 13, 3867, 1980.
34. A. Giusti-Suzor, J.N. Bardsley and C. Derkits. Phys. Rev. A, 28, 682, 1983.
35. M.F.A. Harrison. Brit. J. Appl. Phys. 17, 371, 1966.
36. A.N. Hayhurst and N.R. Telford. Proc. Roy. Soc. A322, 1999, 1974.
37. A. Hazi. Phys. Rev. A27, 1751, 1983.
38. E. Herbst. Ap. J. 222, 508, 1978.
39. G. Herzberg. Electronic Spectra of Polyatomic Molecules. Van Nostrand Reinhold Co., New York, 1966.
40. K. Kawaoka and R.F. Borkman. J. Chem. Phys. 54, 4234, 1971.
41. C.J. Keyser, H.R. Froelich, J.B.A. Mitchell and J.Wm. McGowan, J. Phys. E12, 316, 1979.
42. K.C. Kulander and C. Bottcher. Chem Phys. 29, 141, 1978.
43. K.C. Kulander and E.J. Heller. J. Chem. Phys. 69, 2439, 1978.
44. K.C. Kulander and M.F. Guest. J. Phys. B, 12, L501, 1979.
45. M.T. Leu, M.A. Biondi and R. Johnsen. Phys. Rev. A8, 413, 1973.
46. B.A. Lippman and T.F. O'Malley. Phys. Rev. A2, 2115, 1970.
47. J. MacDonald, M.A. Biondi and R. Johnsen. Planet. Space Sci. 32, 651, 1984.
48. T. Maruyama, Y. Ichikawa, R.M. Hobson, S. Teii, T. Kaneda and J.S. Chang. IEE, Japan. Proc. Symp. 2, 1, 1981.
49. H.N. Maier and R.W. Fessenden. J. Chem. Phys. 62, 4790, 1975.
50. D. Mathur, S.U. Khan and J.B. Hasted. J. Phys. B11, 3615, 1978.
51. F.J. Mehr and M.A. Biondi. Phys. Rev. 181, 264, 1969.
52. J. Wm. McGowan and J.B.A. Mitchell, in Electron-Molecule Interactions and their Applications, Vol.II, ed. L.G. Christophoron. Academic Press, NY, 1984.
53. H.H. Michels, in The Excited State in Chemical Physics. Adv. Chem. Phys., Vol.XLV (ed. J. Wm. McGowan), Wiley Interscience, NY, p.225, 1975.
54. H.H. Michels and R.H. Hobbs. Ap. J. 286, L27, 1984.
55. J.B.A. Mitchell, J.F. Forand, C.T. Ng, D.P. Levac, R.E.

Mitchell, P.M. Mul, W. Claeys, A. Sen and J. Wm. McGowan. Phys. Rev. Lett. 51, 885, 1983.
56. J.B.A. Mitchell and J. Wm. McGowan, in Physics of Ion-Ion and Electron-Ion Collisions (eds. F. Brouillard and J. Wm. McGowan) (NATO ASI, Baddeck, Nova Scotia), Plenum, New York, p.279, 1983.
57. J.B.A. Mitchell, C.T. Ng, L. Forand, R. Janssen and J. Wm. McGowan. J. Phys. B, 17, L909, 1984.
58. C.T. Ng. MSc Thesis, University of Western Ontario, 1982.
59. T.F. O'Malley, Phys. Rev. 150, 14, 1966.
60. T.F. O'Malley, J. Chem. Phys. 61, 322, 1969.
61. T.F. O'Malley in Adv. At. Mol. Phys., Vol.7 (Eds. D.R. Bates and I. Estermann), Academic Press, NY, p.223, 1971.
62. T.F. O'Malley, J. Phys. B, 14, 1229, 1981.
63. H.J. Oskam and V.R. Mittelstadt. Phys. Rev. 132, 1445, 1963.
64. B. Peart and K.T. Dolder. J. Phys. B, 6, L359, 1973.
65. B. Peart and K.T. Dolder. J. Phys. B, 7, 236, 1974a.
66. B. Peart and K.T. Dolder. J. Phys. B, 7, 1948, 1974b.
67. B. Peart and K.T. Dolder, J. Phys. B, 7, 1567, 1974c.
68. B. Peart and K.T. Dolder. J. Phys. B, 8, 1570, 1975a.
69. B. Peart and K.T. Dolder. J. Phys. B, 8, L143, 1975b.
70. R.A. Phaneuf, D.H. Crandall and G.H. Dunn. Phys. Rev. A11, 1983, 1975.
71. K. Rai Dastidar and T.K. Rai Dastidar. J. Phys. Soc. Japan, 46, 1288, 1979.
72. R.E. Rebbert and P. Ausloos. J. Res. NBS A76, 329, 1972.
73. R.E. Rebbert and P. Ausloos. J. Res. NBS A77, 109, 1973.
74. R.E. Rebbert, S.G. Lias and P. Ausloos. J. Res. NBS A77, 249, 1973.
75. I. Schneider, I.N. Mihailescu, L. Nanu and I Iovil-Popescu. J. Phys. B18, 791, 1985.
76. A. Sen, J.B.A. Mitchell and J. Wm. McGowan. Abstracts of Papers, XIV, ICPEAC, 1985, p.667.
77. A. Sen and J.B.A. Mitchell. Abstracts of Papers XIV ICPEAC, 1985, p.600.
78. A. Sen. Ph.D. Thesis, University of Western Ontario, 1985.
79. Y.J. Shiu, M.A. Biondi and D.P. Sipler. Phys. Rev. A15, 494, 1977a.
80. Y.J. Shiu and M.A. Biondi. Phys. Rev. A16, 1817, 1977b.
81. Y.J. Shiu and M.A. Biondi. Phys. Rev. A17, 868, 1978.
82. D. Smith and N.G. Adams in Physics of Ion-Ion and Electron-Ion Collisions (eds. F. Brouillard and J. Wm. McGowan) (NATO ASI, Baddeck, Nova Scotia), Plenum, NY, p.501, 1983.
83. D. Smith and N.G. Adams. Ap. J. 284, L13, 1984.
84. P.O. Taylor, K.T. Dolder, W.E. Kauppila and G.H. Dunn. Rev. Sci. Inst. 45, 538, 1974.
85. E. Teloy and D. Gehrlich. Chem. Phys. 4, 417, 1974.
86. M. Vogler and G.H. Dunn. Phys. Rev. A11, 1983, 1975.
87. F. Von Busch and G.H. Dunn. Phys. Rev. A5, 1726, 1972.
88. J. Von Neumann and E.P. Wigner. Phys. Z. 30, 467, 1929.
89. F.L. Walls and G.H. Dunn. J. Geophys. Res. 79, 1911, 1974.
90. R.L. Wilkins. J. Chem. Phys. 44, 1884, 1966.
91. E. Zipf. J. Geophys. Res. 85, 4232, 1980a.
92. E. Zipf. Geophys. Res. Lett. 7, 645, 1980b.
93. V.P. Zhdanov and M.I. Chibisov. Sov. Phys. JETP 47, 38, 1978.
94. V.P. Zhdanov. J. Phys. B, 13, L311, 1980.

RECENT DEVELOPMENTS IN THE THEORY OF DISSOCIATIVE RECOMBINATION AND RELATED PROCESSES

A. Giusti-Suzor

Laboratoire de Photophysique Moléculaire du C.N.R.S.

Bâtiment 213 - Université Paris-Sud - Orsay, France

Dissociative recombination of molecular ions :

$$AB^+ + e \rightarrow A + B$$

has been presented and discussed by B. Mitchell, in the same volume. The current chapter will comment on some interesting features of this process and present recent theoretical developments in this field. Examples and extension to related processes will close this chapter.

I. INTRODUCTION

As an introduction, I will ask a naïve question : what do we learn from Dissociative Recombination (DR*) studies, experimental or theoretical ? Of course, we hope to obtain reliable values of the cross-section σ or of the rate coefficient $\alpha = \langle v\sigma \rangle$ (v is the incident electron velocity and the average is taken over the velocity distribution, usually Maxwellian). They are very important to understand or predict the electronic density of numerous ionized media, in laboratory plasmas, ionosphere or interstellar medium. But beyond its own interest, DR may shed light on the dynamics of excited molecular states, in particular those which cannot be studied by conventional (i.e. one photon) spectroscopy due to parity or multiplicity selection rules. More generally, the dissociative recombination pertains to a broad class of processes which proceed through the formation of a "molecular complex" (Fig. 1a) in which all the particles (electrons and nuclei) are close together and may interact strongly. All kinds of energy exchanges are thus possible, leading to various decay channels (Fig. 1b). In dissociative recombination, the kinetic energy of the incident electron is transferred to the motion of the nuclei, not directly but via a rearrangement of the whole electronic cloud, resulting in the formation of a dissociative (or predissociated) state of the neutral molecule. It is to be noted that such an excited state, lying above the ionization and dissociation limit of the molecule (they are sometimes called "superexcited states"), generally plays a role in other processes (associative ionization, atomic collisions, photoprocesses) involving the same molecular complex.

---

* Not to be confused with Dielectronic Recombination, denoted also DR by atomic physicists, e.g. Y. Hahn in the same volume ! A comment on "molecular" DR versus "atomic" DR will be made in Section II.2 below.

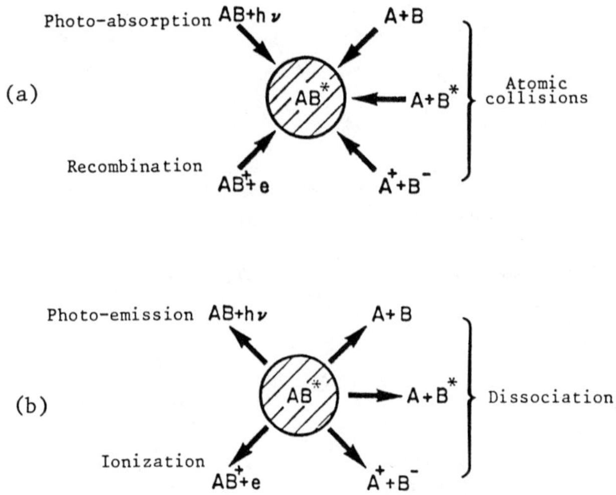

Fig. 1 : Formation (a) and decay (b) of an excited molecular complex

II. COMMENTS ON DISSOCIATIVE RECOMBINATION CROSS-SECTIONS

1. What determines the intensity of the process ?

The strength of the DR process hinges mainly on the existence of a dissociative valence state (doubly excited state) of the neutral molecule, intersecting "favorably" the potential curve of the ion initial state (generally the ground one). Two factors are then determinant : the strength of the electronic interaction $V_{el}(R)$ between the mono- and diexcited electronic configurations, and the size of the overlap between the nuclear wavefunctions of the ion initial level (χ_{v+}) and of the dissociative state (F_d). The basic mechanism for DR, first proposed by Bates [1], has been explained in detail by B. Mitchell in this volume, and by S. Gubermann in Ref. 2. They show that the DR cross-section may be written approximately as the product of a capture cross-section and a survival factor S_F, which represents the probability of dissociation following the electronic capture ($S_F \leq 1$) :

$$\sigma \simeq \sigma_{capt} \times S_F \simeq \frac{4\pi^3}{k^2} g\, V_{el}^2 \, |\langle\chi_{v+}|\, F_d\rangle|^2 \, S_F \tag{1}$$

where g is the ratio of the final and initial state multiplicities and $k = \sqrt{2\varepsilon}$ (in a.u., ε being the energy of the incident electron) is the electron wave number.

Here I just want to make a few remarks about this widely used approximate expression :

i) a more exact expression of the capture cross-section, avoiding the "Frank-Condon approximation" is :

$$\sigma_{capt} = \frac{4\pi^3}{k^2} g \left| \int \chi_{v+} (R)\, V_{el} (R)\, F_d (R)\, dR \right|^2 \tag{2}$$

which shows that not only the size of the electronic interaction but also its variation with the internuclear distance may play a role in the strength of the capture.

ii) the factorization $\sigma_{capt} \times S_F$, which is of course also approximate but generally justified, must not hidden that both the electron capture and the autoionization of the dissociative state (which competes with the dissociation and causes S_F to be generally smaller than 1) are governed by the same interactions, such that one may not "play" with one factor (e.g. for test calculations) without changing the other one. For example, a strong increase of the capture cross-section causes simultaneously a decrease of the survival factor (since autoionization increases), and the actual effect on the DR cross-section depends on the circumstance.

This can be seen more quantitatively on the approximate expression for the DR cross-section, valid for rather weak coupling[3]

$$\sigma_{v_o^+} \simeq \frac{\pi}{k^2}\, g \frac{4\xi^2_{v_o^+}}{\left[1 + \xi^2_{v_o^+}\right]^2} \tag{3}$$

where v_o^+ is the initial ion vibrational level and

$$\xi_{v^+} = \pi V_{v^+ d} = \pi \int \chi_{v^+}(R)\, V_{el}(R)\, F_d(R)\, dR \tag{4}$$

is a dimensionless quantity, since it is a continuum-continuum interaction, coupling the electronic continuum to the nuclear one (the electronic coupling V_{el} and the nuclear wave function F_d have the dimensions $E^{1/2}$ and $E^{-1/2}$, respectively). Note that the capture cross-section (Eq.2) can be viewed as the product of the basic cross-section

$$g \frac{\pi}{k^2} = g \pi ƛ^2$$

generally called the geometrical limit (ƛ is the De Broglie wavelength of the incident electron divided by 2π), by the dimensionless interaction $4\xi^2$. Therefore one gets the expression

$$S_F = \frac{1}{(1 + \xi^2)^2}$$

for the survival factor.

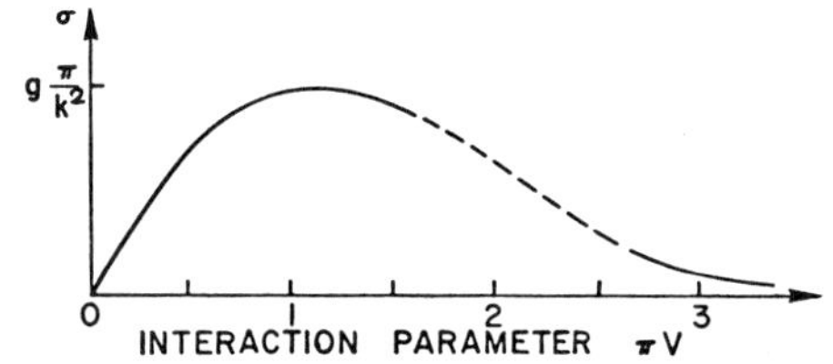

Fig. 2 : variation of the DR cross-section (eq.3) as a function of the dimensionless interaction parameter πV (eq. 4).

Fig. (2) shows the variations of the DR cross-section as a function of the interaction parameter : it begins to decrease when $|\xi|$ increases beyond unity, for which σ reaches its maximum (actually, the expression (3) should not be used for very strong couplings but nevertheless it indicates the general behavior of the cross-section). Note finally that Eq.(3) holds when only one ionization threshold v^+ (corresponding to the initial ion vibrational level) is open for autoionization. It is not realistic at moderate or high electron energy. The more general expression [3]

$$S_F = \frac{1}{\left[1 + \sum_{v^+_{op}} \pi^2 \langle v^+|V_{el}|F_d\rangle^2\right]^2} \qquad (5)$$

where the summation runs over all the open vibrational thresholds, leads to a reduction of the cross-section (with respect to the overall E^{-1} dependance) at high energy. We will see below (section III.5) that the apparent stepdown of the survival factor (5), and hence of the cross-section, when a new threshold opens, is generally shaded off by a series of dips below the threshold, due to resonances on Rydberg states.

iii) a last remark about the strength of the process : it may happen that the curve of the dissociative doubly excited state lies entirely below that of the ion initial state (fig. 3). Although being then classically forbidden (the "capture point" R_c, where in the semi-classical picture the electron is captured into the dissociative state, is located beyond the left turning point), the process can still take place at the expense of energy exchange between atomic and nuclear motion, respecting the conservation of total energy.

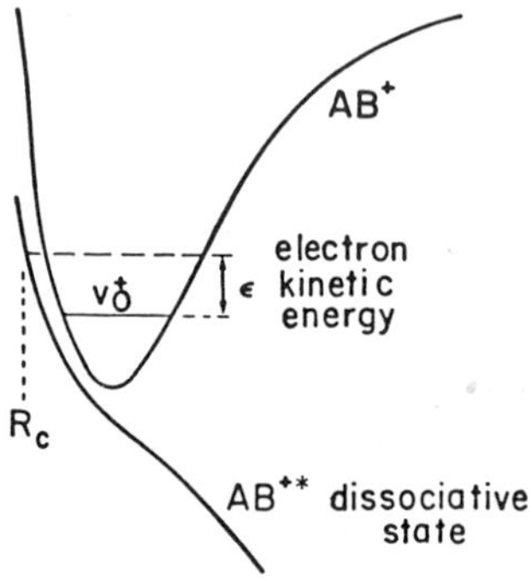

Fig. 3 : DR process via a non-intersecting dissociative curve.

The cross-section is then smaller than usually but not zero. An example of that kind will be given in the last section. Finally, some cases where no dissociative doubly excited state exists in the vicinity of the ion state have been encountered. The only mechanism left is a capture into a high Rydberg state (lying of course below the ion curve), due to (ro)vibrational coupling. The cross-section is then reduced by several orders of magnitude with respect to the electronic mechanism.

2. Is the dissociative recombination a resonant process ?

The two-electron process responsible for the electron capture in most cases reminds strongly the so-called dielectronic recombination (atomic DR !) occuring in atoms and described for example by Y. Hahn in this book. Actually the present molecular process should be called dissociative dielectronic recombination, but the alternative process of radiative dielectronic recombination is so weak compared to the dissociative one that it is never considered. An interesting comparison between the atomic and molecular recombination processes, and between their orders of magnitude, has been made by S. Guberman [2]. Here I will just comment on the resonant character of the dielectronic process : in atoms, the process occurs around definite energies, those of the doubly excited states of the neutral atoms. Hence, the cross-sections consists in a series of resonance peaks, which can be averaged analytically or numerically (see fig. 1 in the recent work by Bell and Seaton [4]). In molecules, the doubly excited state is also "resonant" with respect to the electron capture, at fixed R, but due to the additional degree of liberty (unbound nuclear motion), this resonant capture is not restricted to discrete values of the energy and does not induce "accidents" in the cross-section : in a semi-classical picture, one may say that for each energy ε of the incident electron, there is an internuclear distance $R(\varepsilon)$ which satisfies the energy conservation (see fig. 4)

$$U^+(R) + \varepsilon = U_d(R)$$

where U^+ and U_d represent the potential curves of the ion state and of the dissociative state, respectively. In a quantal description, the capture is not so localized, but it does occur for each energy providing the nuclear overlap is non zero.

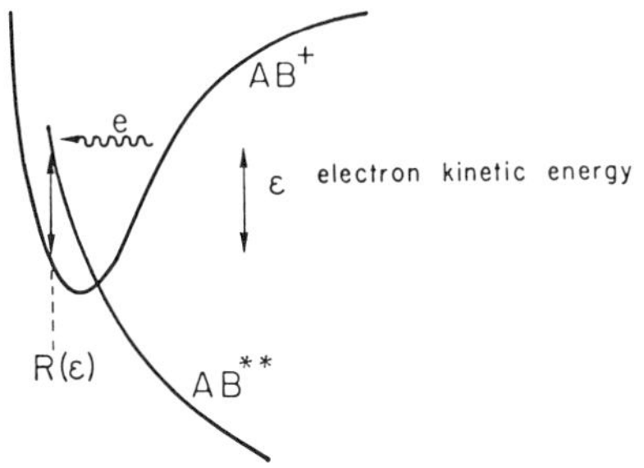

Fig. 4 : Semi-classical picture of the "dielectronic" capture for a given kinetic energy ε of the incident electron.

As a result, the DR cross-section in molecules varies smoothly, basically as E^{-1} with broad oscillations due to the variations of the nuclear overlap, and faster at high energy due to strong autoionization.

Nevertheless, true resonances (i.e. strong variations in a narrow energy range) may also occur in the molecular DR cross-sections, when the energy coincides with that of a true bound state (for electronic and nuclear motion), a vibrational excited level of a Rydberg state : this is the so-called indirect dissociative recombination, for which the capture is not due to the dielectronic process. Analysis of this truly resonant process, and of its interferences with the direct one, is the main purpose of the next section.

III. QUANTUM DEFECT TREATMENT OF DIRECT AND RESONANT DISSOCIATIVE RECOMBINATION

The major step towards the quantitative treatment of the direct DR process is due to Bardsley [5]. He extended to molecules the treatment of Fano [6] describing the mixing between bound and continuous electronic configurations in atoms. Due to the non-locality of the autoionization process in molecules, the nuclear wave-function of the dissociative state satisfies an integro-differential Schrödinger equation which is solved by the Green -function method. At the same time, but using a separate treatment [7], Bardsley considered the "indirect" process, which involves the transitory formation of a vibrationaly excited Rydberg state, predissociated by the repulsive state. He predicted it to induce narrow peaks in the cross-section, with Lorentzian shapes since no interferences between direct and indirect processes, treated separately, were allowed. It is only in 1980 that an unified treatment of the whole process [3] has been proposed, using an extension of the multichannel quantum defect theory (MQDT). I will describe below the main points and the main results of this treatment.

1. The basic concepts of the multichannel quantum defect theory

The quantum defect approach, first introduced by Seaton (1966), rests on two main concepts. First, it unifies the treatment of discrete and continuum states by using the collisional concept of channel, characterized by the bound or free motion of a relative particle in a given long-range potential (an electron in the Coulomb field of an ion, two atoms in a molecular potential... see fig. 5). If the relative motion is bound ($E < 0$ on fig. 5a and 5b), the channel is said to be closed, and open if the system may really be fragmented into two (or more) parts flying away.

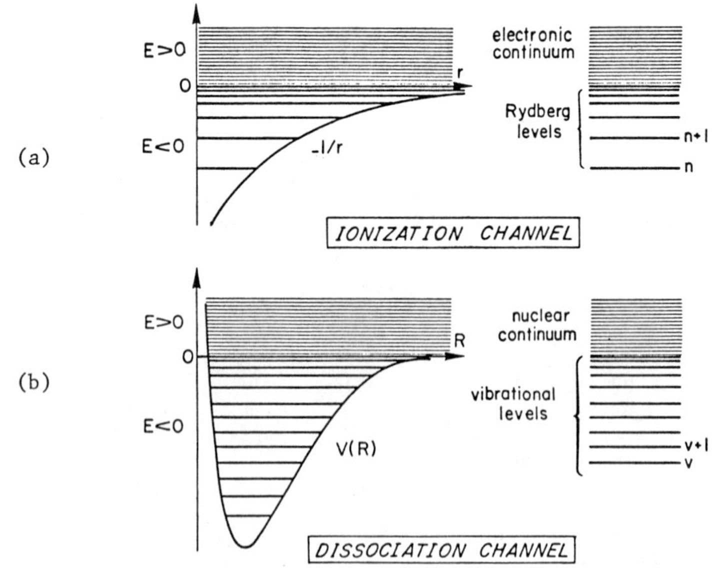

Fig. 5 : Energy levels in an ionization or dissociation channel (left) and their schematic representation (right).

These two kinds of channels may be treated on the same footing as long as one deals with their behavior at short range, where all the particles are close together and interact strongly. Thus, the distinction between short- and long- range interactions, which is the second basic concept of MQDT, and leads to partition the configuration space into (at least) two regions, is the corollary of the unified treatment of bound states (in the closed channels) and continua (open channels). They are distinguished in the external zone only, where their asymptotic behaviors differ radically. Note that this partition of the configuration space relates directly to the notion of molecular complex introduced in section I (fig. 1), and applies to the electronic motion as well as to the nuclear motion.

2. Example of one ionization channel

These two concepts may be illustrated on the simplest case of a single "ionization channel", corresponding to the scattering of one electron by an atomic or molecular "core", concentrated in the inner region $r < r_c$ (fig. 6). (For the sake of clarity, this paragraph will be very schematic. Quantitative details on Coulomb functions and normalization problems can be found for example in the first chapter of Seaton's MQDT review [9]). Outside this ionic core, the electron experiences only the Coulomb attraction and either moves apart if its kinetic energy is sufficient ($\varepsilon > 0$) or stays in a Rydberg orbital if it is not ($\varepsilon < 0$). In the upper part of fig. 6 the radial wavefunction of an electron has been schematically drawn, both for an hydrogen atom ($r_c \simeq 0$) and another atom or molecule : both wavefunctions are Coulomb functions, but they differ by the "boundary condition" at $r = r_c$: the effect of the core, for the complex atom, amounts to a translation of the wave function, which laggs in phase by the quantity δ with respect to the hydrogen wavefunction. This phase-shift δ_ℓ is the crucial quantity of the quantum defect theory, and desserves several remarks :

i) It varies very slowly with the long range kinetic energy ε of the electron, as shown by the lower part of fig. 6 : in the inner region,

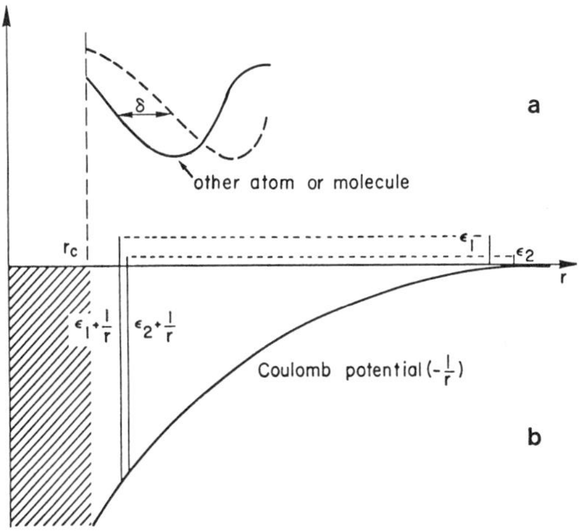

Fig. 6 :

(a) radial wavefunction for $r > r_c$

---- H atom

—— external electron in another atom or molecule

$\delta = \pi\mu$ is the phase-shift between the two functions.

(b) relative kinetic energies for two incident electrons, at long and short range.

where δ is built, the electron has been strongly accelerated by the Coulomb attraction, and its actual kinetic energy $\varepsilon + 1/r_c$ at the core boundary, where $1/r_c >> |\varepsilon|$, is not sensible to moderate changes of ε. This is the great advantage of dividing the space in two regions : the difficult calculations in the inner zone, where strong interactions and exchange effects have to be described, may be performed on a broad mesh of energy. The narrow structures due to resonance effects, on the contrary, must be calculated on a fine energy mesh but this is done in the external zone only, where the asymptotic behavior of the wavefunctions are specified.

ii) Let us see now how the phase-shift δ enters the asymptotic form of the radial wavefunction. Let f denote the hydrogen wavefunction, i.e. the regular ($f(0) = 0$) Coulomb function (for simplicity we treat the case of a s wave, with orbital quantum number $\ell = 0$). Denoting by g the irregular Coulomb function lagging in phase by $\pi/2$ with respect to f, the radial wavefunction in the general case may be written for $r > r_c$ in the form

$$\Phi_\varepsilon(r) = f(\varepsilon,r) \cos \delta - g(\varepsilon,r) \sin \delta . \qquad (6)$$

Indeed, for $\varepsilon > 0$ and $k = (2\varepsilon)^{1/2}$ (in a.u.) we have

$$f(\varepsilon,r) \underset{r\to\infty}{\sim} \sin (kr + \sigma) \qquad g(\varepsilon,r) \underset{r\to\infty}{\sim} - \cos (kr + \sigma)$$

where σ is the Coulomb phase-shift. Hence

$$\Phi_\varepsilon(r) \underset{r\to\infty}{\sim} \sin (kr + \sigma + \delta)$$

which shows that $\Phi_\varepsilon(r)$ is phase-shifted by δ with respect to f.

For $\varepsilon<0$ the wave function (6) behaves asymptotically as

$$\Phi_\varepsilon(r) \underset{r\to\infty}{\sim} A(\varepsilon)\, e^{-r/\nu} + \sin (\pi\nu + \delta) e^{+r/\nu}$$

where $\nu = (2\varepsilon)^{-1/2}$.

This wavefunction corresponds to a physical (bound) state only if the divergent term ($e^{+r/\nu}$) disappears, that is for the ν values satisfying

$$\sin (\pi\nu + \delta) = 0.$$

Each solution $\nu = n - (\delta/\pi)$ corresponds to a discrete value of the energy

$$\varepsilon_n = - \frac{1}{2(n - \delta/\pi)^2} \qquad \text{(a.u.)}$$

This is just the Rydberg formula giving the energies of the successive members of a Rydberg series, measured from the ionization threshold. Therefore the quantity δ/π is nothing else, for negative energy, than the quantum defect μ , known to be almost constant along the whole series. What is pointed out here is the correspondance between the phase-shift resulting, at positive energy, from the electron scattering by the core and the quantum defect which measures the energy shifts of Rydberg levels in complex atoms with respect to the hydrogen levels. The relation $\delta = \pi\mu$ allows fruitfull extrapolations across thresholds and relates the domains of electronic scattering (or ionization processes) and atomic or molecular structure.

3. Multichannel processes

In the multichannel case, the coefficients of the linear combination (6), more precisely their ratio tg δ= tg $\pi\mu$, is replaced by matrix elements of the reactance matrix K, which includes the effets of all the short range interactions and varies slowly with the energy. Numerous applications of MQDT to atoms have been made by Seaton and his coworkers (see e.g. the lecture by D. Moore in this book). MQDT has been applied to molecular process first by Fano [10], then by Jungen and his collaborators [11], who succeded in describing elegantly and precisely the rotational and vibrational interactions in Rydberg states due to coupling between electronic and nuclear motions neglected in the Born-Oppenheimer (B.O.) approximation. Traditionally these couplings are introduced as matrix elements of the centrifugal energy term or the derivation operator with respect to the internuclear distance R, very tedious to calculate. MQDT uses instead a "frame-transformation" technique, that is a change of basis when passing from the inner region, where the electron is strongly linked to the core electrons at each nuclear geometry (B.O. situation) to the outer region, where the external electron "sees" the molecular ion as a whole, in a definite electronical and rovibrational state. To be more explicit, let us write a few simplified equations (without the rotational parts) showing how this frame-transformation works for describing the vibrational interactions, which play a role in resonant dissociative recombination.

Outside the electronic core ($r > r_c$), but not too far away, i.e. in a region A ($r < r_A$) where the electron is still strongly accelerated by the Coulomb attraction (recall fig. 6), the molecular wavefunction may be written, using Eq. 6 :

$$\psi_v^A = \chi_v(R)\ \Phi_{core}\ (q,R) \left[\ f(\varepsilon,r)\ \cos\pi\mu(R)\ -\ g\ (\varepsilon,r)\ \sin\ \pi\mu(R)\ \right] \qquad (7)$$

χ_v is the vibrational wavefunction, $\Phi_c(q,R)$ the core electronic wavefunction, and the phase-shift $\delta = \pi\mu$ is now a function of the internuclear distance, in the B.O. framework.

Further away (region B) the electron is completely disconnected from the core and the wavefunction becomes

$$\psi_{v^+}^B = \chi_{v^+}\ (R)\ \Phi_c\ (q,R) \left[\ f\ (\varepsilon,r)\ a_{v^+} + g(\varepsilon,R)\ b_{v^+}\ \right] \qquad (8)$$

where χ_{v^+} is the vibrational wavefunction of the ion, and the coefficients of the Coulomb functions (f,g) depend now on v+ instead of R.

Performing the "frame transformation" consists here in expanding the B.O. wavefunction (7) on a basis of wavefunctions (8) :

$$\psi_v^A = \Phi_c\ (q,R)\ \sum_{v^+} \chi_{v^+}\ (R) \left[\ \mathcal{C}_{v^+v}\ f\ (\varepsilon,r)\ - \mathcal{S}_{v^+v}\ g\ (\varepsilon,r)\ \right]. \qquad (9)$$

The coupling coefficients

$$\mathcal{C}_{v^+v} = \int \chi_{v^+}(R)\ \cos\ \pi\mu(R)\ \chi_v(R)\ dR \quad \text{and} \quad \mathcal{S}_{v^+v} = \int \chi_{v^+}(R)\ \sin\ \pi\mu(R)\ \chi_v(R)\ dR \qquad (9')$$

which measures the mixing between different vibrational levels, clearly depends on how much the quantum defect varies with R, that is how much the external electron distorts the molecular Rydberg potential curves

$$U_n(R) = U^+(R) - \frac{1}{2[n - \mu(R)]^2}$$

with respect to the ion curve $U^+(R)$.

The actual wavefunction is successively written as a superposition of B.O. wavefunctions (7)

$$\Psi = \sum_v A_v \Psi_v \qquad (10)$$

and of "fragmented" wavefunctions (8) by means of the projection used in (9). As in the one channel case, the asymptotic behavior in each fragmentation channel is then specified, leading to an algebraic linear system which determines the coefficients A_v in (10), sufficient for a "half-collision" problem (e.g. photoionization), or alternatively the matrix elements of the scattering matrix, for a true collisional process.

4. Inclusion of electronic coupling with a dissociative channel

A global treatment of DR requires to describe simultaneously the vibrational coupling, responsible for electron capture in vibrationally excited Rydberg states, and the electronic coupling between the dissociative state and both the electronic continuum (direct process) and the Rydberg states (resonant process). Note that in the inner zone, these two kinds of electron-ion configurations are not distinguished and both correspond to the same electronic coupling $V_{el}(R)$ of Eq. (4), varying very slowly with the electron kinetic energy.

The dissociative channel is plugged into the MQDT treatment through the matrix elements V_{vd} defined in Eq. 4. Only the short internuclear distances ($R < R_c$) contribute to the integral in (4), such that the concept of "reaction zone" may be extended to the R coordinate (Fig. 7). Outside the reaction zone, a further component is now added to the wave function (10) :

$$\Psi = \sum_v A_v \tilde{\Psi}_v + A_d \Phi_d \left[F_d(E,R) + \left(\sum_v \pi V_{vd}\right) G_d(E,R)\right] \qquad (11)$$

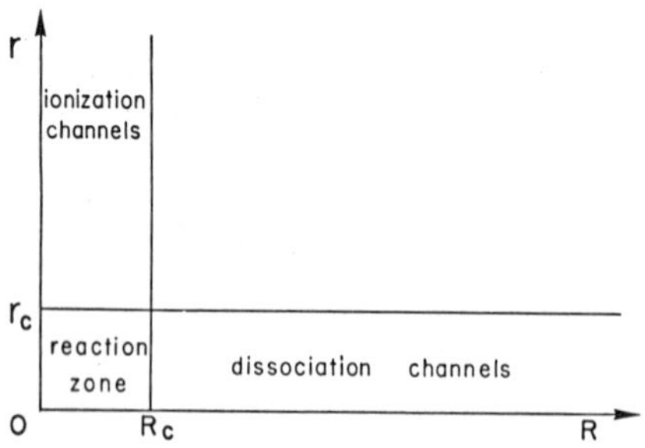

Fig. 7 : Two-dimensional diagram showing the different regions involved in the DR process.

In (11), Φ_d is the electronic wavefunction of the dissociative state, F_d the nuclear radial wavefunction in the "pure" dissociative potential. As in Eq. 6, the short range interaction has added to this function a phase-shifted component G_d (irregular solution in the same potential).

Similarly, a further phase-shifted component, proportional to $\pi\ V_{vd}\ g(\varepsilon,r)$ is added to each electronic radial wavefunction in the Ψ_v of Eq. 10 to yield the "perturbed" wavefunctions $\tilde{\Psi}_v$ of Eq. 11 (after proper normalization). It must be noted that this simple way of combining ionization and dissociation channels, described here for the sake of clarity, is valid only for weak electronic interaction : if the matrix elements V_{vd} are large they must be replaced by the elements of the corresponding partial reactance matrix, associated to this part of the short range interaction [3].

As for pure ionization, the wavefunction (11) is projected on a non B.O. basis set, and the DR cross-section is obtained in a last step by specifying the asymptotic behavior in each channel : ingoing wave in the entrance channel v_o^+, outgoing waves in all the other open channels, including the dissociative one, and convergence in the closed ionization channels. This last boundary condition introduces the resonances in the cross-sections around the energies of the discrete levels in the closed channels.

5. Resonance profile in the DR cross-section :

Figure (8), obtained for the case of H_2^+ + e dissociative recombination [12], shows the typical aspect of DR cross-section when the effect of discrete Rydberg levels (n,v) in the closed channels is taken into account. The most striking point is the shape of the resonances, which appear mostly as dips in the cross-section. This may be understood from the nature of the interactions, summarized on Fig. 9. Out of resonance, recombination proceeds directly via the strong electronic coupling with the dissociative channel. When the energy reaches a discrete level, the

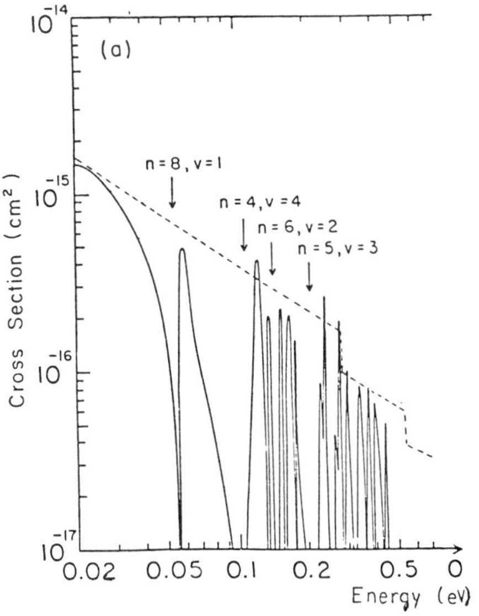

Fig. 8 :

DR cross-section of H_2^+ ions in the initial vibrational level $v_o^+ = 0$, with closed channels included.

(n,v) indicate the Rydberg levels responsible for the most prominent dips. The dashed curve corresponds to the direct process only (no closed channels).

weight of the Rydberg configuration is locally dominant and the electron capture is almost suppressed since it may occur only through vibrational coupling (dotted arrow on Fig. 9), much weaker than the electronic one.

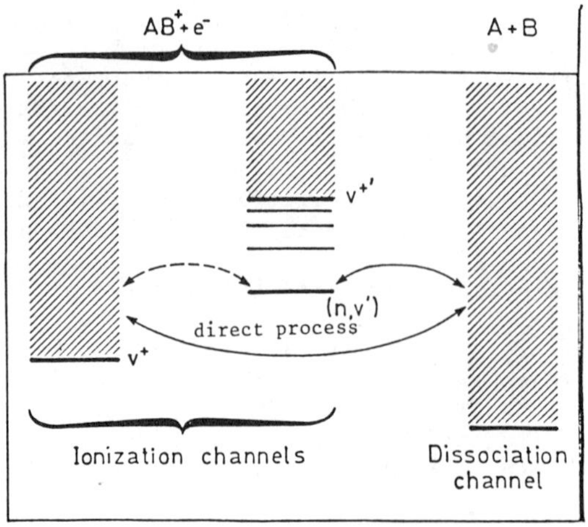

Fig. 9 :

Minimal set of channels for simultaneous direct and resonant process

---- vibrational coupling

—— electronic coupling

This is similar to "window" resonances in a photoionization spectrum, due to discrete states with very small oscillator strengths but coupled with the strongly absorbing direct ionization continuum.

The values of the resonance width may also be understood from Fig.9 : the discrete levels in the closed channel are coupled vibrationally to the ionization continuum, and electronically to the dissociation continuum. Hence their width is roughly proportional to the sum of the corresponding squared coupling matrix elements. Vibrational coupling is strongest for small change in v ($\Delta v = +1$, see Eq. 9'), but in any case the bulk of the width will come from the electronic part for which no "propensity rule" on Δv holds. On the contrary, the levels with higher v correspond to lower values of the principal quantum number n (for total energy conservation), that is to larger widths

$$\Gamma_{n,v} \simeq \frac{2\pi}{(n-\mu)^3} V^2_{vd} , \tag{12}$$

as may be seen on Fig. 8. This fact has been first noted by O'Malley [13], who emphasized the major role of electronic interaction even for the resonant process (in previous work attention were more directed towards levels with low v and high n, due to the "propensity rule" $\Delta v = +1$ for vibrational capture). Note however that whatever the value of n is, the same resonant process is involved (this is not clear in the discussion by Mitchell and Mc Gowan in their lecture at Baddek [14]), but the width is much larger for low values of n. For example, assuming the same magnitude for the electronic coupling in Eq. 12, a 5 meV resonance for n = 8 corresponds to about 40 meV for n = 4.

A last point must be noted on Fig. 8 : the marked steps down in the direct DR cross-section (dashed line) at the opening of each new ionization threshold are shaded off by the numerous resonance dips below the threshold, when the closed channels are included in the calculation. Hence the Rydberg series appear as precursors, for back ionization process, of the adjacent ionization continua. A similar situation occurs in molecular photoionization, where the step at each new threshold is generally masked by autoionization resonances just below.

IV. EXAMPLES

1. H_2^+ dissociative recombination of low energy

The most complete calculation to date - but far from being definitive - concerns the dissociative recombination of H_2^+ ions with slow electrons [12] which has also been intensively studied experimentally (see e.g. the lecture of Mitchell in this volume, and references therein). Before comparing the results shown on fig. 8 with experiment, one has to average them in two ways, for taking into account

i) the experimental resolution (fig. 10)
ii) the repartition of population among the vibrational levels of the recombining H_2^+ ions (assumed to be in the ground electronic state. Fig. 11)

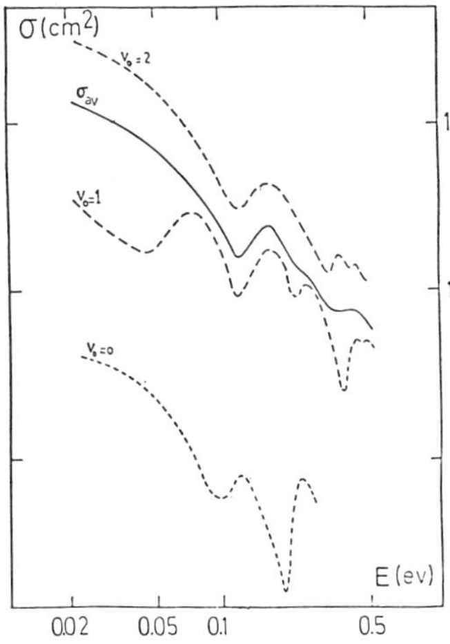

Fig. 10 : DR cross-section for $H_2^+ + e$, convoluted using a triangular apparatus function with a width of 0.04 eV

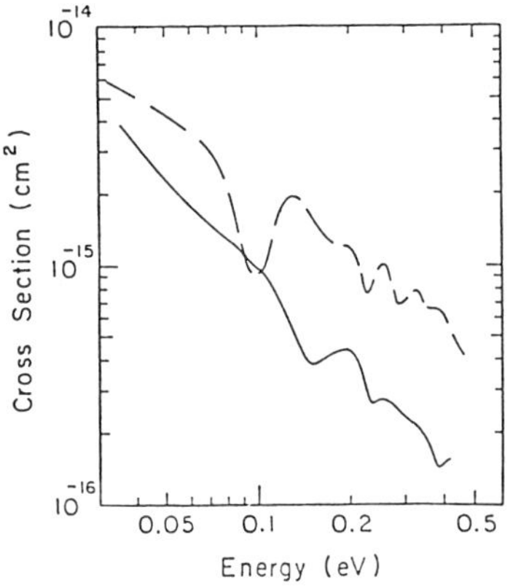

Fig. 11 : DR cross-section for $H_2^+ + e$
---- experiment (ref.15)
—— MQDT calculations, convoluted and averaged assuming $v_o^+ \leq 2$ and the vibrational distribution (1:2:2) for $v_o^+ = 0,1,2$

In both averages the resonance structure get smeared, and probably disappears completely if the experimental resolution is too poor, or if too much vibrational levels are present in the H_2^+ source. Indeed the broad dips resulting from experimental convolution (fig.10) do not coincide for different vibrational levels. This is probably one of the reasons of the experimental difficulties encountered for reproducing the only DR measurements which showed such broad dips [15,14]. The MEIBE technique used for this experiment yields probably the best resolution at low collision energy, but the vibrational population in the H_2^+ ions is very difficult to control.

Fig. 11 shows clearly that the agreement between the present theoretical and experimental results is far from being satisfying. Efforts are needed - and are on the way [17] - to reach a better understanding of this basic example. Investigation of the atomic products (partial cross-sections) is also clearly needed.

2. $H^+ - H^-$ associative ionization

The inverse process of DR - associative ionization - may of course be treated in the same MQDT approach, except that long-range correlations between atomic and molecular states must be carefully studied in this case. The $H^+ - H^-$ association, measured a few years ago at Louvain-La-Neuve [17], has just been theoretically studied in the same laboratory [18]. The dominant path for entering the reaction zone has been selected, and the ionization is assumed to result mainly from the electronic coupling of the doubly excited $(2p\,\sigma_u)^2$ configuration state with the electronic continuum of the H_2^+ ground state. The results shown on Fig. 12 show that this mechanism is certainly the dominant one for this reaction.

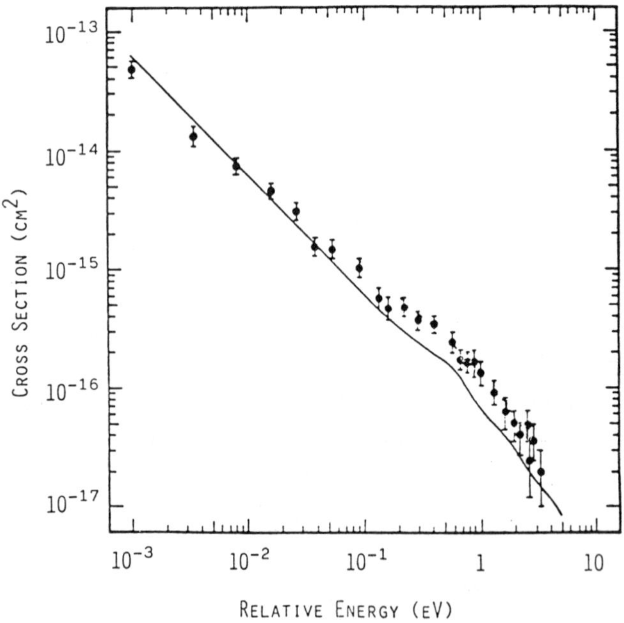

Fig. 12 : cross-section for the associative ionization process $H^+ - H^- \rightarrow H_2^+ + e$ (ref. 18). The experimental values are those of ref. 17.

3. CH^+ dissociative recombination

The case of CH^+ has been extensively discussed due to its importance to interstellar chemistry. The high abundance observed in interstellar molecular clouds might be due to a low rate of destruction by DR. Indeed, a molecular calculation [19] has suggested that the only dissociative state of CH close to the ion ground state stays below the ion curve at all R, as in the example of Fig. 3. This should lead to a low recombination rate, at least one order of magnitude smaller than the common value of $\sim 10^{-7}\ cm^3 \times s^{-1}$ at 100°K. But a short time later, measurements by Mitchell and Mc Gowan [20] yielded on the contrary a high rate ($\sim 5.10^{-7}\ cm^3 \times s^{-1}$) for CH^+ dissociative recombination at 100°K, and specific interstellar models have been developed to explain the CH^+ abundance.

However, it might be well, as for the case of H_2^+ measurements using the same merged beam technique, that vibrationally excited ions were present in the CH^+ source (although buffer gases have been added in order to try to eliminate them). The low rate found in the calculations corresponds to destruction of CH^+ ions in the $v^+ = 0$ vibrational level of the ground state, the only one populated in the low density and low temperature interstellar clouds. Numerical tests have shown that electron capture by vibrationally excited ions ($v^+ = 1,2$) should be faster, due to more favourable nuclear overlaps for non intersecting curves as on Fig. 3. Moreover, recent more sophisticated molecular calculations [21] seem to arrive at the same conclusion that no doubly excited dissociative state of CH intersects the ion ground state in the Frank-Condon region of its lower vibrational level.

REFERENCES

1. D.R. Bates, Phys. Rev. 78, 492 (1950)
2. S.L. Guterman, in "Physics of Ion-Ion and Electron-Ion Collisions", Ed. F. Brouillard and J.W. Mc Gowan, Nato Advanced Study Institutes Series 83, Plenum Press (New-York), p. 167
3. A. Giusti, J. Phys. B 13, 3867 (1980)
4. R.H. Bell and M.J. Seaton, J. Phys. B 18, 1589 (1985)
5. J.N. Bardsley, J. Phys. B 1, 349 (1968)
6. U. Fano, Phys. Rev. 124, 1866 (1961)
7. J.N. Bardsley, J. Phys. B1, 365 (1968)
8. M.J. Seaton, Proc. Phys. Soc. 88, 801, 815 (1966)
9. M.J. Seaton, Rep. Prog.Phys. 46, 167 (1983)
10. U. Fano, Phys. Rev. A 2, 353 (1970)
11. C. Jungen and O. Atabek, J. Chem. Phys. 66, 5584 (1977)
    C. Jungen and D. Dill, J. Chem. Phys. 73, 3338 (1980)
    See also the review by C. Greene and C. Jungen in "Adv. in At. and Molec. Physics" 21, 51 (1985)
12. A. Giusti-Suzor, J.N. Bardsley and C. Derbits, Phys.Rev. A28, 682 (1983)
13. T.F. O'Malley, J. Phys. B 14, 1229 (1981)
14. J.B.A. Mitchell and J.W. Mc Gowan, in "Physics of Ion-Ion and Electron-Ion Collisions", Ed. F. Brouillard and J.W. Mc Gowan, Nato Advanced Institutes series 83, Plenum Press (New-York), p. 279
15. D. Auerbach, R. Cacak, R. Caudano, T.D. Gaily, C.J. Kayser, J.W.McGowan, J.B.A. Mitchell and S.F. Wilk, J. Phys. B 10, 3797 (1977)
16. A. Sen, J.B.A. Mitchell, Abstracts of Papers XIV, ICPEAC (1985), p. 600-667
17. G. Poulaert, F. Brouillard, W. Claeys, H.W. McGowan and G. Van Wassenhove, J. Phys. B11, L 671 (1978).
18. X. Urbain, A. Giusti-Suzor, D. Fussen and C. Kubach, J. Phys. B(1986) in press.
19. A. Giusti-Suzor and H. Lefebvre-Brion, Ap.J. 214, L 101 (1977)
20. J.B.A. Mitchell and J.W. McGowan, Ap.J. 222, L 77 (1978)
21. E. Van Dishoek, private communication (1985).

THEORY OF CHARGE EXCHANGE AND IONIZATION IN ION-ATOM (ION) COLLISIONS

R.K. Janev

Invited Professor at the Université Catholique de Louvain
B1348 Louvain-la-Neuve
Belgium

INTRODUCTION

The present review will deal with the theoretical methods for study of single electron transfer and ionization processes in ion-atom and ion-ion collisions,

$$A + B^{q+} \rightarrow A^{+} + B^{(q-1)+} \quad , \quad q \geqslant 1 \tag{1.1a}$$

$$\rightarrow A^{+} + B^{q+} + e \quad , \tag{1.1b}$$

$$A^{q_1+} + B^{q_2+} \rightarrow A^{(q_1+1)+} + B^{(q_2-1)+} \quad , \quad q_1, q_2 \geqslant 1 \tag{1.2a}$$

$$\rightarrow A^{(q_1+1)+} + B^{q_2+} + e \tag{1.2b}$$

We shall also briefly discuss some conceptual aspects of the transfer ionization process

$$A + B^{q+} \rightarrow A^{2+} + B^{(q-1)+} + e \tag{1.3}$$

which is the simplest example of correlated multi-electron transition processes.

From the point of view of basic electron transition mechanisms, there is no substantial difference between the processes (1.1) and (1.2). The Coulomb repulsion in the initial channel of ion-ion reactions becomes important at low collision energies only, and for charge transfer reactions governed by a localized nonadiabatic transition mechanism it introduces a threshold. Thus, for energies below its maximum, the charge exchange cross section in ion-ion collision systems decreases more rapidly than for ion-atom systems (where the fall-off is exponential). In the case of ion-ion ionization, the Coulomb repulsion introduces an additional factor in the transition probability which exponentially decreases with decreasing collision velocity. Further, the density of interacting states in ion-atom systems is higher than in ion-ion systems, which enhances the possibility for getting an energy quasi-resonance for electron capture in ion-atom systems. With these remarks in mind, we shall not make any further distinction between the ion-atom and ion-ion charge-exchange and ionization processes, and present the theoretical methods for their description in a form identical for both types of collisions.

When presenting the theoretical methods for the inelastic processes in heavy particle collisions, it is convenient to introduce three dynamical regions defined by the ratio of the relative collision velocity v and the characteristic velocity v_0 of active bound electrons. For $v \ll v_0$, the collisional regime is adiabatic. A quasi-molecular description of collisional system is appropriate, and inelastic transitions in the system generally occur due to well localized nonadiabatic couplings between a restricted number of electronic states. For $v \gg v_0$, the collisional regime is such that transition amplitudes of inelastic processes can be expanded in terms of the small parameter V_{int}/E where V_{int} is the interaction inducing electron transitions and E is the collision energy. Finally, for $v \sim v_0$ no small dynamical parameter exists in the problem, and the collisional dynamics has to be solved in its whole complexity. Physically speaking, in this region there is a strong coupling between many (both discrete and continuum) electronic states of the colliding system, and the role of virtual transitions (through intermediate states) may become very pronounced. While in the high velocity region ($v \gg v_0$) a perturbational approach (such as the Born series) may be considered as an appropriate one for description of inelastic ion-atom collision processes (with proper account of continuum states in the case of electron transfer), in the intermediate and adiabatic velocity regions, the methods based on an expansion of total electronic wave function over a set of suitably chosen functions have proved to provide a good description of collisional dynamics. Appart from the perturbation and expansion methods, direct solution of the Schrödinger or classical equations of motion can be used in the intermediate velocity region to treat the charge exchange and ionization ion-atom collisions.

In the present lectures we shall confine ourselves to the methods based on the wavefunction expansion (or the coupled channel) formalism within the framework of semiclassical approximation. In the intermediate and low-energy regions this approach allows to treat the electron capture and ionization processes on the same footing and, under certain circumstances, to obtain approximate analytical solutions for the multistate close-coupling problem. Standard formulations of coupled-channel formalism for charge exchange can be found in the monographs[1-2] and review articles[3-6]. Coupled equations for ionization have been derived by using continuum pseudostates[2,7]. Here we shall develop a formulation of the close coupling method which provides a unified description of charge exchange, ionization and transfer - ionization processes (Sect. 2). A more complete account on the expansion methods for charge exchange at low and intermediate collision velocities will be given in Sects 3 and 4, respectively, while for ionization in Sect. 5.

Throughout these lectures we shall use atomic units ($e = \hbar = m_e = 1$), unless otherwise is explicitely stated.

GENERAL FORMULATION OF EXPANSION METHOD

Within the semiclassical approximation, the electronic motion in a colliding diatomic system is described by the time-dependent Schrödinger equation

$$i \frac{\partial \Psi}{\partial t} = H(\vec{r};\vec{R})\Psi(\vec{r};\vec{R}) \qquad (2.1)$$

where H is the electronic Hamiltonian, $\vec{r}$ is the set of electron position vectors and $\vec{R} = \vec{R}(t)$ is the internuclear position vector, on which H and Ψ depend parametrically. The expansion method starts, with defining a complete basis set $\{\chi\}$ over which the total electronic wavefunction is expanded. If Eq. (2.1) is to describe processes both in the discrete and continuous spectrum of the system $\{\chi\}$ should contain both discrete

(χ_j) and continuum (χ_ω) states. Moreover, the colliding system may contain several continua (χ_ω^α), corresponding to different electronic arrangements. Without loss in generality we shall, however, assume that there is only one continuum, that χ_ω are eigenstates of H, and that χ_j are mutually orthonormalized, but not necessarily eigenstates of H. Thus we assume (for any value of $\vec{R}$)

$$H\chi_\omega = \omega\chi_\omega, \quad < \chi_{\omega'}|\chi_\omega > = \delta(\omega'-\omega), \tag{2.2a}$$

$$< \chi_\kappa|H|\chi_j > = H_{\kappa j}, \quad < \chi_\kappa|\chi_j > = \delta_{\kappa j} , \tag{2.2b}$$

$$< \chi_\omega|H|\chi_j > = H_{\omega j}, \quad < \chi_\omega|\chi_j > = 0. \tag{2.2c}$$

Expanding now Ψ over $\{\chi_j\}$ and $\{\chi_\omega\}$

$$\Psi = \sum_j a_j(t)\chi_j \exp(-i\int_{-\infty}^t H_{jj}dt') + \int d\omega b_\omega(t)\chi_\omega \exp(-i\omega t) \tag{2.3}$$

and inserting into Eq. (2.1), one gets the following system of coupled equations for the state amplitudes $a_j(t)$ and $b_\omega(t)$

$$i\frac{\partial a_k}{\partial t} = \sum_{j\neq k} a_j(t)[H_{kj}-i(\frac{\partial}{\ell t})_{kj}]\exp[-i\int_{-\infty}^t (H_{jj}-H_{kk})dt'] +$$

$$+ \int d\omega b_\omega(t)[H_{k\omega}-i(\frac{\partial}{\partial t})_{k\omega}]\exp[-i(\omega t-\int_{-\infty}^t H_{kk}dt'], \tag{2.4a}$$

$$i\frac{\partial b_\omega}{\partial t} + \int d\omega b_\omega (i\frac{\partial}{\partial t})_{\omega'\omega} \exp[i(\omega'-\omega)t] =$$

$$= \sum_j a_j[H_{\omega' j}-(i\frac{\partial}{\partial t})_{\omega' j}]\exp[i(\omega' t-\int_{-\infty}^t H_{jj}dt'], \tag{2.4b}$$

where

$$(i\frac{\partial}{\partial t})_{k\omega} = i < \chi_k|\frac{\partial}{\partial t}|\chi_\omega >,$$

and the initial conditions are

$$a_j(-\infty) = \delta_{jo}, \quad b_\omega(-\infty) = 0 \tag{2.5}$$

(if at $t \to -\infty$ the state χ_0 is not coupled to the continuum), and in view of the completness of the basis $\{\chi\} = \{\chi_j\}\cdot\{\chi_\omega\}$, the unitarity condition is

$$\sum_j |a_j(t)|^2 + \int d\omega |b_\omega(t)|^2 = 1. \tag{2.6}$$

The probabilities for an electron transition within the discrete spectrum (excitation or charge transfer) and into the continuum (ionization into an interval $\omega+d\omega$), respectively are

$$P_j = |a_j(+\infty)|^2, \quad P_\omega = |b_\omega(+\infty)|^2. \tag{2.7}$$

The system of coupled equations (2.4) contains direct potential coupling between the discrete states (H_{kj}), discrete and continuum states ($H_{k\omega}$), as well as dynamic coupling between all types of states

$$[(\frac{\partial}{\partial t})_{kj}, (\frac{\partial}{\partial t})_{k\omega}, (\frac{\partial}{\partial t})_{\omega'\omega}].$$

By virtue of integral character of Eqs (2.4), terms of the type $H_{k\omega}H_{\omega j}$

are also present in it, describing indirect coupling between discrete states through intermediate continuum states.

In order to achieve further simplifications of the system (2.4), we shall assume that continuum is homogeneous so that $\partial/\partial t|\chi_\omega> = 0$. In this case the matrix elements $(\partial/\partial t)_{k\omega}$ and $(\partial/\partial t)_{\omega'\omega}$ in Eqs (2.4) vanish. If this assumption does not hold, the above matrix elements still vanish in the part of the spectrum far from continuum edge (due to fast oscillations). In both these cases, the system of coupled equations (2.4) is reduced to the form

$$i\,\frac{\partial a_k}{\partial t} = \sum_{j\neq k} a_j[H_{kj}-i(\frac{\partial}{\partial t})_{kj}]\exp[-i\int_{-\infty}^{t}(H_{jj}-H_{kk})dt'] +$$

$$+\int d\omega b_\omega H_{k\omega}\exp[-i(\omega t-\int_{-\infty}^{t}H_{kk}dt'], \qquad (2.8a)$$

$$i\,\frac{\partial b_\omega}{\partial t} = \sum_j a_j H_{\omega j}\exp[i(\omega t-\int_{-\infty}^{t}H_{jj}dt']. \qquad (2.8b)$$

Further reduction of the system of equation (2.8) is possible if the basis $\{\chi_j\}$ is appropriately chosen. If $\{\chi_j\}$ is the adiabatic basis,

$$H_{kj} = E_k\delta_{kj}\ ,\ H_{\omega j} = 0, \qquad (2.9)$$

then equations (2.8) become

$$\frac{\partial a_k}{\partial t} = -\sum_{j\neq k} a_j(\frac{\partial}{\partial t})_{kj}\exp[-i\int_{-\infty}^{t}(E_j-E_k)dt']\ , \qquad (2.10a)$$

$$\frac{\partial b_\omega}{\partial t} = 0. \qquad (2.10b)$$

Thus, in the adiabatic representation, there is no coupling between the discrete and continuous electronic states, if the continuum is assumed homogeneous. The later assumption may, however, be violated for the states lying closely to the continuum edge, and then the operators $\partial/\partial t$ do not vanish. The corresponding system of coupled equations in that case is

$$\frac{\partial a_k}{\partial t} = -\sum_{j\neq k} a_j(\frac{\partial}{\partial t})_{kj}\exp[-i\int_{-\infty}^{t}(E_j-E_k)dt'] -$$

$$-\int d\omega b_\omega(\frac{\partial}{\partial t})_{k\omega}\exp[-i(\omega t-\int_{-\infty}^{t}E_k dt')], \qquad (2.11a)$$

$$\frac{\partial b_{\omega'}}{\partial t} + \int d\omega b_\omega(\frac{\partial}{\partial t})_{\omega'\omega}\exp[i(\omega'-\omega)t] = -\sum_j a_j(\frac{\partial}{\partial t})_{\omega' j}\exp[i(\omega' t-\int_{-\infty}^{t}E_j dt')] \qquad (2.11b)$$

We note that Eqs (2.10a) represents the coupled equations of the perturbed stationary state (PSS) approximation.
If the discrete basis set $\{\chi_j\}$ is chosen in an orthogonalized diabatic representation defined by

$$(\frac{\partial}{\partial t})_{kj} = 0 \quad (\text{or} \simeq 0), \qquad (2.12)$$

then the system of coupled equations (2.8) is reduced to

$$i \frac{\partial a_k}{\partial t} = \sum_{j\neq k} a_j H_{kj} \exp[-i\int_{-\infty}^{t}(H_{jj}-H_{kk})dt'] +$$

$$\int d\omega b_\omega H_{k\omega} \exp[-i(\omega t-\int_{-\infty}^{t} H_{kk} dt')], \qquad (2.13a)$$

$$i \frac{\partial b_\omega}{\partial t} = \sum_j a_j H_{\omega j} \exp[i(\omega t-\int_{-\infty}^{t} H_{jj} dt')] \qquad (2.13b)$$

By solving Eq. (2.13b) formally

$$b_\omega = -i\int_{-\infty}^{t} dt' \sum_j a_j H_{\omega j} \exp[i(\omega t'-\int_{-\infty}^{t} H_{jj} dt')] \qquad (2.14)$$

and substituting it into Eq. (2.13a), one obtains a set of equations which, under the assumption that a_j and $H_{\omega j}$ are slow varying functions of ω, reduces to

$$i \frac{\partial a_k}{\partial t} = -\frac{i}{2}\Gamma_k a_k + \sum_{j\neq k}[H_{kj} - \frac{i}{2}\Gamma_{kj}] \cdot [\exp -i\int_{-\infty}^{t}(H_{jj}-H_{kk})dt'], \qquad (2.15)$$

$$\Gamma_k = 2\pi|H_{\omega k}|^2, \quad \Gamma_{kj} = (\Gamma_k\Gamma_j)^{1/2}, \qquad (2.16)$$

where Γ_k and Γ_j represent respectively the decay widths of the states $|k>$ and $|j>$ into the common continuum, and the term Γ_{kj} represents the coupling of states $|k>$ and $|j>$ through the continuum. The system (2.15) describes the mutual interaction of a set of autoionizing states. When the amplitudes a_j are found from Eqs (2.15), the continuum amplitude b_ω can be calculated from Eq (2.14). Since we assume that $a_j(t)$ and $H_{\omega j}(t)$ are slowly varying functions, the integral in Eq (2.14) can be evaluated by the stationary-phase method.

We note that a system of coupled equations analogous to (2.15) can be obtained by relaxing somewhat the assumption about the slow variation of $a_j(t)$ and $H_{\omega j}(t)$ on ω. In that case one obtains (see section additional energy shifts of the diabatic potentials H_{jj}. At low collision velocities, however, these energy shifts tend to zero.

The system of Eqs (2.15) allows further simplifications. For $|H_{kj}| \gg |\Gamma_k|, |\Gamma_j|$, it reduces to the standard form of close-coupled equations in diabatic representations for transitions within the discrete spectrum,

$$i \frac{\partial a_k}{\partial t} = \sum_{j\neq k} a_j H_{kj} \exp[-i\int_{\infty}^{t}(H_{jj}-H_{kk})dt']. \qquad (2.17)$$

In the opposite case, when Γ_k, Γ_j are much larger than H_{kj}, Eqs (2.15) are reduced to

$$\frac{\partial a_k}{\partial t} = -\frac{1}{2}\Gamma_k a_k - \frac{1}{2}\sum_{j\neq k} a_j \Gamma_{kj} \exp[-i\int_{-\infty}^{t}(H_{jj}-H_{kk})dt'] \qquad (2.18)$$

If one, further, neglects the interaction through the continuum, one is led with the decay approximation

$$\frac{\partial a_k}{\partial t} = -\frac{1}{2}\Gamma_k a_k \qquad (2.19)$$

which describes the exponential decay of the state $|k\rangle$ into continuum.

Equations (2.15) are particularly well suited for describing the coupling of charge exchange and transfer-ionization channels in low-energy ion-atom collisions. Ionization is described in terms of autoionization of discrete states due to their interaction with continuum. We note that not necessarily all of the states $|j\rangle$ become quasi-stationary during the evolution of the system; some of them may become quasi-stationary within an interval $(-t_0, t_0)$ only, but it is also possible that all of the states are quasi-stationary during the entire collision time (as in the case of $A + B^{q+}$ $(q \gg 1)$ systems).

Another approach of describing the coupling of discrete and continuum states is to use pseudostates in the total wavefunction expansion

$$\Psi = \sum_j a_j \chi_j + \sum_k b_k \bar{\chi}_k, \tag{2.20}$$

where the pseudostate functions $\bar{\chi}_k$ are chosen to be orthogonal to all χ_j, and to overlap with the continuum as much as possible. Inserting the expansion (2.20) into Eq. (2.1) leads to a system of coupled (differential) equations for the amplitudes a_j and b_k, which will be discussed in more details in Sect. 4. Here we note that the total ionization probability in this approach is given by

$$P_{ion} = \sum_k |b_k(+\infty)|^2. \tag{2.21}$$

CHARGE EXCHANGE AT LOW COLLISION VELOCITIES

In order to discuss in more detail the specific problems in charge exchange at low velocities $(v \ll v_0)$, we shall neglect the coupling of discrete states to the continuum, and for sake of simplicity we shall confine ourselves to the one-electron systems. Exceptions from these restrictions in our treatments will be stated explicitly.

The Perturbed Stationary State (PSS) method

The PSS method in low-energy heavy particle collisions has been introduced in the early thirties (see e.g. Ref. 8), and it is one of the simplest and the most transparent ways of describing collision dynamics in the quasi-molecular region. It consists in using the electronic adiabatic states

$$\psi_j(\vec{r},\vec{R}) = \Phi_j(\vec{r},\vec{R}) \exp(-i\int_{-\infty}^{t} E_j dt'), \tag{3.1}$$

$$H\Phi_j(\vec{r},\vec{R}) = E_j(R)\Phi_j(\vec{r},\vec{R}), \quad \vec{R} = \vec{R}(t), \tag{3.2}$$

as a complete expansion basis set for the collision problem, which yields the following system of coupled equations for the expansion amplitudes (see Eq. (2.10a)).

$$\frac{da_k}{dt} = - \sum_{j \neq k} a_j \Lambda_{kj} \exp[-i\int_{-\infty}^{t} (E_j - E_k) dt'], \tag{3.3}$$

with

$$\Lambda_{kj} = \langle \Phi_k | \frac{\partial}{\partial t}\Big|_{\vec{r}} | \Phi_j \rangle, \tag{3.4}$$

and the initial conditions $a_k(-\infty)=\delta_{ko}$. (In Eq. (3.4) $\partial/\partial t_{\vec{r}}$ means differentiation with $\vec{r}$ kept fixed). In the asymptotic region ($R \to \infty$), the molecular functions $\Phi_j(\vec{r},\vec{R})$ have to describe atomic states centered either on the target (A) or on the projectile ion (B^+). Consequently, two classes of basis functions have to be distinguished

$$\psi_j^A(\vec{r},\vec{R}) \xrightarrow[R \to \infty]{} \phi_j^A(\vec{r}_A)\exp(-i\varepsilon_j^A t), \tag{3.5a}$$

$$\psi_j^B(\vec{r},\vec{R}) \xrightarrow[R \to \infty]{} \phi_j^B(\vec{r}_B)\exp(-i\varepsilon_j^B t), \tag{3.5b}$$

where $\phi_j^\lambda(\vec{r}_\lambda)$ and ε_j^λ are atomic orbitals and energies for the center $\lambda(=A,B)$. In the case of a symmetrical system (A=B), proper asymptotic behaviour have the combinations

$$2^{-1/2}(\psi_{j,g} \pm \psi_{j,u}) \xrightarrow[R \to \infty]{} \phi_j \exp(-i\varepsilon_j t) \tag{3.5c}$$

where ψ_{jg} and ψ_{ju} are the wavefunctions of gerade (g) and ungerade (u) molecular states. Another obvious asymptotic condition which has to be satisfied is

$$\Lambda_{kj} \xrightarrow[R \to \infty]{} 0. \tag{3.6}$$

However, with the adiabatic functions $\Phi_j(\vec{r},\vec{R})$, the boundary condition (3.6) is not always satisfied. To demonstrate this statement, we refer to Fig. 1 and to the relations

$$\frac{\partial}{\partial t_{\vec{r}}} = \frac{\partial}{\partial t_{\vec{r}_A}} + p\,\vec{v}\,\nabla_{r_A}\Big|_R, \tag{3.7a}$$

$$= \frac{\partial}{\partial t_{\vec{r}_B}} - q\,\vec{v}\,\nabla_{r_B}\Big|_R \tag{3.7b}$$

where $\vec{v}$ is the relative collision velocity, p:q = AO : OB (cf Fig. 1), and the gradients ∇_{r_A} and ∇_{r_B} are calculated with R held fixed.

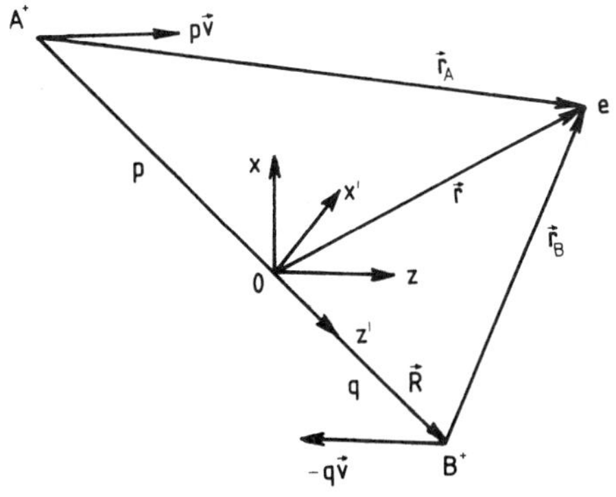

Fig. 1. Space fixed (XYZ) and rotating (X Y Z) coordinate systems for the A + B^+ collision.

$$\Lambda_{kj}^{AA} \xrightarrow[R \to \infty]{} \langle \phi_k^A | \frac{\partial}{\partial t}\Big|_{\vec{r}_A} | \phi_j^A \rangle = 0 \qquad (3.8)$$

$$\Lambda_{kj}^{BB} \xrightarrow[R \to \infty]{} \langle \phi_k^B | \vec{v}\nabla_{r_B} | \phi_j^B \rangle \qquad (3.9)$$

While Eq. (3.8) is consistent with the condition (3.6), the matrix element $\Lambda_{kj}^{BB}(R \to \infty)$ will not satisfy this condition if dipole transitions between the atomic states ϕ_k^B and ϕ_j^B are allowed. Similar violation of the boundary condition (3.6) may appear also in the exchange coupling matrix elements Λ_{kj}^{AB}. If one takes the coordinate origin at the ion B^+, then the condition (3.6) is trivially satisfied for Λ_{kj}^{BB}, but non-vanishing elements appear among Λ_{kj}^{AA} and Λ_{kj}^{BA}. The appearance of such spurious couplings in the asymptotic region signifies that the corresponding adiabatic states tend to "unphysical" asymptotic states, and if the basis $\{\Phi_j\}$ included in the calculations is complete, one can always construct an appropriate linear combination of them by which the boundary condition (3.6) will be satisfied. However, if from practical reasons the basis is restricted to a finite number of states, the above procedure will not remove the spurious couplings at infinity, and the results of calculations will depend on the choice of coordinate origin. This translational non-invariance of the PSS method, as first recognized by Bates and McCarroll[9], originates from the fact that the transfer of momentum, carried out by the translational motion of bound electron during the collision, is not described by the adiabatic wavefunctions ψ_j. Since the electron translational motion in the semi-classical approximation is associated with the heavy particle motion only, it should appear in the channel wavefunctions ψ_j as a phase factor,

$$F_j(\vec{r},\vec{R},\vec{v}) = \exp[i\gamma_j(\vec{r},\vec{R},\vec{v})]. \qquad (3.10)$$

The construction of the phase function $\gamma_j(\vec{r},\vec{R},\vec{v})$ has been subject of many discussions in the past (see e.g. Refs 1-4 for details) and there is still no consensus on the problem. The complexity of the problem may come from the fact that adiabatic states are intrinsically not scattering states of the system. Before going into more detailed discussion of the electron translation factors (ETF's) and the corresponding modifications of the system of coupled equations, we mention that in practice the adiabatic wavefunctions $\Phi_j(\vec{r},\vec{R})$ are usually calculated in the rotating molecular frame (X'Y'Z') with the Z' - axis oriented along the vector $\vec{R}$ (see Fig. 1). The transformation between the space fixed and the rotating frame is

$$X' = x\cos\theta + z\sin\theta, \qquad (3.11a)$$

$$Y' = y, \qquad (3.11b)$$

$$Z' = -x\sin\theta + z\cos\theta\ , \quad \theta = \theta(t) \qquad (3.11c)$$

where θ is the angle of rotation of internuclear axis, and, accordingly, the transition operator $\partial/\partial t_{\vec{r}}$ takes the form

$$\frac{\partial}{\partial t_{\vec{r}}} = v_R(\partial/\partial R_{\vec{r}'}) + \dot{\theta}(\partial/\partial\theta_{\vec{r}'}), \qquad (3.12)$$

$$v_R = dR/dt\ , \quad \dot{\theta} = -\,bv/R^2, \qquad (3.13)$$

with b being the impact parameter and v_R and $\dot{\theta}$ the radial and angular velocities. The matrix element Λ_{kj} now gets the form

$$\Lambda_{kj} = v_R \Lambda^R_{kj} + (bv/R^2)\Lambda^L_{kj} , \tag{3.14a}$$

$$\Lambda_{kj} = < \Phi_k | \frac{\partial}{\partial R}\Big|_{\vec{r}'} | \Phi_j >, \quad \Lambda^L_{kj} = < \Phi_k | iL_{y'} | \Phi_j > \tag{3.14b}$$

The radial coupling operator $\partial/\partial R$ couples states with the same angular symmetry ($\sigma \leftrightarrow \sigma \quad \pi \leftrightarrow \pi$, etc), while the rotational coupling acts between states for which the projection of angular momentum on the Z - axis differs by one ($\sigma \leftrightarrow \pi \quad \pi \leftrightarrow \delta$, etc).

Coupled-channel equations with ETF's

Let us associate with each of the adiabatic basis functions ψ_j (see Eq. (3.1)) a translational factor F_j of the form (3.10). Similarly to the asymptotic classification (3.5), we can, classify the functions of the new basis $\{\chi_j\} = \{F_j\psi_j\}$,

$$\chi^A_j = F^A_j \psi^A_j, \quad F^A_j = \exp[i\alpha_j(\vec{r},\vec{R},\vec{v})] , \tag{3.15a}$$

$$\chi^B_j = F^B_j \psi^B_j, \quad F^B_j = \exp[i\beta_j(\vec{r},\vec{R},\vec{v})] \tag{3.15b}$$

The simplest way to achieve the translational invariance of coupled channel equations is to take α_j and β_j in the form of a Galilean transform (hereafter we take the coordinate origin at the mid-point of internuclear distance)

$$\alpha_j = \frac{1}{2}\vec{v}\cdot\vec{r} - \frac{1}{8}v^2 t, \quad \beta_j = -\frac{1}{2}\vec{v}\cdot\vec{r} - \frac{1}{8}v^2 t. \tag{3.16}$$

It can easily be verified that with the functions $\chi_j = F_j\psi_j$ the boundary condition (3.6) is always satisfied. The form (3.16) for α_j and β_j has first been used by Bates and McCarroll[9] in their analysis of the charge exchange problem at higher energies. It can be seen from Eqs (3.16) that the plane-wave Bates McCarroll ETF's do not depend on the channel j and on the internuclear distance R. The independence of α_j and β_j on j can be justified only at energies considerably larger than the electronic energies of the system, while their independence of R implies that they are correct only at asymptotically large internuclear distances. At low energies, however, the translational motion is determined by the effective channel potentials and, consequently, the ETF's should, in principle, be channel-dependent. By the same token, they have to be also R-dependent. Assuming that $F_j(\vec{r},\vec{R},\vec{v})$ have the general form (3.10), and expanding the total electronic function over the complete basis set $\{\chi_j\} = \{F_j\psi_j\}$, one obtains the following system of coupled equations for the expansion coefficients a_j (in matrix form)

$$i\, \underline{S}\, \dot{\underline{a}} = \underline{M}\, \underline{a}, \tag{3.17}$$

where $\underline{a}$ is a column vector which consists of $a_j(t)$, $\underline{S}$ is the overlap matrix, with elements

$$S_{kj} = < \Phi_k | \exp[-i(\gamma_k - \gamma_j)] | \Phi_j > \exp[-i\int_{-\infty}^{t} (E_j - E_k)dt'], \tag{3.18}$$

and $\underline{M}$ is the coupling matrix, with elements

$$M_{kj} = - < \Phi_k | e^{-i(\gamma_k - \gamma_j)} \lambda_j | \Phi_j > \exp[-i\int_{-\infty}^{t} (E_j - E_k)dt'] \tag{3.19}$$

$$\lambda_j = \frac{\partial}{\partial t_r} + i\frac{\partial \gamma_i}{\partial t_r} + i\nabla_r\gamma_j\nabla_r + \frac{i}{2}\nabla_r^2\gamma_j - \frac{1}{2}(\nabla_r\gamma_j)^2. \tag{3.20}$$

The unitarity condition $\frac{d}{dt} < \Psi|\Psi > = 0$ in case of the system (3.17) can be written as

$$\underline{M} - \underline{M}^+ + i\frac{\partial}{\partial t}\underline{S} = 0. \tag{3.21}$$

It follows from Eq. (3.21) that the matrix $\underline{M}$ will be Hermitian if S_{kj} is time independent, i.e.

$$\frac{\partial S_{kj}}{\partial t} = 0. \tag{3.22}$$

An optimum choice of the phase functions γ_j can be made by using the variational method[11]. A system of Euler-Lagrange equations for γ_j is obtained which has to be solved simultaneously with the coupled equations for the amplitudes. However, this approach meets formidable computational difficulties. Simplifications in the variationally optimized ETF's can be obtained by factoring the $\vec{r}$, $\vec{R}$ and $\vec{v}$ dependences in γ_j[12]. Leaving however, the first principles in determining γ_j, all the possible choices become a matter of practical convenience. In most cases the channel-dependence of γ_j is omitted. In that case a very general form for $\gamma(\vec{r},\vec{R},\vec{v})$ is (omitting the trivial energy term $-v^2t^2/8$)

$$\gamma(\vec{r},\vec{R},\vec{v}) = \vec{v}\cdot\vec{r}f(\vec{r},\vec{R}). \tag{3.23}$$

The form of the function $f(\vec{r},\vec{R})$ is obtained on certain physical grounds. If γ is to satisfy the asymptotic conditions (3.16), then

$$f(\vec{r},\vec{R}) \xrightarrow[R\to\infty]{} \frac{1}{2}, \qquad r_A \ll r_B, \tag{3.24a}$$

$$\xrightarrow[R\to\infty]{} -\frac{1}{2}, \qquad r_B \ll r_A. \tag{3.24b}$$

If one is guided by the idea that at small values of R electronic motion has a truly molecular character, then for R smaller than the characteristic atomic dimensions the effect of γ has to be switched off, i.e.

$$f(\vec{r},\vec{R}) \xrightarrow[(R/r)\to o]{} 0. \tag{3.24c}$$

Many switching functions can be constructed which satisfy the conditions (3.24), but we mention here the simplest one[13]

$$f(\vec{r},\vec{R}) = \frac{1}{2}\cos(\vec{r}\cdot\hat{\vec{R}})\left[\frac{R^2}{R^2+\tilde{a}^2}\right], \tag{3.25}$$

where $\tilde{a}$ is a constant on the order of one. Use of switching functions which satisfy the asymptotic conditions (3.24a-3.24b) makes the basis non-orthogonal and an oscillatory term appears in the exchange matrix element M_{kj}^{AB}, (see Eq. (3.19)). With increasing $\vec{v}$ this oscillatory term leads to a significant decrease of M_{kj}^{AB}. In the case of ETF's in the form (3.23), the dynamic operator λ_j becomes

$$\lambda_j = \frac{\partial}{\partial t_r} + i\vec{v}\frac{\partial}{\partial t_r}(\vec{r}\,f) + i\vec{v}\nabla_r(\vec{r}\,f)\nabla_r + \frac{i}{2}\vec{v}\nabla_r^2(\vec{r}\,f) - \frac{1}{2}v^2[\nabla_r(\vec{r}\,f)]^2, \tag{3.26}$$

from where one can see that with decreasing $\vec{v}$ the effects of electron momentum transfer on the capture process become less and less important.

The experience in cross section calculations with various forms of ETF's shows that in a wide region of collision velocities the best results and the fastest convergence of the close-coupling method are obtained by using either variationally optimized or plane-wave ETF's.

Coupled-channel equations in an orthogonalized diabatic basis.

The construction of an appropriate diabatic molecular basis for charge-exchange ion-atom collisions has been thoroughly discussed many times[1-5,14]. We emphasize here only that the diabatic functions ψ_j^d should be chosen in such a way as to minimize the dynamic coupling $\partial/\partial t_r$ between the states, and should go over into proper atomic orbitals ϕ_j^λ (λ=A,B) or into linear combinations of them, when $R \to \infty$. The problem of ETF's in the basis $\{\psi_j^d\}$ remains the same as in the adiabatic basis. If $\{\psi_j^d\}$ is constructed to form an orthonormalized basis $\{\tilde{\psi}_j^d\}$ (including the ETF's) then the coupled equation for the expansion amplitudes are obtained in the form

$$i\dot{a}_k = \sum_{j\neq k} a_j V_{kj} \exp[-i\int_{-\infty}^{t}(V_{jj}-V_{kk})dt'] \tag{3.27}$$

with

$$\sum_j |a_j(t)|^2 = 1, \tag{3.28}$$

and

$$V_{kj} = \langle \tilde{\psi}_k^d | H | \tilde{\psi}_j^d \rangle \tag{3.29}$$

For the case of plane-wave ETF's, the orthonormalization procedure of diabatic states is demonstrated elsewhere[15].

Approximate treatments of coupled-channel equations. Multichannel models.

Numerical solution of coupled-channel equations is an appropriate approach only when the number of states involved in the charge exchange process is relatively small (below ten). When it is not the case (like in $A + B^{q+}$ ($q \gg 1$) and $A^- + BC^+$ systems, for instance), numerical solutions become to lengthy, and approximate treatments of coupled-channel equations have to be invoked. Although their accuracy generally is not too high (within a factor of two, or so), in the most complex situations they are the only effective way to get information about the capture cross section. Two typical situations are shown in Figs. 2a and 2b, representing the potential energy curves of the interacting states involved respectively in the reactions

$$A + B^{q+} \to A^+ + B^{(q-1)}(n,\ell) \tag{3.30}$$

and

$$A^- + BC^+(\nu) \to A + BC(\nu') \tag{3.31}$$

where ν and ν' are the vibrational quantum numbers.
In the present section we shall review three multichannel models, two of which describe the interaction of one discrete state with a group of other states (cf. Fig. 2a), and the third one describes the interaction of two groups of interacting molecular states (cf Fig. 2b).

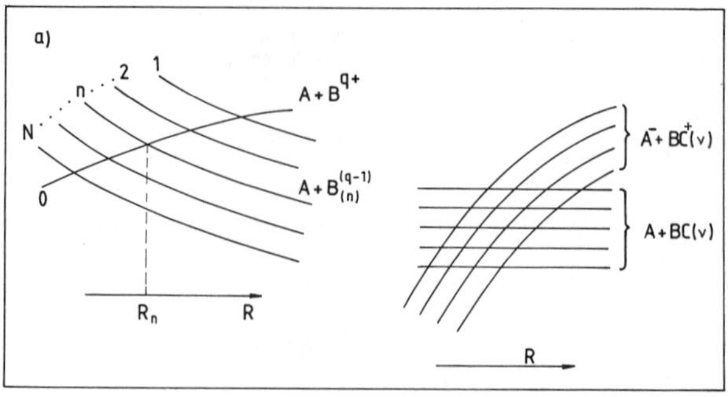

Fig. 2. Potential energy curve-crossings in the $A + B^{q+}$ and $A + BC$ systems.

A. Multichannel Landau-Zener model

Assuming that the curve-crossing points R_n in Fig. 2a are well separated from each other so that the corresponding two-state strong coupling regions δR_n do not overlap, the probability for population of a final charge exchange channel $|n>$ in a collision starting with the initial configuration $|o>$ is obtained by multiplication of independent two-state transition probabilities p_i at each crossing (see the figure for designation of the crossings).

$$P_n = p_1 p_2 \cdots p_n (1-p_n)[1 + (p_{n+1}p_{n+2} \cdots p_N)^2 + (p_{n+1}p_{n+2} \cdots p_{N-1})^2(1-p_N)^2 + (p_{n+1}p_{n+2} \cdots p_{N-2})^2(1-p_{N-1})^2 + \cdots + p_{n+1}^2 (1-p_{n+2})^2 + (1-p_{n+1})^2] \quad , \qquad n \leqslant N, \tag{3.32}$$

where, in the Landau-Zener approximation, the two-state probability p_i is given by [10]

$$p_i = \exp\left| - \frac{\pi \Delta_i^2(R)}{2v_R |\Delta F_i(R)|} \right|_{R=R_i} , \tag{3.33a}$$

$$\Delta(R) = 2V_{oi}, \; \Delta F_i = \frac{\partial}{\partial R} |V_i(R) - V_o(R)|, \tag{3.33b}$$

$$v_R = v \left(1 - \frac{b^2}{R^2} - \frac{V_o(R)}{E} \right)^{1/2}, \tag{3.33c}$$

where V_{oi} is the interaction coupling between the states $|o>$ and $|i>$ $V_i(R)$ are the diabatic potentials, E is the collision energy and b is the impact parameter.

The multichannel Landau-Zener (MLZ) model provides fair predictions only in the energy region where a few channels contribute considerably to the cross section..

Some of the charge exchange channels, for which there are no curve-crossings (e.g. excited product states with energies smaller that the initial state electron binding energy), are completely omitted in this model. Finally, the contribution from rotational transitions is also ignored in the model. As a result, the MLZ model substantially underestimates the total cross section in the energy region below the cross section maximum. The rotational transitions can be accounted for in the MLZ model in the case of hydrogen-atom fully stripped ion (H+Z) collisions. The populated σ - state of the $H^+ + (Z-1)_n$ manifold in the first half of collision is subject to a dynamical (dipole) mixing in the region $R < R_n$ with the other (n^2-1) Stark states. In the outgoing part of the trajectory (the second pass of the crossing point) the population of non-σ-states does not change and therefore the outgoing particle flux along the channel is larger than in the case without Stark mixing. The probability of population of all non-σ-Stark states in the region $R \leqslant R_n$ is given by[16]

$$q_n = 1 - (1 - \sin^2\beta \sin^2\alpha)^{2(n-1)}, \qquad (3.34a)$$

$$\beta = \arctan\left(\frac{2Zbv}{3n} \right), \qquad (3.34b)$$

$$\alpha = \mathrm{arc}\cos\{ \left(\frac{b}{R_n} \right)[1 + \left(\frac{3n}{2Zbv} \right)^2]^{1/2}\} \qquad (3.34c)$$

while the probability for population of the final principal n-shell is[17]

$$\begin{aligned} P_n = {} & p_1 p_2 \cdots p_n [1 + (p_{n+1} p_{n+2} \cdots p_N)^2 + \\ & + (p_{n+1} p_{n+2} \cdots p_{N-1})^2 (1-p_N)^2 (1-q_N) + \\ & + (p_{n+1} p_{n+2} \cdots p_{N-2})^2 (1-p_{N-1})^2 (1-q_{N-1}) + \\ & + \\ & \vdots \\ & + p_{n+1}^2 (1-p_{n+2})^2 (1-q_{n+2}) + \\ & + (1-p_{n+1})^2 (1-q_{n+1})] \\ & + p_1 p_2 \cdots p_{n-1} (1-p_n)^2 q_n , \qquad n \leqslant N \end{aligned} \qquad (3.35)$$

The inclusion of rotational coupling in the MLZ method significantly improves the cross section predictions particularly in the region of the cross section maximum. The results of extensive cross section calculations for the H+Z system for $Z \geqslant 14$ and $0.04 \lesssim v \lesssim 1$ can be fitted to the analytic expression[17] (within a 10% accuracy)

$$\sigma = 2.9\, \pi Z \ln \left(\frac{15}{v}\right) \qquad (3.36)$$

We note that the MLZ model with rotational coupling included gives fair predictions of the partial cross sections σ_n for the states which are dominantly populated in the capture process.

B. Separable interaction model

It has been found recently[18] that the coupling interaction $V_{oi} = \langle o|H|i\rangle$ for the highly excited final states $|i\rangle = |n\ell m\rangle$ of the H+Z system is separable in the quantum numbers n, ℓ and m namely

$$V_{oi} = \delta_{om} f(\ell) U_n(Z,R), \tag{3.37}$$

where, for $n \gg 1$ and $Z \gg 1$, one has

$$f(\ell) = (2\ell+1)^{1/2} \exp[-\frac{\ell(\ell+1)}{2Z}], \tag{3.38a}$$

$$U_n(Z,R) = (\frac{2Z}{\pi n^3})^{1/2} R \exp(-\frac{R^2}{3Z}). \tag{3.38b}$$

For these states, the energy difference of initial and final molecular states $\omega_{io} = V_{ii} - V_{oo}$ depends only on the principal quantum number n of final state

$$\omega_{io} \simeq \frac{1}{2} - \frac{Z^2}{2n^2} - \frac{(Z-1)}{R}. \tag{3.39}$$

The above circumstances allow one to reduce the system of coupled equations constructed on the diabatic molecular basis in which the initial (H+Z) and all final $(H^+ + (Z-1)_i)$ electron states are included. For the expansion

$$\Psi = a_o(t)\tilde{\psi}_o^d(t) + \sum_i b_i(t)\tilde{\psi}_i^d(t)$$

the following system of coupled equations can be obtained (see e.g. Eqs (3.27))

$$i\dot{a}_o = \sum_i b_i(t) V_{oi} \exp[-i\int_{-\infty}^{t} (V_{ii} - V_{oo}) dt'], \tag{3.40a}$$

$$i\dot{b}_i = a_o(t) V_{io} \exp[i\int_{-\infty}^{t} (V_{ii} - V_{oo}) dt']. \tag{3.40b}$$

The basis $\{\tilde{\psi}_j^d\}$ can be made orthonormal and Hermitian[15] $(V_{io} = V_{oi}^{\star})$ and we assume that it is complete, i.e.

$$|a_o|^2 + \sum_i |b_i|^2 = 1. \tag{3.41}$$

The initial conditions for Eqs (3.40) are : $a_o(-\infty) = 1$, $b_i(-\infty) = 0$. Making in Eqs (3.40) the substitutions

$$U_n = \frac{V_n}{[\sum_\ell f^2(\ell)]^{1/2}}, \quad b_i = \frac{f(\ell)}{[\sum_\ell f^2(\ell)]^{1/2}} b_n \tag{3.42}$$

$$i\dot{a}_o = \sum_n b_n V_n \exp(-i\int_{-\infty}^{t} \omega_n dt'), \tag{3.43a}$$

$$i\dot{b}_n = a_o V_n \exp(i\int_{-\infty}^{t} \omega_n dt'), \tag{3.43b}$$

where ω_n is given by Eq. (3.39). The ℓ-dependence of amplitudes is excluded from Eqs (3.43), so that before solving them one can obtain the ℓ-distribution of captured electrons ($i = n\ell$).

$$W_\ell = \frac{|b_i(+\infty)|^2}{|b_n(+\infty)|^2} = \frac{f^2(\ell)}{\sum_\ell f^2(\ell)} \simeq \frac{2\ell+1}{Z} \exp[-\frac{\ell(\ell+1)}{Z}] \tag{3.44}$$

The distribution (3.44) has a maximum at $\ell = \ell_m \lesssim Z^{1/2}$. By solving formally Eq. (3.43b) and inserting the solution into Eq. (3.43a), and using then the fact that for the highly excited states V_n are slow varying functions of $n[V_{n+\Delta n} \simeq V_n$ to within $O(\Delta n/n)$, if $\Delta n \ll n]$, one can transform the infinite system of equations (3.43) into a system of two equations[19]

$$i\dot{a}_o = -\frac{i}{2}\Gamma_o(t)a_o + V_n b_n \exp(-i\int_{-\infty}^{t}\omega_n dt'), \tag{3.45a}$$

$$i\dot{b}_n = V_n a_o \exp(i\int_{-\infty}^{t}\omega_n dt'), \tag{3.45b}$$

where

$$\Gamma_o(t) = 2\pi \frac{ZU_n^2}{(Z^2/n^3)} \equiv 2\pi \sum_\ell |V_{o,n\ell}|^2 \frac{dn}{d\omega} \tag{3.46}$$

is the decay width of the initial state in all the final channels, with $dn/d\omega$ being the density of states. Thus, the interaction of a discrete state $|o\rangle$ with a large group of states is reduced to interaction of a quasi-stationary state $|\tilde{o}\rangle$ with a single final state $|n\rangle$. If the density of final states is large ($\Gamma \gg V_n$), Eqs (3.45) reduce to the decay approximation[20,21]

$$\dot{a}_o = -\frac{1}{2}\Gamma(t)\, a_o,$$

with the charge exchange probability given by

$$P^d_{cx} = 1 - |a_o(+\infty)|^2 = 1 - \exp[-\int_{-\infty}^{\infty}\Gamma(t)dt]. \tag{3.47}$$

The two coupled equations (3.45) can be solved by using, for instance, the Vainstein-Presnyakov-Sobelman (VPS) approximation[22], with the result

$$P^{VPS} = |b_n(+\infty)|^2 = |\int_{-\infty}^{\infty} dt V_n \exp[i\int_{-\infty}^{t}\Omega_n dt' - \frac{1}{2}\int_{-\infty}^{t}\Gamma dt']|^2 \tag{3.48}$$

with

$$\Omega_n(t) = [\omega_n^2 + 4V_n^2]^{1/2}. \tag{3.49}$$

An illustration of the effectiveness of the separable interaction model is given in Fig. 3, where the results of this model for the charge exchange cross section of reaction $H + C^{6+} \to H^+ + C^{5+}$ is shown, in comparison with the 33-molecular orbital close-coupling calculations[23] (with variationally optimized ETF's), with the MLZ model with rotational coupling included[17] and with the experimental data[24]

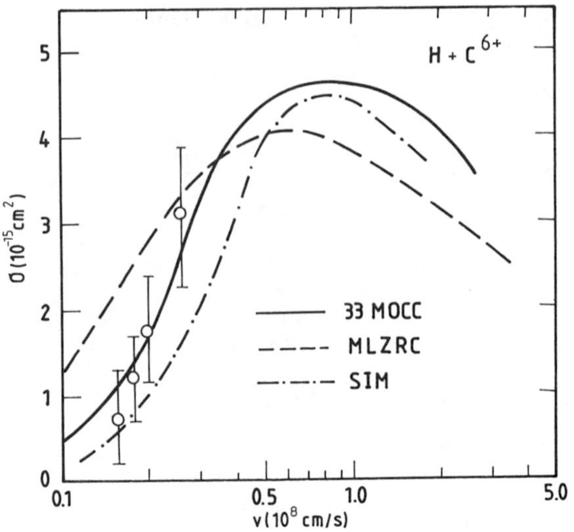

Fig. 3. Comparison of different theoretical cross sections for the reaction $H+C^{6+} \to H^{+}+C^{5}$ with experimental data.

C. Two-dimensional multichannel problem

We now consider two groups of interacting states, {i} and {n} belonging to two different electron arrangements of the system (see Fig. 2a). Let the corresponding basis set {i} • {n} be complete. We shall be interested in the $i_o \to n_o$ transition. Using the Feshbach projection operator formalism (in its time-dependent version), we extract the states $| i_o >$ and $| n_o >$ from the basis by the projector

$$P = u_{i_o} >< u_{i_o} + u_{n_o} >< u_{n_o}, \tag{3.50}$$

and define its complement

$$Q = 1 - P = \sum_{j \neq i_o, n_o} u_j >< u_j . \tag{3.51}$$

P and Q satisfy the relations

$$P^2 = P,\ Q^2 = Q,\ PQ = QP = 0,\ P\, i \frac{\partial}{\partial t} Q = 0. \tag{3.52}$$

The time dependent Schrödinger equation (2.1) now can be written in form of two coupled equations

$$P(i \frac{\partial}{\partial t} - H)P\Psi = PHQ\Psi, \tag{3.53a}$$

$$Q(i \frac{\partial}{\partial t} - H)Q\Psi = QHP\Psi. \tag{3.53b}$$

Introducing the Green's function $G^{(+)}$ with an "outgoing" boundary condition by

$$Q(i \frac{\partial}{\partial t} - H)QG^{(+)} = Q, \tag{3.54}$$

equation (3.53a) can be rewritten as

$$P(i\frac{\partial}{\partial t} - H - H'Q_G^{(+)}QH')P\Psi = 0 \tag{3.55}$$

with

$$H' = H - i\frac{\partial}{\partial t}. \tag{3.56}$$

From Eq (3.55) it follows that to the first-order perturbation theory one has[25]

$$QG^{(+)}Q = -i\theta(t-t') \sum_{j\neq i_o, n_o} u_j><u_j \cdot$$

$$\cdot \exp\{-i[\phi_j(t)-\phi_j(t')-\eta(t-t')]\} \tag{3.57}$$

where $\theta(x)$ is a step function,

$$\phi_j(t) = \int_{-\infty}^{t}[E_j(t')-E_j(\infty)]dt'+E_j(\infty)t, \tag{3.58}$$

and η is a small positive quantity defining the poles of $G^{(+)}$ in the complex t-plane. Using now the expansion

$$P\Psi = a(t)u_{i_o}e^{-i\phi_{i_o}(t)} + b(t)u_{n_o}e^{-i\phi_{n_o}(t)} \tag{3.59}$$

and Eq. (3.57) into Eq. (3.55), one can obtain[25]

$$i\dot{a} = V_{i_oi_o}a - i\sum_{k\neq o} V_{n_ki_o}e^{-i\phi_{n_ki_o}(t)-\eta t}\int_{-\infty}^{t}dt'V^{\star}_{n_ki_o}(t')e^{i\phi_{n_ki_o}(t')+\eta t'}a(t')$$

$$+ V_{n_oi_o}e^{-i\phi_{n_oi_o}(t)}b, \tag{3.60a}$$

$$i\dot{b} = V_{n_on_o}b - i\sum_{k\neq o} V_{i_kn_o}e^{-i\phi_{i_kn_o}(t)-\eta t}\int_{-\infty}^{t}dt'V^{\star}_{i_kn_o}(t')e^{i\phi_{i_kn_o}(t')+\eta t'}b(t')$$

$$+ V_{i_on_o}e^{-i\phi_{i_on_o}(t)}a, \tag{3.60b}$$

where

$$V_{i_kn_j} = < u_{n_j}|H - i\frac{\partial}{\partial t}|u_{i_k}> = V^{\star}_{n_ji_k}, \tag{3.61}$$

$$\phi_{i_kn_j} = \phi_{ik} - \phi_{n_j} = -\phi_{n_ji_k}. \tag{3.62}$$

In Eqs. (3.60) terms coupling states from the same group have been neglected. Assuming that the energy levels in both groups are energetically close (with respective level separations Δ_n and Δ_i), the sums in Eqs (3.60) can be replaced by an integral over the energy. Assuming, further, that the variations of a and b are relatively slow, one obtains from Eqs (3.60)[25]

$$i\dot{a} = (\tilde{V}_{i_o i_o} - \frac{i}{2}\Gamma_{i_o})a + V_{n_o i_o} b\exp(i\phi_{i_o n_o}), \qquad (3.63a)$$

$$i\dot{b} = (\tilde{V}_{n_o n_o} - \frac{i}{2}\Gamma_{n_o})b + V_{i_o n_o} a \exp(-i\phi_{i_o n_o}), \qquad (3.63b)$$

$$\tilde{V}_{j_o j_o} = V_{j_o j_o} + \tilde{\Delta}_{j_o}, \quad (j = i,n), \qquad (3.64)$$

where $\tilde{\Delta}_{j_o}$ and Γ_{j_o} are the energy shift and the width of the state $|j_o>$ due to its interaction with the states of other group. The expressions for $\tilde{\Delta}_{i_o}$ and Γ_{i_o} are

$$\tilde{\Delta}_{i_o} = -\frac{1}{\Delta_n} P \int_{E^{(n)}_{min}}^{E^{(n)}_{max}} dE \frac{\left|V_{E/\Delta_n + (n_o - E_{n_o}/\Delta_n),i_o}\right|^2}{E - E_{i_o} - \dot{\chi}_a}, \qquad (3.65)$$

$$\Gamma_{i_o} = 2\pi \frac{1}{\Delta_n} \int_{E^{(n)}_{min}}^{E^{(n)}_{max}} \left|V_{E/\Delta_n + (n_o - E_{n_o}/\Delta_n),i_o}\right|^2 \delta(E-E_{i_o}-\dot{\chi}_a) \qquad (3.66)$$

where P means the principal value of the integral, $E^{(n)}_{min}$ and $E^{(n)}_{max}$ are the lowest and the upermost levels in the group {n} and χ_a is the phase of the amplitude a. Expressions for $\tilde{\Delta}_{n_o}$ and Γ_{n_o} can be obtained from (3.65) and (3.66) by using the substitutions $i \leftrightarrow n$, $a \rightarrow b$. If the variation of matrix elements $V_{i_k n_j}$ with the energy is slow and the density of states is so high that we can extend the integration limits to $\pm\infty$, then one obtains

$$\tilde{\Delta}_{i_o} = \tilde{\Delta}_{n_o} = 0, \qquad (3.67)$$

and

$$\Gamma_{i_o} = 2\pi |V_{n_o,i_o}|^2 \frac{dn}{dE^{(n)}_n}, \qquad (3.68)$$

Where $dn/dE^{(n)}_n$ is the density of states in the group {n}. Equations (3.63) are valid, of course, for any of the levels from {i} and {n}. Thus, the interaction of two groups of interacting states has been reduced to the problem of two quasi-stationary states coupled with the interaction V_{in}.

The solution of coupled equations (3.63) depends on the form of $\tilde{V}_{j_o j_o}$, Γ_{j_o}, V_{in} and $\phi_{i_o n_o}$. Analytic results for the Landau-Zener, Demkov and Nikitin models, appropriately generalized have already been obtained[25-27]. For the Auger processes, occuring in the $A+B^{q+}$ systems, the decay widths Γ(R) have been calculated in closed form[28,29]. They have usually an exponential dependence on R, but in the models Γ(R) can be appropriately modelled to achieve an analytic solution of coupled equations. Analysis of the probability expressions for the above-mentioned models shows that, apart from the trivial decay during the adiabatic evolution of the system, the inelastic transitions in the strong-coupling region are enhanced with respect to the case of non-decaying states. This is mainly due to the fact that in case of two interacting quasi-stationary states, the transition zone becomes broader.

At intermediate collision velocities ($V \sim V_0$), the active electron spends the overwhelming part of collision time bound either on the target of on the projectile. Therefore, atomic orbitals centered on the target and projectile form a suitable basis for description of electronic motion. Plane-wave ETF's are usually regarded as appropriate in this (and the higher) energy region. In general, the number of states included in the basis depends on how well the resonant condition for the capture process is fulfilled in this region. If the process takes place in a resonant or quasi-resonant manner, then the process can successfully be described by coupling the resonant (qausi-resonant) states and those which are energetically close to them. For non-resonant capture processes (e.g. in the endothermic reactions), the number of states which have to be included in the expansion basis may be very large and has also to include virtual continuum states or pseudostates. Depending on charge symmetry of colliding system, the basis functions may be centered either on one of the centers (strongly charge-asymmetric systems) or on both of the centers (nearly charge-symmetric systems). In the latter case some of the basis functions may also be centered on the center-of-charge of the system (three-center expansions). In the present section we shall give a brief account of the coupled-channel models, based on one- and two-center expansions.

Two-center expansion models

The expansion basis of these models contains functions describing bound and continuum states of the electron around each of the potential centers (the target core and the projectile). With each of the expansion functions a plane-wave ETF is associated. The expansion functions may not necessarily be eigenstates of corresponding atomic Hamiltonians. When working with a finite basis set, the choice depends on the physical situation (resonant condition, role of continuum, etc...) and the requirement on the convergence of the model. Below we consider some of the most frequently used expansion models.

A. Expansion model with travelling atomic orbitals

The basis used in this model is $\{\chi\} = \{\psi_n^A\} \cdot \{\psi_m^B\}$, with

$$\psi_n^A = \phi_n^A(\vec{r}_A)\exp(-i\varepsilon_n^A t + \frac{i}{2}\vec{v}\cdot\vec{r} - \frac{i}{8}v^2 t), \tag{4.1a}$$

$$\psi_m^B = \phi_m^B(\vec{r}_B)\exp(-i\varepsilon_m^B t - \frac{i}{2}\vec{v}\cdot\vec{r} - \frac{i}{8}v^2 t), \tag{4.1b}$$

where $\phi_n^A, \varepsilon_n^A$ and $\phi_m^B, \varepsilon_m^B$ are the eigenfunctions and eigenenergies of atomic Hamiltonians H_0^A and H_0^B. Writing the total electronic Hamiltonian in the form

$$H = H_0^A + V^B = H_0^B + V^A \tag{4.2}$$

and using the expansion

$$\Psi = \sum_n a_n(t)\psi_n^A + \sum_m b_m(t)\psi_m^B, \tag{4.3}$$

in the time-dependent Schrödinger equation (2.1), one obtains the following system coupled channel equations (in matrix form)

$$i(\underline{\dot{a}} + \underline{s}\underline{\dot{b}}) = \underline{K}\,\underline{a} + \underline{W}\,\underline{b} \quad , \tag{4.4a}$$

$$i(\underline{s}^{+}\underline{\dot{a}} + \underline{\dot{b}}) = \underline{\tilde{K}}\,\underline{a} + \underline{\tilde{W}}\,\underline{b}, \tag{4.4b}$$

where

$$s_{nm} = <\phi_m^B|\exp(i\vec{v}\cdot\vec{r})|\phi_n^A>\exp[-i(\varepsilon_n^A-\varepsilon_m^B)t], \tag{4.5}$$

$$K_{nn'} = <\phi_{n'}^A|V^B|\phi_n^A>\exp[-i(\varepsilon_n^A-\varepsilon_{n'}^A)t] \quad , \tag{4.6a}$$

$$\tilde{K}_{mm'} = <\phi_{m'}^B|V^A|\phi_m^B>\exp[-i(\varepsilon_m^B-\varepsilon_{m'}^B)t] \tag{4.6b}$$

$$W_{mn} = <\phi_n^A|\exp(-i\vec{v}\cdot\vec{r})V^B|\phi_m^B>\exp[-i(\varepsilon_m^B-\varepsilon_n^A)t], \tag{4.7a}$$

$$\tilde{W}_{nm} = <\phi_m^B|\exp(i\vec{v}\cdot\vec{r})V^A|\phi_n^A>\exp[-i(\varepsilon_n^A-\varepsilon_m^B)t]. \tag{4.7b}$$

The system of coupled equations (4.4) is subject to the initial conditions

$$a_n(-\infty) = \delta_{no} \quad , \quad b_m(-\infty) = 0,$$

with the target excitation and charge exchange probabilities being given by $|a_n(+\infty)|^2$ and $|b_m(+\infty)|^2$, respectively. The unitarity condition in the present case may be written as

$$\underline{W} - \underline{\tilde{W}}^{+} + i\underline{\dot{s}} = 0 \tag{4.8}$$

and it may be used as a check for the completeness of the basis.

Extensive large base calculations using Eqs (4.4) have been mode for the H+Z(Z $\leqslant$ 3) systems by Bransden and his associates (see e.g. Ref. 4), showing good agreement with more involved calculations, which use pseudostates or a three-center expansion basis. From practical point of view the presence of translational factor $\exp(\pm i\vec{v}\cdot\vec{r})$ in the exchange matrix elements W_{ji} makes their calculations very lengthy (their calculations have to be carried out for each value of $\vec{v}$). Therefore, the choice of important atomic orbitals to be included in the basis for each particular collision system and the rapid convergence of the results with increasing the basis are questions of considerable practical interest. Inclusion of pseudostates in the basis is motivated also from such considerations.

B. Two-center atomic orbital and pseudostate expansion model

The two-center atomic orbital expansion method has been in recent years used in conjunction with pseudostates, introduced in the basis to account approximately for the effects of coupling with the continuum. Inclusion of such "continuum" pseudostates is important especially in the non-resonant case (i.e. for weak electron capture transitions), and at higher energies. If one wishes to extend the method towards lower (but not too low) collision velocities, some features of the quasi-molecular behaviour of electronic states can be taken into account by including atomic orbitals (or pseudostates) of the united atom (UA) centered on A^+ and/or B^{q+} [30,31], or at the center of the charge of the system (three-center expansion)[32]. The expansion (4.3) now reads

$$\Psi = \sum_{n}^{N} a_n \psi_n^A + \sum_{m}^{M} b_m \psi_m^B + \sum_{k}^{K} \bar{a}_k \bar{\psi}_k^A + \sum_{\ell}^{L} \bar{b}_\ell \bar{\psi}_\ell^B \tag{4.9}$$

$$\left(+ \sum_{s}^{S} \bar{c}_s \bar{\psi}_s^{AB}\right) \tag{4.9a}$$

where N,M,K,L and S are the numbers of orbitals (pseudostates, for K,L,S) included in the basis, and $\bar{\psi}_k^A, \bar{\psi}_\ell^B, \bar{\psi}_s^{AB}$ are the pseudostate functions. The pseudostate functions have to satisfy the following requirements : 1) they should be made orthogonal to all the bound state orbitals ϕ_n^A and ϕ_m^B ; 2) if $\bar{\phi}_k^A$ and $\bar{\phi}_\ell^B$ are "continuum" pseudostates, they have to have maximum possible overlap with the continuum wavefunctions of the corresponding atomic Hamiltonian H_o^A and H_o^B, in a given region of positive energies; 3) it some $\bar{\phi}_k^A$ and $\bar{\phi}_\ell^B$ are UA pseudostates, they should have maximum possible overlap with the low-lying bound state of the united atom (or even to be identical with them); 4) the matrix elements involving pseudostate functions should have "good" asymptotic behaviour (i.e. to vanish as $R \to \infty$). In order to satisfy the above conditions, the functions $\bar{\phi}$ have to contain a sufficient number of parameters and to decrease sufficiently rapidly at infinity (exponentially or Gaussian-like). Apart of representing some physical features of colliding system, the main advantage of using the pseudostates is a more rapid convergence of the results toward the correct (experimental) one than in the case of simple two-center atomic orbital expansions having the same number of states in the basis. (In fact, this circumstance is considered as an argument for the physical meaning of the pseudostate functions, which may, however, be a subject for discussion). The most extensive two-center atomic orbital + united atom pseudostate expansion close-coupling calculations have been performed for the one-electron systems H+Z($Z \leqslant 8$)[31], in which up to 48 terms have been included in the basis. The cross section results for H+O^{8+} are shown in Fig. 4, together with the 33-molecular orbital close-coupling calculation[33] and the experiment results for H+O^{8+} and Ne^{8+}+H (from Ref. 34).

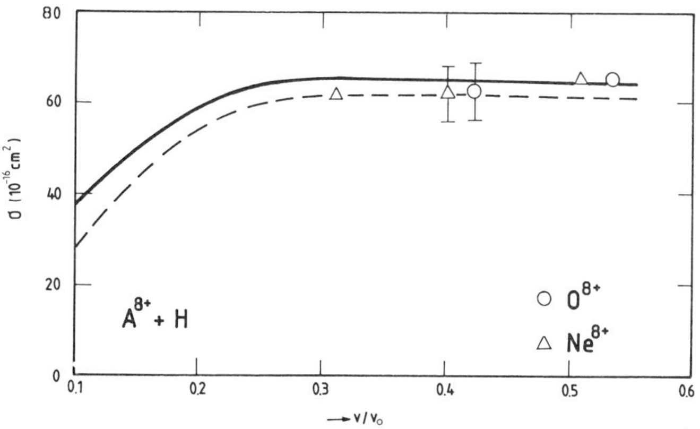

Fig. 4. Total cross section for H+O^{8+} collisions. The full and dashed curves represent the large base molecular[33] and atomic[31] orbital calculations; the symbols are experimental data[34] for O^{8+}+H (open circle) and for Ne^{8+}+H (triangles).

In principle, the method using pseudostate $\bar{\psi}_n^A$ and $\bar{\psi}_m^B$, which are chosen to represent the influence of target and projectile continua to the electron capture process, may also be used to calculate approximately the ionization cross section. The ionization probability is, in an obvious manner, given by

$$P_{ion} = \sum_k^K |\bar{a}_k(+\infty)|^2 + \sum_\ell^L |\bar{b}_\ell(+\infty)|^2 \equiv P_{ion}^A + P_{ion}^B, \tag{4.10}$$

in which the term P_{ion}^A represents the contribution from transitions into the target continuum, whereas the term P_{ion}^B represents the contribution from transitions into the projectile continuum (electron capture to the continuum-ECC). Extensive calculations for H^++H collisions in the 15-200 keV region have been performed[35] using a 35-pseudostate two-center basis in which both the lower excited and low-continuum states have been represented. It has been shown that below ∿ 50 kEv ECC gives the main contribution to the ionization process. The energy spectrum of ECC electrons is sharply peaked in the forward direction. The continuum pseudostates in ion-atom collisions have first been introduced by Cheshire et al[36].

C. Sturmian basis

A special class of pseudostates can be constructed from the solutions of Sturmian equation

$$\left(-\frac{1}{2}\frac{d^2}{dr^2} + \frac{\ell(\ell+1)}{2r^2} - \frac{\alpha_{k\ell}}{r}\right) S_{k\ell}(r) = \varepsilon S_{k\ell} \tag{4.11}$$

where ε is a fixed parameter and $\alpha_{k\ell}$ is an "effective charge", treated in (4.11) as an eigenvalue. Equation (4.11) is Schrödinger-like, and if ε is chosen to be negative, then k takes only discrete values (k=1,2,3,...) and the solutions $S_{k\ell}(r)$ form an infinite and complete discrete set. The boundary behaviour of $S_{k\ell}$ is : $S_{k\ell}(r \to 0) \to 0$ and $S_{k\ell}(r \to \infty) \to 0$. The two-center Sturmian basis has the same form as Eqs (4.1), except that atomic functions ϕ_n^A and ϕ_m^B are now replaced by Sturmian functions

$$\tilde{\phi}_k^{(\lambda)} = \frac{1}{r} S_{k\ell}(r)\, Y_{\ell m}(\theta,\phi) \quad , \quad \lambda = A,B. \tag{4.12}$$

For the special choice of parameter ε[37]

$$\varepsilon = -\frac{1}{2(\ell+1)^2} \tag{4.13}$$

$S_{k\ell}(r)$ become scaled radial hydrogenic functions

$$S_{k\ell}(r) = (\alpha_{k\ell})^{1/2} R_{k\ell}(\alpha_{k\ell} r) \tag{4.14}$$

$$\alpha_{k\ell} = k(-2\varepsilon)^{1/2} = \frac{k}{\ell+1}, \tag{4.15}$$

where the normalization factor is chosen such that

$$< \tilde{\phi}_k^{(\lambda)}(r) | \tilde{\phi}_k^{(\lambda)}(r) > = 1. \tag{4.16}$$

The following remarks have to be made with respect to the use of Sturmian basis in charge exchange cross section calculations. Since $\tilde{\phi}_k^{(\lambda)}$ is not eigenfunction of atomic Hamiltonian H_o^λ, it does not, in general, represent a true physical state. (With the special choice (4.13) for ε. $\tilde{\phi}_{1s},\tilde{\phi}_{2p},\tilde{\phi}_{3d}$ coincide with hydrogenic functions $\phi_{1s},\phi_{2p},\phi_{3d}$, but $\tilde{\phi}_{2s},\tilde{\phi}_{3s},\tilde{\phi}_{3p}$

are different from the corresponding atomic states). If a finite basis set of $\tilde{\phi}_{k\ell}$ is used in the expansion, the amplitudes $\tilde{a}_n(t)$ and $\tilde{b}_m(t)$ corresponding to the "non-physical" Sturmian states have an oscillatory behaviour at $t \to \infty$.

It has been, however, demonstrated that with increasing Sturmian basis, the oscillatory components of transition amplitudes become negligible[38].

One-center expansion models

The one-center expansion models are particularly well suited for description of charge exchange processes in atom-multicharged ion collisions. Usually only one orbital is taken from the target (A) basis, and all other orbitals belong to the projectile (B) basis, so that instead of (4.3) one has the expansion

$$\Psi = a_o \tilde{\psi}_o^A + \sum_m b_m \psi_m^B, \qquad (4.17)$$

where $\tilde{\psi}_o^A$ is the intial state wavefunction ψ_o^A transferred to the projectile and orthogonalized to all ψ_m^B. Then, hermiticity is preserved if all discrete and continuum states are included in $\{\psi_m^B\}$[15]. Assuming that orthogonalization procedure is performed, the system of coupled equations for a_o and b_m is

$$i\dot{a}_o = \sum_m V_m b_m \exp(i \int_{-\infty}^{t} \omega_m dt'), \qquad (4.18a)$$

$$i\dot{b}_m = V_m a_o \exp(-i \int_{-\infty}^{t} \omega_m dt'), \qquad (4.18b)$$

where

$$V_m = \langle \tilde{\psi}_o^A | V^B | \psi_m^B \rangle, \quad \omega_m = H_{oo} - H_{mm}, \qquad (4.18c)$$

and the completeness relation reads

$$|a_o(t)|^2 + \sum_m |b_m(t)|^2 = 1. \qquad (4.19)$$

Equations (4.18) allow approximate treatments. We shall discuss briefly two of them.

A. Multichannel VPS approximation

Introducing a new ("normalized") reaction amplitude

$$K_m(t) = \frac{b_m(t)}{a_o(t)}, \qquad (4.20)$$

Eqs (4.18) can be transformed into a system of non-linear Riccati equations[39]

$$\frac{dK_m}{dt} + i(\tilde{V}_m - \tilde{V}_m^* K_m^2) = Q_m(t) K_m(t) \qquad (4.21)$$

where

$$\tilde{V}_m = V_m \exp(i \int_{-\infty}^{t} \omega_m dt')$$

and

$$Q_m(t) = -i \sum_j K_j(t) \tilde{V}_j(t). \qquad (4.22)$$

From the expression for $Q_m(f)$ it is seen that the term on the right-hand-side of Eq.(4.22) describes capture transitions to state $|m>$ through all virtual ionic states $|j>$.

This second-order term may play a significant role only for exoergic non-resonant $|o> \to |m>$ transitions. Before going into further discussion of Eqs (4.21), we note that from Eqs (4.20) and (4.19) it follows

$$|a_o(t)|^2 = \frac{1}{1+\sum_j |K_j|^2}, \quad |b_m(t)|^2 = \frac{|K_m(t)|^2}{1+\sum_j |K_j|^2} \tag{4.23}$$

Thus, in the considered approach, the unitarity is preserved independent of the approximation to which Eqs (4.22) are solved. Neglecting the $|o> \to |j> \to |m>$ transitions ($Q_m=0$), Eqs (4.2) are reduced to an uncompled system of Riccati equations. The solution $K_m(t)$ at $t \to +\infty$ can be represented as[39]

$$|K_m(+\infty)|^2 = \frac{p_{om}}{1-p_{om}}, \tag{4.24}$$

where p_{om} is the exact two-state ($|o> \to |m>$) transition probability. In Ref. 39 the Vainstein-Presnyakov-Sobelman (VPS) approximation[40] has been adopted

$$p_{om} = \left| \int_{-\infty}^{\infty} dtV_m \exp[i\int_{-\infty}^{t} \omega_m^2+4V_m^2)^{1/2}dt'] \right|^2 \tag{4.25}$$

Using (4.24) in Eq. (4.23) for the electron capture probability one obtains

$$P_m^{cx} = |b_m(+\infty)|^2 = \frac{p_{om}\prod_{j\neq m}(1-p_{oj})}{\prod_j(1-p_{oj}) + \sum_m p_{om}\prod_{j\neq m}(1-p_{oj})} \tag{4.26}$$

The validity conditions for Eq. (4.26) are : $\ell\, |\prod_{j\neq m}(1-p_{oj})| \ll 4$, and $p_{om}^{1/2}\, \ell n|\prod_{j\neq m}(1-p_{oj})| \ll 2$. The partial cross sections σ_m, calculated using Eq. (4.26), have a proper adiabatic behaviour in the low-energy region and a Brinkman-Kramers (v^{-12}) behaviour at high velocities. Thus, although unitarization (important at lower velocities) is achieved, the two-state character of the VPS model remains.

B. Unitarized distorted wave (UDW) approximation

Using the same basis as in expansion (4.17), and neglecting the $|o> \to |j> \to |m>$ transitions, one can use the interaction picture of the S-matrix to achieve unitarization of $|o> \to |m>$ transition probability[41]. In the work of Ryufuku and Watanabe[41], the transition amplitude $b_m(t)$ is taken in the distorted wave (DW) approximation[42].

The UDW electron capture probability is obtained in the form

$$P_m^{UDW} = \frac{1}{p}\, |b_m^{DW}(+\infty)|^2 \sin^2\sqrt{p}\,, \quad p = \sum_m |b_m^{DW}(+\infty)|^2. \tag{4.27}$$

Desprite of performing the unitarization, the use of DW approximation for b_m introduces considerable inaccuracies in partial cross sections both at low ($v \ll v_o$) and ($v \gg v_o$) velocities. The situation in this respect is significantly improved when excitation and ionization channels are included in the treatment[43]. There have been numerous calculations for both partial and total cross sections for the H+Z systems ($Z \leqslant 24$) using the UDW approximation. These calculations, have revealed the following scaling relationship for the total cross section[43]

$$\sigma(E) = Z^{\alpha}\,\tilde{\sigma}(\tilde{E}), \qquad \tilde{E} = EZ^{-\beta}, \tag{4.28}$$

with $\alpha = 1.07$, $\beta = 0.35$. This scaling appears to be a quite successful for $\tilde{E} \gtrsim 0.1$ keV/amu ($Z \geqslant 14$). Another successful prediction of this model is the value of principal quantum number of dominantly populated final state : $n_m^{UDW} = Z^{0.768}$ (but not for too low and high values of E).

In Fig. 5 we show the total cross section for $Li^{3+} + H \rightarrow Li^{2+} + H^{+}$ reaction, calculated in the UDW approximation[43] (the dotted curve), and in the two-center atomic-orbital expansion model using 22 states[44] (the solid curve and 24 states (including united atom pseudostates)[45] (the dashed-line).

The calculations are compared with experimental data[46,47].

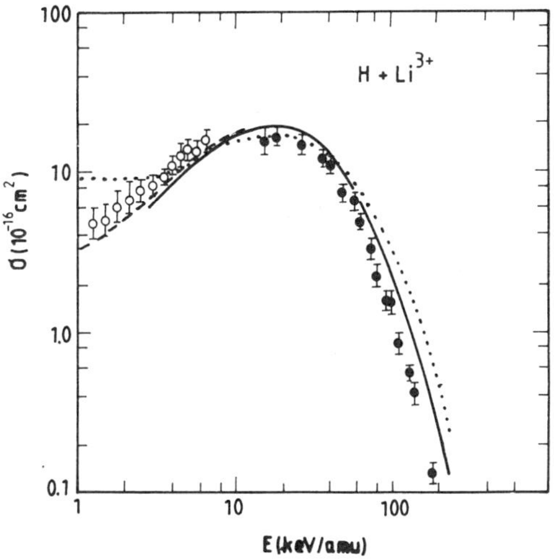

Fig. 5. Total cross sections for $Li^{3+} + H \rightarrow Li^{2+} + H^{+}$ charge exchange. Theoretical calculations : solid curve-22 two-center AO's[44], dashed curve-24 two-center AO's+UA pseudostates[45], dotted curve-UDWA[43]. Experimental data : open circles-Ref 46, closed circles-Ref 47.

IONIZATION IN ION-ATOM (ION) COLLISIONS

The ionization processes in ion-atom (ion) collisions are much less investigated than the electron capture process. This particularly holds for the intermediate - and low - energy regions, where the ionization mechanisms are numerous (diabatic state promotion, Auger-decay, adiabatic ionization). At high energies, the ionization is well described by the first Born approximation, being a correct limit to the Born perturbational series. However, in this region too, other non-potential coupling mechanisms, such as capture to the continuum and shake-off, are also possible for the process. Bellow we shall discuss mainly the perturbational and close-coupling treatments, with some remarks regarding the classical approach to this problem.

Perturbational treatments

A. Semiclassical dipole approximation

The first-order transition probability for ionization of the bound state $|\, o >$ into a continuum state $|\, k >$ is given by

$$a_{ok}(+\infty) = -i \int_{-\infty}^{\infty} dt < k|V_{int}|o > \exp(i\omega_{ok})t, \tag{5.1}$$

where V_{int} is the ion-atom interaction and ω_{ok} is the electron transition energy

$$\omega_{ok} = I_o + \frac{1}{2} K^2, \tag{5.2}$$

with I_o being the ionization potential of the state $|\, o >$.

For the $A+B^{Z+}$ system treated in the "one-electron approximation" (and B^{Z+} as structureless), the electron-ion interaction V_{int}

$$V_{int} = - \frac{Z}{|\vec{R}(t) - \vec{r}|} \tag{5.3}$$

in the dipole approximation has the form

$$V^d_{int} = - \frac{Z\, \vec{r}\cdot\vec{R}}{R^3} , \tag{5.4}$$

where $\vec{r}$ and $\vec{R}$ are the electron and B^{Z+} position vectors (with respect to A^+), respectively. In the straight-line trajectory approximation, $\vec{R} = \vec{b}+\vec{v}t$, the integral (5.1) can be calculated analytically and the ionization probability is

$$W_{ion} = |a_{ok}(+\infty)|^2 = \frac{4Z^2}{3V^2} \sum_k \omega^2_{ok} r^2_{ok} \left[K^2_o\left(\frac{\omega_{ok}b}{v}\right) + K^2_1\left(\frac{\omega_{ok}b}{v}\right)\right] \tag{5.5}$$

where $\vec{r}_{ok}$ is the dipole matrix element, $K_o(x)$ and $K_1(x)$ are the Macdonald functions, and the sum runs over the continuum states. Taking the asymptotic forms of Macdonald functions for large values of their argument, one obtains the following behaviour of W_{ion} for small impact velocities and/or large impact parameters

$$W_{ion} \sim \frac{Z^2}{v} \frac{\omega_{ok}}{b} \exp\left(- \frac{2\omega_{ok}b}{v}\right) . \tag{5.6}$$

Thus, for $(v/b) \ll (2\omega_{ok})^{-1}$, the ionization probability is exponentially small. For a fixed value of v, however, $W_{ion}(b)$ has a singular behaviour as $b \to 0$. The regularization procedure is not unique, but the common prescription is to take $W_{ion} = 1$ for $b \leq b_{min}$, where b_{min} is defined by $W_{ion}(b_{min}) = 1$. For H+Z systems $b_{min} \simeq 2Z/v$, which for high values of Z (or small v) is large, thus giving a large impact parameter region in which W_{ion} remains uncertain. Using this prescription for the unitarized probability, the normalized cross section takes the form

$$\sigma^N_{ion} = \pi b^2_{min} S + 2\pi \int_{b_{min}}^{\infty} W_{ion} b db, \tag{5.7}$$

where $S = \sum_k r_{ok}^2 / \langle r^2 \rangle$. For the case of hydrogen atom S = 0.283.

Another way of normalizing the ionization cross section is to introduce a cut-off in the dipole interaction potential which eliminates the singularity in Eq. (5.5), or to calculate W_{ion} in the region $b \leqslant b_{min}$ by using the sudden approximation. In all cases, the calculated cross sections have a fairly low degree of accuracy.

B. Bethe-Born approximation

The Born approximation for ionization and its Bethe version are well known from the standard text-books on atomic collisions physics (see e.g. Ref. 8) and will not be discussed here. We shall confine ourselves only to mention a recent semi-empirical formula due to Gillespie[48], which improves the Bethe-Born ionization cross sections in the energy region of cross section maximum and below. This formula gives also an adequate scaling of σ_{ion} with the ionic charge Z. The ionization cross section (excluding the region below ∿ 30 keV/amu) is represented as

$$\sigma_{ion} = \exp(-\lambda \frac{Z}{v^2})\, \sigma_{ion}^B (Z,v), \tag{5.8}$$

where $\sigma_{ion}^B(Z,v)$ is the Bethe-Born (or only Bethe) cross section for ionization of an atom (treated in one-electron approximation) by a structureless ion B^{Z+}, and λ is an empirical constant which depends on the atomic properties. For hydrogen λ = 0,76, while for helium $\lambda = 2$. For a given target atom, the scaling relationship (5.8) is valid practically for all ions (independent of charge state and chemical species).

Close-coupling models using pseudostates

In section 4.1B we have already discussed the possibility of obtaining the ionization probability from a two-center atomic expansion model containing also continuum pseudostates in the basis. The corresponding ionization probability (see Eq. (4.10)) consists of two terms, one containing transitions into the target continuum, and the other describes transitions into the projectile continuum (ECC transitions). A numerical application[35] of this method (for the H^++H ionization) has been mentioned in Sect. 4.1B. The simplest realization of the pseudostate expansion method in the ion-atom ionization problem has been proposed in Ref. 49, where besides the ground target state, only two target continuum pseudostates have been included in the basis. The special choice of these pseudostates has allowed to obtain an approximate analytic solution of the three-state close-coupling problem, and to express the ionization cross section for a class of collision partners in form of a scaling relation. The next natural step in pursuing this approach is to include the resonant target level in the basis (which has also been done in Ref. 49) and/or to include a pseudostate from the projectile continuum (not done as yet). Considering an s → p ionizing transition, one expands the electron wavefunction as

$$\Psi = a(t)\psi_s + b_o\bar{\psi}_{p,m=o} + b_1\bar{\psi}_{p,|m|=1}, \tag{5.9}$$

where the two $\bar{\psi}_p$ pseudo-states differ only by the value of magnetic quantum number. Within the dipole approximation (5.4) of the ion-atom interaction potential, the system of coupled equations for the amplitudes a, b_o and b_1 is

$$i\dot{a} = \frac{Z\lambda}{R^2}(b_o\cos\theta + b_1\sin\theta)\exp(-i\omega t), \tag{5.10a}$$

$$i\dot{b}_o = \frac{Z\lambda}{R^2} a \cos\theta \exp(i\omega t), \quad (5.10b)$$

$$i\dot{b}_1 = \frac{Z\lambda}{R^2} a \sin\theta \exp(i\omega t), \quad (5.10c)$$

where

$$\cos\theta = \frac{vt}{R}, \quad \sin\theta = \frac{b}{R}, \quad \lambda = (f_{eff}/2\omega)^{1/2}, \quad (5.11)$$

with ω being the ionization potential and f_{eff} is an effective continuum oscillator strentgh. (For the hydrogen atom f_{eff}=0.6525). Solving Eq. (5.10) by using the "K-matrix" method of section "A. Multichannel VPS approximation, Eq. 4.20" (with the intial conditions : $a(-\infty)=1$, $b_{o,1}(-\infty)=0$), for the ionization probability one obtains

$$P_{ion} = P_{ion}^{m=o} + P_{ion}^{|m|=1} = \frac{p_o + p_1}{1 + p_o + p_1}, \quad (5.12)$$

where

$$p_o = \left| \int_{-\infty}^{\infty} dt \left(\frac{Z\lambda vt}{R^3}\right) \sin\left\{ \int_0^t dt' \left[\omega^2 + 4\left(\frac{Z\lambda}{R^2}\right)^2\right]^{1/2} \right\} \right|^2 \quad (5.13a)$$

$$p_1 = \left| \int_{-\infty}^{\infty} dt \left(\frac{Z\lambda b}{R^3}\right) \cos\left\{ \int_0^t dt' \left[\left(\frac{\omega}{2}\right)^2 + \left(\frac{Z\lambda}{R^2}\right)^2\right]^{1/2} \right\} \right|^2 \quad (5.13b)$$

As seen from Eq. (5.12), P_{ion} is properly normalized.

In this three-state dipole close-coupling approximation, (DCC), the integration of P_{ion} over impact parameters gives

$$\sigma_{s-p}^{DCC} = 2\pi\left(\frac{Z\lambda}{V}\right) D(\beta), \quad (5.14)$$

$$\beta = \frac{Z\lambda\omega}{V^2}, \quad (5.15)$$

$$D(\beta) = \beta \int_0^{\infty} x dx \; \frac{p_o(\beta,x) + p_1(\beta,x)}{1 + p_o(\beta,x) + p_1(\beta,x)} \quad (5.16)$$

The parameter β comprises all the relevant physical parameters of the problem. The universal function $D(\beta)$ has well defined asymptotic properties[49], from where one obtains the cross section behaviour in the adiabatic and high energy regions,

$$\sigma_{s-p}^{ad} \simeq \pi\left(\frac{Z\lambda}{v}\right)^2 \exp\left[-\frac{1}{v}(2Z\lambda\omega)^{1/2}\right], \quad v \ll (Z\lambda v)^{1/2} \quad (5.17)$$

$$\sigma_{s-p}^{B} \simeq 8\pi\left(\frac{Z\lambda}{v}\right)^2 \ln\left(\frac{cV^2}{Z\lambda\omega}\right), \quad v \gg (Z\lambda V)^{1/2}, \quad c = 1.4 \quad (5.18)$$

The function $D(\beta)$ is tabulated in Ref. 49. It defines a "reduced" cross section

$$\tilde{\sigma}_{ion} \equiv \frac{\sigma_{ion}^{DCC}}{2\pi(Z\lambda/\omega)} = D(\beta) \qquad (5.19)$$

for any given target and any "structureless" projectile. For H+Z systems, the reduced ionization cross section σ/Z as function of reduced energy $\tilde{E} = E/Z$ is given in Fig. 6.

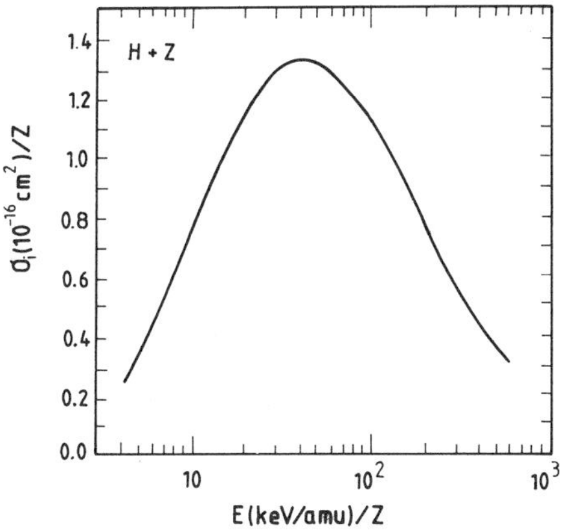

Fig. 6. Reduced ionization cross section vs. reduced energy for H+Z systems.

It should be noted that use of the dipole approximation leads to an overestimation of the DCC cross section in the adiabatic region. Neglect of ECC transitions in three-state DCC model introduces also an uncertainty in the region below the cross section maximum.

Other models for ionization

The unitarized distorted wave (UDW) approximation has also been applied to ionization in H+Z collisions[43], using the prescription (for an $0 \rightarrow k$ ionizing transition)

$$P_{ion}^{UDW} = \frac{1}{p_t} \left| b_{ok}^{DW}(+\infty) \right|^2 \sin^2\sqrt{p_t}, \qquad (5.20)$$

$$p_t = \sum_{j=n,m,k} \left| b_j^{DW}(+\infty) \right|^2 , \qquad (5.21)$$

where the sum includes all the excitation (j=n), charge exchange (j=m) amplitudes, and integration over the ionization (j=k) amplitudes, all calculated in DW approximation. In principle, $|k\rangle$ may belong either to the target or to the projectile continuum. In Ref. 43 the states $|k\rangle$ where taken only from the projectile continuum (ECC transitions), what introduces a considerable uncertainty in computed cross sections for low charged ions.

Another method which has been used for ionization cross section calculations is the classical trajectory Monte Carlo (CTMC) method[50].

This method is appropriate only in a rather limited region of collision energies around the cross section maximum and slightly beyond it. At higher energies, the CTMC method becomes identical with the classical impulse approximation and significantly underestimates the cross section. At the low energies the CTMC method also drastically underestimate the ionization cross section due to neglect of quantum (e.g. barrier penetration) effects.

Attempts have been made to apply higher-order perturbational methods to improve the description of ionization process in the intermediate energy region.

Use of the Glauber approximation for the H+Z systems[51] leads to underestimation of the cross section for energies at and below the cross section maximum. More sophisticated second-order methods, such as the continuum distorted wave approximation, have also proved to be inadequate for description of ionization at intermediate energies. Much theoretical work remains to be done in the field of ionization at these energies.

References

1. B.H. Bransden, Atomic Collision Theory, 2nd ed. (Benjamin, New York, 1982).
2. R.K. Janev, L.P. Presnyakov and V.P. Shevelko, Physics of Highly Charged Ions (Springer, Berlin-Heidelberg-New York, 1985)
3. J.B. Delos, Rev. Mod. Phys. 53, 287 (1981).
4. B.H. Bransden and R.K. Janev, Adv. At. Mol. Phys. 19, 1 (1983).
5. R. Mc Carroll, in : Atomic and Molecular Collision Theory, Ed. by F.A. Gianturco (Plenum, 1982) p. 165.
6. R.K. Janev, Adv. At. Mol. Phys. 12, 1 (1976).
7. E. Pritchard, A.L. Ford and J.F. Reading, Phys. Rev. A16, 1325 (1977).
8. N.F. Mott and H.S.W. Massey, The Theory of Atomic Collisions, 3rd ed. (Oxford University Press, 1965).
9. D.R. Bates and R. Mc Carroll, Proc. Roy. Soc. London, A245, 754 (1958).
10. M. Barat, This volume.
11. M.E. Riley and T.A. Green, Phys. Rev. A4, 619 (1971).
12. T.A. Green, Phys. Rev. A23, 532 (1981).
13. S.B. Schneiderman and A. Russek, Phys. Rev. 181, 311 (1969).
14. E.E. Nikitin and S.Y. Umanskii, Theory of Slow Atomic Collisions, (Springer-Verlag, Berlin-Heidelberg-New York, 1984).
15. R.K. Janev and L.P. Presnyakov, Phys. Rep. 70, 1 (1981).
16. V.A. Abramov, F.F. Baryshinikov and V.L. Lisitsa, Sov. Phys. JETP 47, 469 (1979).
17. R.K. Janev, D.S. Belic and B.H. Bransden, Phys. Rev. A 28, 1293 (1983).
18. F.L. Duman and L.I. Men'shikov, Sov. Phys. JETP 50, 433 (1979).
19. L.P. Presnyakov, D.B. Uskov and R.K. Janev, Sov. Phys. JETP 525 (1982).
20. M.I. Chibisov, JETP Lett. 24, 46 (1976).
21. T.P. Grozdanov and R.K. Janev, Phys. Rev. A 17, 880 (1978).
22. L.A. Vainstein, L.P. Presnyakov and I.I. Sobelman, Sov. Phys. JETP, 18, 1983 (1964).
23. T.A. Green, E.J. Shipsey, I.C. Browne, Phys. Rev. A 25, 1364 (1982).
24. R. Phaneuf et al. Phys. Rev. A 26, 1892 (1982).
25. R.K. Janev and P. Krstic, J. Phys. B (in press).
26. V.A. Bazylev, N.K. Zhevago, M.I. Chibisov, Sov. Phys. JETP 42, 436 (1975).
27. A.Z. Devdariani, V.N. Ostrovskii and Yu N. Sebyakin, Sov. Phys. JETP 44, 477 (1976).
28. Kishinevskii and E.S. Parilis, Sov. Phys. JETP 28, 1020 (1969).
29. R.K. Janev, V.Yu. Lazur and T.P. Grozdanov, J. Phys. B (in press).

30. W. Fritsch and C.D. Lin, J. Phys. B 15, 1255 (1982); Phys. Rev. A 26, 762 (1982).
31. W. Fritsch and C.D. Lin, Phys. Rev. A 29, 3039 (1984).
32. D.J.M. Anderson, M.J. Antal and M.B. McElroy, J. Phys. B 7, 918 (1974); ibid, 8, 1513 (1974); 14, 1707 (E) (1981).
33. E.J. Shipsey, T.A. Green and J.C. Browne, Phys. Rev. A 27, 821 (1983).
34. D. Ciric, D. Dijkkamp and F.J. de Heer, J. Phys. B 18 (1985).
35. R. Shakeshaft, Phys. Rev. A 18, 1930 (1978).
36. I.M. Cheshire, D.F. Gallaher and A.J. Taylor, J. Phys. B 3, 813 (1970).
37. D.F. Gallaher and L. Wilets, Phys. Rev. 169, 139 (1968).
38. R. Shakeshaft, Phys. Rev. A 14, 1626 (1976).
39. L.P. Presnyakov and A.D. Ulantsev, Sov. J. Quantum Electron 4, 1320 (1975).
40. L.A. Vainshtein, L.P. Presnyakov and I.I. Sobelman, Sov. Phys. JETP 18, 1383 (1964).
41. H. Ryufuku and T. Watanabe, Phys. Rev. A 18, 2505 (1978).
42. D.R. Bates, Proc. Roy. Soc. A 247, 294 (1958).
43. H. Ryufuku, Phys. Rev. A 25, 720 (1982).
44. B.H. Bransden and C.J. Noble, J. Phys. B 15, 451 (1982).
45. W. Fritsch and C.D. Lin, J. Phys. B 15, L281 (1982).
46. W. Seim, A. Muller et al., J. Phys. B 14, 3475 (1981).
47. M.B. Shah, T.V. Goffe and H.B. Gilbody, J. Phys. B 11, L233 (1978).
48. G. Gillespie, J. Phys. B 15, L729 (1982).
49. R.K. Janev and L.P. Presnyakov, J. Phys. B 13, 4233 (1980).
50. R.E. Olson and A. Salop, Phys. Rev. A 16, 531 (1977).
51. J.H. McGuire, Phys. Rev. A 26, 143 (1982).

ELECTRON CAPTURE IN ION ATOM AND ION-ION COLLISIONS

M. Barat

Laboratoire des Collisions Atomiques et Moléculaires *

Université Paris-Sud, Bat. 351, 91405 Orsay Cedex, France

I - INTRODUCTION

Electron capture (EC) by positive ions in collision with various targets has remained one of the most important subjects of research since the early 30's. From a theoretical point of view, EC is obviously a coupled 3-body problem : at least two cores and an *active* electron that jumps between them. Practical interest in EC arose in a variety of fields. In aeronomy and astrophysics electron capture processes play a crucial role in the equilibrium of the various ionic and atomic species. More recently a renewed interest arose from the physics of thermonuclear fusion, where capture by highly charged ionic impurities were found to be an important process in tokamak devices. For that reasons, a number of reviews were devoted to this subject during the past years, including lectures given in various NATO advanced science institutes. The aim of this lecture is not at all to add a new review to this list, but (i) to sumarize the very basis of the present theoretical approaches at low and moderate collision energy, (ii) to pinpoint some crucial difficulties in the theoretical treatment, (iii) to select specific examples which, to the taste of the author, reflect some present practical interest, or some significant advances.

I.1 - Three simple ideas

Fig. 1 shows the behaviour of the cross section for various ions colliding with hydrogen atoms in a large range of collision velocities.

1st idea : translational energy and collision time

For most ions the EC cross section exhibits a flat maximum for collision velocities $V < v$ the electron velocity, and then drops drastically at larger V. This feature is related to the increasing difficulty for the electron to jump to a *fast* moving ion.

* Associated with CNRS (UA 281).

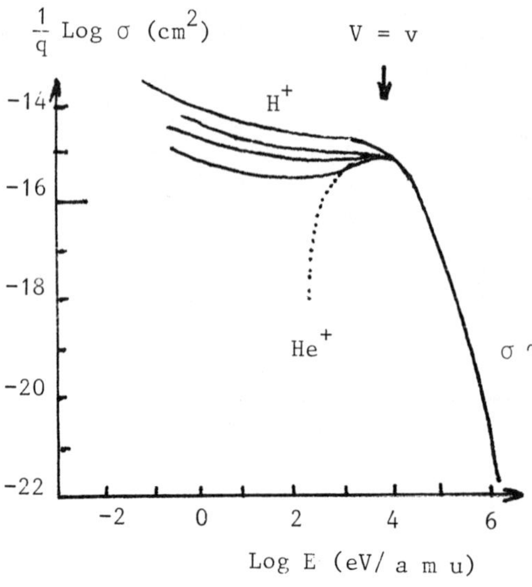

Fig. 1 - Scaled cross-section for EC by various ions on H target.

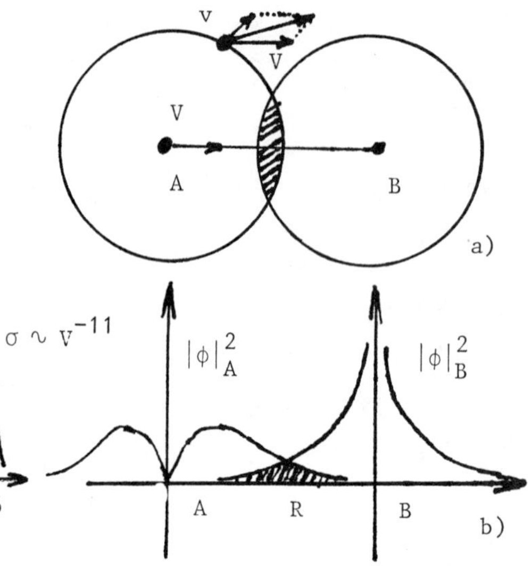

Fig. 2

2nd idea : electron velocity matching

EC will be favoured if the electron velocity is about the same when it is bound either to the initial or to the final core. Or, in other words, *resonant* or quasi resonant processes will display larger cross-sections specially at low velocity. But this condition is not sufficient. Lets take for example a K-K capture in process as

$$Ar^{18+} + Ar \rightarrow Ar^{17+}(1s) + Ar^{*}(1s^{-1})$$

Although the reaction is quasi-resonant, the cross-section would be small ($\sim 10^{-22}$ cm²). This case introduces a 3rd idea : space matching of the wave function (fig. 2).

Actually, the above EC cross section is small because the 1s electron of Ar is "concentrated" very close to the nucleus and therefore the *geometrical* cross section is small. The EC cross section will be closely related to the overlap of the wave functions (fig. 2b) describing the electron bound to A or B.

I.2 - Theoretical approaches relevant to various collision velocity domains

A) $v \gg V$ *Quasi-molecular region*

Following idea (1), the type of theoretical treatment of EC will depend primarily on the relative value of the electron velocity v and the collision velocity V. At low collision energy, when v is much larger than V, the electron motion adapts nearly adiabatically to the nuclear motion as in a molecule thereby suggesting a theoretical treatment in terms of the quasi-molecular electronic states formed by the two colliding partners during the collision. The cross-section can be very large (10^{-14} cm²) provided some quasi-resonant condition be fulfilled. Usually only *few* molecular states are concerned simplifying the theoretical treatment. This approach will be developed in section II.2.

B) *$v \simeq V$ Intermediate region*

As usual, intermediate regimes are the most difficult to deal with. For EC, neither a molecular treatment, nor an atomic treatment is really appropriate. However, when carefully used both approaches may sometimes converge towards identical results (see section III.2.3). This usually requires consideration of *many* states. Description of the *atomic* expansion method will be outlined in section II.1.

C) *$v << V$ High energy region*

As already mentioned, the cross-sections drastically drop for asymptotically large collision velocities. This allows the use of *perturbative* theories. During the last decade , a wealth of theoretical treatments based on improvements of the "Born approximation" has emerged (see e.g. / 1 /). It should also be pointed out that in this domain, collision processes are dominated by excitation and ionisation. In this context, an interesting aspect is the discovery of a new process, the capture into the continuum where the active electron is captured into an unbound orbital of the fast moving ion.

An interesting feature in this collision region, is provided by the so called "Thomas peak". In 1929 Thomas / 2 / pointed out that for large V the EC process could be described in terms of two successive classical binary encounters (fig. 3). In a first step, the electron assumed at rest with respect to the energetic projectile A elastically scatters towards the target nucleus B. In the second step the electron bounces off B with a final velocity equal and parallel to that of A. The α and θ angles are easily determined by momentum and energy conservation :

$$\alpha = \cos^{-1} \frac{1}{2} \sqrt{1 + \frac{m}{M}} \simeq 60 \text{ deg}$$

$$\theta \simeq \frac{m}{M} \frac{\sqrt{3}}{2}$$

(m and M are the electron mass resp, A nucleus mass).

It is amazing to point out that the experimental confirmation of the Thomas prediction awaited more than 50 years. The corresponding experiment has been performed by Hordsdal-Pedersen *et al.* / 3 / who measured the differential cross-section for EC in H^+ + He collisions at few MeV energies and very small scattering angles. The Thomas peak clearly shows up at $\theta \simeq 0.5$ mrad (fig. 4).

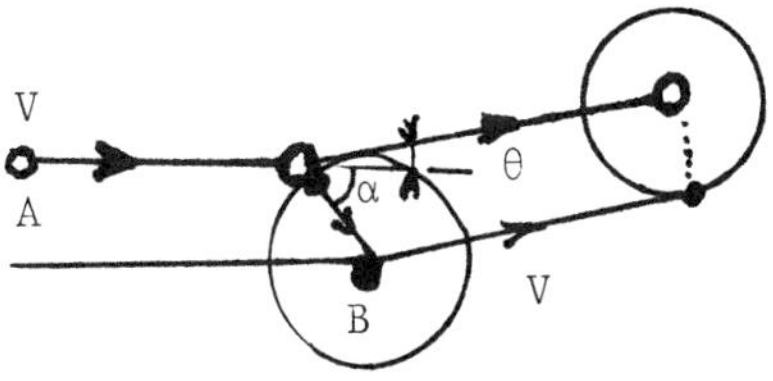

Fig. 3 - Double scattering mechanism.

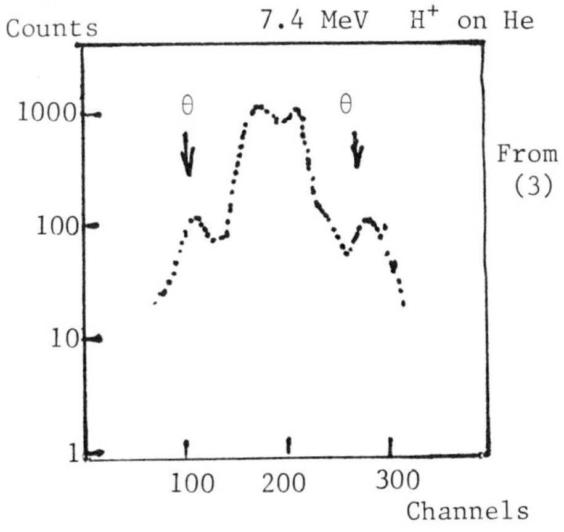

Fig. 4

I.3 - Two simple models

The body of this lecture shall be devoted to the quantum treatment of EC processes in the case of atomic and molecular approaches. However, as a preamble, it is worth describing two simple models which are quite useful in the understanding of the EC mechanisms and in the derivation of instructive scaling laws.

A) *The classical Bohr-Lindhardt model*

Bohr and Lindhardt / 4 / developed a classical model which is well adapted for medium and high collision energies. The electron initially attached to nucleus B is captured by A if *two* conditions are fulfilled :

(i) Release condition

EC might occur at a distance R (fig. 5) such that the force exerted on the active electron by A would at least compensate the force which binds it to B

$$\frac{q}{2} = \frac{v^2}{a} \rightarrow R < R_1 \quad \frac{\sqrt{qa}}{v} \quad \text{(in a.u.)}$$

where :

q = charge of the incident ion A
v = electron orbital velocity
a = Bohr radius of the electron bound to B.

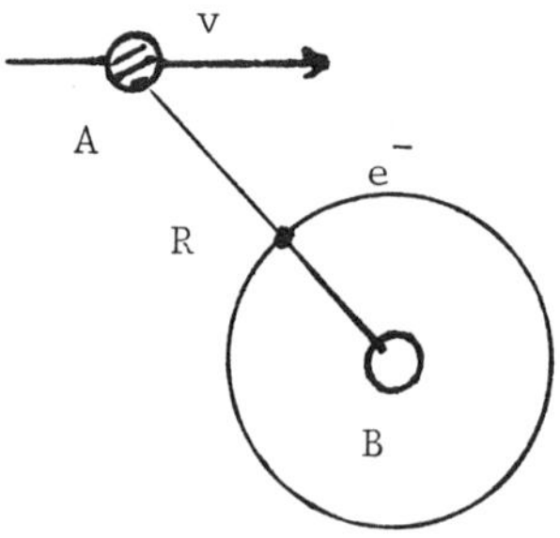

Fig. 5

(ii) Capture condition

The potential energy of the electron in A is larger than its kinetic energy with respect to A (traveling energy)

$$\frac{1}{2} V^2 < \frac{q}{R} \, , \; R < R_2 = \frac{2q}{V^2}$$

Two cases happen :

- if $R_1 < R_2$, capture always occurs for $R < R_1$, then the cross-section is determined by :

$$\sigma_1 = \pi R_1^2 = \frac{\pi q a}{v^2}$$

- if $R_2 < R_1$, condition (ii) determines the cross-section. However, for $R_2 < R < R_1$, the released electron is lost. So, the geometrical cross-section $\sigma_2 = \pi R_2^2$ should be weighted by the probability of loosing the electron before capture. This loss corresponds to ionisation process. The loss rate is $\simeq$ v/a and the collision time R_2/V then

$$\sigma_2 = \pi R_2^2 \left(\frac{v}{a} \frac{R_2}{V}\right)$$

$$\sigma_2 = \frac{8\pi}{a} q^3 \, v \, V^{-7}$$

Finally, the Bohr model of the atom provides a relationship between a and v

$$a = \frac{Z^{1/3}}{v}$$

where Z is the charge of nucleus B, this leads to :

$$\sigma_1 = \pi\, q\, Z^{1/2}\, v^{-3} \qquad \sigma_2 = 8\,\pi\, q^3\, Z^{-1/3}\, v^2\, V^{-7}$$

It can be readily seen that this simple model predicts a constant cross-section for moderate energy (σ_1) and a rapid fall off at high energy (σ_2). This model was developed further by Knudsen *et al.* / 5 /.

B) *The over Barrier model (obm)*

Lets consider the single electron capture,

$$A^{q+} + B \rightarrow A^{(q-1)+} + B^+$$

From a static point of view, for a given internuclear distance R, the active electron moves in the potential shown on fig. 7, that is a superposition of the potential well associated with the cores A and B

$$U(r) = \frac{-1}{r} - \frac{q}{|R-r|} \quad \text{(along the internuclear axis)}$$

EC by A will be favored if two conditions are fulfilled.

(i) A quasi-resonance between the two perturbed levels $I_B^* \simeq I_A^*$ where

$$I_B^* = I_B + \frac{q}{R} \quad , \quad I_A^* = I_A + \frac{1}{R}$$

I_A and I_B being the ionization potential of A and B respectively. This equality determines the favorable distance

$$R_C = \frac{q-1}{I_A - I_B}$$

(ii) The two levels should lie close to the top of the barrier $U_M = -U(r_M)$

$$\left.\frac{dU}{dr}\right|_{r=r_M} = 0 \quad \text{leads} \quad r_M = \frac{R}{1+q^{1/2}} \; , \; U_M = \frac{(q^{1/2}+1)^2}{R}$$

with the condition $U_M = I_{B*}$ one gets

$$R_{obm} = \frac{1 + 2\,q^{1/2}}{I_B}$$

Capture will be favored for energy levels n for which $R_{obm} \simeq R_c$. For hydrogenic ions one gets :

$$n = q\left[\frac{2\,q^{1/2} + 1}{2\,|I_B|\,(2q^{1/2} + q)}\right]^{1/2}$$

for large q, $n \simeq q^{3/4}\, |I_B|^{-1/2}$.

This model has been recently extended to multiple capture processes / 6 /. In these cases, the model implies that the electrons are captured, successively (see discussion in section III.2). Once the first electron is captured, the second one actually moves in a potential given by :

$$U\ (r) = \frac{-2}{r} - \frac{q-1}{|R-r|}$$

Using similar arguments one gets easily :

$$R'_{obm} = \frac{2\ |1 + 2\sqrt{2\ (q-1)}|}{I'_B}$$

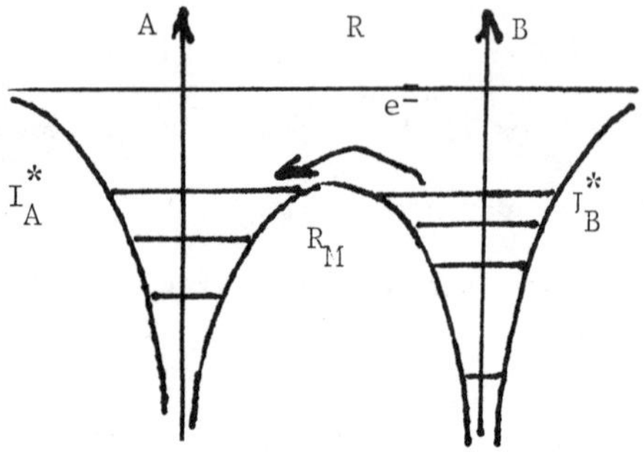

Fig. 6

I'_B is the ionization potential for the second electron. From the calculation of the mostly populated n value, it is possible to determine the most probable variation of the internal energy, or the energy gained by the projectile for one and two electron capture processes respectively

$$Q_1 = \frac{I_B(q-1)}{1+2q^{1/2}} \quad Q_2 = Q_1 + \frac{I'_B\ (q-3)}{2+2[2(q-1)]^{1/2}}$$

As seen from fig. 7, energy gain spectra obtained for Ar^{q+} + B collision compare nicely with these predicted values.

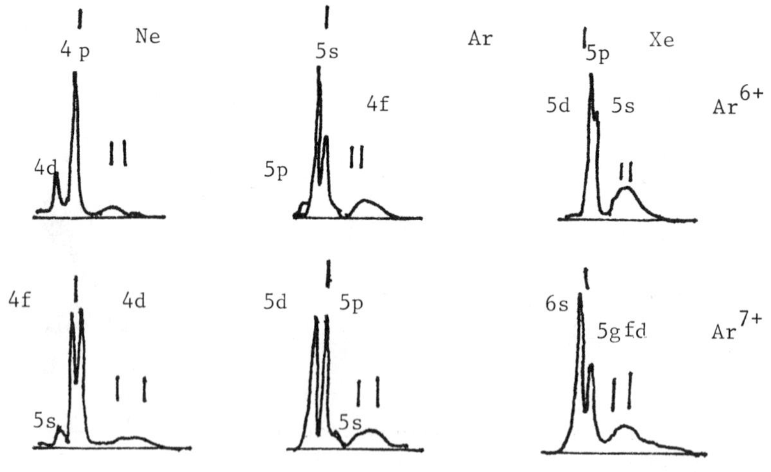

Fig. 7

Energy gain spectra for electron capture by Ar^{9+} on rare gas targets at 1keV - The markers show the values predicted by the o b m model for 1,2 and 3 capture. From /6/

II - BASIC FORMALISM ATOMIC AND MOLECULAR EXPANSIONS

(time dependent treatment)

At very low energy (say thermal energy) a full quantum treatment of both the nuclear and electronic motions may be necessary. However, for most applications, the collision energy is high enough (> 10 eV/amu) so that the De Broglie wave length associated with the nuclear motion is much smaller than the size of the atom. In such a case the nuclear motion can be treated semi-classically in terms of *trajectories*. The simplest trajectory : a straight line is suitable to describe small angle scattering. This trajectory characterizes the so-called Impact Parameter Method (IPM). However, for ion-ion collisions a coulomb trajectory is more appropriate. More generally current calculations aiming at the determination of differential cross-sections make use of an average common trajectory / 7 /.

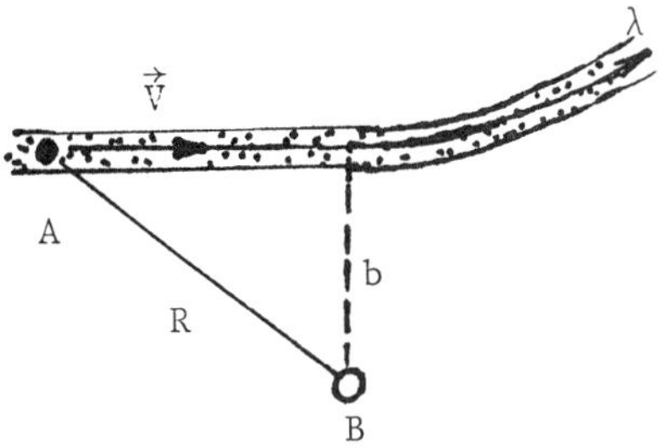

Fig. 8

On the other hand the motion of the electrons is treated quantum mechanically. So, in the most general formulation we have to solve the Schrödinger equation for a fixed value of the internuclear distance R. If $r(r_1, r_2, ..., r_n)$ describe collectively the motion of the electron with respect to the CM.

The time-dependent Schrödinger equation writes as :

$$(1) \qquad H_{el}\ \psi(r,t) = \left[\sum_i \left(-\frac{1}{2}\nabla^2_{r_i} - \frac{Z_A}{r_{Ai}} - \frac{Z_B}{r_{Bi}}\right) + \sum_{j<i} \frac{1}{r_{ij}}\right]\psi = i\,\frac{\partial\psi}{\partial t}$$

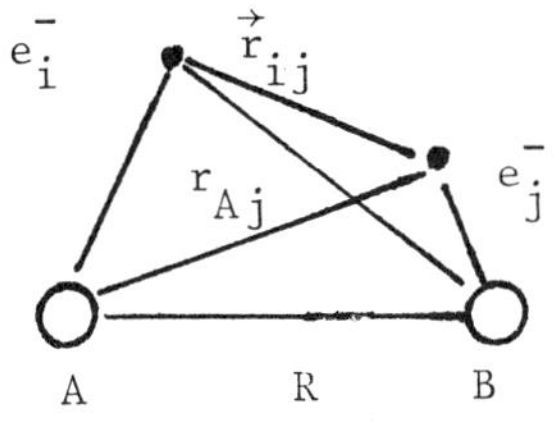

Fig. 9

Z_A charge of nucleus A

Z_B charge of nucleus B

Then ψ is expanded over a *suitable* basis set of function $\phi_i(r_i,t)$ as :

$$(2) \qquad \psi = \sum_i a_i(t)\ \phi_i(r_i,t)$$

Obviously most of the physics will be put in the choice of the $|\phi_i\rangle$ *basis set.*

Replacing ψ by the expansion (2) in (1), multiplying by ϕ_j^* and integrating over the electronic coordinates, one gets the set of coupled equations :

$$(3) \qquad \Sigma\, a_i \langle\phi_i|H_{el}|\,\phi_j\rangle = i\Sigma\, a_i\langle\phi_i|\frac{\partial}{\partial t}|\phi_j\rangle + i\,\Sigma\,\dot{a}_i\,\langle\phi_i|\phi_j\rangle$$

which can be written in the matrix form :

(4) $$i\mathbb{S}\dot{A} = \mathbb{M}A$$

where :

$\mathbb{S}$ is the overlap matrix $S_{ij} = \langle i|j\rangle$

$\mathbb{M}$ is the potential and coupling matrix :

$$M_{ij} = \langle i|H_{el} - i\frac{\partial}{\partial t}|\ j\rangle$$

A and $\dot{A}$ are the column matrix a_j and $\frac{da_j}{dt}$. This coefficient vector provides the probability to find the system in a given state j :

(5) $$P_j(E,b) = |a_j(+\infty)|^2$$

integrating over the impact parameters, one obtains the total cross-section

(6) $$\sigma_j(E) = 2\pi \int P_j(b)bdb$$

II.1 - Atomic expansion

A first obvious choice of the $|\phi\rangle$' s is provided by an *atomic basis*. We have then to face a number of problems that I will try to discuss. Let's start with the simplest EC problem in the $H_A^+ + H_B(1s) \rightarrow H_B(1s) + H_A^+$ one electron system. Because it is a basic example that often serves as a reference the corresponding mathematics will be presented in a pedestrian way.

- *Resonant EC*

For simplicity, let's consider a two state problem

$$|\phi_i\rangle = |1s_A\rangle \quad |\phi_j\rangle = |1s_B\rangle ,$$

then eq. (2) reduces to :

(7) $$\psi = a\ |1s_1\rangle + b\ |1s_B\rangle \equiv a|A\rangle + b|B\rangle$$

H_{el} writes now as :

(8) $$H_{el} = -\frac{1}{2}\nabla_r^2 - \frac{1}{r_A} - \frac{1}{r_B} = h_A - \frac{1}{r_B} = h_B - \frac{1}{r_A}$$

where h_A and h_B are the hamiltonians of atoms A and B respectively :

(9) $$h_A|A\rangle = \varepsilon_A|A\rangle \quad , \quad h_B|B\rangle = \varepsilon_B|B\rangle$$

the coupled equations (4) write as :

(10) $$i\left[\dot{a} < A|A > + \dot{b} < A|B >\right] = a < A|M|A > + b < A|M|B >$$

$$i\left[\dot{b} < B|B > + \dot{a} < B|A >\right] = b < B|M|B > + a < B|M|A >$$

The wave functions $|A\rangle$ and $|B\rangle$ are assumed to be normalized ($\langle A|A\rangle = \langle B|B\rangle = 1$). However $|A\rangle$ and $|B\rangle$ are not necessarily orthogonal. Let $S = \langle A|B\rangle$ be the corresponding overlap. Eq. (10) writes in the matrix form :

$$(11) \qquad i\,\mathbb{S}\begin{pmatrix}\dot a\\ \dot b\end{pmatrix} = \mathbb{M}\begin{pmatrix}a\\ b\end{pmatrix} \qquad \text{where} \quad \mathbb{S} = \begin{pmatrix}1 & S\\ S & 1\end{pmatrix}$$

Multiplying by the inverse matrix S^{-1} we get :

$$i\begin{pmatrix}\dot a\\ \dot b\end{pmatrix} = \mathbb{S}^{-1}\mathbb{M}\begin{pmatrix}a\\ b\end{pmatrix} \qquad \text{with} \quad \mathbb{S}^{-1} = \frac{1}{1-S^2}\begin{pmatrix}1 & -S\\ -S & 1\end{pmatrix}$$

that is :

$$(12) \qquad i\begin{pmatrix}\dot a\\ \dot b\end{pmatrix} = \frac{1}{1-S^2}\begin{pmatrix}1 & -S\\ -S & 1\end{pmatrix}\cdot\begin{pmatrix}M_{AA} & M_{AB}\\ M_{BA} & M_{BB}\end{pmatrix}\cdot\begin{pmatrix}a\\ b\end{pmatrix}$$

- *Expression of M_{ij}* :

$$M_{ij} = \langle i|H_{el}|j\rangle - i\,\langle i|\frac{\partial}{\partial t}|j\rangle$$

let us first consider the second term :

$$(13) \qquad \frac{\partial}{\partial t}\Big|_A |A\rangle = 0 \quad \text{then} \quad \langle i|\frac{\partial}{\partial t}\Big|_A|A\rangle \equiv 0$$

and similarly for B : $\frac{\partial}{\partial t}\Big|_B |B\rangle = 0$.

Nota : here we boldly jump over a *subtle* problem since we use *two different coordinate origins* to perform the derivative of $|A\rangle$ and $|B\rangle$ which is of course incorrect ! we will see later on how this difficulty should be solved. Then, M_{ij} reduce to $M_{ij} = \langle i|H_{el}|j\rangle$ with

$$(14) \qquad M_{AA} = \langle A|h_A - \frac{1}{r_B}|A\rangle = \varepsilon_{1s} - \langle A|\frac{1}{r_B}|A\rangle = \varepsilon_{1s} - h_{AA}$$

$$M_{AB} = \langle A|h_A - \frac{1}{r_B}|B\rangle = \varepsilon_{1s}\,S - \langle A|\frac{1}{r_B}|B\rangle = \varepsilon_{1s}\,S - h_{AB}$$

Similarly for M_{BB} and M_{BA}. Let's call $\varepsilon = \varepsilon_{1s} - h_{AA} = \varepsilon_{1s} - h_{BB}$ eq. (12) becomes :

$$(15) \qquad i\begin{pmatrix}\dot a\\ \dot b\end{pmatrix} = \frac{1}{1-S^2}\begin{pmatrix}1 & -S\\ -S & 1\end{pmatrix}\cdot\begin{pmatrix}\varepsilon & h\\ h & \varepsilon\end{pmatrix}\cdot\begin{pmatrix}a\\ b\end{pmatrix}$$

developing the matrix product :

$$(16) \qquad i\begin{pmatrix}\dot a\\ \dot b\end{pmatrix} = \frac{1}{1-S^2}\begin{pmatrix}\varepsilon - sh & h - s\varepsilon\\ h - s\varepsilon & \varepsilon - sh\end{pmatrix}\cdot\begin{pmatrix}a\\ b\end{pmatrix}$$

Neglecting the small S^2 terms (see below) one gets :

(17)
$$i\dot{a} = \varepsilon_A a + h' b$$
$$i\dot{b} = \varepsilon_B b + h' a$$

with $h' = h - S\varepsilon = Sh_{AA} - h_{AB}$, $\varepsilon_A = \varepsilon_B = \varepsilon - Sh$

This form can be made even simpler by introducing the following transformation :

(18)
$$c = a\, e^{-i \int^t \varepsilon_a dt'} \qquad d = b\, c^{-\int^t \varepsilon_b dt'}$$

e.g. $\frac{dc}{dt} = e^{-i \int \ldots} (\frac{da}{dt} - i\varepsilon_a a) \ldots$ which cancels the $\varepsilon_a a(\varepsilon_b b)$ term in eq. (17)

(19)
$$i\dot{c} = h'd$$
$$i\dot{d} = h'c$$

Adding and subtracting these two equations one easily obtains after integration :

(20)
$$c = A\, e^{-i \int_{-\infty}^t h'dt'} + B\, e^{i \int_{-\infty}^t h'dt'}$$
$$d = A\, e^{-i \int_{-\infty}^t h'dt'} - B\, e^{i \int_{-\infty}^t h'dt'}$$

A and B are determined by the initial conditions :

$$c\ (t = -\infty) = 1 \qquad d\ (t = -\infty) = 0$$

One then obtains the resonant charge exchange probability :

(21)
$$P(b) = |d(+\infty)|^2 = \sin^2 \int_{-\infty}^{+\infty} h(R)\, dt$$

In the IPM approximation :

(22)
$$R^2 = b^2 + v^2 t^2 , \qquad \frac{dt}{dR} = \frac{1}{v} \frac{r}{(R^2 - b^2)^{1/2}}$$

one gets :

(23)
$$P(b) = \sin^2 \frac{2}{v} \int_b^\infty \frac{h'(R)\, RdR}{(R^2 - b^2)^{1/2}}$$

where : $h' = Sh_{AA} - h_{AB}$

- *Evaluation of the exchange interaction h'*

One has to calculate : $h_{AB}(R) = \langle A| \frac{1}{r_A} |B\rangle$

The hydrogenic ground state wave function is :

(24) $$|A> = |B> = \frac{2}{\sqrt{4\pi}} e^{-r_{A,B}}$$

(25) $$h_{AB}(R) = \int \frac{4}{4\pi} e^{-r_A} \frac{1}{r_A} e^{-r_B} d^3 r_A$$

Let's introduce the prolate spheroidal coordinates

(26) $$\xi = \frac{r_A + r_B}{R} \qquad \eta = \frac{r_A - r_B}{R}$$

(27) $$h_{AB} = \frac{1}{\pi} e^{-(\xi+\eta)\frac{R}{2}} e^{-(\xi-\eta)\frac{R}{2}} \frac{2}{\xi + \eta} \dots$$

$$\dots \frac{R^3}{8} (\xi^2 - \eta^2) \, d\xi \, . \, d\eta \, . \, d\Phi$$

After integration we get :

$$h_{AB} = - (1 + R) e^{-R}$$

the evaluation of $S = <A\,B> = \frac{4}{4\pi} \int e^{-r_A} e^{-r_B} d^3 r_A$ and h_{AA} are obtained with a similar procedure :

$$S = e^{-R} (1 + R + \frac{R^2}{3})$$

$$h_{AA} = \frac{1}{R} - e^{-2R} (\frac{1}{R} + 1) \simeq \frac{1}{R} \text{ for large R}$$

finally

(28) $$h' = \frac{2}{3} R e^{-R}$$

- *Cross-sections*

Replacing eq. (28) in eq. (23) one gets :

(29) $$P(b) = \sin^2 \frac{1}{v} \int_b^\infty \frac{R^2 e^{-R} dR}{R^2 - b^2}$$

$$= \sin^2 \frac{b^2}{v} K_0(b) \simeq \sin^2 (\frac{2}{v} \frac{\pi b}{e} b.e^{-b})$$

where K_0 is the McDonald function. P(b) is represented on fig. 11.

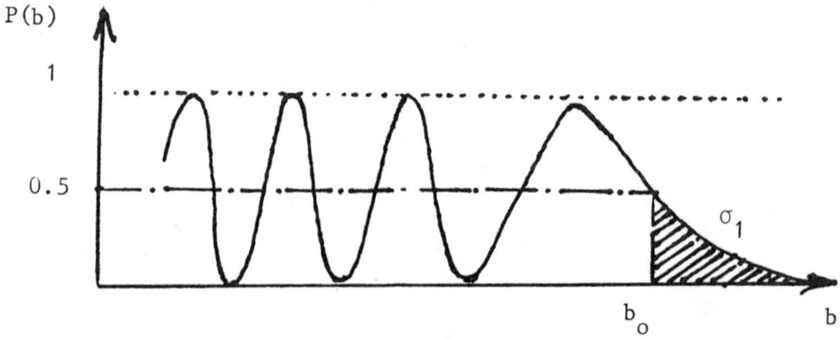

Fig. 11

the total cross-section can be calculated after splitting the integral into two parts (fig. 11) :

$$\sigma = \sigma_o + \sigma_1 = 2\pi \left(\int_o^{b_o} P(b)\, db + \int_{b_o}^{\infty} P(b)\, db\right)$$

in a first approximation σ_1 can be neglected : since P(b) strongly oscillates, one can average the integral $\int \sin^2 = 1/2$

(30) $$\sigma \simeq \frac{\pi\, b_o^2}{2} \text{ (geometrical cross section)}$$

Firsov showed that :

$$b_o \simeq \ln \frac{2}{0.28} \sqrt{\frac{\pi}{2}} - \ln v$$

with replaced in eq. (30) entails

$$\sigma = \frac{\pi}{2} \left(\ln \frac{2}{0.28} \sqrt{\frac{\pi}{2}} - \ln v\right)^2$$

- *GENERAL CASE : PLANE-WAVE ELECTRON TRANSLATIONAL FACTOR (ETF)*

Lets consider a more general case, where only one electron is active in the EC process and moves in the field of the A and B ion-cores. The electronic hamiltonian writes now as :

(31) $$H_{el} = \frac{-1}{2} \nabla^2 + V_A (r_A) + V_B (r_B)$$

where $V_{A,B}$ are respectively the interaction of the electron with the central potential of cores A and B. The derivation of P(b) and are handled in a similar way as already done for H^+H.

However, remember that we have jumped an important difficulty : namely, the time derivation in eq. (13). This a *key point* in the treatment of electron capture processes which will be shown to correspond to a basic physical aspect of the problem. Actually, we have to describe the electron cloud as moving with the nucleus to which it is bound. This is achieved by introducing some "traveling wave functions" $|\chi_A\rangle$ instead of the $|\phi_A\rangle$. We have first to choose a unique coordinate origin (for instance the midpoint of R which will be held fixed when differentiating the wave-functions. Then it can be shown that :

$$\left.\frac{\partial|\chi_A\rangle}{\partial t}\right|_C = \left.\frac{\partial|\chi_A\rangle}{\partial t}\right|_A + \vec{\nabla}\chi_A \cdot \left.\frac{d\vec{r}_A}{dt}\right|_C = \nabla\vec{\chi}_A \cdot \left.\frac{d\vec{r}_A}{dr}\right|_C$$

In order to keep the treatment as simple as presented above this term should be somehow cancelled by a proper choice of the function χ.

Bates and McCarroll / 8 / have solved this problem in the IPM approximation (V = const.) by multiplying the atomic wf by a phase factor :

$$\chi_A = \phi_A \, e^{-i\left(\varepsilon_A t + \frac{\vec{V}.\vec{r}}{2} + \frac{1}{8}V^2 t\right)} \tag{32}$$

The exponential term is called the plane-wave ETF. Similarly

$$\chi_B = \phi_B \, e^{-i\left(\varepsilon_B t - \frac{\vec{V}.\vec{r}}{2} + \frac{1}{8}V^2 t\right)}$$

($\varepsilon_{A,B}$ are the energies of the electroniques ϕ_A and ϕ_B states).

Nota : This is true only if V = const. (IPM app) as already mentioned. If more realistic trajectories are needed, acceleration terms show up requiring more sophisticated ETF / 9 /.

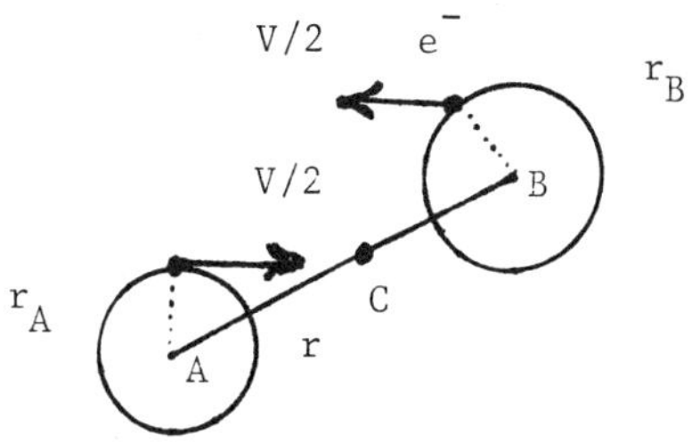

Fig. 12

Then, following the same derivation as in eqs. (15)-(19), we get finally the set of coupled equations :

$$ic = d\,U_{AB}\,e^{i\int_{-\infty}^{t}(U_A-U_B)\,dt} \tag{33}$$

$$id = c\,U_{BA}\,e^{-i\int_{-\infty}^{t}(U_A-U_B)\,dt}$$

where :

(potential terms)

$$U_A = \varepsilon_A + \frac{h_{AA} - S\,h_{BA}}{1 - S^2}\ , \quad U_B = \varepsilon_B + \frac{h_{BB} - S\,h_{AB}}{1 - S^2}$$

(interaction terms)

$$U_{AB} = \frac{h_{AB} - S\,h_{BB}}{1 - S^2}\ , \quad U_{BA} = \frac{h_{BA} - S\,h_{AA}}{1 - S^2}$$

the matrix elements h_{ij} are very similar to those discussed above. However, their calculation involve the velocity dependent ETF factors. Owing to be complexity of the related expressions the reader is referred to a paper by Lin / 10 /.

The potentials U_A, U_B correspond to some molecular energy curves of the (AB) system, but they are not the *adiabatic* potentials since the off-diagonal U_{AB} terms are not zero. They are *diabatic* potential curves : some sort of molecular curves better adapted to the collision problem. But we are anticipating the next section. This theoretical approach has been applied successfully to electron capture by *bare nuclei* colliding with hydrogen atoms (see III.2.3).

II.1.1 - Multi-electron problem

The presence of the electron-electron repulsion term in H_{el}, $\frac{1}{r_{ij}}$ in eq. (15), considerably complicates the treatment of multi-electron systems. In fact, no satisfactory treatment have been proposed to date although some tentative approaches has been suggested by a few authors / 11 /. In some case it is possible to reduce the *many* electron problem to the *one* electron case discussed above. For instance, two cases can easily fit in the previous framework :

(1) The active electron can easily be described using a one electron *model potential* or *pseudo potential* approximation. Collisions involving alkali-atoms and alkali-earth ions are obviously very good candidates for such a treatment.

(2) The electron correlation can also be often neglected in electron capture from *inner shells*. Using for instance a Hartree-Fock description of the atoms individual electrons are constrained to be in *atomic orbitals* (AO). Electron transfer between AO can be described within the one electron model. In these cases the χ_A, χ_B of the above treatment are traveling orbitals defined by eqs. (32) (33). Such treatments which have been extensively developped during the past years for K shell-K shell, K-L, ... electron transfer will not be discussed here (see e.g. the very recent review by U. Wille and R. Hippler / 12 /).

II.1.2 - Recent improvements

The previously discussed two-state atomic approach is well adapted when the capture process occurs at *large R* where the two atoms do not perturb each other too much. However, in many cases, electron capture takes place at small R values where the atomic wf (or AO) of the atoms *strongly overlap*. In that case the separated atom basis set is not well appropriated and may give very poor results.

At small R the AB system ressembles the *united atom* (UA) thereby suggesting to add AO of the UA to each center in order to better describe the small R behaviour. Such *pseudo-state* methods are also called the AO+ approximation / 13 / (for transitions taking place at very small R, it is also possible to use a *one* center expansion.

The natural extension of the AO+ method is the *three center expansion*. The 3rd center is placed at the center of charge of the two atoms and AO'S of both separate atoms (SA) and UA are placed on the 3rd center / 15 /. Although very appealing this method turns out to be very time consuming. A comparison with other methods will be discussed at the end of the next section. The extension of these pseudo-state method to *multi-electron* cases is extremely difficult and time consuming.

II.2 - Molecular expansion

The discussion of the last section, leads quite naturally to seek a *quasi-molecular* description of the collision. When the AO of both atoms overlap strongly one cannot say that the electron belong specifically to one of the two atoms and specially for low collision velocities ($V \ll v$) one can look for a description of the collision process in term of some *molecular state* $\phi_i\,(R,r) \equiv |i\rangle$.

Let's consider a complete set of orthonormalized "molecular" wf $\langle \phi_i | \phi_j \rangle = \delta_{ij}$. The collision wavefunction is expanded as

$$\psi = \Sigma\, a_i(t)\ \phi_i(R(t),r)\ e^{-i\int^t \varepsilon_i(R(t))\,dt'} \tag{34}$$

As in eqs. (17)-(18) the phase factor is introduced to remove the diagonal term in the time dependent close-coupling equations which write :

$$i\,\frac{da_j}{dt} = \sum_{i\neq j}\left[\varepsilon_{ij}(R) - i\,\langle j\,|\frac{\partial}{\partial t}|\,i\rangle\right]\ a_i\ e^{-i\int^t(\varepsilon_j - \varepsilon_i)\,dt'} \tag{35}$$

$|a_i(t)|^2$ represents the probability to find the system in the *molecular* state $|i\rangle$ at time t.

It is seen that in the general formulation (35) the interaction responsible for the electronic transition (electron capture process for instance) is composed of two terms

$\varepsilon_{ij}(R) = \langle i\,|H_{el}|\,j\rangle$ often called the *electrostatic coupling*

$\langle j\,|\frac{\partial}{\partial t}|\,i\rangle$ the *dynamical coupling*

(actually both couplings terms are dynamical !)

$\varepsilon_i(R) = \langle i\,|H_{el}|\,i\rangle$ is the molecular energy terms.

A - ADIABATIC REPRESENTATION

Up to this point some flexibily is still left in the choice of the function $\phi_i(R)$. Let's first consider a molecular basis which diagonalizes H_{el}. This is the so called *adiabatic basis* in the Born-Oppenheimer approximation in Molecular Physics.

$$\langle i\,|H_{el}|\,j\rangle = \delta_{ij}\ \varepsilon_i(R) \tag{36}$$

eq. (35) writes now :

$$\frac{da_j}{dt} = -\,\Sigma\,\langle i\,|\frac{\partial}{\partial t}|\,j\rangle\ a_i\ e^{-i\int^t \varepsilon_i(R) - \varepsilon_j(R)\,dt'} \tag{37}$$

The coupling terms $\langle i\,|\frac{\partial}{\partial t}|\,j\rangle$ can be rewritten using the "classical trajectory" relation R(t) (fig. 14).

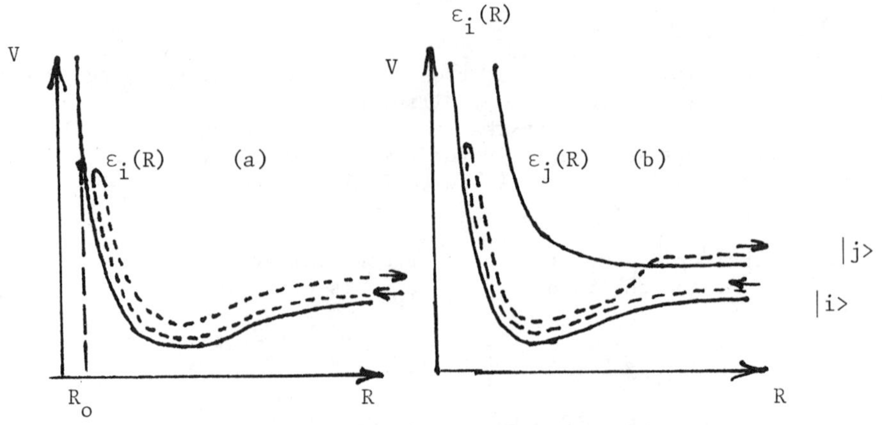

In a (semi-) classical description R_o is the distance of closest approach.

---- classical part describing the elastic collision.

Representation of a transition between channel |i> and |j> described respectively by potential $\varepsilon_i(R)$ and $\varepsilon_j(R)$.

Fig. 13

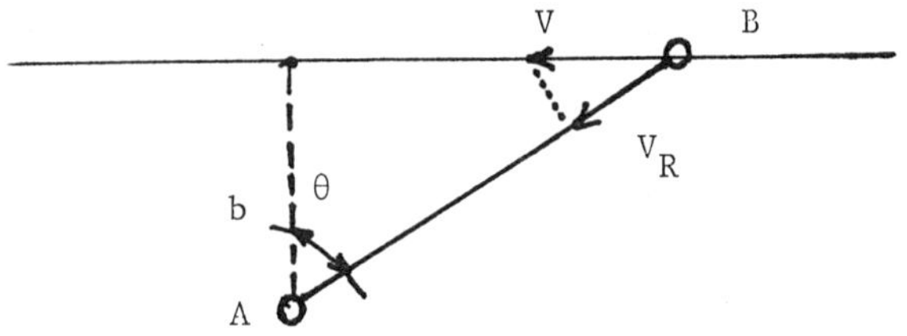

Fig. 14

(38) $$\frac{\partial}{\partial t} = \frac{dR}{dt} \cdot \frac{\partial}{\partial R} + \frac{d\theta}{dt} \cdot \frac{\partial}{\partial \theta} = v_R \frac{\partial}{\partial R} + \frac{V_o b}{R^2} \frac{\partial}{\partial \theta}$$

b is the impact parameter and V_o the incident collision velocity, V_R the "radial" velocity the coupling term is now written as the sum of two terms :

(39) $$i \left|\frac{\partial}{\partial t}\right| j> = v_R <i \left|\frac{\partial}{\partial R}\right| j> + \frac{V_o b}{R^2} <i \,|iLy|\, j> \qquad \text{with } \frac{\partial}{\partial \theta} = iLy$$

The first term $v_R <i \left|\frac{\partial}{\partial R}\right| j>$ called *radial coupling* couples molecular states having the same symmetry ($\Sigma \leftrightarrow \Sigma \quad \Pi \leftrightarrow \Pi \ldots$).

The second term called *rotational coupling* couples molecular states, obeying the selection rule $\Delta\Lambda = \pm 1$ (Λ being the projection of the angular momentum along the internuclear axis).

B - DIABATIC REPRESENTATION

As an alternative to the adiabatic representation, one can look for states $|\ i>$ which do *not* diagonalize H_{el} : $<\phi_i|H_{el}|\phi_j> \neq 0$ but minimize the dynamical coupling terms. Such states are called diabatic states. Diabatic states are usually defined with respect to radial coupling.

In fact, there exists no rigorous definition of diabatic states. However, diabatic states should follow the "prescription" :

$$(40) \qquad v_R <i\ |\frac{\partial}{\partial R}|\ j> \quad << \varepsilon_{ij}(R)$$

Remarks :

(i) The formal definition proposed by F.T. Smith / 16 / $<i|\frac{\partial}{\partial R}|\ j> = 0$ leads, if the $|\phi_i>$ basis set is complete, to $\frac{\partial}{\partial R}|\phi_i> = 0$, $\phi_i(R) = \text{const.}$ (e.g. $= \phi\ (R \to \infty)$. This is the case for a common *atomic* basis which then constitutes the crudest diabatic basis (§ II.1).

(ii) Through the condition (40), it is readily seen that the choice of diabatic states may depend on the collision velocity range.

For each category of collision problems the underlying physics should be the guide to research for the most appropriate diabatic representation. In this spirit, one can look for diabatic states within perturbation theory. ε_{ij} can be considered as a small perturbation, inducing the inelastic transitions. Quasi-molecular diabatic states are calculated by diagonalizing *all* interactions except those responsible for the transition. For example, in a one-electron capture process, all interactions (coulombic, polarization, etc...) should be diagonalized *except the exchange interaction*.

C - AN IMPORTANT EXAMPLE : THE MOLECULAR CURVE CROSSING

The Wigner-Von Neuman / 17 / non crossing rule tells us that two adiabatic potential curves of same symmetry do *not* cross (fig. 15). At the vicinity of an avoided crossing ($R \simeq R_c$), the wave functions suffer a sudden change of character, which is reflected in the strongly peaked behaviour of the $<|\partial/\partial R|>$ matrix element. In such a case, diabatic states are those which cross at R_c and merge with the adiabatic states outside the crossing. The *smooth* behaviour of these states results in a very weak $<|\partial/\partial R|>$ term. The coupling is given by $\varepsilon_{12}(R)$ which also displays a smooth behaviour (an interesting feature for practival calculations).

D - COUPLED EQUATIONS WITH DIABATIC STATES

The set of eq. (35) reduces to

$$(41) \qquad i\frac{da_j}{dt} = \sum_{i\neq j} a_i(t)\ \varepsilon_{ij}(R)\ e^{-i\int^t(\varepsilon_i - \varepsilon_j)\,dt'}$$

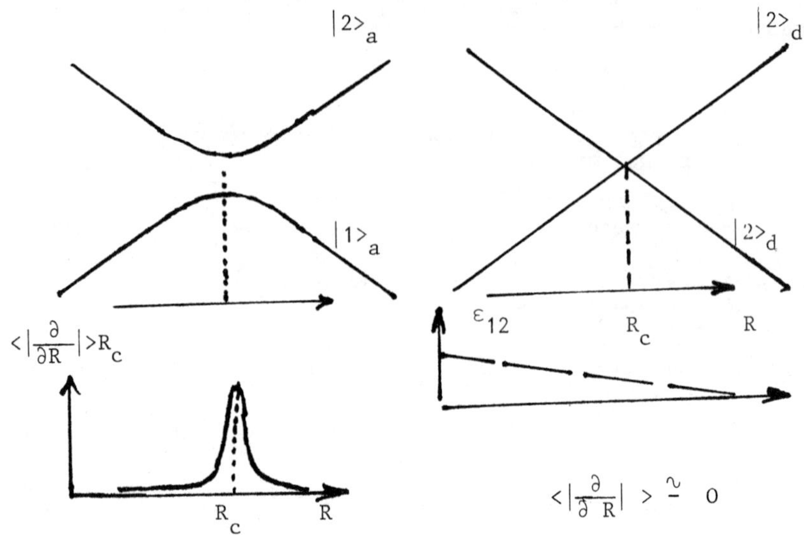

Fig. 15

Fortunately, in some cases of practical interest, the coupling between many states can be seen as a succession of couplings between two states. The problem is then reduced to solving a succession of two state close coupling equations :

$$(42) \qquad i\,\frac{da_1}{dt} = a_2\,\varepsilon_{12}(R)\,e^{i\int^t \Delta\varepsilon\,dt'}$$

$$i\,\frac{da_2}{dt} = a_1\,\varepsilon_{12}(R)\,e^{-i\int^t \Delta\varepsilon\,dt'}$$

Let's now compare the set of coupled equations (33) obtained with the atomic expansion with eqs. (42) obtained with the molecular expansion. The similarity is striking, justifying *a posteriori*, why U_A and U_B were named *diabatic potentials*.

E - MOLECULAR ETF

As in the case of the atomic expansion (§ I.1, eq. (13)) we have jumped over the problem of the coordinate origin for the evaluation of the coupling term $<|d/dt|>$ or $<|\nabla_R|>$. It can be shown that $<\nabla_R>$ terms considerably differ according to whether they are calculated with the origin of electronic coordinates on the projectile, the target, the center of mass or the center of charge. Removing these inconsistancies requires the introduction of molecular ETF. Unfortunately this problem is much more difficult than in the atomic expansion case. Even now, the problem is not completely solved. Since a decade, this problem has been attacked by many investigators and several practical "recipes" have been proposed. A detailed discussion of this problem is outside the scope of this lecture and only the basic ideas and some of these recipes will be mentionned. Although this problem has also been formulated in a fully quantal treatment of the collision / 18 / most of the works have been carried out within the semi-classical treatment. All attempts to solve this problem start with a "molecular" basis set of the form :

$$\psi = \Sum a_n(t)\, \phi_n(\vec{R},\vec{r}) \tag{43}$$

$$\phi_n(\vec{R},\vec{r}) = \chi_n(\vec{R},\vec{r})\, e^{i S_m(\vec{R},\vec{r})}\, e^{-i \int^t \varepsilon_n dt'}$$

with $\varepsilon_n = \langle\phi_n|H_{el}|\phi_n\rangle$.

The collision problem consists then of solving a set of coupled equations similar to eq. (35) with the new set of wave functions ϕ_n. The choice of the best $S_n(\vec{R},\vec{r})$ is, of course, the preoccupation of the theoretical approaches. In any case, $S_n(\vec{R},\vec{r})$ should asymptotically match the atomic ETF of Bates and McCarroll :

$$S_n(\vec{R},\vec{r}) \to \alpha_n \,.\, \vec{V}.\vec{r}$$

where the proportionality factor depends on the coordinate origin. One simple choice is to neglect the ETF and select the most appropriate coordinate origin (O). In particular Piacentini and Salin / 19 / showed that, if the origin is taken at the nucleus to which the active electron is initially attached (incident channel), the total (all channels included) electron capture cross-section is correctly reproduced but the individual charge exchange cross-sections cannot be estimated correctly. Alternatively if the origin is taken on the other nucleus, the total cross-section is incorrect but the branching between the various final channels is correctly reproduced. Guided by the asymptotic behaviour of the ETF, several authors look for ETF of the form

$$S'_n = f(\vec{R},\vec{r})\, \vec{V}.\vec{r} \tag{44}$$

with f being a "switching function" obeing the asymptotic condition

$$f \to \beta \qquad \text{for } r_A >> r_B$$

$$f \to -1 + \beta \qquad \text{for } r_B >> r_A$$

where β depends on the coordinate origin. For example, in their study of the $H^+ + H$ resonant charge exchange, Schneiderman and Russek / 20 / proposed the arbitrary form :

$$f = \frac{\cos\theta}{2} \Big/ \left(1 + \left(\frac{R}{a}\right)^2\right) \tag{45}$$

θ (see fig. 14), is defined with the origin at the center of mass of the nuclei and a is an arbitrary parameter : f is common to all wave functions and is seen to easily fulfill the asymptotic conditions

$$f \to \frac{1}{2} \qquad \text{for } r_A >> r_B$$

$$f \to -\frac{1}{2} \qquad \text{for } r_B >> r_A$$

Furthermore f vanishes when $R \to 0$ to account for the fact that, near the "united atom limit", the electrons belong to both atoms A and B. The parameter a is characteristic of the relevant united atom wavefunctions.

A more elaborate choice of the switching function has been proposed by Vaaben and Taulbjerg / 21 / in the case of one electron systems. The semi-classical scattering equations take a form similar to eq. (35) if f is chosen as :

$$f = \frac{1}{2} \left(\frac{Z_B - \alpha Z_A}{Z_B + \alpha Z_A} + \frac{Z_A - Z_B}{Z_A + Z_B} \right) \tag{46}$$

with $\alpha = \left(\frac{r_B}{r_A}\right)^3$.

Z_A and Z_B are the charges of nuclei A and B respectively.

An interesting feature of both approaches described above, is provided by the fact that a *common* ETF is used for all wavefunctions. This property automatically insures the orthogonality of the wavefunctions and therefore greatly simplifies the calculations : the coupled equations are those discussed in II.2 with additional dynamical coupling terms of the form :

$$< \phi_i \quad (\vec{v}.\vec{r}) \; (\vec{\nabla}_r f.\vec{\nabla}_r) \; + \; (\vec{v}.\vec{r}) \; (\frac{1}{2} \vec{\nabla}_r^2 f) + \vec{v}.\vec{\nabla}_r \; f \; |\phi_j > \tag{47}$$

More rigourous attemps / 22 / to define ETF specific to each wavefunction ϕ_i leads to much heavier calculation, since the orthogonality of the wave functions is lost.

F - INTERFERENCE EFFECTS

The oscillations, seen (fig. 11) in the charge exchange probability for $H^+ + H$, is actually a manifestation of some interference effects. This can easily be seen using the quasi-molecular approach.

Due to the potential symmetry of the H_2^+ ion, two molecular wave functions can be constructed which are respectively symmetric and antisymmetric with respect to the center of charge ("u" and "g" molecular states resp.). For a 1s electron, the corresponding σ_u and σ_g states asymptotically behave as :

$$\sigma_g \; (R \to \infty) \to (|1s_A> + |1s_B>) \; / \sqrt{2} \tag{48}$$

$$\sigma_u \; (R \to \infty) \to (|1s_A> - |1s_B>) \; / \sqrt{2}$$

In a two state approximation and a time-independent treatment:

$$\psi = F_u(R)|\sigma_u> + F_g(R)|\sigma_g>$$

with the asymptotic behaviour (j = g or u):

(49) $$F_j\ (R \to \infty) \sim \frac{1}{\sqrt{2}} \left[e^{ikz} + f_j(\theta)\ \frac{e^{ikR}}{R} \right]$$

$$\psi \to \frac{1}{\sqrt{2}}\ (\sigma_u + \sigma_g)\ e^{ikz} + \frac{e^{ikR}}{R} \left[f_u(\theta)\ \sigma_u + f_g(\theta)\ \sigma_g \right]$$

using eq. (48)

$$\psi \simeq \frac{1}{\sqrt{2}} \left\{ |1s_A\rangle e^{ikz} + \frac{1}{2}\ e^{ikr} \left[f_u(\theta) + f_g(\theta) \right] |1s_A\rangle + \frac{1}{2} e^{ikr} \left[f_u(\theta) - f_g(\theta) \right] |1s_B\rangle \right\}$$

↑ incident channel ↑ elastic scattering channel ↑ electron capture channel

$$\sigma_{el}(\theta) \propto |f_u + f_g|^2 \quad \sigma_{ec}(\theta) \propto |f_u - f_g|^2$$

with the partial wave expansion :

(50) $$f_j(\theta) = \frac{1}{2ik} \sum_l (2l+1)\ (e^{i\delta_l} - 1)\ P_l\ (\cos\theta)$$

which becomes in the JWKB approximation

(51) $$f_j(\theta) = \sqrt{S_j}\ e^{i(\alpha_j + \frac{\pi}{4})}$$

S_j is the classical elastic cross-section calculated along the potential $U_j(R)$ and α_j is a phase shift developed along this trajectory. We finally obtain the electron capture probability :

(52) $$P_{ec} = \frac{\sigma_{ec}}{\sigma_{ec} + \sigma_{el}} = \frac{1}{2} \quad 1 - \frac{2\sqrt{S_u S_g}}{S_u + S_g} \cos\ (\alpha_g - \alpha_u)$$

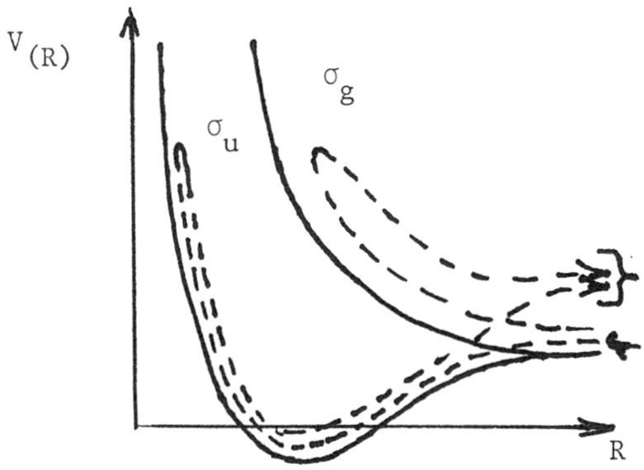

(a) with the potential curves.

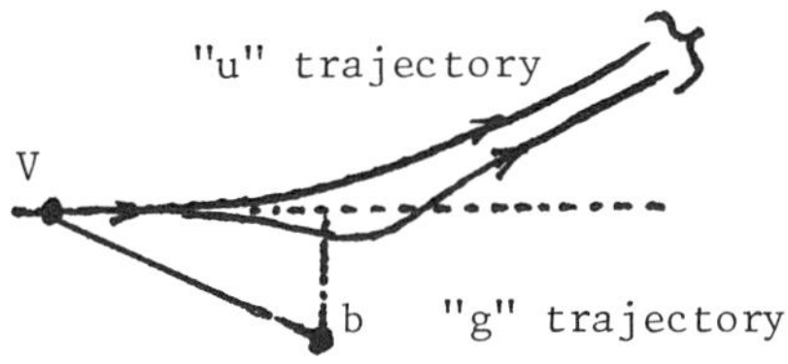

(b) with the trajectories.

Fig. 16 - "view" of the interference mechanism.

This type of interference has a very general character and has been observed in many circumstances :

Two examples :

Curve crossing — Non-resonant electron capture

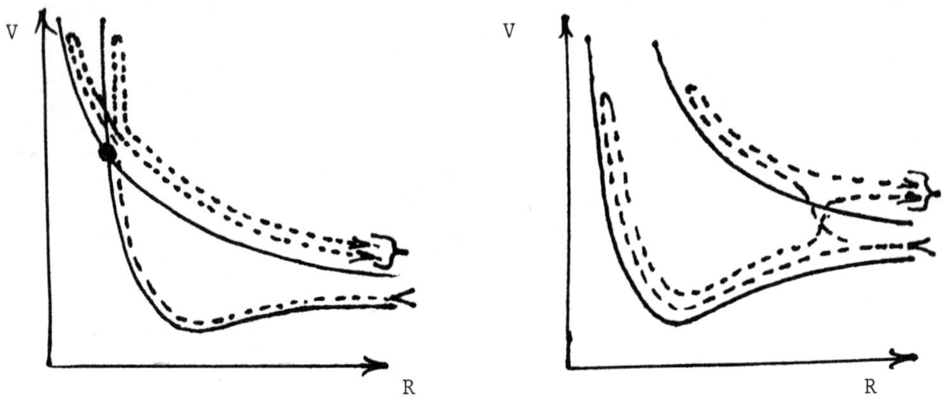

Fig. 17

G - MODELS

It may be useful to have at hand, simple models which allow a rapid estimate of the transition probability. First of all, such models can be developed in the case of *two-state* problems. If more than 2 states are involved simultaneously a numerical solution of the coupled equations must be considered.

G.1 - The Landau-Zener model (LZ) / 23 /

Approximate solutions of the two coupled equations (42) can be derived in some cases, leading to analytical expressions of the transition probabilities. In particular in the case of a curve crossing Landau and Zener have derived a simple formula which is very useful to estimate the transition probability. In the vicinity of the crossing, the potential energies $\varepsilon_1(R)$ and $\varepsilon_2(R)$ are approximated by their tangents at R_c. The first important parameter is then the difference of slopes :

$$\alpha = \left| \frac{d\varepsilon_1}{dR} - \frac{d\varepsilon_2}{dR} \right|_{R = R_c} \qquad (53)$$

Furthermore, the interaction is supposed to be constant and equal to its values at R_c

$$\varepsilon_{12}(R) = \varepsilon_{12}(R_c) \qquad (54)$$

With (53) and (54) equation (42) can be solved approximately and one gets the transition probabilities (fig. 18a).

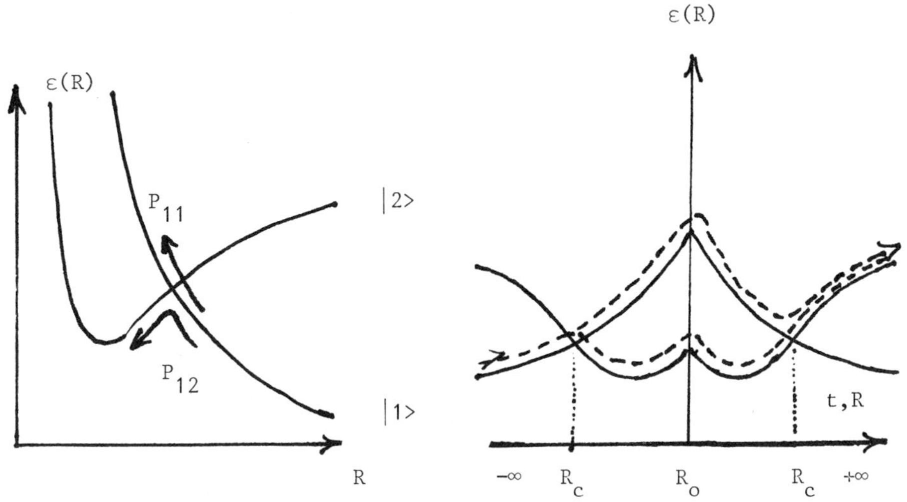

Fig. 18

(55)
$$P_{11} = |a_1(+\infty)|^2 = e^{-\gamma}$$
$$P_{12} = |a_2(+\infty)|^2 = 1 - e^{-\gamma}$$

with $\gamma = \frac{2\pi}{v_{R_c}} \quad \frac{\varepsilon^2_{12}(R_c)}{\alpha}$

where $v_R = \frac{dR}{dt} = V\left[1 - \frac{U(R_c)}{E} - \frac{b^2}{R^2}\right]^{1/2}$ is the radial velocity

P_{11} and P_{12} correspond to transitions 1 → 1 and 1 → 2, respectively.

For a *real* collision, the system passes twice through the crossing point (fig. 18 b), before and after the turning point R_o has been reached. The total probability corresponds to the squared modulus of the sum of the amplitudes developed along "paths" 1 and 2 :

(56)
$$P = |\, a_1(1 - a_1)_{(1)} + a_1(1 - a_1)_{(2)}|^2$$
$$P = 4p_{11}\ (1 - p_{11})\ \sin^2\ (\Delta\phi + \Delta\omega_{LS})$$

$\Delta\phi$ is the difference of semi-classical phases developed along paths 1 and 2. $\Delta\omega_{LZ}$ is an additional phase introduced at the crossing. This interference effect, introduced by stukelberg gives rise to an oscillatory behaviour of the differential cross-section. However when P(b) is integrated over impact parameter this effect is washed out. So, if only the total cross-section is desired the phase in (56) is averaged :

(57)
$$\tilde{P} = 2p_{11}\ (1 - p_{11})$$

and the total cross-section is :

(58) $$\sigma_{LS} = 2\pi \int \tilde{P}(b).b\,db = 2\pi \int_0^\infty 2\, e^{-\gamma}\,(1 - e^{-\gamma})\, b\, db$$

which can be written with $v_{R_c} \underset{-}{\sim} V\,(1 - \frac{U(R_c)}{E})^{1/2}$

(59) $$\sigma_{LZ} = 4\pi\, R_c^2\, (1 - \frac{U(R_c)}{E})\, G(\gamma)$$

where $G(\gamma)$ is an universal function shown in fig. 19

$$G(\gamma) = \int_1^\infty e^{-\gamma x}\,(1 - e^{-\gamma x})\, \frac{dx}{x^3}$$

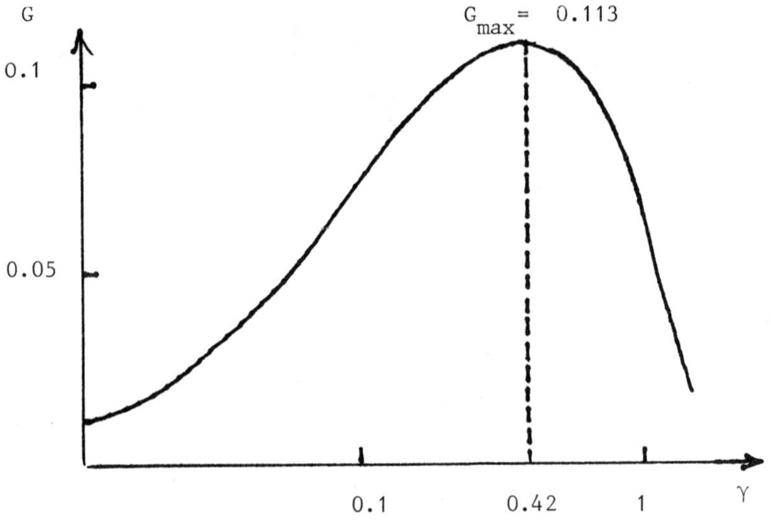

Fig. 19

It is worth noting that $\sigma_{max} = 0.452\, R_c^2$ and roughly corresponds to half of the geometrical cross-section for $R = R_c$.

G.2 - The Demkov model / 24 /

Let us consider the quasi-resonant charge exchange process

$$A^+ + B \rightarrow A + B^+$$

Neglecting the effect of polarisation forces one has $\Delta\varepsilon \underset{-}{\sim} \Delta\varepsilon(\infty)$ = Const. The Demkov model is particularly well adapted to this problem since it assumes that :

(60) $$\Delta\varepsilon = \text{Const.}$$
$$\varepsilon_{12} = A\, e^{-\alpha R}$$

Solving the coupled equation (42), one gets the transition probability (fig. 16) :

(61) $$p_{12} = |a_2(+\infty)|^2 = \frac{e^{-\gamma}}{1+e^{-\gamma}}$$

with $\gamma = \frac{\pi\,\Delta\varepsilon}{v_R\,\alpha}$

For an estimation of the transition probability, α can be expressed in terms of the ionization potentials I_A and I_B of atoms A and B, respectively :

(62) $$\alpha = \frac{1}{2^{1/2}}\,(I_A^{1/2} + I_B^{1/2}) \quad \text{(atomic units)}$$

and $\Delta\varepsilon = I_A - I_B$

(63) $$\gamma \simeq \frac{\pi}{V}\,2^{1/2}\,(I_A^{1/2} - I_B^{1/2})$$

For a real collision, as in the Landau-Zener case, one has to take into account that the system passes twice accross the transition region R_c, R_c being determined by : $\Delta\varepsilon = 2\,\varepsilon_{12}(R_c)$. Averaging the oscillations caused by Stueckelberg-like interferences one gets the following formula for the transition probability :

(64) $$P = 2p_{12}\,(1 - p_{12}) = \frac{2}{(e^{\gamma/2} + e^{-\gamma/2})^2} = \frac{1}{2}\,\mathrm{Sech}^2\,\frac{\gamma}{2}$$

The total cross-section is given by

(65) $$\sigma = 2\pi \int_0^\infty P(b)\,db = \pi \int_0^\infty \mathrm{Sech}^2\,\frac{\gamma}{2}\,b\,db$$

which can be expressed in term of a universal function :

(66) $$\sigma = \frac{\pi R_c^2}{2}\,\sigma^*(\delta^{-1}) \qquad \sigma^* = \int_1^\infty \frac{8\,e^{-\delta x}}{(1+e^{-\delta x})^2}\,\frac{dx}{x^3} \quad \delta = \gamma(v_R = v_o)$$

σ^* is represented in fig. 20. Solving numerically the coupled equation (15), Olson / 25 / found a better σ^* function also represented in fig. 20 .

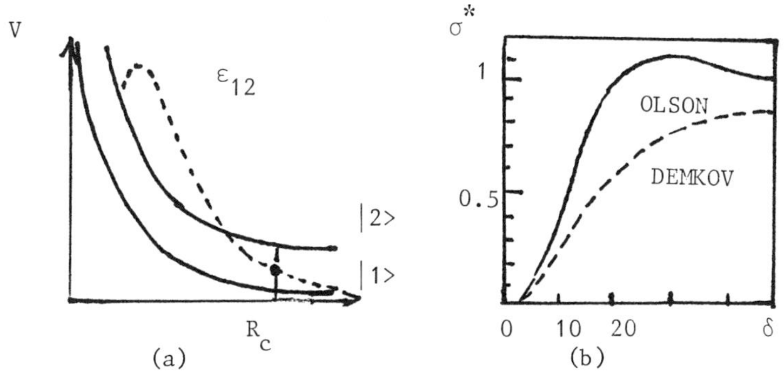

Fig. 20

III - CASES STUDIES

Electron capture by multiply charged ions (ion-atom, ion-ion collisions). Neutralisation.

III.1 - Introduction

In view of eq. (42), one sees that the EC probability is primarily governed by two parameters : the interaction h(R) and the difference between the energy curves corresponding to the active channels $\Delta\varepsilon(R) = \varepsilon_2(R) - \varepsilon_1(R)$. For most EC collision of interest, i.e. having large cross-sections, the transition occurs at *large R* where h(R) presents a exponential behaviour (*) (see eq. (28)). On the other hand, the energy curves $\varepsilon(R)$ may have very different long range behaviours : coulombic, polarisation. The following situations are often found :

(i) Singly charged ion-atom collisions (fig. 21)

At large R, the behaviour of the *diabatic* potential curves is essentially governed by the polarisability of the neutral partner.

(67) $$\varepsilon_1(R) = -\frac{\alpha_B}{R^4} \qquad \varepsilon_2(R) = -\frac{\alpha_A}{R^4} + \Delta E \qquad \Delta\varepsilon \simeq \frac{\alpha_B - \alpha_A}{R^4} + \Delta E$$

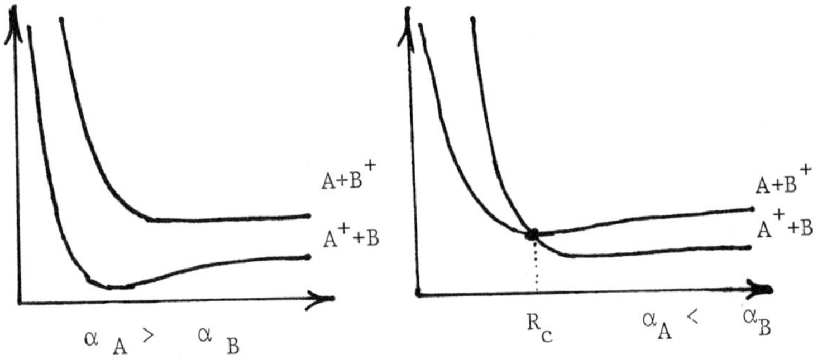

Fig. 21 The curves may cross.

(ii) Neutralisation $A^+ + B^- \rightarrow A + B$

The attractive Coulomb interaction in the incoming channel leads often to curve crossings (fig. 22a)

$$\varepsilon_1 \simeq \cdots \frac{A}{R^6} \qquad \varepsilon_2 \simeq -\frac{1}{R} + \Delta E \qquad R_c \simeq \frac{1}{\Delta E}$$

This is the well known problem of ionic-covalent interactions in chemistry.

(*) the exchange interaction asymptotically behaves as $\simeq A r^{\alpha} e^{-\beta R}$.

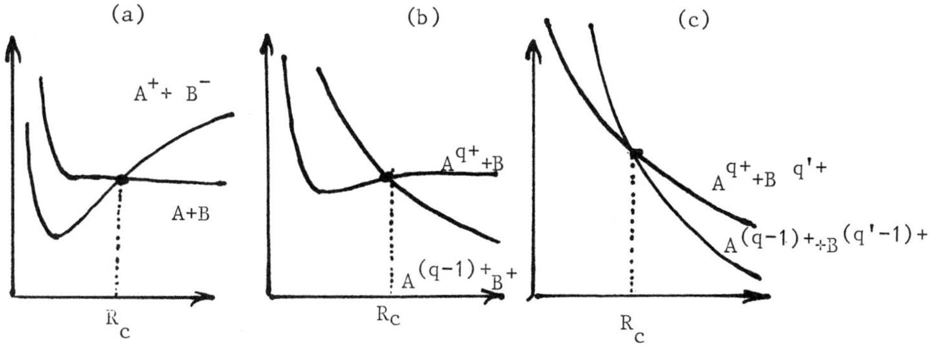

Fig. 22

(iii) Multiply charged ion-atom collisions $A^{q+} + B \rightarrow A^{(q-1)+} + B^{+}$

The exit channel is dominated by the Coulomb repulsion (fig. 22b)

(68) $$\varepsilon_1 = -\frac{\alpha}{R^4} \qquad \varepsilon_2 = \frac{q-1}{R} - \Delta E \qquad R_c \simeq \frac{q-1}{\Delta E}$$

(iv) Ion-ion collisions. $A^{q+} + B^{q'+} \rightarrow A^{(q-1)+} + B^{(q'+1)+}$

Both channels have a repulsive character (fig. 22c).

(69) $$\varepsilon_1 = \frac{qq'}{R} \qquad \varepsilon_2 = \frac{(q-1)\ (q'+1)}{R} + \Delta E \qquad R_c = \frac{q' - q+1}{\Delta E}$$

In the spirit of this school. I will choose selected examples or current problems related to EC involving multicharged ions and ion-ion collisions.

III.2 - Electron capture, involving multicharged ions

Let's consider the pure one electron case where a bare nucleus collides with an hydrogen atom in the ground states

$$A^{q+} + H(1s) \rightarrow A^{(q-1)+}\ (n,l,m) + H^{+}$$

The qualitative behaviour of the potential curves is given on fig.18b. The energy defect and the crossing distance are given by :

(70) $$\Delta E \simeq \frac{1}{2}\left(\frac{q^2}{n^2} - 1\right) \qquad R_c = \frac{1}{\Delta E} = \frac{2}{\frac{q^2}{n^2} - 1} > 0$$

It is well-known that in the case of a one electron-two center problem the solution of the Schrödinger equation is obtained using the prolate spheroidal coordinates, in which case this equation becomes separable in ξ and η. The corresponding quantum number n_ξ, n_η, m easily connect with the quantum number of the *united atom* n_u, l_u, m).

$$n_\xi \leftrightarrow n_r = n_u - l_u - 1$$
(71) $$n_\eta \leftrightarrow l_u = n_u - |m|$$

m is always a good quantum number.

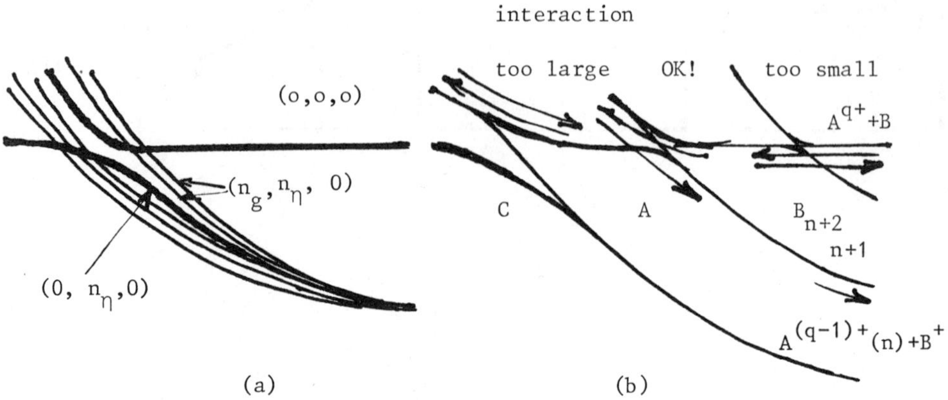

Fig. 23

However, the correspondence with the separate atom states is not as simple. For $R \to \infty$ the prolate spheroidal coordinates becomes *parabolic* coordinates and $n_\xi = n_1$, $n_\eta = n_2$, $m = m$. Still, each parabolic state has not a unique correspondence, with a state expressed in spherical coordinate. We will come back to this point later on.

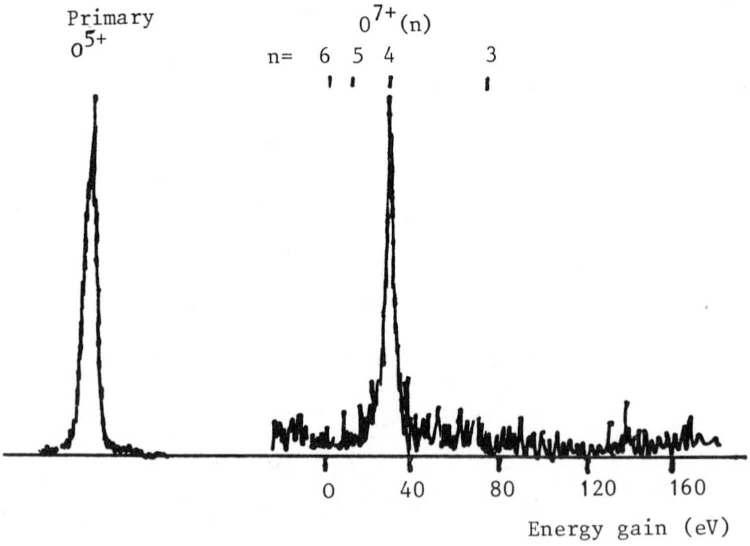

Fig. 24

Energy spectrum of scattered O^{7+} ions O^{8+} + He collision at a collision energy of 8 keV.

$$O^{8+} + He \to O^{7+}\ (n = 4) + He^+$$

III.2.1 - *n* distribution

An interesting property of the separability of the Schrödinger equation is that only one σ state within the manifold of molecular states correlated with a given final channel (fig. 23a) actually interacts with the incident channel (0,0,0) namely (0, n_η, 0) state which differs by the single quantum number ($n_\eta \neq 0$). The problem then reduces to solve the coupled equation for the two relevant states (for a given n value). However, as q increases, an increasing number of states crosses the incident channel (fig. 23b). Nevertheless, due to the exponential decrease of the interaction (see eq. (28)) only a few crossings effectively populate EC channels. In fig. 23b only the B crossing is efficient. This leads to a *high selectivity* in the n-quantum number population. This feature has been first experimentally verified by translational (or energy gain) spectroscopy at the "NICE" facility (Japan) / 25 / see fig. 24.

III.2.2 - *l, m* distribution. Influence of rotational coupling

- *TOTAL CROSS-SECTION*

Up to this point, for the sake of clarity, the picture used to describe the EC process has been over simplified. In particular, the π states have not been considered. Considering for instance fig. 25, within the π manifold the mate 4f π of the 4fσ final channel plays an important role that limits the validity of the aforementioned 2 state model. According to ref. / 28 / (fig. 25) : (i) Half of the flux following the V_0 path down to R_{00} is not significantly affected by the rotational coupling. (ii) On the contrary, the flux corresponding to a transition at R_A, on the way in, will be strongly affected, by the strong 4fσ-4fπ rotational coupling. This effect corresponds to the partial unability of the electron to follow rotation of the nuclear axis.

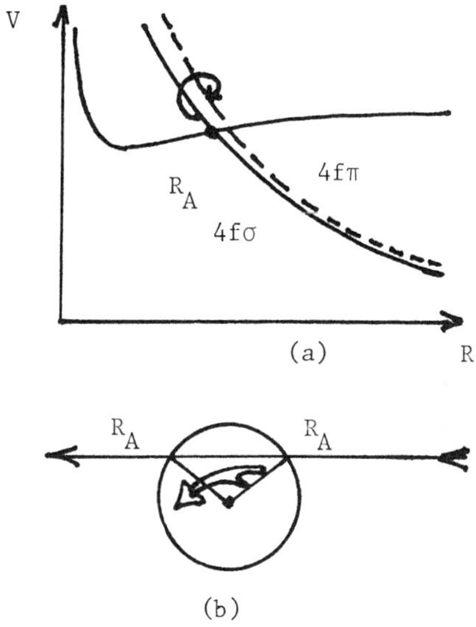

Fig. 25

Let's suppose, for instance, a strong mixing, i.e. : 4fσ population $\simeq$ 4fπ population. On the way out, at the R_A crossing, part of the "σ" flux will transfer back to the incident channel whereas the total "π" flux goes though R_A to the EC channel. One then expects an increase of the total cross-section by $\simeq$ 50 % as compared to LZ theories a figure that is not too far from the results given by detailed numerical close-coupling calculations / 26 /.

- *m, l DISTRIBUTION*

Let's first neglect the effect of the rotational coupling. Only the (o,n,o) state is initially populated. At large R this state will interact with the other σ states correlated with the same final channel (fig. 26) *(dynamical stark effect)*.

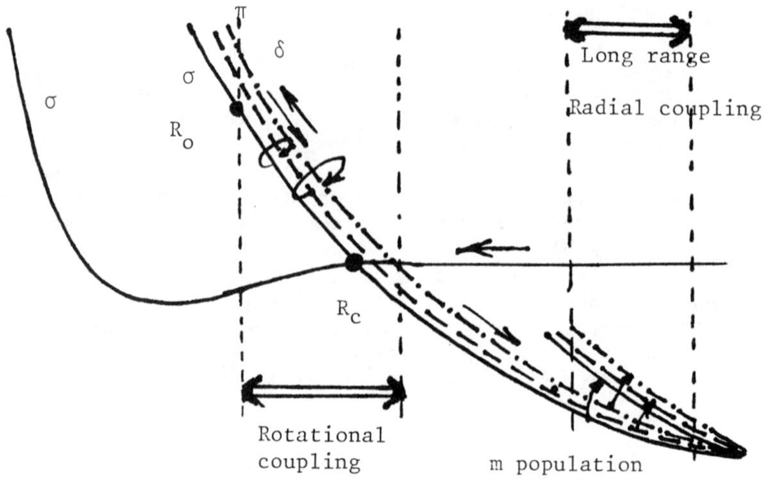

Fig. 26

Two extreme cases can be considered :

* In the limit of *low collision velocity : this dynamical coupling is negligible* and the (o,n_2,o) state adiabatically dissociates into the various spherical states weighted by $|C_{nl}|^2$:

(72) $$\phi(o,n_2,o) = \sum_{\ell} C_{nl}\,\phi(n,l,o)$$

with the population distribution given by

(73) $$W_{nl} = |C_{nl}|^2 = \frac{(2l+1)\,(n-1)!}{(n+l)!\,(n-l-1)!}$$

* Another limit would correspond to a *strong* radial coupling between the various states. In that case, a statistical distribution would be invoked

(74) $W_{nl} = \frac{1}{n}$ (statistical with only m = o states populated)

* Neglect of the rotational coupling, as already discussed, is a poor approximation. We may then assume *both radial and rotational couplings are strong*, we then get a statistical distribution on the complete 2l+1 manifold

(75) $$W_{nl} = \frac{2l+1}{n}$$

* *Complete l mixing model*

This model was recently proposed by Salin / 27 /. The author first solved the system of coupled equations including all the couplings (radial and rotational) and then analysed the results. He found that the actual $|lm\rangle$ population is obtained as a *statistical distribution* among the various l states of a given initial m population :

(76) $$\sigma_{nl} = \sum_{m=o}^{1} \frac{\sigma_{nm}}{n-m}$$

this implies that the l distribution is not the most important source of information about collision mechanisms. More important is the determination of the *polarization* of the light emitted which provides information on the population of the m sublevels/ 28 /.

III.2.3 - A comparison between atomic and molecular calculations for one electron system

During the past years, two groups have indertaken close-coupling calculations using large basis sets and appropriate ETF for EC in one electron systems. Such calculations can be considered as the "state of the Art" to date. The systems that have been investigated so far are : C^{6+}, (N^{7+}), O^{8+} + H(1s). The calculations concern (nl) the total cross section for a given sublevel.

* *Green, Shipsey and Browne* / 29 / have chosen a *molecular* treatment with very sophisticated ETF / 30 / in large basis set (e.g. 33 states for C^{6+} + H). The mathematical derivations are by far too complicated to be given here.

* *Fritsch and Lin* / 31 / used a two center atomic expansion eventually including atomic states of the united atoms (AO+ approximation) and appropriate plane wave ETF. They also used large basis sets (35 AO for C^{6+} + H). A comparison of the results is given in fig. 27. The agreement between both calculations is excellent for the total EC cross section as well as EC into the dominant channel (n = 4). As good an agreement is found for the l distribution within this dominant n-channel. On the other hand both calculation strongly disagree for the minor n-channels.

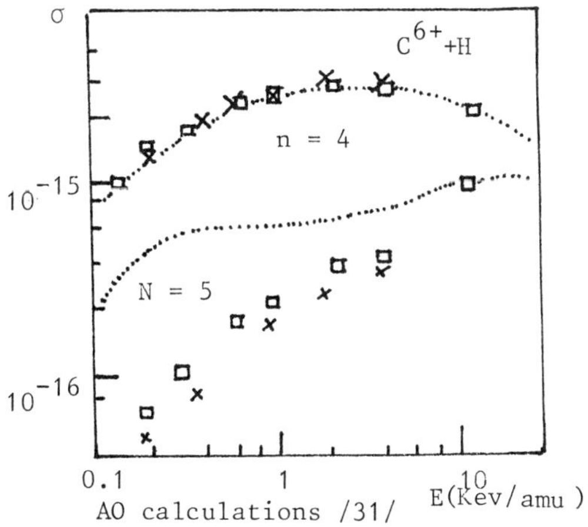

AO calculations /31/
MO calculations /29/
AO-MO calculations /33/

Fig.27

A THEORETICAL APPROACH WHICH RECONCILES THE MOLECULAR AND THE ATOMIC APPROACH

Many problems in atomic collision theory arise because one wants to *apply* the same framework, the same approximation for *all* values of R. A given approximation can be appropriate for a given range of R whereas another approximation better applies to another R-range : an atomic expansion better describes large R-values whereas a molecular expansion is more appropriate for small R-values. In order to join both descriptions, when treating a given problem, one may make use of the R-matrix technique which was first devised in Nuclear Physics.

The space is shared into various subspaces. In each subspace, the problem is solved with the most suitable approximation. The matching of the solution at a border between two subspaces is obtained through continuity conditions on the logarithmic derivative of the wavefunction (fig. 28).

$$R_M \ \Psi_I(R_M) = \Psi_n(R_M)$$

$$\frac{\partial \Psi_I}{\partial R}\Big|_{R_M} = \frac{\partial \Psi_2}{\partial R}\Big|_{RM}$$

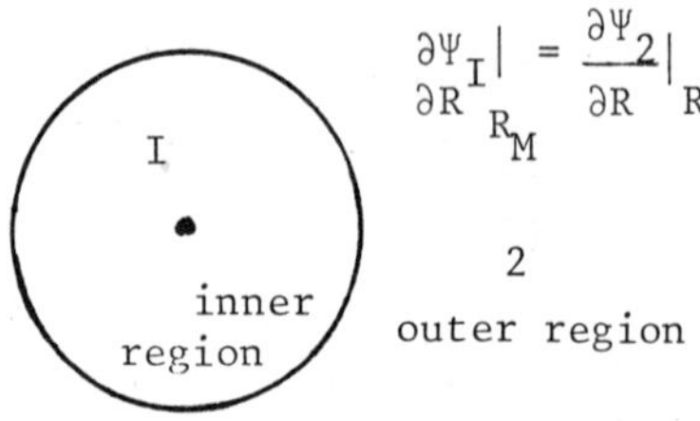

Fig. 28

In this spirit, Kimura and Lin / 32 / have recently proposed a similar approach for EC problem :

* In the *inner region* the molecular treatment is well suited , and the conceptual difficulty for introducing molecular ETF is removed

* In the *outer region* the atomic treatment is used, allowing the use of the well defined plane wave ETF.

The choice of the radius R_M which is somewhat arbitrary, is governed by an *a priori* knowledge of the collision mechanism. Sensitivity of the final results to R_M values is a good check of the initial guess. The authors have recently applied this method to the C^{6+} H system / 33 /. The results are in excellent agreement with the MO and AO methods for the EC into the n = 4 dominant channel and favor the AO results as concerning the population of the n = 5 channel (fig. 27). However, the main interest of this method will show up in *many* electron problems : in the inner region the electron does not belong specifically to a given center, therefore one has not to bother with ETF : standard methods of quantum chemistry can be used. In the outer region where ETF are required, plane-wave ETF can easily be introduced *(at least in the AO approximation)*.

III.2.4 - Extension to multi-electron systems

Theoretical work is not as advanced for multielectron systems as for one-electron systems and only some aspects will just be outlined in this lecture. Two cases merit some special attention.

* *Partially stripped ions colliding with H atoms*

Only one electron is *active* in this problem, thereby simpliying again the theoretical treatment. The other electrons just screen the nuclear charge of the multiply charged ion which allows the use of model potential methods / 34 /. In the same spirit : the fact that capture takes place into high Rydberg states suggest the use of *quantum defect theory* / 26 /.

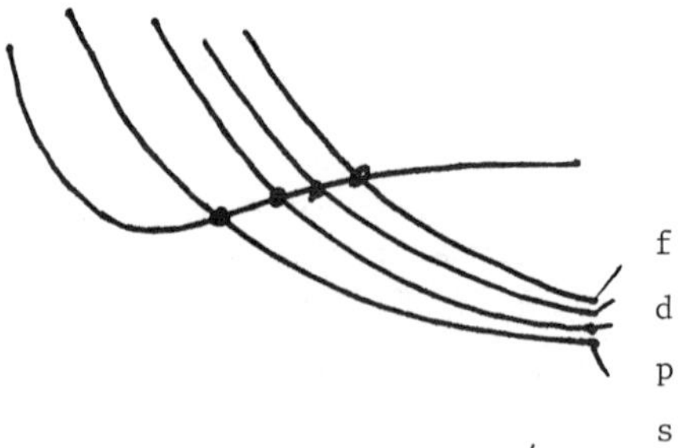

Fig. 29

However, compared to the simple one-electron systems departure from a pure two center Coulomb problem, hinders the separability in ξ and η coordinates. As a consequence, all σ states correlated with the final n levels (fig. 29) have *a priori* a non vanishing interaction with the incident σ state. Furthermore the degeneracy of the final nl states is lifted up. Interesting results as the "complete l-mixing" / 27 / should be reexamined in this context.

* *Fully stripped ions. Colliding with many electron atoms*

- The final states being hydrogenic one can expect that the "complete l-mixing" model of Salin / 27 /, should be appropriate to treat the l distributions.

- *Multiple* electron capture can also occur this point will be discussed below.

III.2.5 - Capture of several electrons by multiply-charged

This subject being covered by an other lecture / 35 /, I will consider only the simplest case : a two electron capture in the collision between a bare nucleus and an He target. The potential curves of relevance for this problem are schematized on fig. 30.

In this very simplified scheme with only three channels, no less than 5 "paths" are involved for the population of the doubly excited Ne^{5+} states. In fig. 30, the most significant states have been selected :

a) the capture occurs in two steps : a first electron is captured at the A crossing, and then a 2nd electron is captured at the C-crossing.

b) both electrons are capturel *simultaneously* at the B crossing.

It is expected that interaction at A and C crossings is much larger (one electron interaction) than at B crossing (electron correlation). In that case process (a) would be favored (when crossing B is efficient,

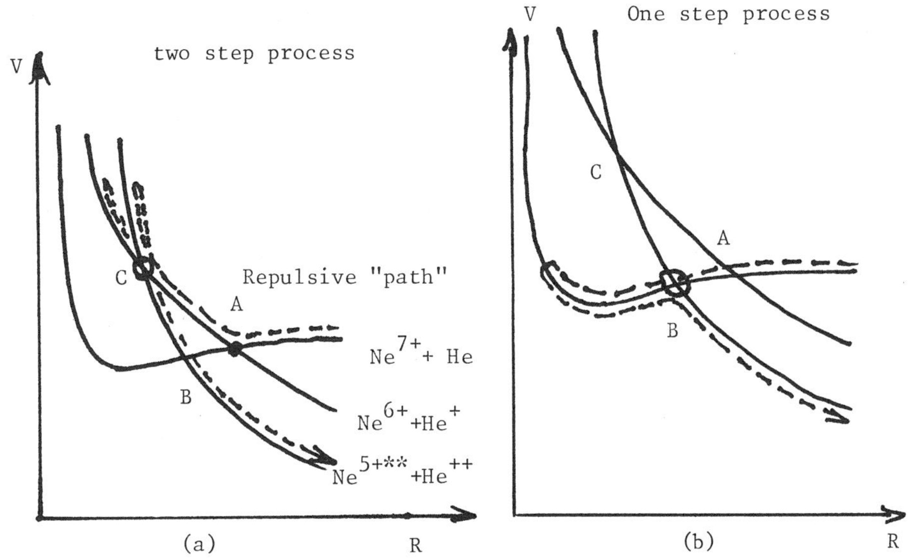

Fig. 30

the crossing A is very adiabatic). Measurements of differential cross-section provides good checks of this hypothesis : mechanism (b) occurs in forward scattering whereas mechanism (a) requires a deflection along the repulsive AC "path" leading always to scattering at *some finite* angle. Investigation of angular scattering in EC by multicharged ions has been recently undertaken by Roncin *et al.* / 36 /. The experiment makes use of a parallel plate electrostatic analyser associated with a two-dimension position sensitive detector (fig. 31a). The results (fig. 31b) clearly show that capture of two electron correspond to scattering at finite angle ($\theta \underset{-}{\sim} 0.25$ deg. and $\theta \underset{-}{\sim} 0.35$ deg.). On the contrary, the single electron capture processes are peaked at 0 deg.

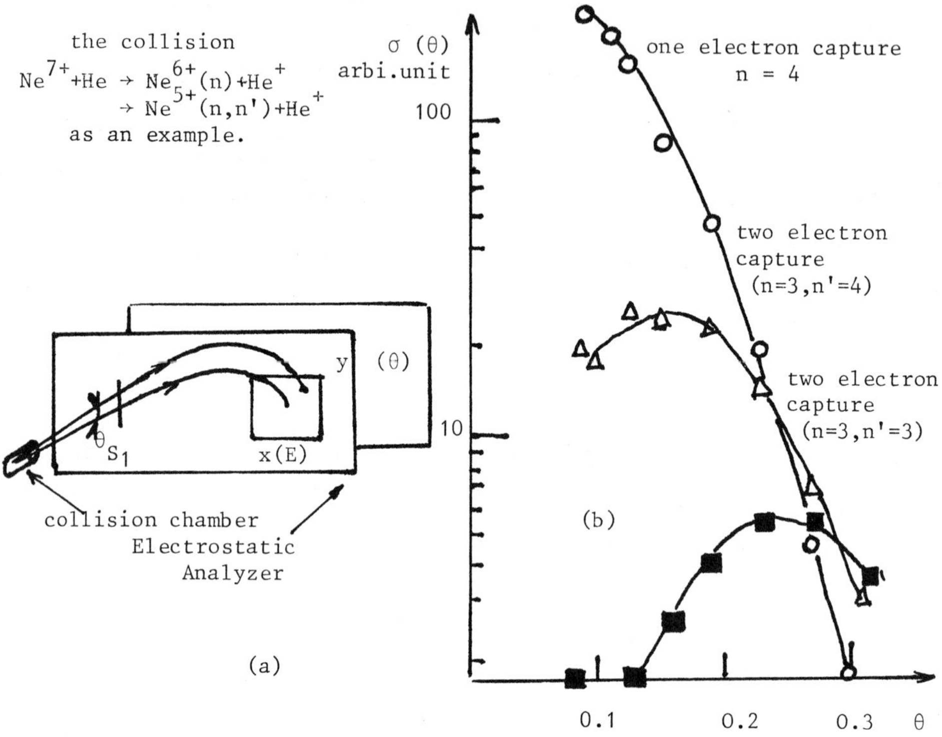

Fig. 31

As a final remark, it is noteworthy that the double capture process is a particularly efficient tool to create *doubly excited states of multiply charged ions* double Rydberg states. Their study opens a new area in spectroscopy / 37 /, the discussion of which is outside the scope of this lecture.

III.3 - Electron capture in ion-ion collision

From a theoretical point of view / 14 /, the problem is not significantly different fro the ion-atom case, except from one aspect. The potential curves of both the *incident* and the final channels have a repulsive behaviour primarly governed by the Coulomb interaction (fig. 32). As a consequence, the EC process will have an *energy threshold* given by the energy of the crossing point $E(R_c)$.

Lets' consider the reaction : $A^{q+} + B^{(q-1)+} \rightarrow A^{(q-1)+} + B^{q'+}$ (ΔE). R_c is given by :

$$R_c \underset{-}{\sim} \frac{q - q'}{\Delta E} \tag{77}$$

for a pure hydrogenic system :

$$(78) \quad R_c = \frac{(Z_A - Z_B)}{Z_A^2/2n_A^2 - Z_B^2/2n_B^2}$$

At present, most calculations have been devoted to collision between low charge states (q = 1, 2) for which experimental results are available. Let's take some examples :

A) $H^+ + He^+ \rightarrow H + He^{++}$

This is the inverse reaction of the doubly charged ion-atom case discussed previously. Recent experimental results (see ref. / 38 / and ref. thereincluded) are in good agreement with bot MO based calculation (E > 30 keV) and AO based calculation (in the whole range 10-100 keV).

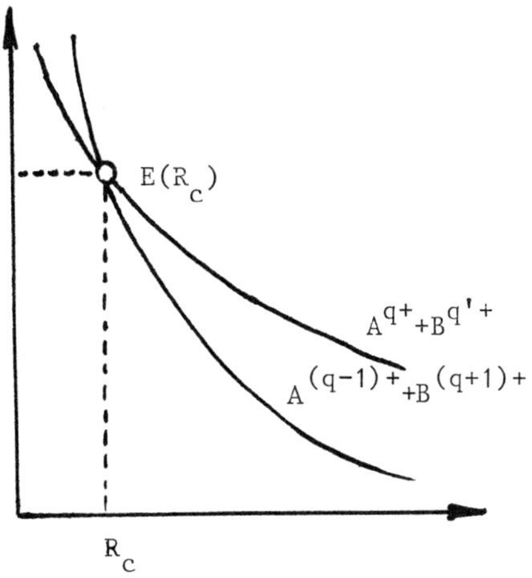

Fig. 32

B) Resonant charge exchange $He^+ + He^{++} \rightarrow He^{++} + He^+$

The problem is quite similar to the H^+-H collision discussed above except that curved trajectories have to be used to describe the motion along the repulsive Coulomb curves. Experiments were performed by Jognaux, Brouillard and Szucs / 39 /. Nevertheless the desagreement between theory / 40 / and experiment / 39 / (fig. 33) is surprising in view of the simplicity of the system.

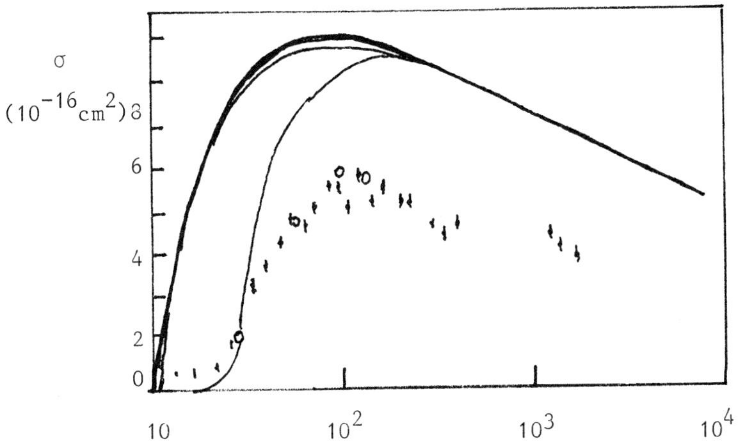

Fig. 33

Fig. 33. Cross section for $He^{2+} + He^+ \rightarrow He^+ + He^{2+}$. Thin curve calculation of Bates and Boyd (1962) for the $^4He^+$ - $^4He^{2+}$ system. Bold curve : present calculation for the $^4He^+$ + $^3He^{2+}$ system. Broken curve : present calculation, taking into account the experimental limits on the accepted scattering angle : present experiment results from measurements where $^3He^{2+}$ is faster than $^4He^+$. : present experimental results from mesurements where $^3He^{2+}$ is slower than $^4He^+$.

C) Quasi resonant charge exchange / 41 /

$$A^{(q-1)+} + B^{q+} \rightarrow A^{q+} + B^{(q-1)+}$$

In that case, the two potential curves have the same coulomb repulsion (parallel curves). Therefore the transition should be well described by the Demkov model. Again due to the strong coulomb repulsion, the equations have to be solved along coulomb trajectories at least for low collision energy. This leads to a drastic reduction of the cross section / 42 /.

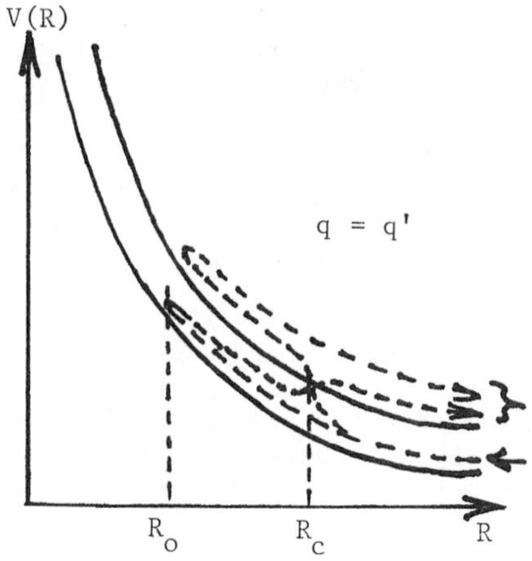

Fig. 34

The probability is given by eq. (53). Actually we should reintroduce the phase $\eta = \int_{-\infty}^{+\infty} \Delta\varepsilon \, dt'$ which was averaged over in eq. (52) as $\sin^2 \eta \rightarrow 1/2$

(79) $$P = \frac{1}{2} \mathrm{Sech}^2 \frac{\gamma}{2} \sin^2 \eta$$

where η is calculated along the trajectory as :

(80) $$\eta = 2 \int_{R_1}^{\infty} \frac{\Delta\varepsilon(R)}{v_R} dR$$

(81) $$v_R = v \left[1 - \frac{q(q-1)}{RE} - \frac{b^2}{R^2} \right]^{1/2}$$

The distance of closest approach R, is given by

$$R_1 = \frac{1}{2} \frac{q(q-1)}{E} + \left[\frac{1}{4} \frac{q^2(q-1)^2}{E^2} + b^2 \right]^{1/2}$$

Belic and Janev have integrated the probability with the following assumption : for *low velocity* η oscillates rapidly thus allowing the use the $\sin^2 \eta \rightarrow 1/2$ approximation. For large velocity $\mathrm{sech}^2 \sigma/2 \sim 1$ and $P \sim 1/2 \sin^2 \eta$. One should note the high energy interferences which are due to a maximum in the η phases.

D) Asymmetric systems

Contrarily to the previous case, curve crossings may occur (fig. 36) for asymmetric system. In the absence of experiments and accurate calculations a rough estimate of the cross sections can be obtained using the Landau-Zener approximation. Actually, there are two extreme cases to be considered.

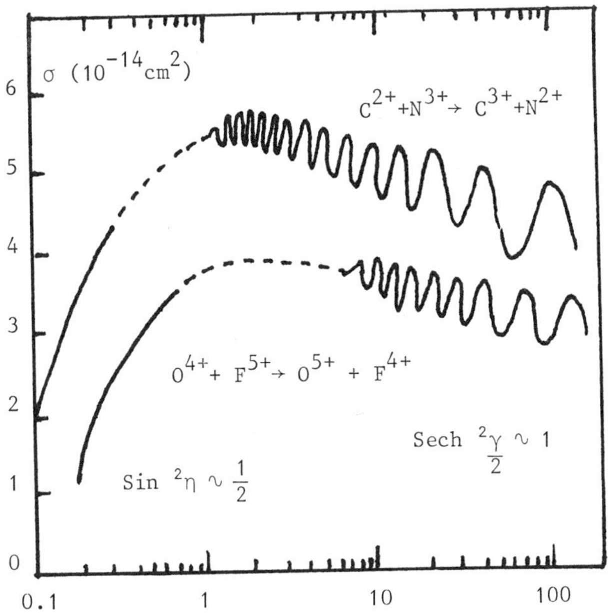

From /42/

Fig. 35

(i) q' << q the active electron is initially attached to the highly charged ion, ex. : $He^{++} + Fe^{25+}$ (1s) → He^{+} (1s) + Fe^{26+} (case (a) in fig. 36) the crossing region occurs at small R values and the geometrical cross sections are very small

(82) $$R_c^{Max} = \frac{2}{q+q'} < \frac{2}{3}\, a_o$$

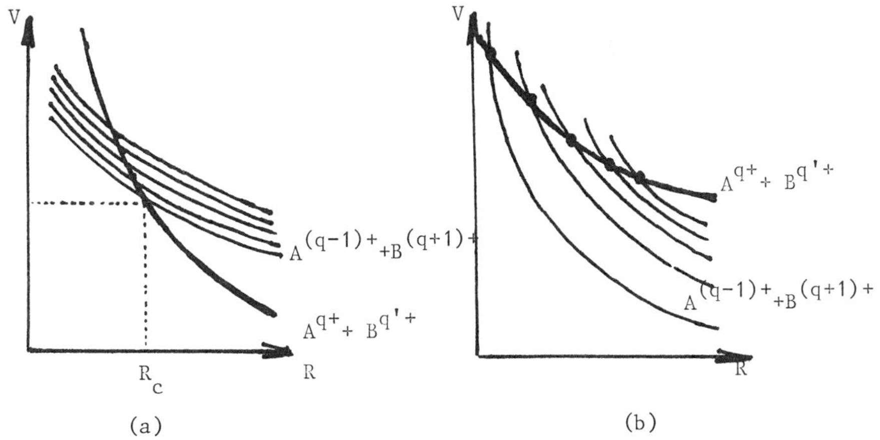

Fig. 36

(ii) q' >> q the active electron is initially attached to the low-charge ion, ex. : He^{+} (1s) + Fe^{26+} → He^{++} + Fe^{25+} (*) (nl), case (b) in fig. 36. The crossing radium may be large and highly excited states may be populated. Therefore *cross sections can be large as indicated* by the rough estimate, shown in table I. (These estimates were obtained / 43 /

from the LZ approximation). The interaction ε_{12} (R) was obtained by an extrapolation of the formula of Olson and Salop :

TABLE I - Rough estimate for EC in $He^+(1s) + A^{Z+} \rightarrow He^{++} + A^{(Z-1)+}$ (n) (R_c crossing radius, H_{12} interaction at the crossing point, E_{Th} thershold energy (in eV), E^M_{LZ} (in eV) energy for maximum cross-section (Q_{max} in 10^{-16} cm^2).

A^{q+}	n	R_c	H_{12}	E_{Th}	E^M_{LZ}	Q_{MAX}
O^{8+}	3	3.85	8.7 (-2)	56	6000	6.5
Fe^{26+}	9	11	5.7 (-3)	64	70	50
	8	7.3	4 (-2)	96	435	23
	7	4.9	1.4 (-1)	144	8300	10
	6	3.2	3.3 (-1)	217	48000	4.5
Mo^{42+}	11	7.5	6.4 (-2)	150	1000	25
	9	4.5	2.2 (-1)	250	14000	9
	8	3.4	3.5 (-1)	336	27000	5

From Barat et Sidis, Ann. Phys. (1983), 8, 133.

In all these examples the active electron was in the ground-state. In some special conditions, a large population of ions can be initially into an excited states as found in inertial plasmas (*laser, electron or ion created plasmas*). In such cases EC cross-sections in ion excited-ion collision can also be high. Corresponding estimates can in principle be derived in a similar way.

- POSITIVE ION-NEGATIVE ION COLLISION NEUTRALISATION

I cannot end this series of lectures without talking about the $H^+ + H^-$ mutual neutralisation process :

$$H^+ + H^- \rightarrow H^*(n) + H$$

which has remained an intriguing challenge for collision physicists until very recently.

Historically, the challenge started with the experiment of Peart, Grey and Dolder / 44 /. These authors measured the total cross-section for this reaction found a "Cusp" at $E_{CM} \simeq 100$ eV (see fig. 38). Furthermore, these experimental results agreed very well with those obtained by Moseley *et al.* / 45 / below $\simeq$ 100 eV. Comparison with various theories was very poor and stood very poor until the 80's.

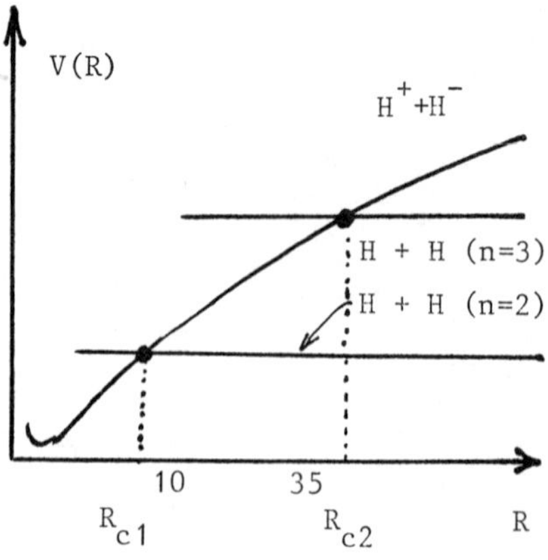

Fig. 37

On fig. 37 are represented the schematic curves relevant to this

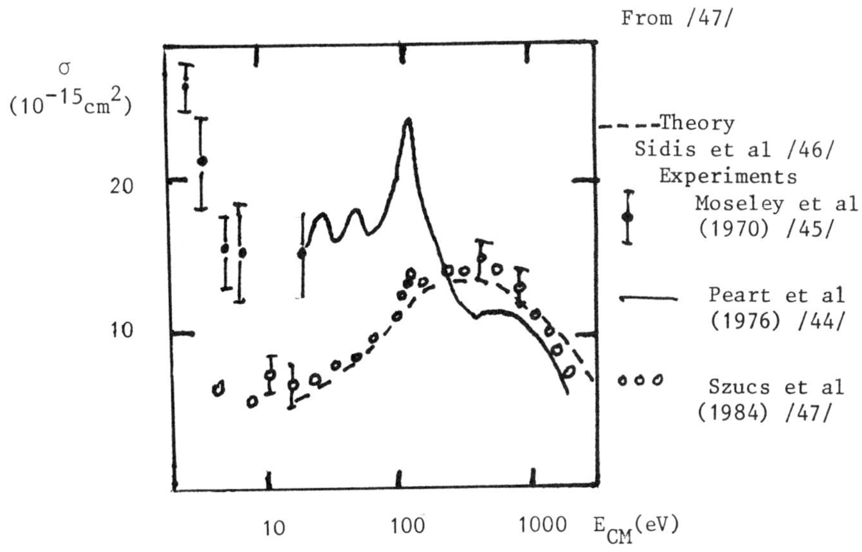

Fig. 38

problem. The attractive coulomb curve describing the incident channel crosses flat curves corresponding to capture into n = 2,3 (it also crosses the curve leading to n = 4 at too large a distance (≃ 270 Å) to be efficient.

This problem looks simple *a priori*. It is well known that H^- should be described by 1s 1s' configuration at least and in this problem, only the diffuse 1s' orbital (whose radius is about $4a_0$) should be active. These characteristic features enable to reduce the actual problem to a one-electron problem. At a first sight a LZ theory should be adequate. As expected such theory predicts that capture into n = 4 is unlikely and the capture into n = 3 would dominate for $E < 10$ eV; at higher energy the capture should take place into n = 2. The comparison with experiment was very poor. Any improvement upon the LZ approximation did not results in better agreement with experiment.

The first surprise came in 1981 when Sidis Kubach and Fussen / 46 / reinvestigated the problem theoretically. They used a one-electron model potential method to describe H^- but they did include ETF. The results changed drastically. They found that EC into n = 3 *is still important in the 100-1000 eV energy range* (fig. 38) contrarily to previous theories. Agreement with experiments somewhat improved. However, they did not found the "Cusp" and claimed that they do not see any physical improvement to fill the gap between experiment and theory.

The second surprise came with the new experimental results of Szücs et al. / 47 / obtained at about the same time, and which did not display the "cusp" anymore (fig. 38). Furthermore, a very good agreement was found with the theory of Sidis et al. As a third surprise, Peart, Bennet and Dolder / 48 / reinvestigating the problem experimentally also found good agreement with the theoretical prediction of Sidis *et al.*!

This example clearly illustrates that care should be exercised when using simple theories like LZ not so much because LZ is wrong but rather because other transitions than those occuring at curve crossings may take place. In the $H^+ + H^-$ case, the additional effect found in ref. / 46 / arises from a transition at R _ 15 $a_0 < R_c$ (n = 3). The coupling parameters characterizing this transition are extremely dependent on the choice of electron coordinate origin. In such pathological cases the use of ETF is unavoidable.

ACKNOWLEGDEMENTS

I would like to thank Dr. V. Sidis for suggestions and improvements for the written version of this lecture.

REFERENCES

1. See e.g. K. Taulberg in "Fundamental Processes in energetic atomic collisions. Nato ASI series V v. 103 : 349).
   M. Kimura in invited lectures XIV ICPEAC (Palo Alto) (1985).
2. L.H. Thomas, Proc. Roy. Soc. A114 : 561 (1927).
3. E. Horsdal Pedersen, C.L. Cocke and M. Stockli, Phys. Rev. Let. 50 : 1910 (1983).
4. N. Bohr and J. Lindhard, K. Dan. Vid. Sel. Mat. Phys. Medd 28, n. 7 (1954).
5. H. Knudsen, H.K. Haugen and P. Hvelplund, Phys. Rev. A23 : 597 (1981).
6. P. Hvelplund et al., The Physics of highly ionized atoms (eds. Silver and Peacock, North Holland) 421 (1985).
7. C. Gaussorgues et al., J. Phys. B 8 : 239 (1975).
8. D.R. Bates and R. McCarroll, Proc. Roy. Soc. A245 : 175 (1958).
9. W. Fritsch, J. Phys. B 15 : L389 (1982).
10. C.D. Lin, Phys. Rev. A17 : 1646 (1978).
11. See e.g. M. Kimura, Phys. Rev. A31 : 2153 (1985) and ref. there inclosed.
12. U. Wille and R. Hippler, Physics Report to be published (1986).
13. W. Fritsch and C.D. Lin, J. Phys. B 15 : 1255 (1982).
14. R.K. Janev, L.P. Prestniakov and V.P. Shevelko, Physics of highly charged ions, Springer-Verlag, Berlin (1985).
15. T.G. Winter and C.L. Lin, Phys. Rev. A29 : 567 (1984).
16. F.T. Smith, Phys. Rev. 179 : 111 (1969).
17. J. von Neuman and E.P. Wigner, Phys. Z. 30 : 457 (1927).
18. W.R. Thorson and J.B. Delos, Phys. Rev. A18 : 135 (1978).
19. R.D. Piacentini and A. Salin, J. Phys. B 7 : 1966 (1974) ; 9 : 563 (1976) ; 10 : 1515 (1977).
20. S.B. Schneiderman and A. Russek, Phys. Rev. 181 : 311 (1969).
21. J. Vaaben and K. Taulberg, J. Phys. B 14 : 1815 (1981).
22. V.H. Ponce, J. Phys. B 12 : 3731 (1979).
23. L. Landau, Phys. Z. Sov. 246 : 14 (1932).
24. Yu N. Demkov, Sov. Phys. JETP 18 : 138 (1964).
25. S. Ohtani et al., J. Phys. B 15 : L533 (1982).
26. O.G. Larsen and K. Taulberg, J. Phys. B 17 : 4523 (1984).
27. A. Salin, J. de Physique 45 : 671 (1984).
28. D. Vernhet et al., Phys. Rev. A 32 : 1256 (1985).
    L.J. Lembo et al., Phys. Rev. Let 55 : 1874 (1985).

29. T.A. Green, E.J. Shipsey and J.C. Browne, Phys. Rev. A25 : 1364 (1982) ; A27 : 821 (1983).
30. T.A. Green, Phys. Rev. A23 : 519 (1981) and following papers.
31. W. Fritsch and C.D. Lin, Phys. Rev. A19 : 3039 (1984).
32. M. Kimura and C.D. Lin, Phys. Rev. A31 : 590 (1985).
33. M. Kimura and C.D. Lin, Phys. Rev. A32 : 1357 (1985).
34. M. Gargaud, I. Hanssen, R. McCarroll and P. Valiron, J. Phys. B 14 : 1359 (1981).
35. E. Salzborn (see lecture in this volume).
36. P. Roncin et al., XIV ICPEAC (Palo Alto) : 677 (1985).
37. A. Bordenave et al., J. Phys. B 17 : L127 (1984).
38. K. Rinn, Melchert and E. Salzborn, J. Phys. B, to be published.
39. A. Jognaux, F. Brouillard and S. Szücs, J. Phys. B 11 : L669 (1978).
40. D.R. Bates and A.M. Boyd, Proc. Phys. Soc. 80 : 1301 (1962).
41. R. Janev, D. Belic, J. Phys. B 15 : 3479 (1982).
42. R. Janev, Phys. Lett. 89A : 190 (1982).
43. M. Barat and V. Sidis, Ann. Phys. (Paris) 8 : 133 (1983).
44. B. Peart, R. Grey and K.T. Dolder, J. Phys. B 9, 3047 (1976).
45. J.T. Moseley, W. Aberth and J.R. Peterson, Phys. Rev. Let. 24 : 435 (1970).
46. V. Sidis, C. Kubach and D. Fussen, Phys. Rev. Lett. 47 : 1280 (1981); Phys. Rev. A27 : 2431 (1983).
47. S. Szücs, M. Karema, M. Terao and F. Brouillard, J. Phys. B17 : 1613 (1984).
48. B. Peart, M.A. Bennett and K. Dolder : J. Phys. B 18 : L439 (1985).

ION-ION COLLISIONS

K. Dolder

Department of Atomic Physics
The University
Newcastle upon Tyne, NE1 7RU, U.K.

1. INTRODUCTION

Let us start by asking three questions. What types of ion-ion reaction have been studied? Why are these studies worthwhile? Where can more detailed information be found?

Experimentalists have so far looked at the following types of reaction :

Mutual neutralization,

$$A^+ + B^- \rightarrow A + B \qquad (1)$$

Detachment by positive ion impact,

$$A^+ + B^- \rightarrow A^+ + B + e \qquad (2)$$

Double charge transfer,

$$H^+ + H^- \rightarrow H^- + H^+ \qquad (3)$$

Association,

$$H^+ + H^- \rightarrow {}^*H_2 \rightarrow H_2^+ + e \qquad (4)$$

Ionization by positive ion impact,

$$A^+ + B^+ \rightarrow A^+ + B^{2+} + e \qquad (5)$$

Charge transfer between positive ions,

$$A^+ + B^+ \rightarrow A + B^{2+} \qquad (6)$$

Reactions (3) and (4) were confined to H^+ and H^-, but for the other processes a number of ion pairs have been studied. In several cases the symbols A and B represent either atomic or molecular ions and the positive ions were not always singly-charged.

Let us now ask what prompted these studies. Mutual neutralization is important in the upper atmosphere, and in various cool ionized gases where positive and negative ions co-exist. Since this process destroys free charges it can profoundly influence the principal properties of cool plasmas, but in the hotter thin plasmas encountered in thermonuclear devices or stellar atmospheres, interactions between positive ions are important. Consider, for example, charge transfer between deuterium fuel (D^+) and impurity ions (X^{n+}) in a thermonuclear device, i.e.

$$D^+ + X^{n+} \rightarrow D + X^{(n+1)+} \qquad (7)$$

The charge on the impurity is enhanced so that energy radiated by bremmsstrahlung increases. Simultaneously, the neutralized fuel escapes from the containing fields and when it strikes the walls of the apparatus it sputters further impurities into the plasma; these reactions therefore cause the loss of both energy and fuel from the plasmas.

The role of positive ion collisions in astrophysics has so far received little attention but there is clearly scope for much more work. Baliunas and Butler(1) have already demonstrated that

$$Si^+ + H^+ \rightarrow Si^{2+} + H \qquad (8)$$

and

$$Si^{2+} + He^+ \rightarrow Si^{3+} + He \qquad (9)$$

very significantly influence the distribution and concentrations of silicon ions in stellar coronae.

There are numerous other practical applications, but ion-ion interactions are particularly fascinating because they also provide experimentalists with opportunities to study collision theory at the most fundamental level. We will, for example, describe reactions involving He^+ (H-like, but with the attraction that each ion is "labelled" by its positive charge), H_2^+ (the simplest of all molecular structures and the only one for which Schrödinger's equation can be exactly solved), or H_3^+ (the simplest polyatomic system). If you are intrigued by these possibilities and by the realization that ion-ion collisions is a relatively new field with considerable potential, you may wish to discover more.

The best introduction to experimental aspects of mutual neutralization was given by Moseley et al(2) in 1975 and, although it is now somewhat dated, it is a good place to start. A more recent and extensive review of ion-ion collisions was prepared by Dolder and Peart(3) (1985) and there are several valuable articles in the proceedings of an earlier NATO ASI, particularly the accounts by Brouillard and Claeys(4) and Mitchell and McGowan[5].

2. GENERAL FEATURES OF THE EXPERIMENTAL METHOD AND COMMENTS ON TECHNICAL MATTERS OF CURRENT INTEREST

The experiments require two ion beams of specified type and energy which are made to intersect normally, obliquely or they can be "merged" to move in confluence over a path of about 10 cm. Details of specific techniques and apparatus were given in the reviews already cited, and so we will merely draw attention, very briefly, to some more important topics and indicate articles where further information can be found.

The beams are necessarily tenuous (particle densities do not usually exceed 10^6 cm^{-3}) because they would otherwise be spread by space charge forces(6) and this often limits the types of experiment which can be undertaken. The interpretation of results will be obscured if the initial internal energy of the ions is unknown and this problem is particularly severe for molecular ions where vibrational excitation is difficult to specify or control. The limited progress made in this field has recently been reviewed(3),(5). It is, for example, often possible to prepare atomic ion beams free from metastable contamination(3) by judiciously chosing the ion source and by operating it under certain conditions. Development is also in progress(5) on sources in which molecular ions are vibrationally de-excited before extraction.

The relative merits of arrangements in which beams are made to merge or, alternatively to collide at a finite angle have recently been discussed[3] and so we will only stress that beams moving in confluence can provide access to interaction energies in the thermal range (≦ 0.1eV) with correspondingly good energy resolution, but it is sometimes overlooked that ion beams are not composed of bundles of perfectly parallel rays and imperfect collimation therefore degrades the energy resolution. An excellent account[4] exists on the equations which quantify these arguments. Enhanced resolution can also be obtained with inclined beams, which usually intersect at an angle between about 8° and 20°.

An intriguing possibility which has scarcely been exploited arises if we consider the interaction of two fast beams. If a collision occurs which produces fragments (e.g. part of a dissociated molecular ion) with energy of several eV in the centre of mass frame, it may happen that corresponding energy measured in the laboratory frame may be one or two orders of magnitude larger. Unfortunately the enhancement of energy will depend upon the angle at which the fragment is formed and this will limit or, at best, complicate the application of this technique to the detailed study of inelastic ion-ion collisions. Equations which can provide numerical values have been derived and discussed by Brouillard and Claeys[4].

We will now consider some experimental results.

3. MUTUAL NEUTRALIZATION

The simplest example is,

$$H^{+} + H^{-} \rightarrow H + H \qquad (10)$$

and this process must be explained theoretically before confident predictions can be made about the more complicated processes occurring in nature. Unfortunately, the understanding of this reaction was delayed by some strange and unexplained experimental results. Briefly, three independent measurements of reaction made in different laboratories with different apparatus produced results in good mutual accord, but at variance with all theoretical predictions. Eventually the anomalies grew so puzzling that there arose what Szűcs et al[7] aptly described as an "amazing situation". The position has recently become much clearer in the wake of new measurements by Szűcs et al[7] and Peart et al[8] which are mutually compatible and agree quite well with theory. But no explanation has yet been discovered for the earlier results.

The symbols in fig. 1 show measurements by Peart et al obtained with H^+ and H^- beams inclined at 8 ° and at 15°. The H^+ ion energies are also indicated by the inset. The continuous curve (Sz) illustrates the measurements by Szücs et al and the broken curves (S and B) show theoretical results by Sidis et al(9) and Borondo et al(10). The agreement is quite good.

Measurements have also been made recently in this laboratory on the mutual neutralization of He^+/H^-, H_2^+/H^- and one-electron transfer between $^3He^{2+}/H^-$. All the results are compared in fig. 2 which shows the measurements plotted on a common velocity scale. We see at once that results for all three singly-charged ions are very similar, especially at the higher velocities. For $^3He^{2+}$ the cross sections are, as one might expect, considerably larger because of the stronger attraction of the double charge.

This kind of behaviour was predicted by Massey(11) who argued that mutual neutralization occurs essentially through long-range crossings of

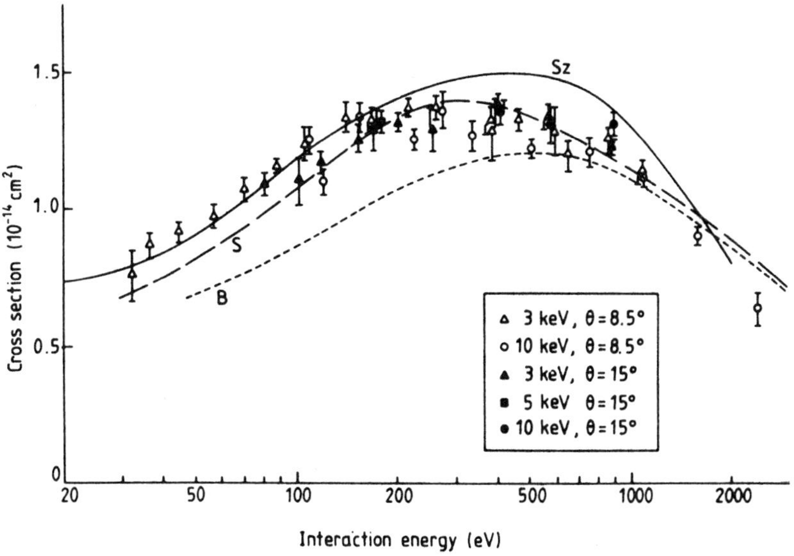

Fig. 1. The symbols show measurements by Peart et al of mutual neutralization of H^+ and H^-. The H^+ energies and angles of beam intersection are shown by the inset. The continuous curve (Sz) illustrate similar measurements by Szücs et al. Theoretical results by Sidis et al (S) and Borondo et al (B) are represented by broken curves.

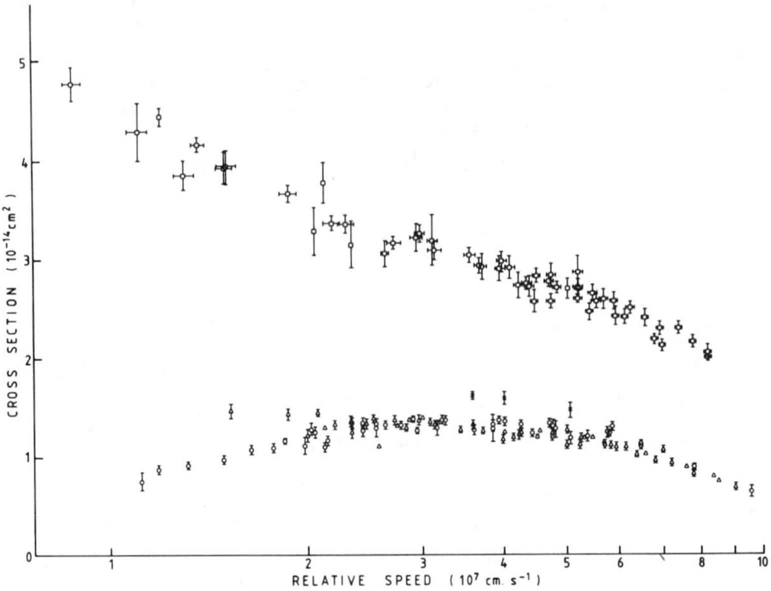

Fig. 2. Measurements of mutual neutralization cross sections for H^+/H^- (circles), He^+/H^- (triangles) and H_2^+/H^- (crosses) and one-electron transfer between $^3He^{2+}/H^-$ (squares) plotted against relative speed of the reactants.

potential energy curves and so one would expect the cross sections to be very similar because H^+, He^+ and H_2^+ "look much the same" to a distant H^- ion.

It can be seen that fig. 2 includes only three measurements of H_2^+/H^- neutralization and these were all obtained with 20 keV H_2^+ ions. It was found[12] that when ions with energies between 3 and 10 keV were used, the apparent cross section increased with ion beam energy. This almost certainly implies that the neutralization of H_2^+ forms products with transverse energies which are large enough to prevent some of them from reaching the neutral particle detector. This result may have far-reaching implications because a great deal of our knowledge about the dissociative recombination of molecular ions has come from beam experiments where the ion beam energy was typically 3keV. It is therefore likely that at least some of these cross sections are too low and further work is clearly needed.

Another avenue for new work is state selection of collision products. When mutual neutralization occurs the neutral atoms produced are not normally formed in their ground states and, by measuring the relative numbers of atoms in the various excited states, one could obtain much more insight into the recombination process and provide more searching tests for theory. This aspect of mutual neutralization is unexplored with the exception of an attempt by Weiner et al[13] to look at sodium atoms formed in 3^2D and 4^2P states by,

$$Na^+ + O^- \rightarrow Na + O \qquad (11)$$

Their method relied upon absolute measurements of the intensity of radiation emitted by sodium atoms at the appropriate wavelengths. Unfortunately, absolute radiometric calibrations are notoriously difficult and their result for the formation of $Na(3^2D)$ was about three times that measured by conventional methods[14] for the function of sodium in <u>all</u> states. Clearly the results of Weiner et al are wrong, but they point the way for experiments to develop in future.

4. COLLISIONS BETWEEN POSITIVE IONS

We have already mentioned the application of positive ion collisions to astrophysics and magnetically confined fusion devices. Interest was also roused in connection with "heavy ion fusion" (HIF) which offers a more speculative route to fusion. Considerable work has been performed on "laser fusion" in which lasers very rapidly heat small pellets of deuterium and tritium. Unfortunately, contemporary lasers operating at suitable wavelengths,are not efficient enough to produce net power gain. It has therefore been proposed that laser light might be replaced by extremely intense, pulsed, energetic beams of heavy ions. But the high ion particle densities required would encourage charge transfer between positive ions, e.g.

$$Cs^+ + Cs^+ \rightarrow Cs + Cs^{2+} \qquad (12)$$

Ions would then be lost from the beams and the fast escaping particles would probably damage the machines used for acceleration and storage. Since the cost of this type of apparatus is of order $\$10^9$ careful design is necessary. This requires knowledge of heavy ion collisions, similar to reaction (12), at energies of order 100 keV, which corresponds to the relative velocities of ions in the particle accelerators and storage rings.

Before we discuss individual reactions beween positive ions it is worthwhile to draw a distinction between "resonant" and "non-resonant" charge transfer. Consider

$$He^{+} + He^{2+} \rightarrow He^{2+} + He^{+} \qquad (13)$$

Here, the commuting electron needed an energy of 54 eV to be extracted from He^{+}, but it gained 54 eV on being absorbed in the He^{2+} ion. The net energy change (ΔE) is therefore zero and the reaction is said to be resonant. By contrast, $\Delta E = 62$ eV for he non-resonant process,

$$H^{+} + Li^{+} \rightarrow H + Li^{2+} \qquad (14)$$

The distinction between resonant and non-resonant reactions is not quite so clear-cut as for ion-neutral collisions because mutual repulsion between two positive ions tends to keep them apart at low energies and therefore reduces the cross section. Nevetheless, at lower energies charge transfer cross sections (σ_c) are typically two orders of magnitude larger than for non-resonant processes.

It is also customary to distinguish three energy regimes for ion-ion reactions. At low energies interacting ions can be considered to form a short-lived molecular complex and so molecular-type calculations are often used. At high energies impact theories are more appropriate, but theory is most difficult at the intermediate energies where the relative speed of the ions is comparable to the orbital speeds of the active electrons. For light ions this corresponds to energies of several tens of keV, which is in the range most often encountered in stellar atmospheres and thermonuclear devices.

It is logical to start by discussing interactions between H^{+} and He^{+} ions since these involve only one electron and the reaction inevitably occurs in thermonuclear devices where He^{+} is a product of the fusion of hydrogen isotopes. The reactions,

$$H^{+} + He^{+} \rightarrow H + He^{2+} \qquad (15)$$

and

$$H^{+} + He^{+} \rightarrow H^{+} + He^{2+} + e \qquad (16)$$

have therefore attracted considerable attention from experimentalists and theoreticians and they have, in the words of Fritsch and Lin[15] developed into a "testing ground" for theoretical models. Reactions

(15) and (16) will be called charge transfer and ionization, respectively, and their cross sections will be denoted by the σ_c and σ_i. Details of earlier work can be found in the review by Dolder and Peart[3] but here we will stress recent developments.

Figure 3 illustrates measurements by Peart et al[16] (open circles) and Salzborn's group[17] (diamonds) of σ_c obtained by counting, in coincidence, H and He^{2+} formed by reaction (15). At these energies ionization is negligible ($\sigma_i \simeq 0$) and so an earlier measurement[18] of cross sections (σ_t) for the production of He^{2+} ($\sigma_t = \sigma_c + \sigma_i$) are also included (closed circles). The agreement between the three measurements and theory is impressive. The open squares, crosses and full squares are results of a molecular state model,[19] a Sturmian calculation[20] and pseudo-state expansion[21]. The two-centre atomic expansion model of Bransden and Noble[22] also agrees with experiment below 15 keV. This agreement is gratifying because the understanding of this one-electron reaction provides a firm base from which theory can advance to more complicated reactions. Moreover, the results in fig. 3 cover the low-energy and intermediate-energy ranges where theory is most difficult.

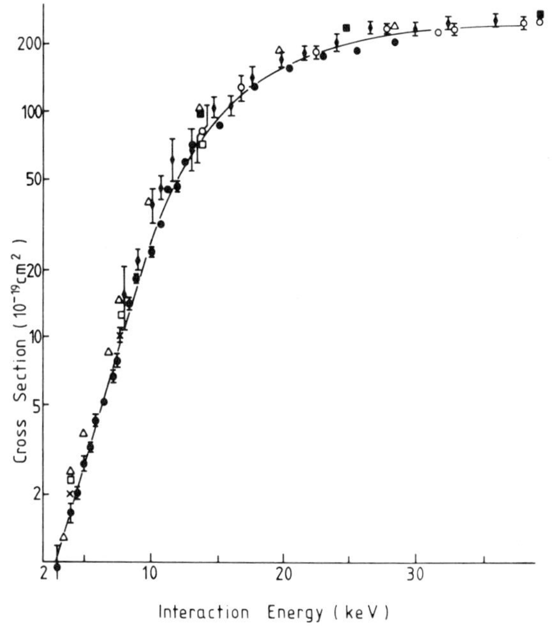

Fig. 3. Experimental and theoretical results for H^+/He^+ collisions. Measurements of charge transfer by Peart et al (open circles) and Rinn et al (diamonds) and of He^{2+} production by Peart et al (closed circles) compared with three theoretical results identified in the text (open and closed squares and crosses).

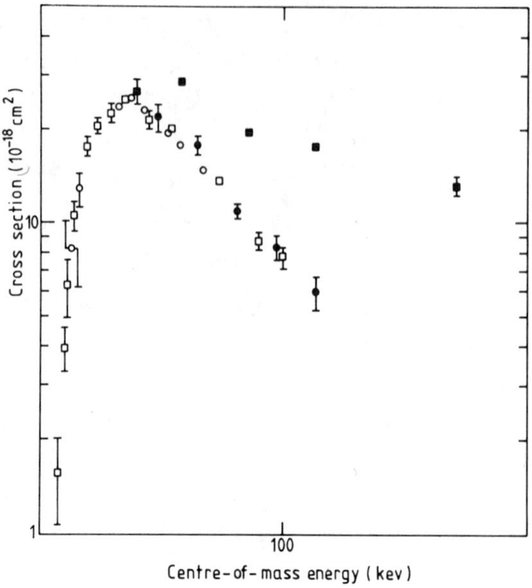

Fig. 4. Experimental results for collisions between H^+ and He^+ at higher energies. The symbols denote measurements of charge transfer by Peart et al (open circles), Rinn et al (open squares) and Watts et al (closed circles). The closed squares illustrate cross sections for He^{2+} formation by Angel et al.

Turning to higher energies, there was previously a difference between measurements of σ_c by Angel et al[23] and Peart et al[16]. This, however, has recently been resolved. It transpired that there was an arithmetic error in the reduction of data by Angel et al and the corrected values are illustrated in fig. 4. New have recently been reported by Rinn[25] et al and these are also included in fig. 4.

Measurements have also been obtained[26],[27],[28] for σ_t which, it will be remembered, is the sum of σ_c and σ_i. It follows that the ionization cross section can immediately be deduced since $\sigma_i \simeq \sigma_t - \sigma_c$, but the results will necessarily only be approximate because we are dealing with the difference between two relatively large quantities. Fig. 5 illustrates the estimate by Peart et al (curve N) which is in fair agreement with theoretical predictions by Reading et al[29] (triangles) Fritsch[30] (crosses) and Miraglia[31] (MSI and MSF).

Recent measurements of σ_t and σ_c by the Belfast group[32] for collisions between protons and Tl^+, Al^+, In^+ and Ga^+ have enabled a number of ionization cross sections to be deduced and compared with a simple scaling law suggested by Neill et al[33]. This takes the form,

$$\sigma_i = R^2 Z_p^{\ 2}\ \overline{\sigma_i} \quad \sum_{n=1}^{n} n_i/u_i^{\ 2} \qquad (17)$$

where R is Rydberg's constant, Z_p the projectile charge, and u_i represents the ionization energy of the n_i electrons in the i th subshell. Figure 6 illustrates these scaled cross sections for several ion pairs and the results are seen to be compatible with equation (17), within a factor two.

Charge transfer between H^+ and He^+ is a non-resonant, one-electron process and it is equally interesting to consider resonant, one-electron transfer. The most obvious example is,

$$He^+ + He^{2+} \rightarrow He^{2+} + He^+ \qquad (18)$$

but here the experimental situation is much less satisfactory. A problem arises because both collision products are positively charged and so they repel. Both experiments(34),(35) which have been reported used detectors which had insufficient angular acceptance to collect all of the ions formed and so corrections were derived from theoretical differential cross sections and applied to the measurements. When this was done the results were in broad agreement with semi-classical, two-state approximation by Dickinson and Hardie,(36) but it is clearly unsatisfactory for experimentalists to appeal to theory for corrections exceeding 30%. New measurements are needed. It is worth adding that the maximum charge transfer cross sections for reaction (15) and (18) were about 2 x 10^{-17} cm^{-2} and 5 x 10^{-16} cm^2, respectively. As one would expect, resonant reactions have much larger cross sections than non-resonant processes.

Another point to stress is that resonant processes tend to proceed via the ground states of the reactants. If excited states were involved there would be an energy defect and so these channels are relatively unattractive. One might therefore expect cross sections for forward and reverse reactions to be similar. An example is,

$$H^+ + Ti^+ \rightarrow H + Ti^{2+} \qquad (19)$$

and

$$H + Ti^{2+} \rightarrow H^+ + Ti^+ \qquad (20)$$

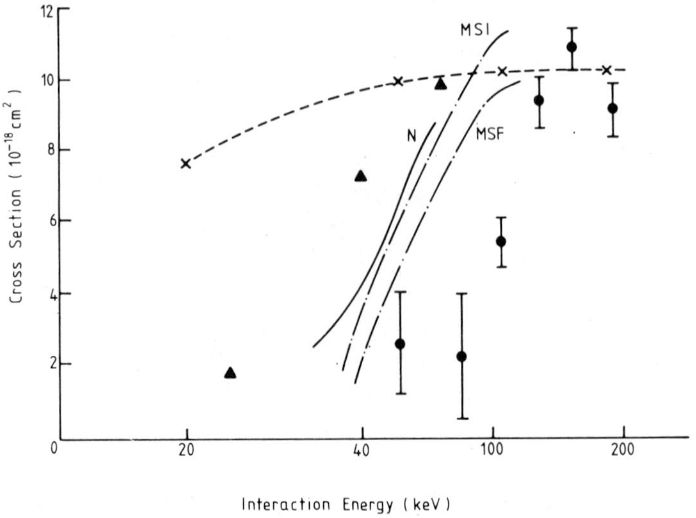

Fig. 5. Cross sections for the ionization of He^+ by protons deduced from measurements by Peart et al (curve N) and Angel et al (closed circles) compared with theoretical predictions by Fritsch (crosses) and Miraglia (curves MSI and MSF).

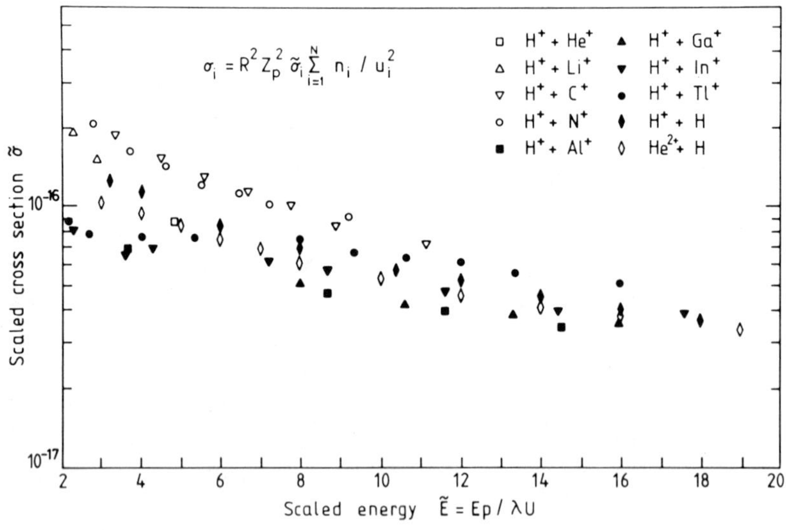

Fig. 6. Comparison of scaled measurements of ionization cross sections for ten ion pairs with the scaling law proposed by Neill et al.

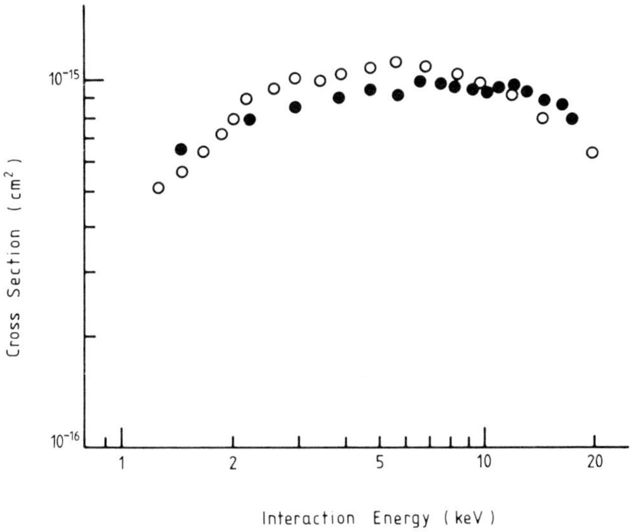

Fig. 7. Comparison of measured cross sections for charge transfer between H^+/Ti^+ (closed circles) and H/Ti^{2+} (open circles).

which are very good examples of accidental resonance because the second ionization potential of titanium happens to be 13.57 eV so that ΔE is only 0.025 eV. Figure 7 compares cross sections measured[(38),(39)] for reactions (19) and (20) and it can be seen that they are almost identical.

Other near-resonant reactions that have been studied in the forward[(40)] and reverse[(37)] directions include,

$$H^+ + Mg^+ \rightarrow H + Mg^{2+} \qquad (21)$$

and

$$H + Mg^{2+} \rightarrow H^+ + Mg^+ \qquad (22)$$

for which $\Delta E = 1.4$ eV. Again there is very close agreement between the two cross sections. Measurements reported for,

$$H^+ + C^+ \rightarrow C^{2+} + \ldots.. \qquad (23)$$

$$H^+ + N^+ \rightarrow N^{2+} + \ldots.. \qquad (24)$$

$$H^+ + Fe^+ \rightarrow Fe^{2+} + \ldots.. \qquad (25)$$

are particularly interesting because the ions are all likely to be found in thermonuclear reactors. The results are illustrated by figs. 8, 9 and 10. Very recently the Belfast group measured charge transfer cross sections for H^+/C^+ and H^+/N^+ reactions and these should be published by mid 1986.

The closed circles in fig. 8 represent measurements by Neill et al[41] (see Gilbody, 1981) for reaction (2). These can be compared with measured[42] cross sections for the charge transfer process,

$$H + C^{2+} \rightarrow H^+ + C^+ \qquad (26)$$

illustrated by the broken curve. The chain curve shown the electron impact cross sections[43]

$$e + C^+ \rightarrow C^{2+} + 2e \qquad (27)$$

for electrons scaled to the same relative speed. A reasonable interpretation suggests that charge transfer and ionization both contributed appreciably to the formation of C^{2+} throughout the energy range studied by Neill et al since, at the highest and lowest energies their measurements appear to converge respectively towards the ionization and charge transfer cross sections.

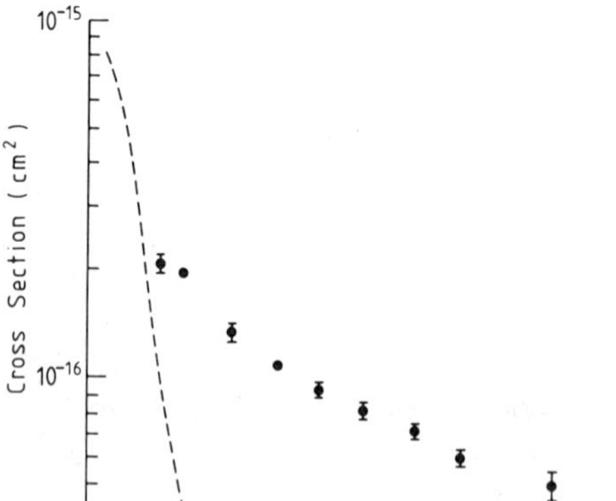

Fig. 8. Measured cross sections for C^{2+} formation by H^+/C^+ collisions and charge transfer between C^{2+} and H (broken curve).

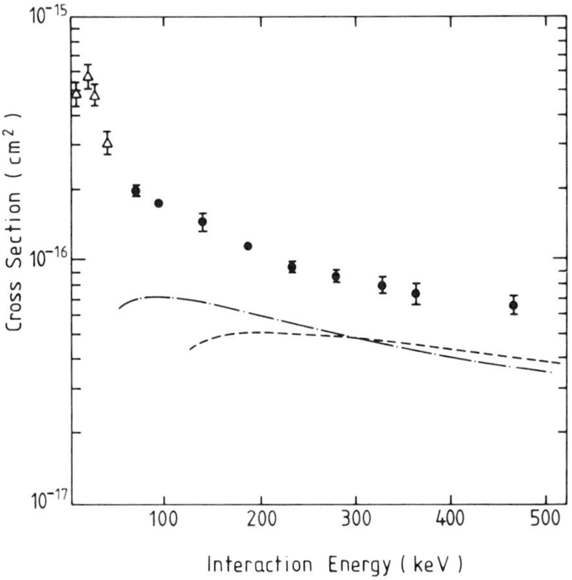

Fig. 9. Measured cross sections for N^{2+} formation by H^+/N^+ collisions (closed circles) and for charge transfer between N^{2+} and H (triangles). The broken and chain dashed curves respectively show measurements and calculations for the ionization of N^+ by electrons with the same relative speed.

One might expect a similar situation to prevail for H^+/N^+ collisions because the energy defects for charge transfer between H^+/C^+ and H^+/N^+ are not too different (10.8 eV and 16.0 eV, respectively). This view is supported by fig. 9 which compares measurements[(41),(42)] for reaction (24) (solid) circles with those by Phaneuf[(44)] et al (open triangles) for,

$$N^{2+} + H \rightarrow N^+ + H^+ \qquad (28)$$

The two results appear to converge at lower energies. At higher energies they approach the broken and chain dashed curves which show measurements and calculations for the ionization of N^+ by electrons with the same relative speed.

Iron, in various stages of ionization is another probable impurity and reaction (25) is the only ion-ion reaction involving this element for which results are available[(38)]. They should be regarded as preliminary because the cross section appears to depend upon the Fe^+ ion energy, presumably because not all the Fe^{2+} ions were collected. The

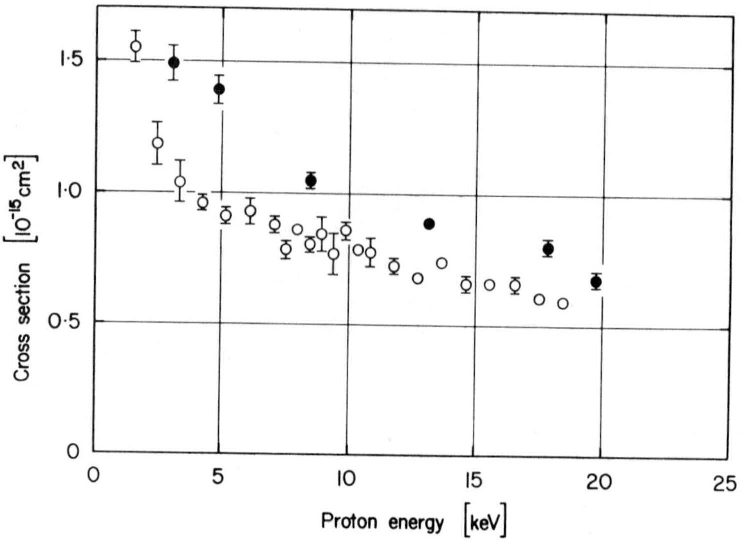

Fig. 10. Preliminary measurements for Fe^{2+} formation by H^+/Fe^+ collisions. The solid and open circles respectively represent measurements with 4keV and 2keV Fe^+ ions. They can be taken to indicate lower bounds for the cross section.

solid and open symbols in fig. 10 respectively refer to measurements witxh 4 keV and 2 keV ions; they can be taken to represent a reliable lower limit.

Lithium is less likely to be found in thermonuclear or astrophysical plasmas but its relative simplicity makes it an attractive test for theory. The two-electron processes,

$$H^+ + Li^+ \rightarrow H + Li^{2+} \tag{29}$$

and

$$H^+ + Li^+ \rightarrow H^+ + Li^{2+} + e \tag{30}$$

are a natural progression from the one-electron H^+/He^+ and He^+/He^+ reactions already discussed. These processes (in addition to reactions between H^+/Li^{2+} and H^+/Li^{3+}) have been studied by Ford et al(45) in the 'perturbed-one-and-a-half-centre' (POHCE) formulation. The sum (σ_t) of cross sections (σ_c and σ_i) for reactions (28) and (29) can be directly compared with measurements by Sewell et al(46) for,

$$H^+ + Li^+ \rightarrow Li^{2+} + \ldots\ldots \tag{31}$$

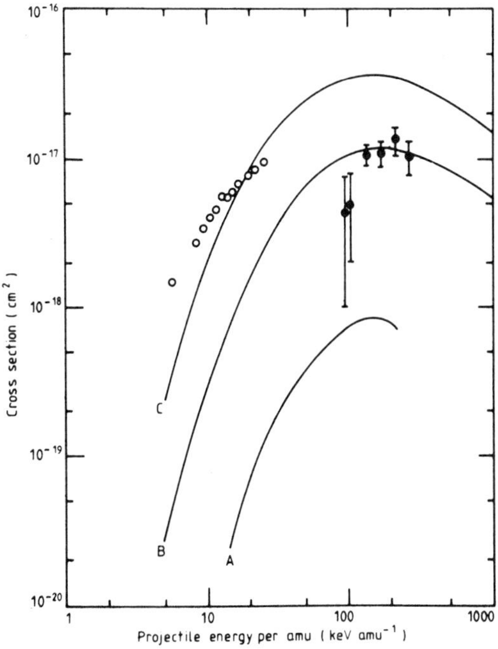

Fig. 11. Ionization and charge transfer in H^+/Li^+ and Li^+/Li^+ collisions. Curves A and C respectively illustrate Born calculations for charge transfer and ionization in Li^+/Li^+ collisions. Open circles denote measurements for Li^{2+} formation, which is the sum of both processes. Curve B and the full circles illustrate results of similar calculations and measurements for Li^{2+} formation by H^+/Li^+ collisions.

which are represented by full circles in the upper part of fig. 11. There is fair agreement with the POHCE calculation (crosses) and a classical Monte Carlo calculation(47) (triangles). The figure also illustrates the theoretical and experimental results for σ_c and σ_i and the result of a Coulomb distorted-wave calculation(49) at 200 keV. amu^{-1}.

It was mentioned previously that an interest in ion-ion collisions arose in connection with heavy ion fusion, where cross sections were needed for interaction between heavy ions of the same type, e.g.

$$Cs^+ + Cs^+ \rightarrow Cs^{2+} + \ldots.. \qquad (32)$$

The study of these heavy ions poses considerable theoretical and experimental difficulties and so it was decided to approach the problem by looking at ions of increasing complexity. Measurements have therefore been reported for

$$He^+ + He^+ \rightarrow He^{2+} \ldots\ldots \qquad (33)$$

and for similar reactions between Li^+/Li^+, Na^+/Na^+, K^+/K^+, Rb^+/Rb^+, Cs^+/Cs^+ and Tl^+/Tl^+.

Measurements[50] of charge transfer and He^{2+} formation by He^+/He^+ collisions are illustrated, respectively, by the open and closed circles in fig. 12. There is no theory with which these results can be directly compared but a comparison can be made with calculations[51] (full curve) and measurements (crosses) by Afrosimov et al[52] for,

$$He(1s^2) + He^{2+} \rightarrow He^+(1s) + He^+(1s) \qquad (34)$$

The energy must first be scaled to allow for the fact that Afrosimov et al used the $^3He^{2+}$ isotope and their cross sections must then be multiplied by 0.25 to take account of the total spin of the system (note: in the revese reaction to (33) the electrons in the $He^+(1s)$ ions can have either spin up or down, whereas for the forward reaction they must both have opposite spins; therefore the forward reaction is four times less probable). When this is done there is good agreement beweeen the results for (33) and or He^+/He^+ charge transfer (open circles).

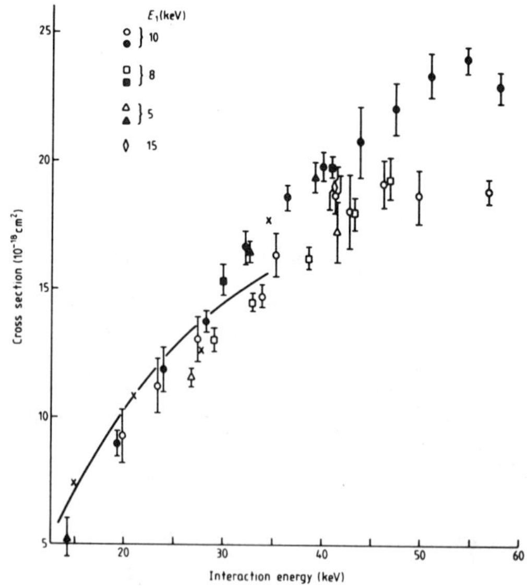

Fig. 12. Measured cross sections for charge transfer (open circles) and He^{2+} formation (closed circles) by He^+/He^+ collisions, compared with measurements (crosses) and calculations (full curve) for He^+ (1s) formation by He^{2+}/He collisions.

The situation is less satisfactory for Cs^+/Cs^+ collisions where measurements for reaction (31) by Peart et al[53] and Neill et al[54] differ at lower energies. It has been persuasively argued[54] that the experiments by Peart et al failed to collect all of the Cs^{2+} ions at lower energies. This is a very valid consideration because at energies of order 10 keV Cs^+ ions move quite slowly and so they may be appreciably scattered by reactive collisions. If this occurs the angular acceptance of the particle detector used to detect scattered ions may have been insufficient and the error would be even greater for the heavier Tl^+ ions[55]. The matter is, however, not entirely resolved because Stalder[56] et al subsequently studied Cs^+/Cs^+ collisions and applied correction factors, based on the work of Neill et al, for losses due to angular scattering. When this was done their results were closer to those of Peart et al, although they had "difficulty in estimating these effects" and they concluded that their "results do not resolve this discrepancy".

REFERENCES

1. S. Baliunas and S.K. Butler, Astrophys. J. Letts, 235. L45, 1980.
2. J. Moseley, R.E. Olson and J.R. Paterson, Case Studies in Atomic Colln. Phys. 5. 1, 1975 (Eds. E.W. McDaniel & M.R.C. McDowell).
3. K. Dolder and B. Peart, Repts. Prog. Phys. 48, 1283, 1985.
4. F. Brouillard and W. Claeys, Physics of ion-ion and electron-ion collisions. Eds F. Brouillard and J.W. McGowan (Plenum: New York) 1983.
5. J.B. Mitchell and J.W. McGowan, Physics of Ion-Ion and Electron-Ion collisions, Eds. F. Brouillard and J.W. McGowan (Plenum: New York) 1983.
6. K. Dolder, Case Studies in Atomic Colln. Phys. 1, 249, Eds. E.W. McDaniel and M.R.C. McDowell (N. Holland Amsterdam) 1969
7. S. Szücs, M. Karemera, M. Terao and F. Brouillard, J. Phys. B. 17, 1613, 1984.
8. B. Peart, M. Bennett and K. Dolder J. Phys. B. 18, L439, 1985.
9. V. Sidis, C. Kubach and D. Fussen, Phys. Rev. Lett. 47, 1280, 1981; Phys. Rev. A. 27, 2431, 1983.
10. F. Borondo, A. Macias & A. Riera, Phys. Rev. Lett. 46, 420, 1981.
11. H.S.W. Massey, "Negative Ions", 3rd Edition, Cambridge University Press, 1976.
12. B. Peart, M. Bennett and K. Dolder, J. Phys. B. (in course of publication) 1985.
13. J. Weiner, M.B. Peatman & R.S. Berry, Phys. Rev. A.4, 1824, 1971.
14. J. Moseley, W. Aberth and J.R. Peterson, J. Geophys. Res. 77, 255, 1972.
15. W. Fritsch and C.D. Lin, J. Phys. B. 15, 1255, 1982.
16. B. Peart, K. Rinn and K. Dolder, J. Phys. B. 16, 1461, 1983.
17. E. Salzborn (private communication) 1985.
18. B. Peart, R. Grey and K. Dolder, J. Phys. B., 10, 2675, 1977.
19. T.G. Winter, G.J. Hatton and N.F. Lane, Phys. Rev. A, 22, 930, 1980.

20. T.G. Winter, Phys. Rev. A 25, 697, 1982.
21. W. Fritsch and C.D. Lin, J. Phys. B. 15, 1255, 1982.
22. B.H. Bransden and C.J. Noble, J. Phys. B. 14, 1849, 1981.
23. G.C. Angel, E.C. Sewell, K.F. Dunn and H.B. Gilbody, J. Phys. B. 11, L297, 1978.
24. K.F. Dunn (private communication) 1985.
25. K. Rinn, F. Melchert and E. Salzborn, J. Phys. B, 18, 3783, 1985.
26. J.B.A. Mitchell, K.F. Dunn, G.C. Angel, R. Browning and H.B. Gilbody, J. Phys. B. 10, 1897, 1977.
27. G.C. Angel, K.F. Dunn, E.C. Sewell, H.B. Gilbody, J. Phys. B. 11. L49, 1978.
28. B. Peart, R. Grey and K. Dolder, J. Phys. B. 10, 2675, 1977.
29. J.F. Reading, A.L. Ford and R.L. Becker, J. Phys. B. 15, 625, 1982.
30. W. Fritsch (private communication) 1984.
31. J.E. Miraglia, J. Phys. B. 16, 1029, 1983.
32. K.F. Dunn, M.F. Watts, G.C. Angel and H.B. Gilbody, Proc. XIV ICPEAC, Palo Alto, Calif, 1985.
33. P.A. Neill, G.C. Angel, K.F. Dunn and H.B. Gilbody, J. Phys. B. 16, 2185, 1983.
34. A. Jognaux, F. Brouillard and S. Szücs, J. Phys. B. 11, L669, 1978.
35. B. Peart and K. Dolder, J. Phys. B., 12, 4155, 1979.
36. A.S. Dickinson and D.J.W. Hardie, J. Phys. B. 12, 4147, 1979.
37. R.W. McCullough, W.L. Nutt and H.B. Gilbody, J. Phys. B. 12, 4159, 1974.
38. M.F.A. Harrison (private communication) 1985.
39. R. McCullough, W.L. Nutt and H.B. Gilbody, J. Phys. B. 12, 4159, 1979.
40. B. Peart, D. Gee and K. Dolder, J. Phys. B. 10, 2683, 1977.
41. H.B. Gilbody, Proc XII ICPEAC Gatlinburg, 1981.
42. T.V. Goffe, M.B. Shah and H.B. Gilbody, J. Phys. B. 12, 2763, 1979.
43. K.L. Aitken, M.F.A. Harrison and R.D. Rundel, J. Phys. B. 4, 1189, 1971.
44. R.A. Phaneuf, F.W. Meyer, and R.H. McKnight, Phys. Rev A 17, 534, 1978.
45. A.L. Ford, J.F. Reading and R.L. Becker, J. Phys. B 15, 3257, 1982.
46. E.C. Sewell, G.C. Angel, K.F. Dunn and H.B. Gilbody, J. Phys. B. 13, 2269, 1980.
47. R.E. Olson and A. Salop, Phys. Rev. A. 16, 531, 1977.
48. K.E. Banyard and G.W. Shirtcliffe, J. Phys. B. 12, 3247, 1979.
49. B. Peart, K. Rinn and K. Dolder, J. Phys, B. 16, 2831, 1983.
50. C. Harel and A. Salin, J. Phys. B. 13, 785, 1980.
51. V.V. Afrosimov, A.A. Basalaev, C.A. Leiko and M.N. Panov, Sov. Phys. - JETP 47, 837, 1978.
52. B. Peart, R.A. Forrest and K. Dolder, J. Phys. B. 14, 1655, 1981.
53. P.A. Neill, G.C. Angel, K.F. Dunn and H.B. Gilbody, J. Phys. B. 15, 4219, 1982.
54. K.R. Stalder, K.H. Berkner and R.V. Pyle, Phys. Rev. A 29, 3052, 1984.

CHARGE EXCHANGE AND IONISATION IN COLLISIONS BETWEEN POSITIVE IONS

K F Dunn

Department of Pure and Applied Physics
The Queen's University of Belfast
Belfast BT7 1NN UK

INTRODUCTION

It is now twenty years since the first reported studies of collisions between positive ions. The first such investigation was carried out by Guidini et al (1965) and Sinda et al (1967) who obtained data of limited accuracy for $H_2^+ - N_2^+$ collisions at keV energies. In subsequent work Brouillard and Delfosse (1967) described a merged beam method for the study of $He^{2+} - He^+$ charge transfer collisions at low energies and obtained a tentative result. However, since 1977 reliable experimental data on charge transfer and ionisation for a number of different collision processes have become available. Apart from our group in Belfast (Gilbody), groups in Culham (Harrison), Newcastle (Dolder), Louvain-la-Neuve (Brouillard) and Giessen (Salzborn) are now active in this field.

Much of the interest in charge transfer and ionisation in collisions between positive ions stems from their relevance to controlled thermonuclear fusion research (cf Gilbody 1979). For example, in a Tokamak device, particle loss and plasma cooling can occur through charge transfer processes of the type

$$H^+ + X^{q+} \rightarrow H + X^{(q+1)+} \qquad (1)$$

involving collisions between hydrogen isotope fuel ions and impurity species. Fast H atoms formed in this way escape from the magnetic confinement while the charge state of X is increased leading to increased energy loss through electron-ion recombination, line radiation and free-free bremsstrahlung.

There has also been considerable interest in schemes to promote fusion by the heating and inertial confinement of DT pellets by intense pulsed beams of GeV energy heavy ions (Godlove 1979). The design of suitable beam drivers is greatly influenced by ion-ion collisions within the beam which occur at energies of approximately 0.5 keV a.m.u.$^{-1}$ (Kim 1976) due to the lateral and transverse motion of the beam particles. If ions which change their charge as a result of charge transfer

$$X^+ + X^+ \rightarrow X + X^{2+} \qquad (2)$$

or ionisation

$$X^+ + X^+ \rightarrow X^+ + X^{2+} + e \qquad (3)$$

are lost from the beam, this may result in not only an unacceptable reduction in beam intensity but in serious wall-effect secondary processes within the storage ring.

In the past decade ion beam probing has been demonstrated to be a versatile technique for measuring plasma density, space potential and particle temperature in high energy, magnetically confined plasma (Reinovsky et al 1975). A monoenergetic beam of heavy singly charged ions is incident on the plasma. Ions which undergo charge changing collisions with the plasma particles will emerge from the plasma with an increase in energy proportional to the space potential at the point of collision. The number of multiply charged ions emerging is proportional to the cross sections for the charge changing processes and the density of the target particles.

EXPERIMENTAL APPROACH

An excellent detailed account of the general principles of beam-beam interactions has been given by Brouillard and Claeys (1983). All the experiments we will discuss have been carried out using the intersecting beams method and only an outline of the general experimental approach will be given here. A typical reaction is of the type

$$X^+ + Y^+ \rightarrow X + Y^{2+} \qquad (4a)$$

$$\rightarrow X^+ + Y^{2+} + e \qquad (4b)$$

The measurements directly provide cross sections $\sigma(Y^{2+})$ for the production of Y^{2+} ions from the combined processes of charge transfer (4a) and ionisation (4b) and, using a coincidence method, the cross section σ_c for charge transfer (4a). The ionisation cross section is obtained from the relation $\sigma_i = \sigma(Y^{2+}) - \sigma_c$.

The angle at which the two beams intersect may be chosen to provide a wide range of CM energies. At Belfast we have used a beam intersection angle of 90° to provide CM energies in the range 38-600 keV while the Newcastle group have used inclined beams at acute angles to give lower CM energies or obtuse angles to enhance their CM energy. At Louvain-la-Neuve a merged beam apparatus has given data down to interaction energies of 10 eV.

The main experimental problems arise from the low densities of the primary ion beams and the comparatively small signal count rates in relation to the background signals arising from the interaction of both beams with the residual gas. Efficient separation of all the scattered collision products formed in the beam intersection region from the primary beam components, which may be up to 10^{12} times larger, requires very carefully designed electrostatic or magnetic analysers. A comprehensive range of experimental checks have been developed for beam-beam experiments and the interested reader is referred to papers by Neill et al (1982) and Harrison (1966).

Unlike many other types of experiments the inclined beams approach yields absolute results since, in principle, all the necessary parameters can be accurately measured.

Table 1 summarises the experimental data available from groups in Belfast (B), Culham (C), Giessen (G), Louvain-la-Nueve (L) and Newcastle (N).

TABLE 1. Experimental data available from groups in Belfast (B), Culham (C), Giessen (G), Louvain-la-Neuve (L) and Newcastle (N).

Process	CM Energy (keV)	Measured cross section	Group	References
H^++He^+	40-402	$\sigma(He^{2+}),\sigma_c,\sigma_i$	B	Mitchell et al (1977) Angel et al (1978a, b)
	40-120	σ_c,σ_i	B	Watts et al (1985)
H^++He^+	3-29	$\sigma(He^{2+})$	N	Peart et al (1977a)
	14-67	$\sigma(He^{2+}),\sigma_c,\sigma_i$	N	Peart et al (1983)
H^++He^+	8-100	σ_c	G	Rinn et al (1985)
H^++Li^+	62-350	$\sigma(Li^{2+}),\sigma_c,\sigma_i$	B	Sewell et al (1980)
H^++C^+	70-462	$\sigma(C^{2+})$	B	Neill et al (1981)
H^++N^+	70-465	$\sigma(N^{2+})$	B	Neill et al (1981)
	50-150	σ_c,σ_i	B	Hopkins et al (1985)
H^++Mg^+	1-45	$\sigma(Mg^{2+})$	N	Peart et al (1977b)
H^++Ti^+	1.5-18	$\sigma(Ti^{2+})$	C	Hobbis et al (1979)
H^++Fe^+	1.5-18	$\sigma(Fe^{2+})$	C	Hobbis et al (1979)
H^++Al^+	59-580	$\sigma(Al^{2+}),\sigma_c,\sigma_i$	B	Dunn et al (1985)
H^++Ga^+	60-592	$\sigma(Ga^{2+}),\sigma_c,\sigma_i$	B	Dunn et al (1985)
H^++In^+	60-595	$\sigma(In^{2+}),\sigma_c,\sigma_i$	B	Dunn et al (1985)
H^++Tl^+	50-598	$\sigma(Tl^{2+}),\sigma_c,\sigma_i$	B	Dunn et al (1985)
H^++Ba^+	52-497	$\sigma(Ba^{2+})$	B	Dunn et al (1985)
He^++He^+	14-58	$\sigma(He^{2+}),\sigma_c$	N	Peart et al (1983)
	5-55	σ_c	G	Melchert (1985)
$^4He^{2+}+^3He^+$	0.1-1.7	$\sigma(^3He^{2+})$	L	Jognaux et al (1978)
$^3He^{2+}+^4He^+$	0.1-20	$\sigma(^4He^{2+})$	N	Peart and Dolder (1979)
Li^++Li^+	19-88	$\sigma(Li^{2+})$	N	Peart et al (1981a)
	53-240	$\sigma(Li^{2+})$	B	Watts et al (1984)
Na^++Na^+	19-88	$\sigma(Na^{2+})$	N	Peart et al (1981c)
Ar^++Ar^+	50-370	$\sigma(Ar^{2+})$	B	Angel et al (1980)
K^++K^+	19-88	$\sigma(K^{2+})$	N	Peart et al (1981c)
Rb^++Rb^+	19-88	$\sigma(Rb^{2+})$	N	Peart et al (1981c)
Xe^++Xe^+	38-303	$\sigma(Xe^{2+})$	B	Angel et al (1980)
Xe^++Xe^+	10-50	$\sigma(Xe^{2)}),\sigma_c,\sigma_i$	G	Melchert (1985)
Cs^++Cs^+	40-280	$\sigma(Cs^{2+})$	B	Dunn et al (1979) Neill et al (1982)
	19-88	$\sigma(Cs^{2+}),\sigma_c$	N	Peart et al (1981b)
Tl^++Tl^+	33-92	$\sigma(Tl^{2+})$	N	Forrest et al (1982)

RESULTS

(a) Collisions involving protons

The simplest ion-ion reaction to study is the one electron system

$$H^+ + He^+ \rightarrow H + He^{2+} \quad (5a)$$
$$\rightarrow H^+ + He^{2+} + e \quad (5b)$$

Measurements of $\sigma(He^{2+})$ have been reported by Mitchell et al (1977), Angel et al (1978a) and Peart et al (1977a) at CM energies ranging from 3-380 keV. Charge transfer cross sections σ_c have been measured by Angel et al (1978b), Peart et al (1983) and Rinn et al (1985). The results from the Newcastle and Giessen groups are in excellent agreement but are approximately 25% below those of Angel et al (1978b). However the results of Angel et al (1978b) have been found to contain an arithmetical error and should be multiplied by 0.75 to obtain their correct results. Although the corrected results of Angel et al (1978b) are in good agreement with the other measurements the Belfast group remeasured σ_c (Watts et al 1985) and these results, shown in figure 1, are in good agreement with both the measurements of Peart et al and Rinn et al. Indeed the high level of agreement among the three laboratories is most striking.

When this reaction was first measured the only charge transfer calculation was that of Olson (1978) who used the classical trajectory Monte-Carlo (CTMC) method which in this case overestimates the cross section by up to 60%. Since then 'electron transfer in the collision system H^+ + He^+ has evolved into a testing ground for different models since calculated cross sections are rather model sensitive and experiments are available' (Fritsch and Lin 1982).

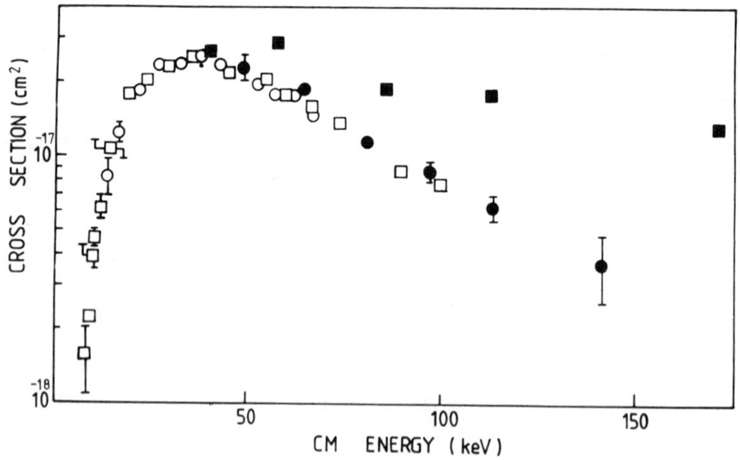

Fig 1. Cross sections σ_c for charge transfer in H^+-He^+ collisions (all results are shown at 68% confidence level): ●, Watts et al (1985); o, Peart et al (1983); □, Rinn et al (1985). Cross section $\sigma(He^{2+})$ for He^{2+} formation: ■, Angel et al (1978).

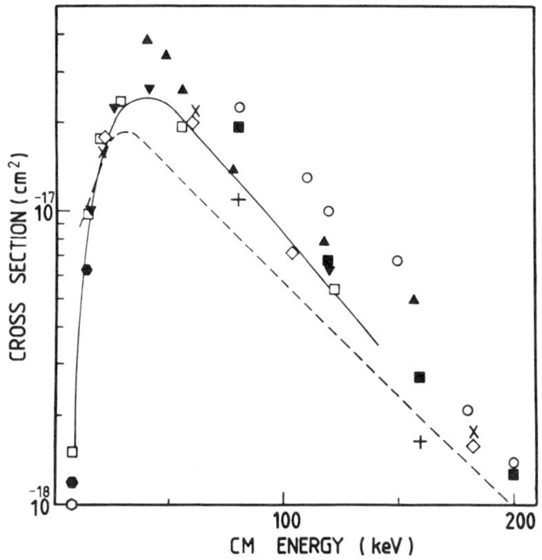

Fig 2. Cross sections σ_C for charge transfer in H^+-He^+ collisions: ____, the combined experimental results of figure 1; o, Olson (1978); ▲, Belkic et al (1979); +, Mukherjee and Sil (1980), - -, Sinha and Sil (1979); ■, Datta et al (1984); ◇, Reading et al (1982); □, Fritsch and Lin (1982); ▼, Winter (1982); x, Bransden et al (1983) ⬣, Winter et al (1980); ⬡, Kimura and Thorson (1981).

In figure 2 the experimental data for σ_C is represented as a curve and compared with a number of theoretical predictions. In the lower energy range the coupled-molecular state approach of Winter et al (1980) and Kimura and Thorson (1981) are in excellent agreement with experiment. The coupled-state calculations of Fritsch and Lin (1982), Winter (1982) and Bransden et al (1983) using basis sets of atomic wavefunctions all agree well with experiment over almost the entire energy range. Impact parameter coupled state calculations by Reading et al (1982) using a perturbative version of a previously developed one and a half centred expansion method (POHCE) yield values of σ_C which are in good accord with the experimental values. The continuum-distorted-wave (CDW) approximation used by Belkic (1979) for capture into all states of H lies about 50% above the experimental results but the CDW approximation of Mukherjee and Sil (1980) for capture into the ground state underestimates the experimental results. At the highest energies the continuum-intermediate state approximation of Datta et al (1984) agrees with the experimental results whereas the results of the Coulumb-Born calculations by Sinha and Sil (1979) considerably underestimate the cross section.

A theoretical estimate of the ionisation cross section σ_i for $H^+ + He^+$ collisions may be obtained by scaling the cross sections for $H^+ + H$ collisions based on the Born approximation (Bates and Griffing 1953). Calculations by Bates and Boyd (1962a) indicate that for interactions of a bare nucleus with a hydrogenic ion, the effects of Coulomb repulsion in ionisation are negligible except in a region on the low energy side of the cross section maximum. At the higher impact energies they suggest that cross sections for a process

$$A^{Z_p+} + B^{(Z_t-1)+} \rightarrow A^{Z_p+} + B^{Z_t+} + e \qquad (6)$$

characterised by projectile and target atomic numbers Z_p and Z_t may be obtained from the Born cross sections for $H^+ + H$ collisions by multiplying the cross section scale by Z_p^2/Z_t^4 and the projectile energy scale by $M_p Z_t^2$ where M_p is the projectile mass. Cross sections for (5b) obtained by scaling in this way are in generally poor accord (figure 3) with our experimental values of σ_i. At the highest CM energy of 386 keV the experimental value is only (79 ±6)% of the value scaled from the $H^+ + H$ Born cross sections at the equivalent projectile energy of 121 keV. However this discrepancy is not surprising in view of recent high precision measurements in our laboratory (Shah and Gilbody 1981) of σ_i for $H^+ + H$ collisions which, when normalised to the Born approximation at 1.5 MeV, are only about 83% of the Born value at 121 keV. Furthermore, Shah and Gilbody (1981) through studies of ionisation in $He^{2+} + H$ collisions, have confirmed the inadequacy of the Born scaling procedure at such low collision energies.

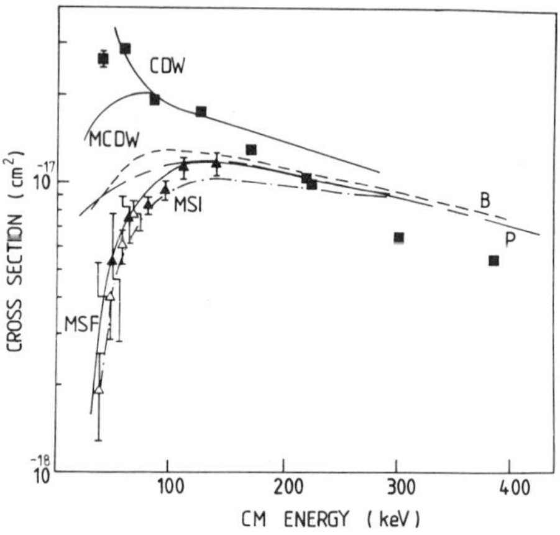

Fig 3. Cross sections σ_i for ionisation in H^+-He^+ collisions: ▲, Watts et al (1985); △, Peart et al (1983); B, Bates and Griffing (1953); P, Reading et al (1982); CDW, Belkic et al (1980)·, MCDW, Miraglia (1983); MSI and MSF, Miraglia (1983); total cross sections $\sigma(He^{2+})$: ■, Angel et al (1978).

The CDW calculation of Belkic (1980) is in poor agreement with experiment while the POHCE method of Reading et al (1982) overestimates the cross section at both high and low energies. Miraglia (1983) has applied both the multiple scattering approximation (MS) and the CDW approximation to the problem. His evaluation of the T-matrix element in the MS approximation did not neglect the internuclear potential and his CDW approximation is modified to provide a better description of the binary electron-projectile collision in the entrance channel through momentum conservation. This results in an improved low energy behaviour but still agrees with Belkic above 80 keV. Multiple scattering approximations MSF, taking account of internuclear potentials, and MSI, neglecting internuclear potentials, both agree with the measured cross sections within experimental uncertainties.

The first simple two-electron collision system studied was $H^+ + Li^+$. In our laboratory Sewell et al (1980) measured cross sections $\sigma(Li^{2+})$ for Li^{2+} formation, σ_c for charge transfer and $\sigma_i = [\sigma(Li^{2+}) - \sigma_c]$ for

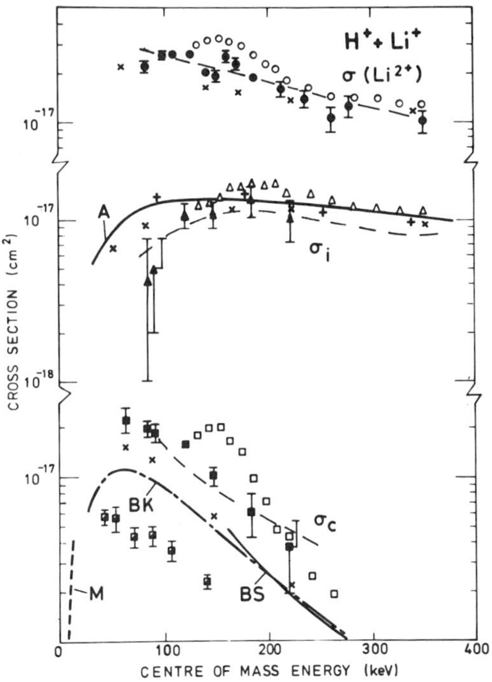

Fig 4. Data for H^+-Li^+. o, △, □ , Sewell et al (1980); ●, ▲, ■ , Olson (1980); X, Forde et al (1982); – –, Reinhold and Falcon (1985); +, McGuire (1984); ◻, Bogdanov et al (1965); Curve M, Mapleton et al (1975); Curve BS, Banyard and Shirtcliffe (1979); Curve BK, Todd (1981).

ionisation. These results (figure 4) show that as the interaction energy decreases, charge transfer makes an increasing contribution to $\sigma(Li^{2+})$ and becomes dominant below about 140 keV. At the higher impact energies values of σ_i are approaching $\sigma(Li^{2+})$. The results in Curve A were obtained by scaling the K shell ionization cross sections of Merzbacker and Lewis (1958) calculated using simplified wave functions. This procedure is equivalent to the classical scaling of ionisation cross sections described by Vriens (1969) and for isoelectronic systems by Thompson (1912). The calculated ionisation cross sections of McGuire (1984), who used the plane-wave Born approximation, are in very good agreement with the experimental results.

Ionisation and charge transfer cross sections have been calculated using the CTMC method by both Olson (cited by Sewell et al 1980) and Reinhold and Falcon (1985). The difference between the two sets of results cannot readily be explained since for other systems such as H^+ + H and H^+ + He there is almost exact agreement between the two groups. The results of Reinhold and Falcon, obtained with both Coulumb effective potentials and model potentials, are in excellent agreement with the experimental results. Values of σ_c calculated by Banyard and Shirtcliffe (1979) using a continuum distorted wave method and by Todd (1980) using the Brinkman-Kramers approximation are approximately a factor of two smaller than experiment.

Coupled channel calculations of both ionisation and charge transfer in H^+ + Li^+ collisions have been carried out by Forde et al (1982) using their POHCE formulation. Their ionisation cross sections are in good agreement with experiment but their capture cross sections are lower by a factor of two. Figure 4 also shows low energy theoretical estimates of σ_c obtained by Mapleton et al (1975) using a two-state atomic expansion method. In addition, it is interesting to note that experimental values obtained by Bogdanov et al (1965) from a study of the scattering of protons in a lithium arc, while about four times smaller, exhibit an energy dependence similar to our experimental values.

Figures 5 and 6 show our measured cross sections (Neill et al 1983, Hopkins et al 1985) $\sigma(C^{2+})$ for $H^+ + C^+$ collisions and $\sigma(N^{2+})$, σ_c for the $H^+ + N^+$ system. In the latter case charge transfer increasingly dominates with decreasing energy. An estimate of the charge transfer contribution to the measured cross section is provided by those for the corresponding reverse processes

$$C^{2+} + H \rightarrow C^+ + H^+$$

and $$N^{2+} + H \rightarrow N^+ + H^+$$

provided that the excited state channels are not greatly different. Such cross sections for both C^{2+} and for N^{2+} impact (Goffe et al 1979, Phaneuf et al 1979) are plotted in figures 5 and 6 respectively. The magnitudes of the forward and reverse reactions for the $H^+ - N^+$ system suggests that the excited state populations are not greatly different.

At high velocities, where ionisation is assumed to dominate our cross sections, $\sigma(C^{2+})$ tends to converge to the cross section for ionisation of C^+ by equivelocity electrons measured by Aitken et al (1971). In the case of electron impact ionisation of N^+ (Harrison et al 1963) this convergence is not as marked. No evidence of significant populations of long lived metastable C^+ and N^+ could be found in parallel experiments by Nutt et al (1979) using an identical ion source operating under similar conditions. In spite of the findings of these beam attenuation studies other authors, such as Moran and Wilcox (1978, 1979) have suggested that a small metastable content may be present. However, Harrison et al (1963) using a similar type source found no evidence of ionisation of N^+ by electron impact at electron energies less than the ionisation threshold of the ground state ion. There is a further possibility that the N^+ beam contained a small admixture of impurity N_2^{2+} ions (McGowan and Kerwin 1963) which could not be resolved in the analysis. This may contribute to the cross section through the reaction

$$H^+ + N_2^{2+} \rightarrow H^+ + N^{2+} + N$$

where the N^{2+} ions are indistinguishable in our experiment from those produced by $H^+ + N^+$ collisions.

The first Born approximation calculations of Peach (1971) for the cross sections for proton impact ionisation of N^+ are seen in figure 6 to be consistently about 50% of the measured cross section. McGuire (1984) has used the plane-wave Born approximation to calculate ionisation cross sections for C^+ and N^+ targets. He calculated both the direct ionisation cross sections from the outer shell and the contribution from 2s excitation followed by autoionisation from the excited states $1s^2 2s^1 2p^m(nl)$. Excitation-autoionisation enhances the total ionisation cross section by approximately 10% for both targets over the present energy range and the total ionisation cross sections agree, above 100 keV CM, with the measurements to within 25% or better.

The concept of resonance cannot strictly be applied to reactions between positive ions since Coulomb repulsion dominates at low energies. Nevertheless it is observed that near resonance reactions have large cross

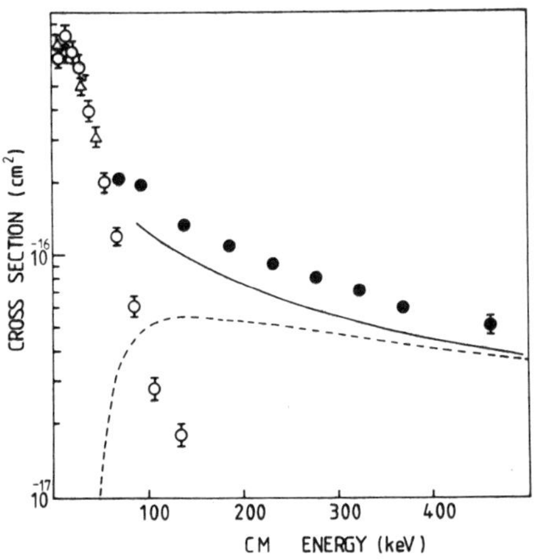

Fig. 5 Data for H^+ - C^+ collisions. ●, $\sigma(C^{2+})$, Neill et al (1981); ___, σ_i, McGuire (1984); ○ , cross sections for the inverse charge transfer process, Goffe et al (1979); Δ , Phaneuf et al (1979); - -, cross sections for ionisation of C^+ by equivelocity electrons, Aitkin et al (1971).

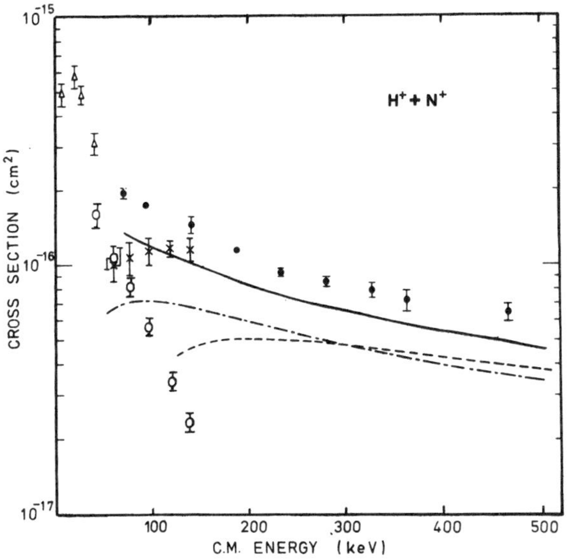

Fig 6. Data for H^+ - N^+ collisions. ●, $\sigma(N^{2+})$, Neill et al (1981); X , σ_i, Hopkins et al (1985); __ ., σ_i, Peach (1971); ___ , σ_i, Maguire (1984); - - , cross section for ionisation of N^+ by equivelocity electrons, Harrison et al (1963); ○ , σ_c, Hopkins et al (1985), Δ , cross section for the inverse process, Phaneuf et al (1979).

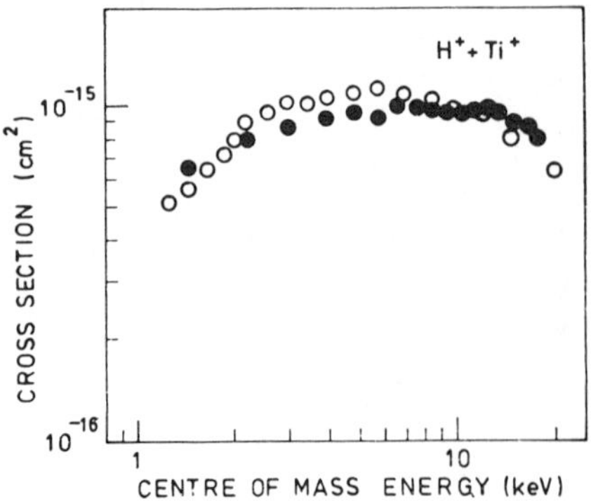

Fig 7. Data for H^+-Ti^+ collisions
$\sigma(Ti^{2+})$: ●, Hobbis et al (1979). Cross section for inverse charge transfer: o, McCullough et al (1979).

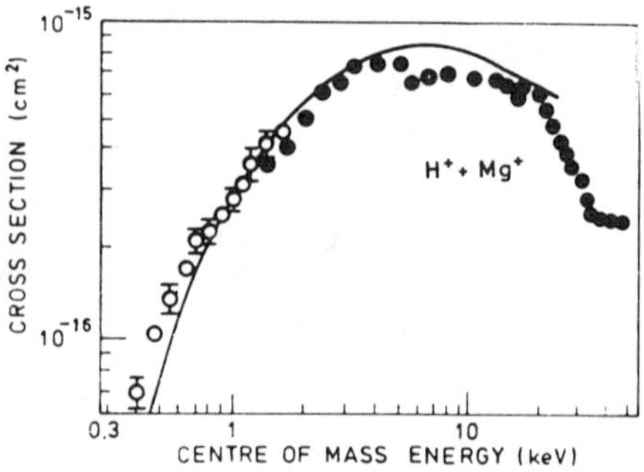

Fig 8. Data for H^+-Mg^+ collisions
$\sigma(Mg^{2+})$: ●, Peart et al (1977b). Cross section for inverse charge transfer process: o, McCullough et al (1979); ___, Bates et al (1964).

sections. Figures 7 and 8 show cross sections $\sigma(Ti^{2+})$ and $\sigma(Mg^{2+})$ for H^+ + Ti^+ and $H^+ + Mg^+$ collisions measured by Hobbis et al (1979) and Peart et al (1977b). If the measured cross sections are dominated by the charge transfer process,

$$H^+ + Ti^+ \rightarrow H + Ti^{2+} \quad (\Delta E = 0.025 \text{ eV}) \tag{7}$$

$$H^+ + Mg^+ \rightarrow H + Mg^{2+} \quad (\Delta E = -1.4 \text{ eV}) \tag{8}$$

which involves small energy defects of 0.025 eV and -1.4 eV respectively for the H(1s) formation channel, close agreement with cross sections for the inverse process might be expected if ground state species are predominant. Cross sections for the reverse reactions measured in this laboratory (McCullough et al 1979) for Ti^{2+} + H and Mg^{2+} + H and semi-classical two-state calculations by Bates et al (1964) for Mg^{2+} + H collisions are seen to be in excellent agreement with the corresponding ion-ion cross sections. This agreement between the measurements is gratifying especially in view of the disparity of the experimental methods.

Hobbis et al (1979) has also obtained data for Fe^{2+} production in H^+ - Fe^+ collisions in the CM energy range 1.5 - 18 keV. At energies above 6 keV these cross sections, figure 9, are comparable in shape and magnitude to those of both Ti^+ and Mg^+. However, at lower energies $\sigma(Fe^{2+})$ increases with decreasing energy, reaching a value of 1.5 x 10^{-15} cm^2 at the lowest energy of the experiment. When the metallic ion energy was varied no significant effect could be observed with either Ti^+ or Mg^+ but in the case of Fe^+ the measured cross section increases with increasing Fe^+ energy suggesting that a long lived excited state or states is present in the ion beam. The charge exchange reaction is not energy resonant so that the agreement with the inverse charge transfer reaction, measured by Shah (1985), is somewhat surprising.

Recently we have studied charge transfer and ionisation in collisions between protons and the ions Al^+, Ga^+, In^+ and Tl^+ (Dunn et al 1985). Aluminium is an impurity in Tokamak devices and Tl^+ is being used as a heavy ion probe for plasma diagnostics. This study of the group IIIB elements was designed to explore the extent to which measured cross sections relate to other relevant data and to simple scaling relations. Measurements of both

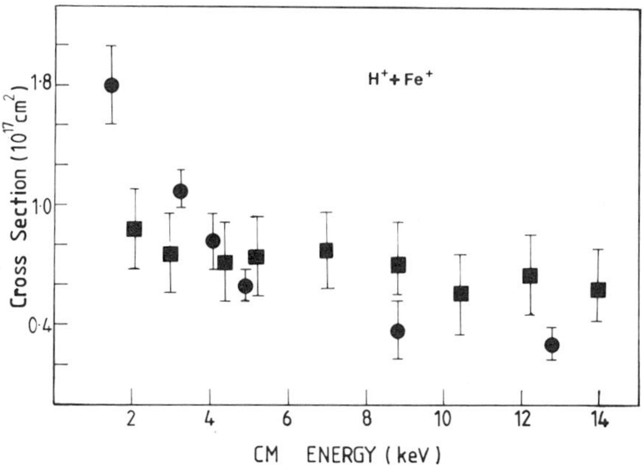

Fig 9. Data for H^+-Fe^+ collisions.
$\sigma(Fe^{2+})$: ●, Hobbis et al (1979). Cross sections for inverse charge transfer process: ■, Shah (1985).

$\sigma(X^{2+})$ and σ_c have been carried out at CM energies in the range 50-580 keV. In each case the target ions were produced from a thermionic source thereby ensuring that only ground state species were present. Figure 10 shows that the cross section $\sigma(X^{2+})$ increases with increasing mass of the target ion. Over the energy range investigated the charge transfer cross sections (figure 11) increase with decreasing energy becoming 30-50% of $\sigma(X^{2+})$ at the lowest CM energies. Conversely the ionisation cross sections are the dominant contribution to $\sigma(X^{2+})$ at CM energies above ~ 200 keV.

In the absence of detailed theoretical estimates of ionisation cross sections for many-electron ions it is interesting to investigate to

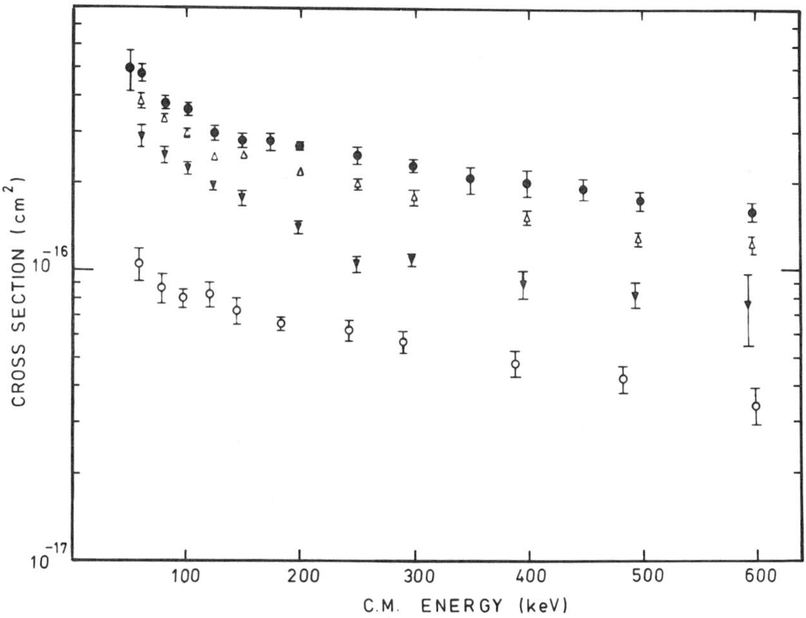

Fig 10. Cross sections $\sigma(X^{2+})$ for X^{2+} formation in H^+-X^+ collisions. ● Tl △ In ▽ Ga o Al: Dunn et al (1985).

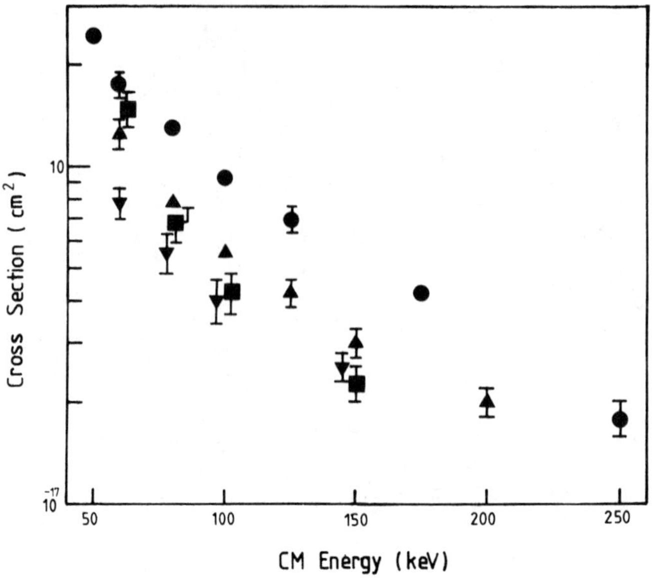

Fig 11. Charge transfer cross sections σ_c for $H^+ + X^+$ collisions. ▼, $H^+ + Al^+$; ■, $H^+ + Ga^+$; ▲, $H^+ + In^+$; ●, $H^+ + Tl^+$. (Dunn et al 1985).

what extent the experimental values can be described in terms of a general scaling relation. For isoelectronic systems A and B classical scaling based on the Thomson (1912) theory show that ionisation cross sections $\sigma_i(A)$ and $\sigma_i(B)$ should be related by the expression

$$\sigma_i(A)/\sigma_i(B) = |u(B)/u(A)|^2 \qquad (9)$$

for corresponding reduced energies where u(A) and u(B) are the ionisation energies of A and B. In a more general scaling law described by Vriens (1969) the ionisation cross section σ_i should scale to a universal form when expressed as

$$\tilde{\sigma}_i = u^2\sigma_i/R^2Z_p{}^2n \qquad (10)$$

where Z_p is the projectile charge, u is the ionisation energy, R is the Rydberg constant and n is the number of electrons responsible for the process.

The scaled energy is expressed as $E = E_p/\lambda u$ where λ is the projectile mass expressed in units of electron mass. We have modified the relationship to take account of inner shell contributions to the ionisation cross section and obtain a universal scaling law of the form

$$\sigma_i = R^2 Z_p^2 \tilde{\sigma}_i{}^2 \sum_{i=1}^{N} n_i/u_i{}^2 \qquad (11)$$

where u_i is the ionisation energy of the electrons in the ith subshell and n_i is the number of such electrons. Figure 12 shows experimental data for proton ionisation of He^+, Li^+, C^+, N^+, Al^+, Ga^+, In^+, Tl^+ and H and He^{2+} ionisation of H all scaled in this way. As can be seen the agreement is

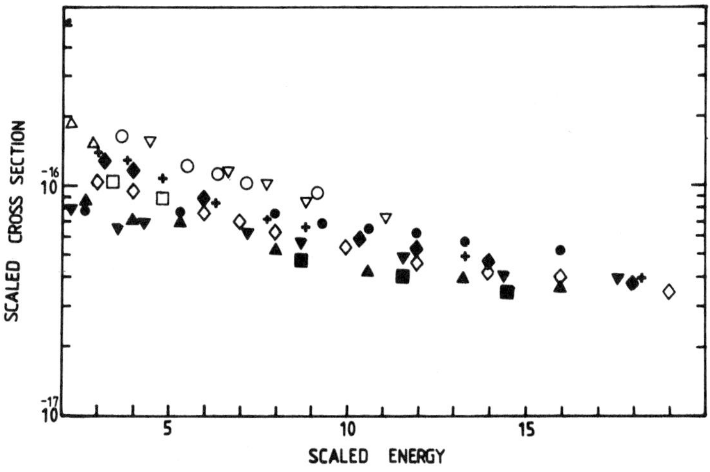

Fig 12. Classically scaled ionisation cross sections.
: H^+ + He, de Heer et al (1966), Hooper et al (1962); □, H^+ + He^+, Watts et al (1985); ▽ : H^+ + C^+ ○ : H^+ + N^+, Neill et al (1983); △ : H^+ + Li^+, Sewell et al (1980); ■ , ▲, ▼ , ●: H^+ + Al^+, Ga^+, In^+, Tl^+, Dunn et al (1985); ✚ : H^+ + H, Shah and Gilbody (1981); ◇ : He^{2+} + H, Shah and Gilbody (1981).

remarkably good, to within about 40% for all the measured ionisation cross sections.

(b) Collisions between helium ions

A second simple one electron system which has been studied experimentally by both Jognaux et al (1978) and Peart and Dolder (1979) is the resonance charge transfer interaction.

$$He^+ + He^{2+} \rightarrow He^{2+} + He^+$$

The measurements of Jognaux et al were carried out over the CM energy range 0.01-1.7 keV using merged beams and those of the Newcastle group over the range 0.1-20 keV using inclined beams. In this charge transfer process both products are positively charged so that they repel each other. Unfortunately in both experiments the angular acceptance of the collision product analysers and detectors was too small to accommodate all the scattered He^{2+} particles so that precise comparison with theory is precluded. Cross sections for charge transfer have been calculated by Bates and Boyd (1962b) and Dickinson and Hardie (1979) using perturbed stationary state methods (figure 13). Cross sections measured by Peart and Dolder are only about 50% and 70% of the theoretical values at 106 keV and 20.5 keV respectively. However, Dickinson and Hardie, by considering the differential cross sections, have applied a correction to their total cross sections appropriate to the upper and lower estimates of the angular acceptance of the apparatus of Peart and Dolder. Curves AL and AS show these corrected cross sections for 5 keV He^+ target ions while curves BL and BS show similar predictions for 3 keV He^+ target ions. Agreement with the cross sections measured by Peart and Dolder is then within 18%. Curve J

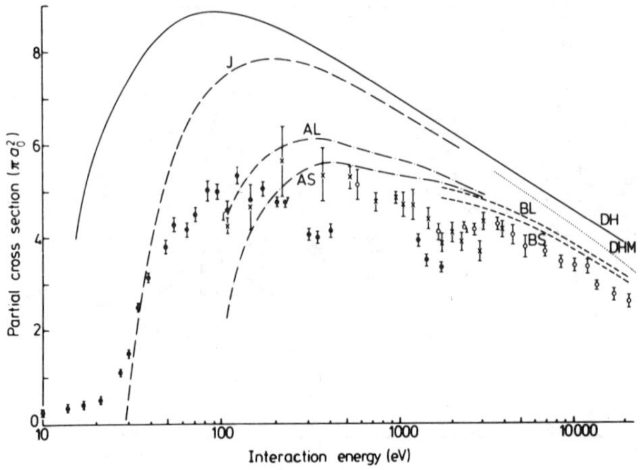

Fig 13. Data for He^{2+}-He^+ collisions. $\sigma(He^{2+})$: ●, Jognaux et al (1978); O, X, Peart et al (1979). Curve DH, Dickinson and Hardie (1979). Curve J, Jognaux et al (1979). Curves AL, AS, BL, BS; Modified theory of Dickinson and Hardie (see text).

shows a semi-classical calculation by Jognaux et al, corrected for their experimental angular acceptance, which agrees well with Dickinson and Hardie's calculations at high energies.

Peart and Dolder (1983) have also studied both charge transfer and ionisation in collisions between He^+ ions for CM energies between 14 and 58 keV. The charge transfer reaction has also been studied by Melchert (1985) over the CM energy range 8-55 keV and these results are in excellent agreement with the Newcastle work. Both the charge transfer and ionisation cross sections have been calculated by Willis et al (1985) using the CTMC method. Their calculated values for σ_c (figure 14) show quite good agreement with the measurements.

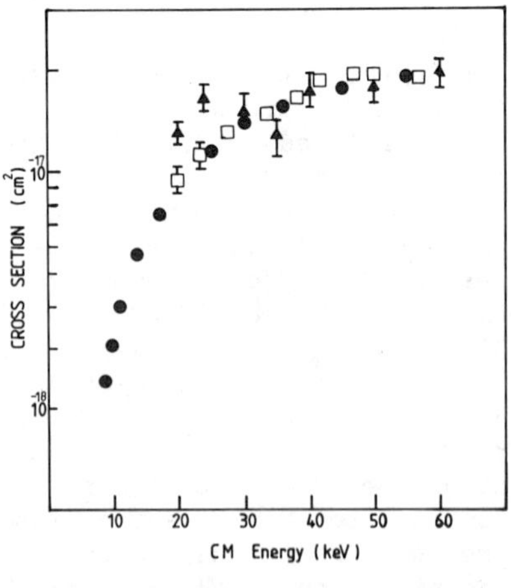

Fig 14. Data for He^+-He^+ collisions. σ_c: □ , Peart et al (1983b); ● , Melchert (1985); ▲ , Willis et al (1985).

Comparison can also be made with the reaction

$$He(1s^2) + He^{2+} \rightarrow He^+(1s) + He^+(1s) \qquad (12)$$

The calculations of Harel and Salin (1980) for this process agree well with the experimental results of Afrosimov et al (1975, 1978) and, when their results are corrected to allow for the total spin of the system and the use of the $^3He^{2+}$ isotope, they agree with the measured charge transfer cross sections in the CM energy range 5-35 keV. This close agreement suggests that at low energies charge transfer proceeds almost exclusively into the ground state of helium.

(c) Collisions between heavy ions

Since the main motivation to study collisions between multielectron ions has come from schemes for Heavy Ion Fusion it is not surprising that only identical ion pairs have been investigated experimentally. The first experimental study of total Cs^{2+} production in Cs^+ - Cs^+ collisions was reported by Dunn et al (1979) in the CM energy range 40-280 keV (0.3-2.1 keV/amu). We measured cross sections $\sigma(Cs^{2+})$ for the combined processes

$$Cs^+ + Cs^+ \rightarrow Cs + Cs^{2+} \qquad \text{Charge Transfer} \qquad (13a)$$

$$\rightarrow Cs^{m+} + Cs^{2+} + me \qquad \text{Ionisation} \qquad (13b)$$

for $m > 1$ (where m is unspecified). This was followed by similar measurements by the Newcastle group (Peart and Dolder 1981b) who used an apparatus in which two Cs^+ beams intersected at 160° to obtain cross sections in the overlapping CM energy range 19-79 keV. At their high energy limit their values of $\sigma(Cs^{2+})$ are tending to converge to our higher energy values but below this limit, down to our lowest energy of 40 keV, the two sets of data exhibit an increasing divergence.

In order to explain this large discrepancy two different but complimentary experiments were carried out at Belfast (Neill et al 1982). Firstly we used a simple beam-static-gas target configuration to measure electron loss cross sections σ_{12} for the process

$$Cs^+ + Xe \rightarrow Cs^{2+} + Xe(\Sigma) + e \qquad (14)$$

(where Σ denotes all final bound and continuum states) in the CM energy range 35-100 keV. Since ionisation provides the predominant contribution to $\sigma(Cs^{2+})$ (Olson and Liu 1981) σ_{12} for (14) (in which Xe and Cs^+ are isoelectronic) might be expected to be not greatly different from $\sigma(Cs^{2+})$. This was indeed shown, figure 15, to be the case. Also shown are two values of σ_{12} measured by the Lawrence Berkeley Group (Tanis et al 1979, Stalder et al 1984) at 50 keV and 75 keV CM which are in good accord with our results. We then investigated the influence of scattering on the measured cross sections σ_{12}. The apparent cross sections $\sigma_{12}{}^A$ corresponding to particular detector acceptance angles are shown in figure 16. It can be seen that in each case, when the angular acceptance is sufficiently large $\sigma_{12}{}^A$ approaches a plateau value corresponding to the true value of σ_{12}. The minimum half angle of acceptance required to attain the plateau value can be seen to increase from about 1.1^o at 100 keV CM to at least 1.9^o at 35 keV CM.

The results of Neill et al for $\sigma(Cs^{2+})$, which are in reasonable accord with the earlier work of Dunn et al, are shown in figure 17 along with the results of Peart et al (1981). The large discrepancy between the two sets of data is clearly apparent. However if we assume that the angular distribution of Cs^{2+} products from reactions (13b) and (14) are not greatly

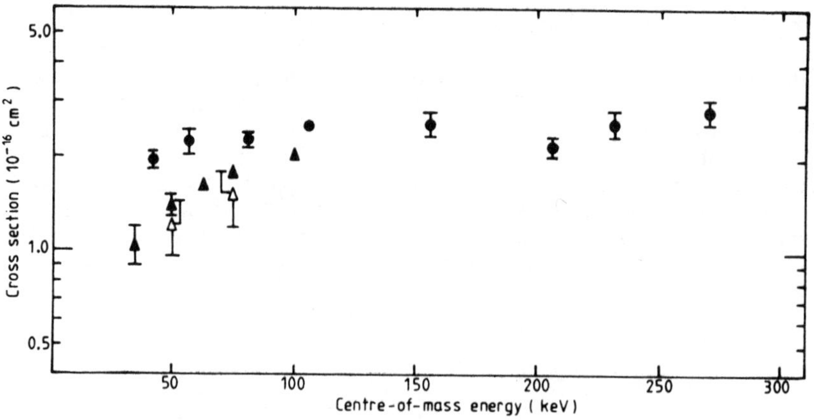

Fig 15. Cross sections $\sigma(Cs^{2+})$ for Cs^{2+} production in Cs^+-Cs^+ collisions: ●, Neill et al 1983. Cross sections σ_{12} for one electron loss by Cs^+ in Xe: ▲, Neill et al (1983); △, A S Schlachter (private communication) and Tanis et al (1979).

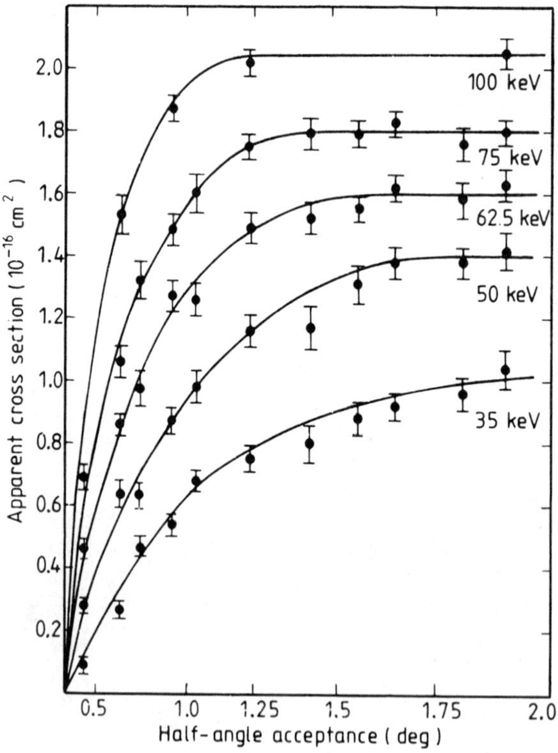

Fig 16. Dependence of measured one electron loss cross section $\sigma_{12}{}^{A}$ on the laboratory half-angle acceptance of the Cs^{2+} product detector for the indicated CM energies.

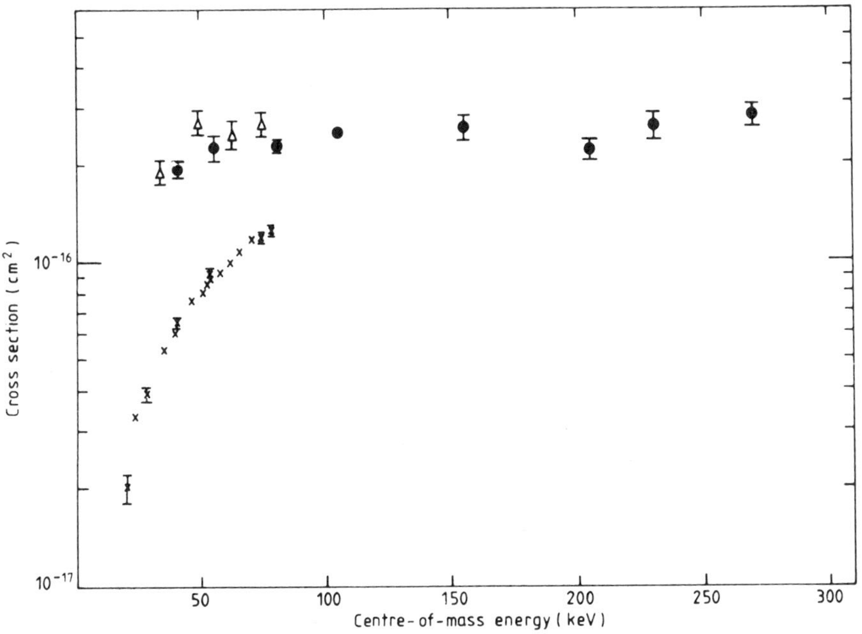

Fig 17. Cross sections $\sigma(Cs^{2+})$ for Cs^{2+} product ion in Cs^+ - Cs^+ collisions: ●, Neill et al (1983); X, Peart et al (1981b); △, values of Peart et al corrected as described in text

different we can use the scattering data of Neill et al for Cs^+ - Xe collisions to obtain a rough estimate of the fraction of Cs^{2+} that would be recorded within a particular angular acceptance in a Cs^+ - Cs^+ intersecting beam experiment. We applied this analysis to the conditions of the Newcastle experiment (Neill et al 1983) and their revised values are seen in figure 17 to be in satisfactory accord with our data. The Newcastle group had shown to their satisfaction that the measured cross sections, at a given CM energy, did not vary with target beam energies in the range 9.9 -14 keV. However, our analysis shows that the approach to such a plateau is very gradual and would not be attained in their case for target beam energies less than about 100 keV.

An alternative measurement of $\sigma(Cs^{2+})$ has been provided by Stalder et al (1984) who used a cesium plasma as a target for a beam of Cs^+ ions. Like the intersecting beams method this experimental approach suffers from problems with scattering of the product Cs^{2+}. Stalder et al thus conclude 'because of the large angular scattering effects at the lower energies, ... the present results do not resolve the discrepancy between the results of Peart et al and those of Neill et al'.

Theoretical studies of processes of the type (13) are difficult since they involve the coupling between the nuclear and electronic motions. A molecular approach has been applied to slow $Cs^+ + Cs^+$ collisions by Das et al (1978), Olson (1978) and Olson and Liu (1981). The latter have used the Fano-Lichten molecular orbital description to show that collisions will be dominated by ionisation rather than charge transfer. This takes place (figure 18) through a pseudo-crossing of the $5p\sigma_u$ and $6s\sigma_g$ states at an internuclear separation of about 2.1 a_0. Here the dominant mechanisms are

$$Cs^+ + Cs^+ \rightarrow Cs^{2+}(5p\sigma_g{}^2\ 6s\sigma_g{}^2) \rightarrow Cs^{+*}(6s^2) + Cs^+ \qquad (15a)$$

$$\rightarrow Cs^{+*}(6s) + Cs^{+*}(6s) \qquad (15b)$$

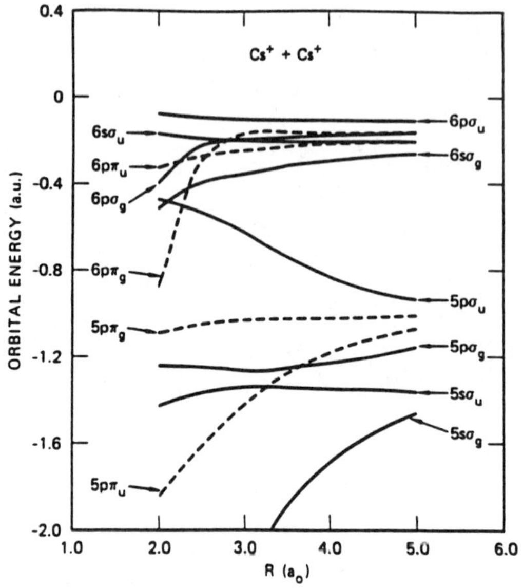

Fig 18. Orbital energy variation with internuclear separation R for the 5s, 5p, 6s and 6p levels of Cs^+-Cs^+.

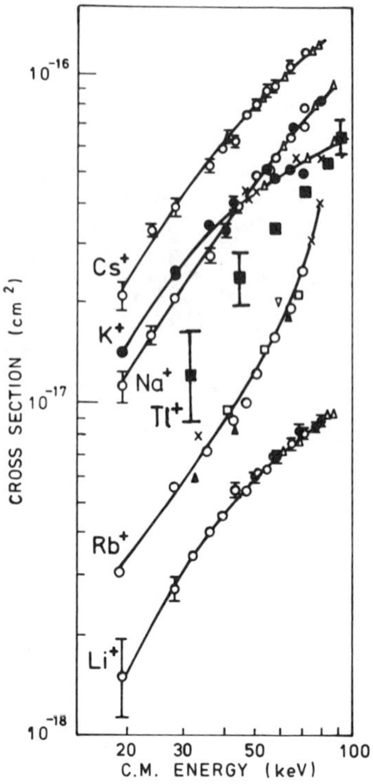

Fig 19. Cross sections for X^{2+} formation in X^+-X^+ collisions for Li, Na, K, Rb, Cs and Tl ions.
(Peart et al 1981; Forrest et al 1982)

where (15a) leads to the ionisation channel through autoionisation of the doubly excited product ion and (15b) can also lead to a small contribution to (13b) via molecular ionisation. On the basis of this model and by reference to known data for the Ar-Ar system, Olson and Lin estimated that at 50 keV $\sigma(Cs^{2+}) \simeq 3 \times 10^{-16}$ cm², to within a factor of 2, a value which compares favourably with the experimental value. An experimental estimate of σ_c for (13a) by Peart et al (1981) in the CM energy range 28-47 keV is at least an order of magnitude smaller than $\sigma(Cs^{2+})$ and is thus in general agreement with the theoretical model. Ermolaev et al (1982) have used a two state approximation in the impact formalism to calculate σ_c in the CM energy range 300 keV-5 MeV. At their lowest energy their cross-section is an order of magnitude lower than our measured values.

Peart et al (1981) have measured $\sigma(X^{2+})$ for X^+ - X^+ collisions for the alkali ions Li, Na, K and Rb in the CM energy range 19-88 keV and $\sigma(Tl^{2+})$ for Tl^+ - Tl^+ collisions have been measured by Forrest et al (1982) in the CM energy range 33-92 keV, figure 19. Our measurements on scattering have important implications for these and other heavy ions, particularly at the lower impact energies. Indeed an analysis of the Rb^+ - Rb^+ results shows that the measured cross sections are not independent of the target beam energies, 8-15 keV, used in the experiment. Unless the angular acceptance of the product ion detector is very large, comparatively high laboratory energies of the target beam must be used to ensure that the loss of scattered collision products is small.

The contribution of charge transfer to the formation of doubly charged product ions might be expected to be greater in the case of the open-shell $Xe^+ - Xe^+$ system than for the closed-shell $Cs^+ - Cs^+$ system. Figure 20 shows results for $\sigma(Xe^{2+})$ for Xe^{2+} production in $Xe^+ + Xe^+$ collisions measured by Angel et al (1980) in the CM energy range 38-303 keV and Melchert (1985) in the CM energy range 10-50 keV. The measurements from the two laboratories are in good accord. The Giessen group has also investigated the charge transfer cross section σ_c. When they attempted to measure the charge transfer they found that the apparent cross section $\sigma_c{}^A$ abruptly increased with decreasing CM energy below about 20 keV. A gas cell was then inserted inside the accelerator analysing magnet of the target Xe^+ beam and $\sigma_c{}^A$ measured as a function of the pressure in this cell. As shown in figure 21 $\sigma_c{}^A$ decreased with increasing pressure and attains the same equilibrium value for both H_2O and SF_6 attenuating gases suggesting that in the magnetic field metastable Xe^+ states, which lie approximately 12-14 eV above the ground state, are preferentially removed from the ion beam by one-electron loss. When metastable ions were similarly removed from the projectile beam there was, as expected, no further observable change in the cross section. This metastable effect is only dominant at energies below 20 keV and has no apparent effect on either $\sigma(Xe^{2+})$ or σ_c at higher energies. The equilibrium values of σ_c are plotted in figure 20 and can be seen to exhibit the expected energy dependence. The magnitude of the charge transfer cross section is, as expected, much larger than in the case of the $Cs^+ - Cs^+$ system and contributes to about 50% of $\sigma(Xe^{2+})$. For the $Ba^+ - Ba^+$ system, where each ion has a loosely bound outer electron, Sramek et al (1980) have used a classical trajectory method with basis functions obtained from a multi-configuration valence bond method to estimate charge transfer

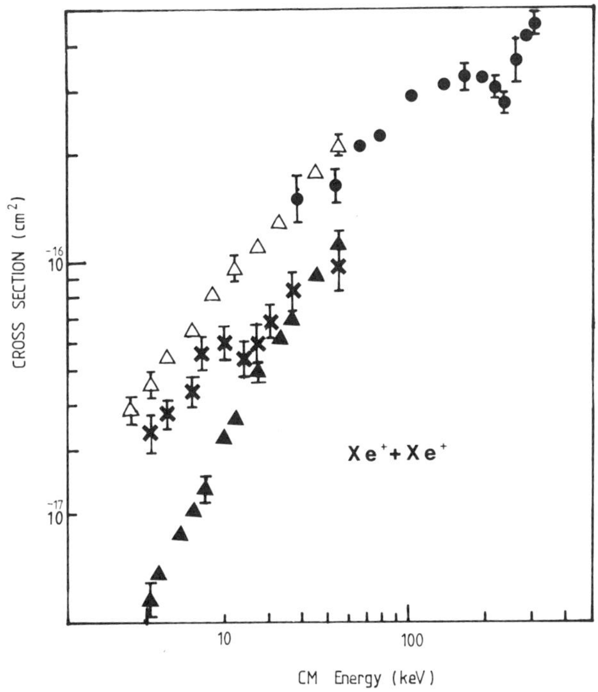

Fig 20. Data for the Xe^+-Xe^+ system
$\sigma(Xe^{2+})$: ●, Angel et al (1980); △, Melchert et al (1985). σ_c: ▲, Melchert et al (1985). σ_i: X, Melchert et al (1985).

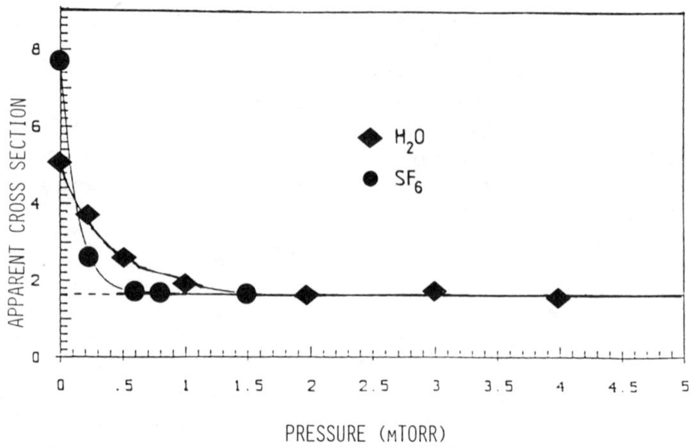

Fig 21. Dependence of the measured capture cross section σ_c^A on pressure in the attenuation cell for two gases, H_2O and SF_6.

cross sections of up to 2.4 x 10^{-15} cm^2 in the CM energy range 25-500 keV.

Both theoretical and experimental work has been carried out on the Li^+ - Li^+ system. The experimental results of Peart et al (1981) and Watts et al (1984) are shown in figure 22 and are in good agreement within the maximum combined uncertainties. This agreement indicates that in this case

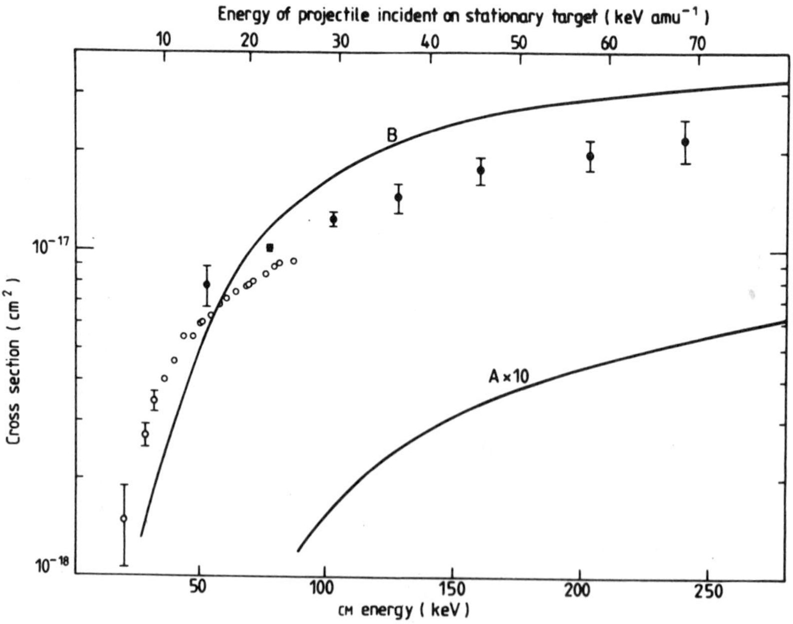

Fig 22. Data for Li^+-Li^+ collisions.
$\sigma(Li^{2+})$: o, Peart et al (1981c); ●, Watts et al (1984). Curve A, theoretical charge transfer x 10, Ermolaev et al (1982b). Curve B, theoretical ionisation, Ermolaev et al (1982a).

there is no serious loss of scattered product ions in either of the two experiments. Cross sections for charge transfer in $Li^+ - Li^+$ collisions calculated by Ermolaev et al (1982a) using a two-state approximation in the impact parameter formalism are about 50 times smaller than the experimental values of $\sigma(Li^{2+})$. Theoretical estimates of the ionisation cross section were calculated by Ermolaev et al (1982b) using the first Born approximation and exhibit an energy dependence different from the measured values of $\sigma(Li^{2+})$.

CONCLUSION

We have seen the maturity of the experimental approach for the measurement of ionisation and charge transfer in both simple one electron and complex multielectron systems. Future experimental advances, such as the determination of the initial and final states of the collision partners, depends on overcoming the many formidable technical problems which exist at present. Reliable calculations have until recently been available only for one and two electron systems or systems which approximate to this. However although theory is now capable of looking at more complex systems a knowledge of cross sections for multielectron/multielectron ion collisions must continue to rely on experiment.

REFERENCES

Afrosimov V V, Basalaev A A, Leiko G A and Panev M N, 1978, Sov Phys JETP 47: 837.

Afrosimov V V, Leiko G A, Mamev Yu A and Panev M N, 1975, ibid 40: 661.

Aitken K L, Harrison M F A and Rundel R D, 1971, J Phys B 4: 1189.

Angel G C, Dunn K F, Neill P A and Gilbody H B, 1980, J Phys B 13: L391.

Angel G C, Dunn K F, Sewell E C and Gilbody H B, 1978a, J Phys B 11: L49.

Angel G C, Sewell E C, Dunn K F and Gilbody H B, 1978b, J Phys B 11: L297.

Banyard K E and Shirtcliffe G W, 1979, J Phys B 12: 3247.

Bates D R and Boyd A H, 1962a, Proc Phys Soc 79: 710.

Bates D R and Boyd A H, 1962b, Proc Phys Soc 80: 1301.

Bates D R and Griffing G, 1958, Proc Phys Soc A66: 961.

Bates D R, Johnston H C and Stewart A L, 1964, Proc Phys Soc 84: 517.

Belkic Dz, 1980, J Phys B 13: L589.

Belkic Dz, Gayet R and Salin A, 1979, Physics Reports 56: 279.

Bell K L and Kingston A E, 1969, J Phys B 2: 653.

Bogdanov G F, Karkhov A N and Kucheryaev, Ya A, 1965, Atomnaya Energiya 19: 1316.

Bransden B H, Noble C J and Chandler J, 1983, J Phys B 16: 4191.

Brouillard F and Claeys W, 1983, 'Physics of Ion-Ion and Electron-Ion

Collisions', Brouillard and McGowan, ed., Plenum Press, New York and London.

Brouillard F and Delfosse J M, 1967, Abstracts Proc 5th ICPEAC Leningrad: 159.

Das G, Raffenetti R C and Kim Y-K, 1978, Argonne Nat Lab Report,NL-79-41: 195.

Datta S, Mandal C R and Mukherjee S C, 1984, Can J Phys 62: 307.

Dickinson A S and Hardie D J W, 1979, J Phys B 12: 4147.

Dunn K F, Angel G C and Gilbody H B, 1979, J Phys B 12: L623 and unpublished data.

Dunn K F, Watts M F, Angel G C and Gilbody H B, 1985, Abstracts Proc XIV ICPEAC Palo Alto: 613.

Ermolaev A M, Miraglia J E and Bransden B H, 1982a, J Phys B 15: L677.

Ermolaev A M, Noble C J and Bransden B H, 1982b, J Phys B 15: 457.

Forde A L, Reading J F and Becker R L, 1982, J Phys B 15: 3257.

Forrest R A, Peart B and Dolder K, 1982, J Phys B 15: L45.

Fritsch W and Lin C D, 1982, J Phys B 15: 1255.

Gilbody H B, 1979, Advances in Atomic and Molecular Physics 15: 293.

Godlove T F, 1979, IEEE Trans Nucl Sci 26: 2997.

Goffe T V, Shah M B and Gilbody H B, 1979, J Phys B 12: 3763.

Guidini J, Manus C, Sinda T and Watel G, 1965, Abstracts Proc 4th ICPEAC Quebec: 450.

Harel C and Salin J, 1980, J Phys B 13: 785.

Harrison M F A, 1966, Brit J Appl Phys 17: 371.

Harrison M F A, Dolder K T and Thoneman P C, 1963, Proc Phys Soc 82: 368.

Hobbis D A, Nicholson P, Harrison M F A and Montague R G, 1979, private communication.

Hopkins C J, Watts M F, Dunn K F and Gilbody H B, 1985, private communication.

Jognaux A, Brouillard F and Szucs S, 1978, J Phys B 11: L669.

Kim Y-K, 1976, US Dept of Commerce Report LBL-5543: 11.

Kimura M and Thorson W R, 1981, Phys Rev A 24: 3019.

McCullough R W, Nutt W L and Gilbody H B, 1979, J Phys B 12: 4159.

McGowan W and Kerwin L, 1963, Proc Phys Soc 82: 357.

McGuire E J, 1984, Phys Rev A29: 3429.

Melchert F, PhD Thesis, 1985, University of Giessen.

Merzbacher E and Lewis H W, 1958, Handb Phys 34: 166.

Miraglia J E, 1983, J Phys B 16: 1029.

Mitchell J B A, Dunn K F, Angel G C, Browning R and Gilbody H B, 1977, J Phys B 19: 1897.

Moran T F and Wilcox J B, 1978, J Chem Phys 68: 2855.

Moran T F and Wilcox J B, 1979, ibid 70: 1467.

Mukherjee S and Sil N C, 1980, J Phys B 13: 3421.

Neill P A, Angel G C, Dunn K F and Gilbody H B, 1982, J Phys B 15: 4219.

Neill P A, Angel G C, Dunn K F and Gilbody H B, 1983, J Phys B 16: 2185.

Nutt W L, McCullough R W and Gilbody H B, 1979, J Phys B 12: L157.

Olson R E, 1978, J Phys B 11: L227.

Olson R E, Argonne Nat Lab Report, 1978, ANL-79-41: 195.

Olson R E and Liu B, 1981, J Phys B 14: L279.

Peach G, 1971, J Phys B 4: 1670 and private communication.

Peart B and Dolder K T, 1979, J Phys B 12: 4155.

Peart B, Forrest R A and Dolder K T, 1981a, J Phys B 14: 3457.

Peart B, Forrest R A and Dolder K T, 1981b, J Phys B 14: 1655.

Peart B, Forrest R A and Dolder K T, 1981c, J Phys B 14: L603.

Peart B, Gee D M and Dolder K T, 1977b, J Phys B 10: 2683.

Peart B, Grey R and Dolder K T, 1977a, J Phys B 10: 2675.

Peart B, Rinn K and Dolder K T, 1983a, J Phys B 16: 1461.

Peart B, Rinn K and Dolder K T, 1983b, J Phys B 16: 2831.

Phaneuf R A, Meyer F W and McKnight R H, 1979, Phys Rev A 17: 534.

Reading J F, Ford A L and Becker R L, 1982, J Phys B 15: 625.

Reinhold C O and Falcon C A, 1985, private communication.

Reinovsky R E, Glowienka J C, Jennings W C and Hickok R L, 1975, IEEE Trans Plasma Sci PS-3, 1975: 194.

Rinn K, Melchert F and Salzborn E, 1985, J Phys B 18: 3783.

Sewell E C, Angel G C, Dunn K F and Gilbody H B, 1980, J Phys B 13: 2269.

Shah M B, 1985, private communication.

Shah M B and Gilbody H B, 1981, J Phys B 14: 2361.

Sinha C and Sil N C, 1979, Phys Lett 71A: 201.

Sramek S J, Macek J H and Gallup G A, 1980, Phys Rev A 22: 1467.

Stalder K R, Berkner K H and Pyle R V, 1984, Phys Rev A 29: 3052.

Tanis J A, Stalder K R, Stearns J W and Schlachter A S, 1979, Bull Am Phys Soc 24: 1196.

Thomson J J, 1912, Phil Mag 23: 449.

Todd N, 1981, private communication.

Vriens L, 1969, Case Studies in Atomic Collision Physics, 1: 337.

Watts M F, Angel G C, Dunn K F and Gilbody H B, 1984, J Phys B 17: 1631.

Watts M F, Dunn K F and Gilbody H B, 1985, Submitted to J Phys B.

Willis S L, Peach G, McDowell M R C and Banerji J, 1985, J Phys B to be published.

Winter T G, 1982, Phys Rev A 25: 697.

Winter T G, Hatton G J and Lane N F, 1980, Phys Rev A 22: 930.

TRANSFER IONIZATION IN COLLISIONS OF MULTIPLY CHARGED IONS WITH ATOMS

Erhard Salzborn and Alfred Müller

Institut für Kernphysik
Justus-Liebig-Universität Giessen
D-6300 Giessen, Fed. Rep. Germany

1. INTRODUCTION

One of the important fundamental collision processes in hot plasmas containing multiply charged ions is the transfer of electrons in ion-ion or ion-atom collisions:

$$A^{q+} + B^{r+} \rightarrow A^{(q-k)+} + B^{(r+i)+} + (i-k)e \qquad (1)$$

Because of the strong Coulomb attraction of electrons by highly charged ions one can expect large cross sections even for ionic collisions with $r \geq 1$ as long as the charge state of one collision partner strongly exceeds that of the other, i.e. when $q >> r$. Moreover it is possible then to set free additional electrons, which means that the less charged ion looses more electrons in a collision of type (1) than will be found attached to the highly charged ion when detected long after the collision. The basic features of this phenomenon of electron transfer with additional ionization will be the principle topic of the present paper.

Although most important for laboratory or astrophysical plasmas the process of electron capture with ionization in ion-ion collisions did not yield yet to direct measurements. Because of low reaction rates and problems to eliminate background in experiments with colliding ion beams[1], studies of ion-ion processes (1) are limited so far[2] to $q = 1,2$; $r = \pm 1$, and

$k = i = 1$. Thus the present report will deal with cases where $r = 0$, i.e. with collisions of multiply charged ions with neutral atoms. Nevertheless it is not unreasonable to assume that the basic physical mechanisms of electron transfer with ionization are similar for atomic and ionic targets.

In recent years processes (1) with $i > k$ were often called transfer ionization thus extending the original meaning of the term to a wide variety of mechanisms. Transfer ionization (TI) in its simplest form is an exoergic collision process in which a single electron is transferred from an atom B to an ion A^{1+} while a doubly charged target ion B^{2+} is produced. The recombination energy $I_A^{(0)}$ of the ion A^{1+} exceeds the sum of the first and the second ionization potentials of B: $I_A^{(0)} > I_B^{(0)} + I_B^{(1)}$. Hence, TI in its original meaning does not require kinetic energy of the colliding particles and may occur at vanishing relative velocity. In the present paper, however, we will follow the generally adopted habit to use the term transfer ionization for any collision process which involves capture of electrons from a target atom into the projectile and the loss of electrons to the continuum by whatever mechanism this may happen. This wide reaching definition hence includes high-energy endoergic collisions where the ejection of target electrons proceeds at the expense of kinetic rather than potential energy of the projectile. Thus it appears reasonable to divide the discussion of TI into three parts dealing with low, intermediate and high kinetic energies. The criterion for this division is the ratio v_{proj}/v_B of projectile velocity v_{proj} and the Bohr orbit velocity $v_B = 2.2 \cdot 10^8$ cm s^{-1}.

Depending on the projectile velocity range and the related experimental conditions various techniques are applied to obtain information about TI processes in the general sense. Energy analysis of ejected electrons, ion energy gain spectroscopy, angular scattering analysis, measurements of charge states of projectile and target after the collision and combination of these techniques in coincidence experiments have been employed to study TI. The experimental methods used will be discussed in some detail in connection with the discussion of results obtained on TI in the different energy regimes.

Since TI belongs to the large family of electron-capture processes we will first briefly review the fundamental features of electron capture in ion-atom collisions and thus provide a basis for the further discussion and a more detailed introduction to the problems involved in TI experiments.

2. MAIN CHARACTERISTICS OF ELECTRON-CAPTURE PROCESSES IN ION-ATOM COLLISIONS

2.1. Experimental Method

The simplest experiment to get information about capture of electrons in ion-atom collisions is the measurement of projectile ion charge states after the collision without taking notice of the subsequent fate of the target particle. A schematic picture of such a measurement is shown in Figure 1.

Behind an ion source with a subsequent accelerator a beam consisting of ions A^{n+} (n = ... q-2, q-1, q, q+1...) is momentum-analzed by a magnet. One charge state is selected by a proper magnet setting and ions A^{q+} are passed through a gas cell at target thickness π. Ions with reduced charge state (q-k)+ produced by the capture of k electrons into projectiles A^{q+}, are separated by a second magnet. Under single-collision conditions the flux I_{q-k} of ions $A^{(q-k)+}$ is directly proportional to the total cross section $\sigma_{q,q-k}$ for the capture of k electrons in one collision process. The total cross section can be obtained from $\sigma_{q,q-k} = I_{q-k}/(I_q \cdot \pi)$ where

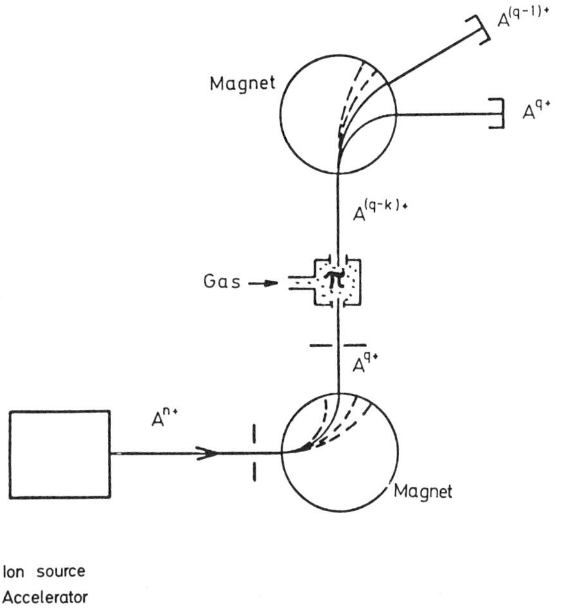

Fig. 1: Schematic picture of an experiment for the measurement of total electron-capture cross sections.

I_q is the incident-ion flux. The single collision condition can be tested experimentally by checking the linear dependence of I_{q-k} on π. Since this procedure may often be too time-consuming the experimentalist needs a mathematical criterion to decide at which target thickness the assumption of single collisions is still safe. Such a criterion considering possible two-step processes is

$$\sum_n \left(\frac{\sigma_{q,n} \cdot \sigma_{n,q-k}}{\sigma_{q,q-k}}\right) \cdot \pi \ll 1 \quad . \tag{2}$$

Generally, $\sigma_{m,n}$ stands for the total cross section of electron capture (or loss) by the projectile with initial and final charge states m and n, respectively. The application of (2) requires knowledge about all relevant cross sections $\sigma_{m,n}$ indicating that the measurement of multiple electron capture may be difficult and require iterative approaches to find the range of "allowed" target thicknesses. As will be shown later more detailed experiments employing coincidence techniques with simultaneous analysis of projectile and target ion charge states allow to decide whether or not single collision conditions are fulfilled. When charge transferred projectile ions $A^{(q-2)+}$ are detected in coincidence with B^{i+} recoil ions the minimum charge state of the target after transfer of 2 electrons is i = 2. The presence of B^{1+} recoil ions would indicate that $A^{(q-2)+}$ ions are formed by multiple collision processes such as $A^{q+} \rightarrow A^{(q-1)+} \rightarrow A^{(q-2)+}$,i.e. the capture of one electron in each of <u>two</u> subsequent collisions.

2.2. Results: Single and Multiple Electron Capture in Slow or Fast Collisions

Experimental data on single and multiple electron capture have been reviewed in a large number of articles. The reader is referred to Invited Papers and Progress Reports at ICPEAC XI and ICPEAC XII[3,4] and to reviews by de Heer[5] and Schlachter et al.[6].

At low collision energies, i.e. when the projectile velocity v_{proj} is below the Bohr orbit velocity v_B, total electron-capture cross sections are generally independent of the kinetic energy. An example with multiple capture in Xe^{q+}+ Kr collisions is shown in Fig. 2.
Müller and Salzborn[7] derived scaling rules for cross sections $\sigma_{q,q-k}$ in collisions of slow multiply charged ions with $q \leq 10$ and the capture of k = 1,2,3 and 4 electrons in a single ion-atom encounter. From a great number of experimental data on rare gas targets they inferred a dependence

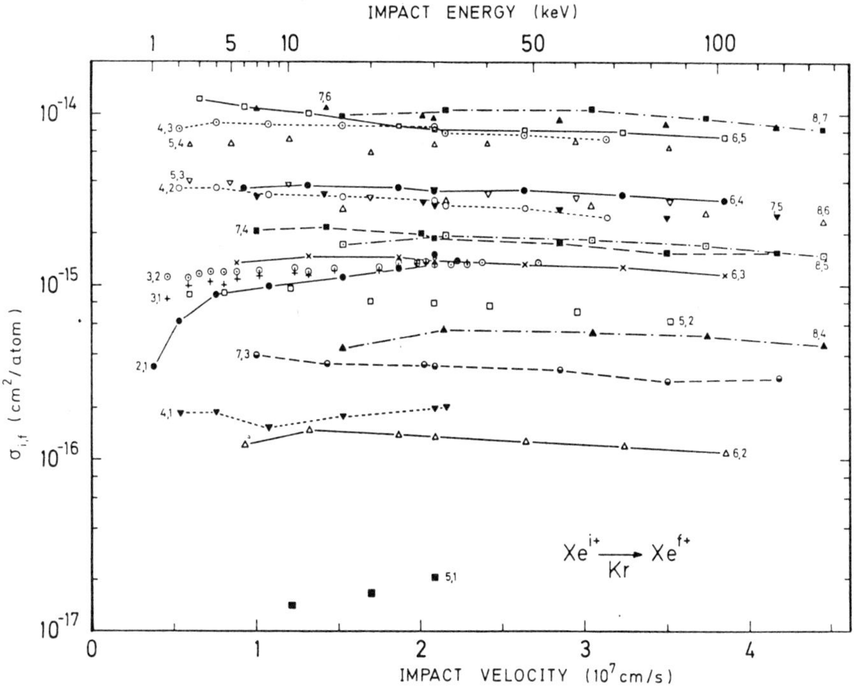

Fig. 2: Electron capture cross sections $\sigma_{q,q-k}$ for Xe^{q+} ions incident on krypton[3]. The initial and final charge states of the projectile ions q and q-k, respectively, are indicated. Lines are drawn to guide the eye.

$$\sigma_{q,q-1} = 1.4\cdot10^{-12}\ \mathrm{cm}^2 \cdot q^{1.17}\left(I_B^{(0)}\right)^{-2.76} \tag{3}$$

which fitted most of their data within deviations of less than 30 %. Recently it was shown that this empirical formula is useful for projectile charge states up to q = 40 (see Fig. 3).

For an extended range of target atoms (from Cs with $I_B^{(0)} = 3.89$ eV to He with $I_B = 24.58$ eV) Müller, Achenbach and Salzborn[9] found a dependence $\sigma_{q,q-1} \propto \left(1/I_B^{(0)}\right)^{\alpha}$ with α close to 2 (Fig. 4).

At high collision velocities $v_{proj} > v_B$ electron capture cross sections rapidly decrease with energy. Macdonald and Martin[10] have measured $\sigma_{q,q-k}$ for O^{q+} ions (q= 3,...,8) incident on Ar atoms with k = 1,2,3,4. Their results are displayed in Fig. 5.

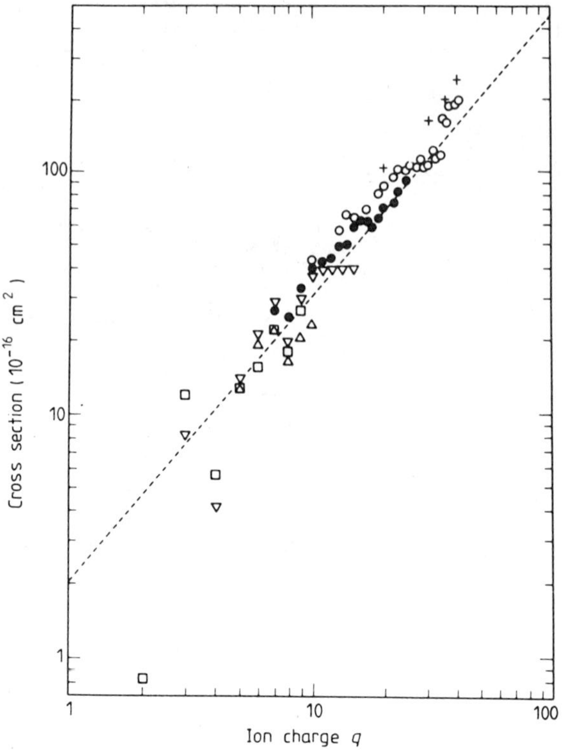

Fig. 3: Cross sections for one-electron capture from He as a function of the charge q of ions I^{q+} and Kr^{q+} at low collision velocity $v < v_B$ (from ref. 8). The dashed line represents the empirical formula (3) by Müller and Salzborn[7].

Schlachter et al.[11] have derived an empirical scaling rule for electron capture by fast highly charged ions in gases. Their prescription to calculate $\sigma_{q,q-1}$ involves the knowledge of the projectile charge state q, the collision energy E and the atomic number Z_2 of the target only. With reduced parameters

$$\tilde{\sigma} = \sigma_{q,q-1} \, Z_2^{1.8}/q^{0.5} \text{ and } \quad \tilde{E} = E/(Z_2^{1.25} \cdot q^{0.7}) \tag{4}$$

where $\sigma_{q,q-1}$ is in cm² and E in keV/amu they find

$$\tilde{\sigma} = \frac{1.1 \cdot 10^{-8}}{\tilde{E}^{4.8}} \left(1-\exp(-0.037 \, \tilde{E}^{2.2})\right) \cdot \left(1-\exp(-2.44 \cdot 10^{-5} \tilde{E}^{2.6})\right) \tag{5}$$

as a good fit to the available data. Approximately 70 % of the data lie within a factor of 2 of the scaling rule. A comparison of measured cross sections with this rule is shown in Figure 6 for a number of different targets.

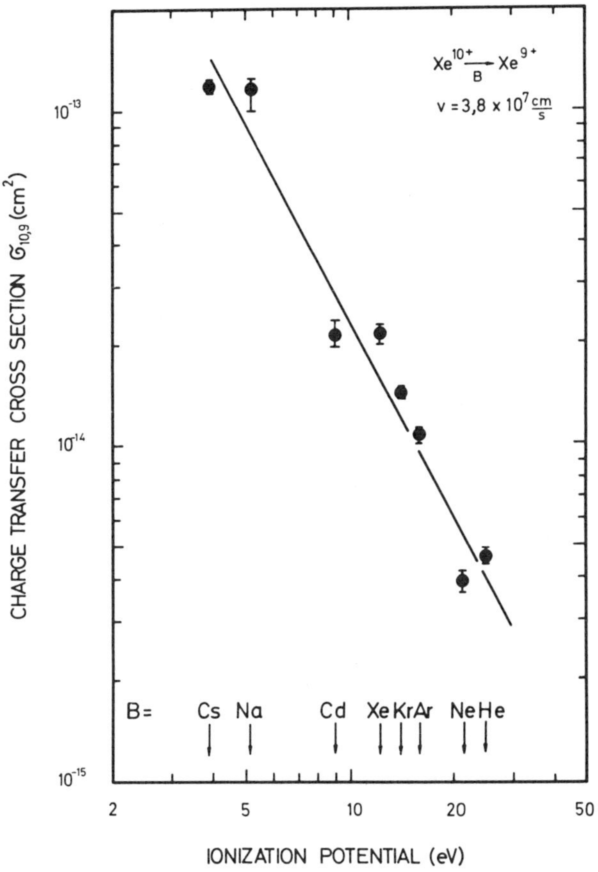

Fig. 4: Cross sections $\sigma_{10,9}$ for one-electron capture by 100 keV Xe^{10+} ions from various target atoms B as a function of the ionization potential $I_B^{(0)}$ of the atoms[9]. The solid line represents a fit of the experimental data with $\sigma_{10,9} \sim (I_B^{(0)})^{-1.94}$.

On the basis of the classical Bohr-Lindhard theory[28] Knudsen, Haugen and Hvelplund found a slightly different scaling behaviour of electron capture cross sections[12]. Their resulting formulas are compared to experimental data in Figure 7, which also includes the region of low energies covered by Equation (3).

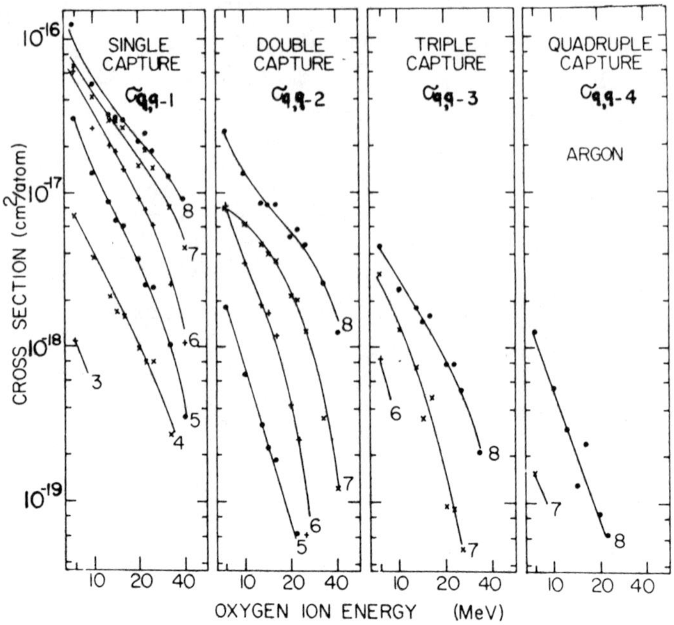

Fig. 5: Electron capture cross sections $\sigma_{q,q-k}$ (k = 1,2,3,4) for O^{q+} + Ar collisions[10]. The initial charge state of the oxygen projectiles is indicated with the data.

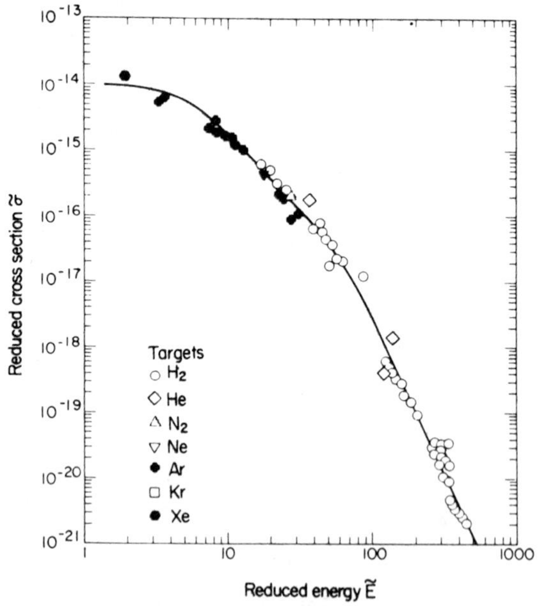

Fig. 6: Reduced plot of single-electron capture cross sections[11] for fast highly charged ions with charge state q incident on gas targets with atomic number Z_2. Cross sections in molecules are divided by 2 and plotted with the atomic Z_2. The line is an empirical fit (Eq.(5)) to the cross sections. The symbols used to represent the targets are shown in the figure. $\tilde{E} = E/(Z_2^{1.25}\ q^{0.7})$ and $\tilde{\sigma} = \sigma Z_2^{1.8}/q^{0.5}$.

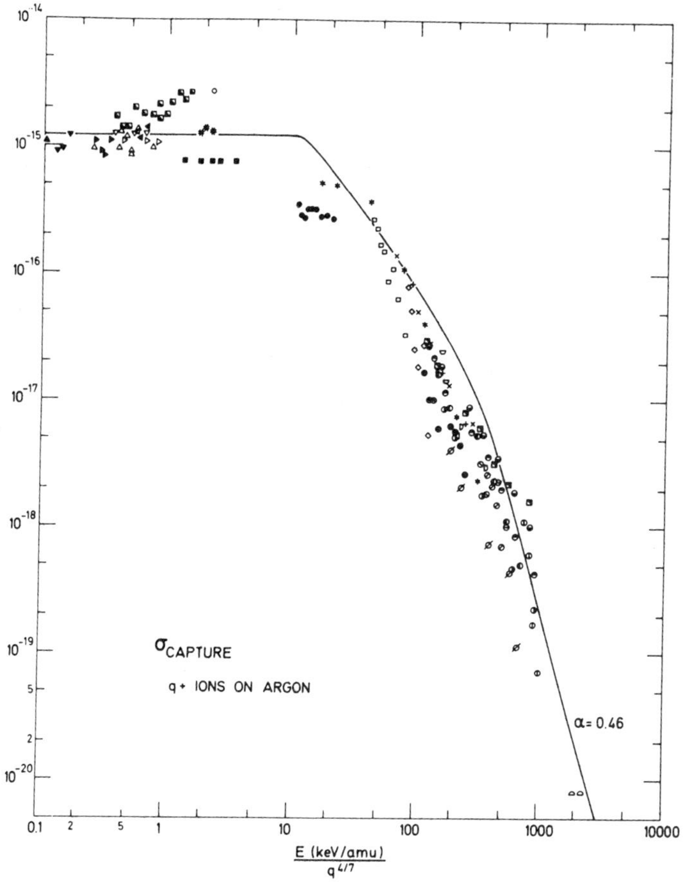

Fig. 7: Experimental data for single-electron capture cross sections of various ions with charge states $q \geq 4$ colliding with Ar atoms. The solid line represents the theoretical estimate of Knudsen et al[12] on the basis of the Bohr-Lindhard model. For the origin of the data see ref. (12).

Fig. 7 clearly shows the general energy-dependence of electron capture cross sections. Apart from very low velocities (energies below 1 eV/amu) total cross sections $\sigma_{q,q-1}$ are large for $v_{proj} < v_B$ and do not depend much on the collision energy. Beyond $v_{proj} = v_B$ there is a rapid decrease of electron capture probabilities as the energy increases. As can be inferred from Fig. 2 and Fig. 5 multiple electron capture shows the same energy dependence. Scaling behaviour was also found[13] for net target charge production by fast highly charged ions incident on atoms. The cross sections σ_q measured refer to the total charge collected on condenser plates across the target region where the gas pressure is low enough to provide single collision conditions. Thus σ_q includes pure ionization (k = 0 in the nomenclature of Eq. (1)), pure electron capture (k > 0, k-i = 0) and transfer ionization (k > 0, i-k > 0). Fig. 8 shows scaled cross sections σ_q/q versus E/q. Data points for a wide range of projectiles are grouped near curves predicted for specific target species by classical trajectory-Monte-Carlo calculations.

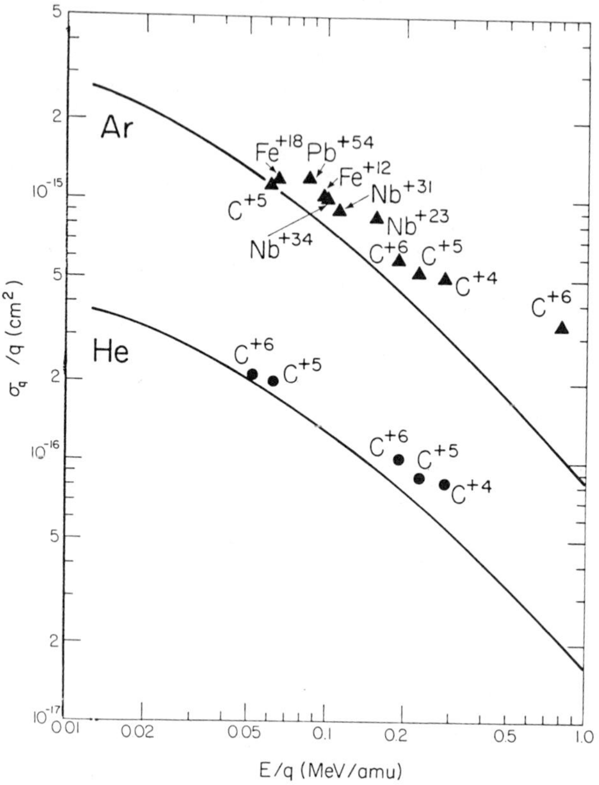

Fig. 8: Reduced plot of net-ionization cross sections for a highly stripped ion in charge state q in He (●) and Ar (▲) targets. The closed symbols are experimental results, and the curves are classical-trajectory-Monte-Carlo calculations (from ref. 13).

2.3. Theoretical Treatment

Theoretical approaches to electron capture processes are reviewed in some detail in another contribution to the present volume[13]. Therefore we will not discuss the different models and calculational techniques applied in different ranges of collision energy. We will rather point out some principle aspects in regard to comparisons between theory and experiment.

Somewhat exaggerating one can say that theory essentially can treat charge transfer collisions where only one electron is involved, i.e. when a completely stripped ion captures the electron from a hydrogen atom or a hydrogenic ion. Every step beyond this problem requires some kind of modeling and makes the predictive power of theory uncertain. So far there is no general theoretical treatment of multi-electron capture processes available and especially the treatment of the outer shells of many-electron targets in collision processes is the theoreticians' nightmare.

When a cross section $\sigma_{q,q-1}$ for the capture of one electron is calculated theory refers to a process $A^{q+} + B \rightarrow A^{(q-1)+} + B^{+}$. The electron is captured from one of the target atom shells into an empty level of the projectile ion. In an experiment as it is shown in Figure 1, the experimentalist detects ions $A^{(q-1)+}$ and measures a total cross section $\sigma^{exp}_{q,q-1}$. It is not necessarily reasonable to compare the theoretical and experimental results since both cross sections generally have a different meaning. In the experiments the time of flight of the ions $A^{(q-1)+}$ between the collision region and the charge analyser is at least a couple of nanoseconds. This time is usually long enough to allow a doubly excited ion $\left(A^{(q-2)+}\right)^{**}$ to autoionize and be detected with the charge state (q-1). Thus, $\sigma^{exp}_{q,q-1}$ may include double-capture processes followed by autoionization. In principle any multi-electron capture process finally ending with a projectile ion $A^{(q-1)+}$ will contribute to $\sigma^{exp}_{q,q-1}$. This is of course also true for mechanisms where the capture of one electron from the target simultaneously leads to additional ionization of the target. Hence, apart from inner shell capture, a theoretical cross section for single electron capture should only be compared to measured cross sections obtained by identification of projectile <u>and</u> target charge states q-1 and 1, respectively, after the collision process.

It is useful in this context to introduce a more precise nomenclature for electron capture cross sections describing collision processes of type (1). Rather than giving only the initial and final charge state of the projectile (q and q-k, respectively) one should also provide the initial and final charge state of the target particle (r and r+i, respectively). In this convention the theoretical single-electron capture cross section is denoted $\sigma^{0,1}_{q,q-1}$. In general, electron capture and TI processes (1) are characterized by cross sections $\sigma^{r,r+i}_{q,q-k}$ where k= 1,2,3... is the number of transferred electrons found attached to the projectile when detected long after the collision and (i-k) = 0,1,2,... is the number of electrons set free during or after the collision. The one-electron capture cross section obtained from an experiment as diplayed in Fig. 1 is then given by $\sigma^{exp}_{q,q-1} = \sum_i \sigma^{o,i}_{q,q-1}$. Net-ionization cross sections σ_q discussed in section 2.2. can be represented by $\sigma_q = \sum_{i,k}(i \cdot \sigma^{o,i}_{q,q-k})$.

3. THE PHENOMENON OF TRANSFER IONIZATION

Transfer ionization (TI) includes every charge transfer process leading to additional target ionization:

$$A^{q+} + B \rightarrow A^{(q-k)+} + B^{i+} + (i-k)e \tag{6}$$

This means $k \geq 1$ and $(i-k) \geq 1$ for a TI process. Similar to the treatment of total electron capture cross sections TI phenomena can be characterized by the ratio of the projectile velocity v_{proj} and the Bohr orbit velocity v_B.

3.1. Slow Collisions ($v_{proj} < v_B$)

3.1.1. Global Experimental Methods The most direct assessment of the relative importance of TI processes (i.e., i > k)compared to pure capture (i.e., k=i) can be accomplished through the identification of projectile and target charge states after the collision. In an experiment this requires both charge state analysis and detection of $A^{(q-k)+}$ and B^{i+} ions in coincidence. In principle one can use electrostatic or magnetic charge-state analysis for the charge transferred projectiles $A^{(q-k)+}$. The target recoil ions generally have low kinetic energies in the eV range. It is necessary therefore to extract the ions B^{i+} from the collision region by an electric field and produce a "beam" of ions B^{i+}. The charge state distribution of ions in this beam can then be analyzed for instance by using a separating magnet. With the two analysers set to transmit ions B^{i+} and $A^{(q-k)+}$ the required coincidence detection can be performed and for another combination of charge states, e.g. (i-1) and (q-k), a new analyser setting is chosen.

Because of the low recoil ion energy, which is due to the fact that electron transfer predominantly occurs in peripheral collisions, the energy width of the B^{i+} ion beam after acceleration is so small that a much more elegant method of charge state detection can be employed. The principle of this method is demonstrated in Fig. 9. Rather than using an analyzing magn for the ions B^{i+} one can make use of the well defined time of flight t_{TOF} of a target ion with given charge state i and mass m between its creation in a collision and its detection by a proper detector ($t_{TOF} \sim \sqrt{m/i}$). The information about the time when the collision occured is carried also by the charge-transferred projectile ion $A^{(q-k)+}$. When the projectile ions A^{q+} are fast they do not change their velocity v_{proj} much by TI collisions. Hence, the produced ions $A^{(q-k)+}$ all take about the same time t_{proj} to travel the distance L from the collision region to the detector behind the analyzing magnet ($t_{proj} = L/v_{proj}$).

When the pulses of the recoil ion detector are delayed by t_{proj} the time spectrum of coincidences between B^{i+} and $A^{(q-k)+}$ ions, measured by a time-to-amplitude converter, directly gives information on the time of flight t_{TOF} of the target ions B^{i+}. Since t_{TOF} is proportional to $\sqrt{m/i}$ one can immediately extract the target ion charge state from the measurement. In addition all ions B^{i+} with different charge states i produced by TI collisions are collected and analyzed at a time so that the relative intensities obtained from the time-of-flight spectrum do not depend on fluctua tions of experimental parameters during a measurement.

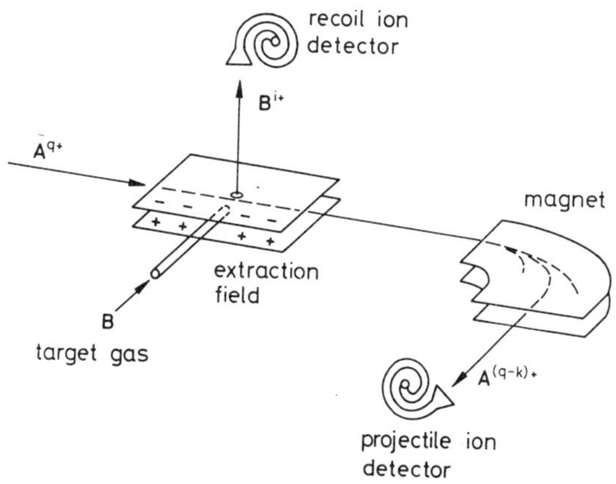

Fig. 9: Schematic view of an experimental setup for the coincident detection of charge-state-analyzed projectile ions $A^{(q-k)+}$ and target ions B^{i+} produced by TI collisions of A^{q+} ions with gas atoms. For explanations see text.

A series of three time-of-flight spectra taken with the setup described above is displayed in Fig. 10. The projectiles are 80 keV Xe^{8+} ions and the target is Xe. The first spectrum was taken with the analyzing magnet set to transmit $Xe^{(8-1)+}$ ions, the second with $Xe^{(8-2)+}$ transmitted and the third with $Xe^{(8-3)+}$ ions transmitted. The results show that the capture of k electrons in the meaning of section 2.3. to a great extent is accompanied by additional target ionization. If pure transfer of 2 electrons to the Xe^{8+} projectile were the only possible process producing fast Xe^{6+} ions in single collisions, the target atom would be detected as Xe^{2+}. Instead the production of Xe^{3+} target ions is the dominant process belonging to k=2 and even Xe^{5+} ions are observed. TI processes in this case dominate pure transfer of 2 electrons by a factor $\sum_{i>2} \sigma^{o,i}_{q,q-2}/\sigma^{0,2}_{q,q-2} = 17.8$. The resulting average recoil ion charge state $\langle i \rangle = \sum_i F_i \cdot i$ is 3.4 compared to 2 for pure electron transfer. The relative fractions F_i of target ions Xe^{i+} are determined from the areas of the peaks in the

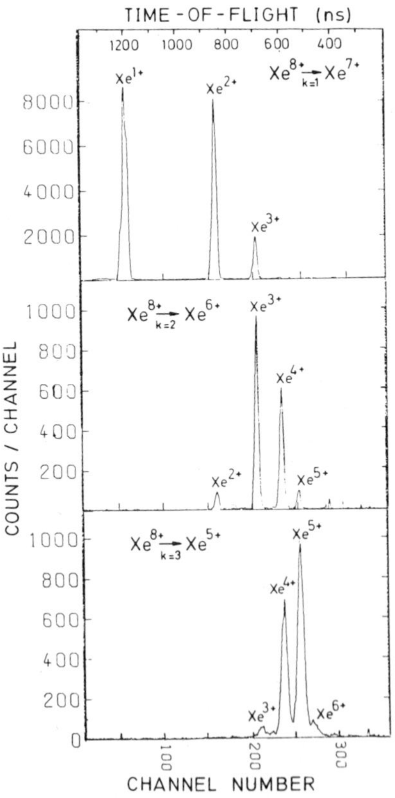

Fig. 10: Recoil-ion charge-state spectra for 80 keV Xe^{8+} projectiles in a thin Xe gas target from which k = 1,2 or 3 electrons are captured[15].

time-of-flight spectrum normalized to the sum of all peak areas and correspond to $F_i = \sigma^{o,i}_{q,q-k}/\sum_i \sigma^{o,i}_{q,q-k}$.

Equivalent information on the ion charge states produced in slow TI collisions was obtained in experiments similar to the one described above. Cocke et al. were the first to produce a beam of slow ions by bombardement of atoms with fast highly charged ions from an accelerator[16] and to use the multiply charged recoil ions for the measurement of absolute cross sections $\sigma^{o,i}_{q,q-k}$ in subsequent collisions of the slow ions in a gas target[17]. Apart from the recoil ion source their experimental technique is similar to the one described. The same is true for measurements by Astner et al.[18] who used the time-of- flight technique to identify not only the charge states of ions B^{i+} but also those of the projectiles after the collision. Their experimental setup is shown in Fig. 11. The charge state analysis of the fast ions is made possible by an acceleration of these ions behind the collision cell so that their flight time becomes dependent on their charge state. Fig. 12 shows a typical result of such an experiment. By the projection of selected intensities in the obtained contour plot one can get spectra equivalent to the ones displayed in Fig. 10.

Global information on TI can also be obtained from energy gain spectroscopy. It was mentioned above that fast projectiles do not change their velocity much in TI collisions. With a high-resolution energy analyzer

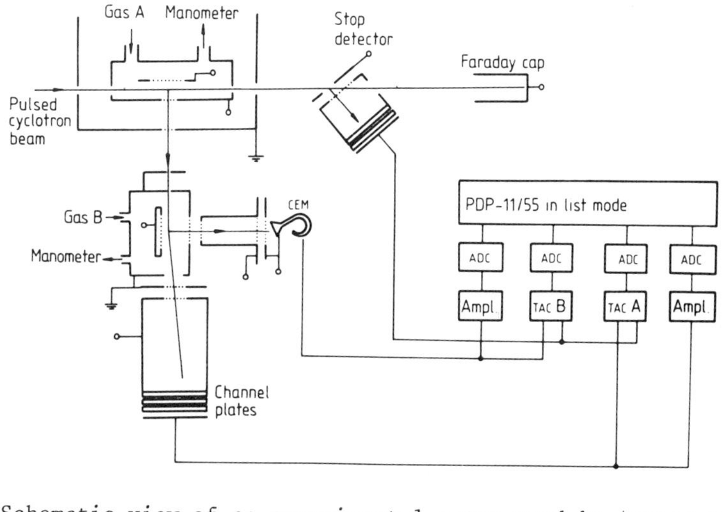

Fig. 11: Schematic view of an experimental setup used by Astner et al.[18] to study electron capture and TI.

(e.g. the magnet in Fig. 9), however, one can detect small energy differences in the projectile energy depending on the change of internal energy Q in an inelastic charge transfer collision (6). Assuming that the projectile A^{q+}(mass M_A, energy E_{proj}) is scattered into an angle Θ with energy $E'_{proj}(A^{(q-k)+})$ while the target atom B (mass M_B) recoils with an energy $E_{recoil}(B^{i+})$ then the total inelastic energy gain Q is defined as $Q=(E'_{proj}+E_{recoil})-E_{proj}$ and the energy gain of the projectile $\Delta E_{proj} = E'_{proj} - E_{proj}$ is given on the basis of kinematic considerations[19] by

$$\Delta E_{proj} = \frac{2M_A M_B}{(M_A+M_B)^2} E_{proj} \left\{ 1+ \frac{M_A}{M_B} \sin^2\Theta + \frac{M_A + M_B}{M_A} \frac{Q}{2E_{proj}} - \cos\Theta\left(1- \left(\frac{M_A}{M_B}\right)^2 \sin^2\Theta - \frac{M_A + M_B}{M_B} \frac{Q}{E_{proj}} \right)^{1/2}\right\} \qquad (7)$$

At an angle $\Theta = 0°$ the interpretation of Eq. 7 becomes easy. Under the assumption of $x = Q/E_{proj} \ll 1$ one can apply $(1-x)^{1/2} \approx 1-1/2\ x$ and therefore one measures $\Delta E = Q$ (at this angle $E_{recoil} = 0$).

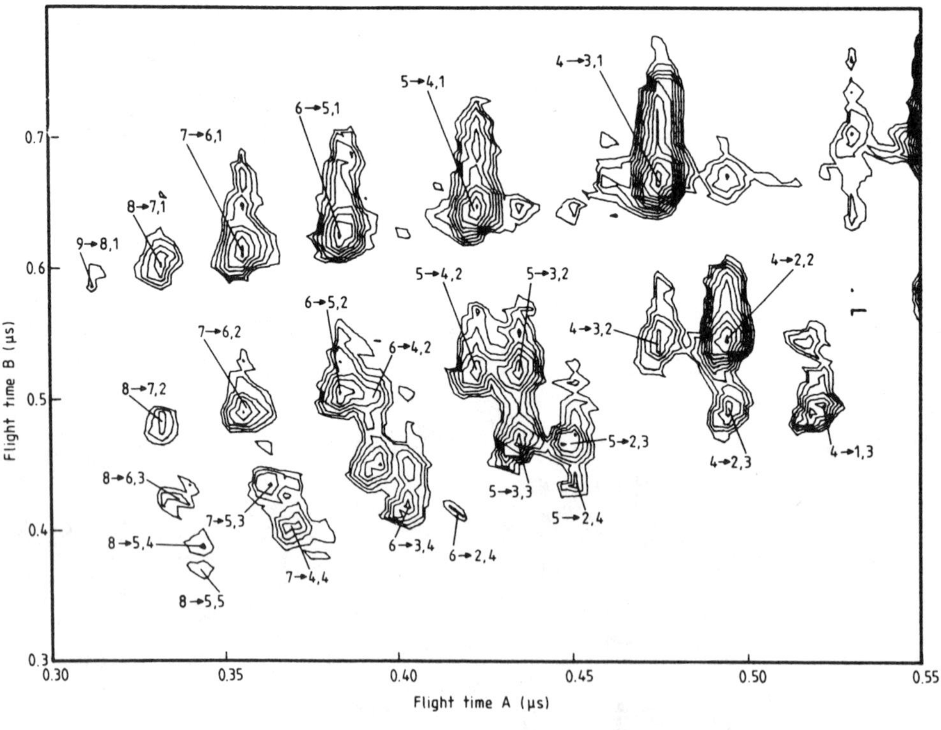

Fig. 12: Logarithmic contour plot of the coincidence intensity as a function of the times of flight in channels A and B. A peak corresponding to the reaction $Ar^{q+}+Ar \rightarrow Ar^{(q-k)+} + Ar^{i+} + (i-k)e$ is denoted $q \rightarrow (q-k),i$ (from ref. 18).

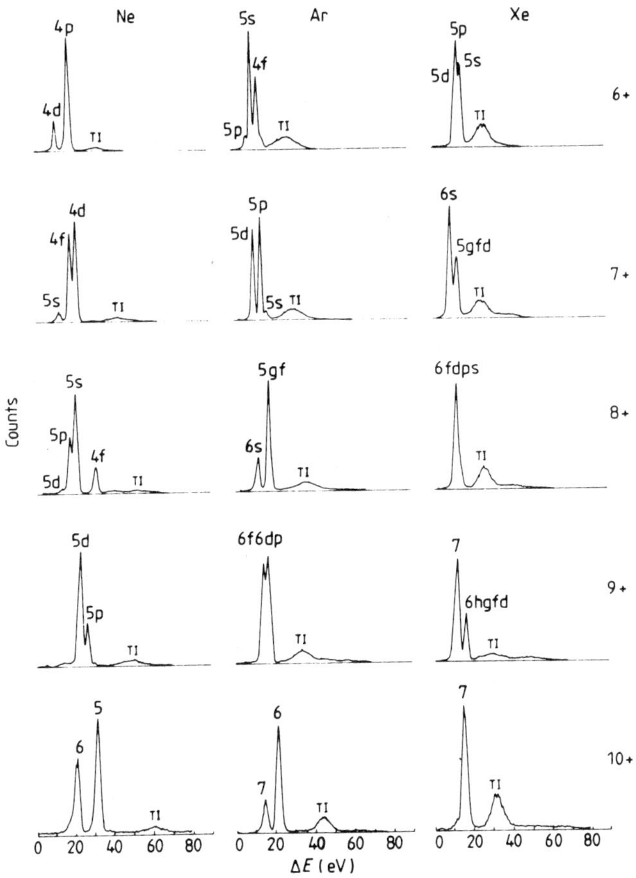

Fig. 13: Energy-gain spectra for 1000 eV $Ar^{q+} \rightarrow Ar^{(q-1)+}$ (q= 6-10) on Ne, Ar and Xe. The narrow peaks are identified by their (n,l) values, whereas the broad peaks are identified as corresponding to transfer ionization (from ref. 21).

Projectile energy gain spectroscopy has been widely used to obtain information on the population of excited states in charge transfer collisions[20]. In a recent publication by Nielsen et al.[21] energy gain spectra were obtained for one-electron capture collisions of 1 keV Ar^{q+} ions on rare gas atoms (Fig. 13). The authors could identify capture of one electron into various (n,l) states depending on the projectile charge state q and the target species B. At the high-energy side of the corresponding narrow peaks broad intensity distributions are found which are ascribed to TI collisions. With increasing projectile ion charge state the relative intensity in the TI peak increases and TI is more important for the more complex atom Xe than it is for Ne. The increased width of the TI peak is due to the large number of intermediate excited states which could not be resolved in these experiments. Also, TI involves the emission of electrons which carry kinetic energy, so that ΔE is no longer equal to Q.

3.1.2. Main Features of TI Processes The Figures 10 and 13 indicate already the main trends common to most of the TI collision systems: the degree of target ionization found in electron capture processes increases with the charge state q of the projectile, with the number k of electrons captured, with increasing atomic number Z_T of the target and decreasing atomic number Z_P of the projectile e.g. in the series of rare gases.

The energy dependence of partial cross sections $\sigma^{o,i}_{q,q-k}$ for multiply charged projectile ions is usually flat giving rise to the assumption that it is not the kinetic energy of the projectile but the potential energy available in the collision which determines TI. Fig. 14 shows measured target-ion charge state fractions F_i for single-electron capture by Xe^{9+} ions from Xe atoms in an energy range from about 20 to 150 keV[22]. Over this entire range of E_{proj} there is no dependence on the kinetic energy of the projectile.

From the relative fractions $F_i = \sigma^{o,i}_{q,q-k}/\sum_i \sigma^{o,i}_{q,q-k}$ measured with an experimental setup as displayed in Fig. 9 and total electron capture cross

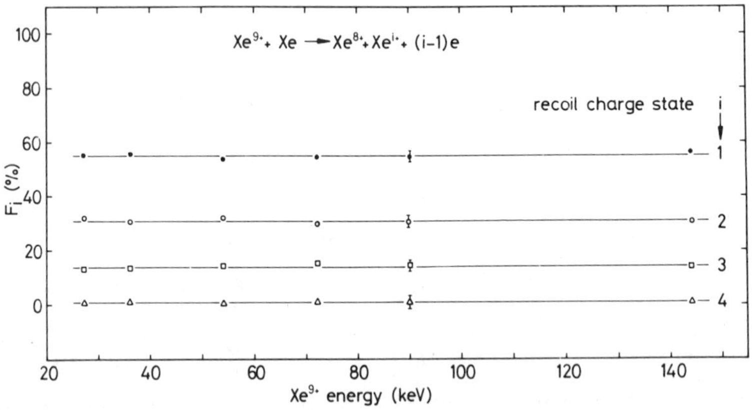

Fig. 14: Charge state fractions F_i of recoil ions as a function of projectile energy in one-electron capture collisons of Xe^{9+} ions in a thin Xe gas target[22].

sections $\sigma_{q,q-k} = \sum_i \sigma^{o,i}_{q,q-k}$ obtained from measurements like those displayed in Fig. 2 one can immediately extract absolute partial cross sections $\sigma^{o,i}_{q,q-k} = F_i \cdot \sigma_{q,q-k}$. Since the total cross sections are generally independent of energy and this is also found for the charge state fractions F_i also the partial cross sections do not depend on E_{proj}. Such partial cross sections obtained from separate measurements of F_i and $\sigma_{q,q-k}$ for the capture of k = 1,2 and 3 electrons, are displayed in Fig. 15 as a function of projectile charge state q for collisions of 10·q keV Ar^{q+} ions with Xe atoms. While pure electron capture (i=k) is still the dominant contribution to $\sigma_{q,q-1}$ it does not play a significant role for $\sigma_{q,q-3}$ although the cross section $\sigma^{0,3}_{q,q-3}$ is still of order 10^{-16} cm².

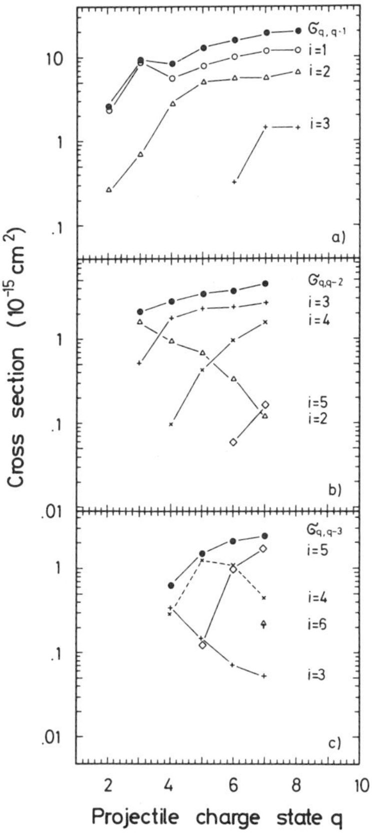

Fig. 15: Absolute partial cross sections $\sigma^{o,i}_{q,q-k}$ for collisions $Ar^{q+} + Xe \rightarrow Ar^{(q-k)+} + Xe^{i+} + (i-k)e$ at 10·q keV projectile energy.

In exceptional cases when not too many intermediate excited states are involved in TI charge state fractions F_i may depend on the projectile energy. Such examples are shown in Fig. 16 where measured values F_i[23] are plotted for collisions $He^{2+} + B \rightarrow He^{+} + B^{i+} + (i-1)e$. In the case of the Ar and Kr targets an interesting result is obtained at He^{2+} energies below 6 keV and especially for Kr there is a strong energy dependence in F_1 and F_2, with $F_2 \gg F_1$ below 4 keV. Single-electron capture in this case is by far dominated by the TI channel. An explanation for this behaviour can be constructed along the following lines. It is well known[3] that endoergic channels with $Q \ll 0$ do not significantly contribute to charge transfer at low energies. Hence, only three final states are likely to contribute to the capture of an electron in $He^{2+} + Kr$ collisions. The "allowed" channels are

$$He^{2+} + Kr \rightarrow He^{1+}\ (n=1) + Kr^{1+} + 40.4\ \text{eV} \qquad \text{I}$$

$$He^{1+}\ (n=2) + Kr^{1+} - 0.4\ \text{eV} \qquad \text{II}$$

$$He^{1+}\ (n=1) + Kr^{2+} + e + 15.8\ \text{eV} \qquad \text{III}$$

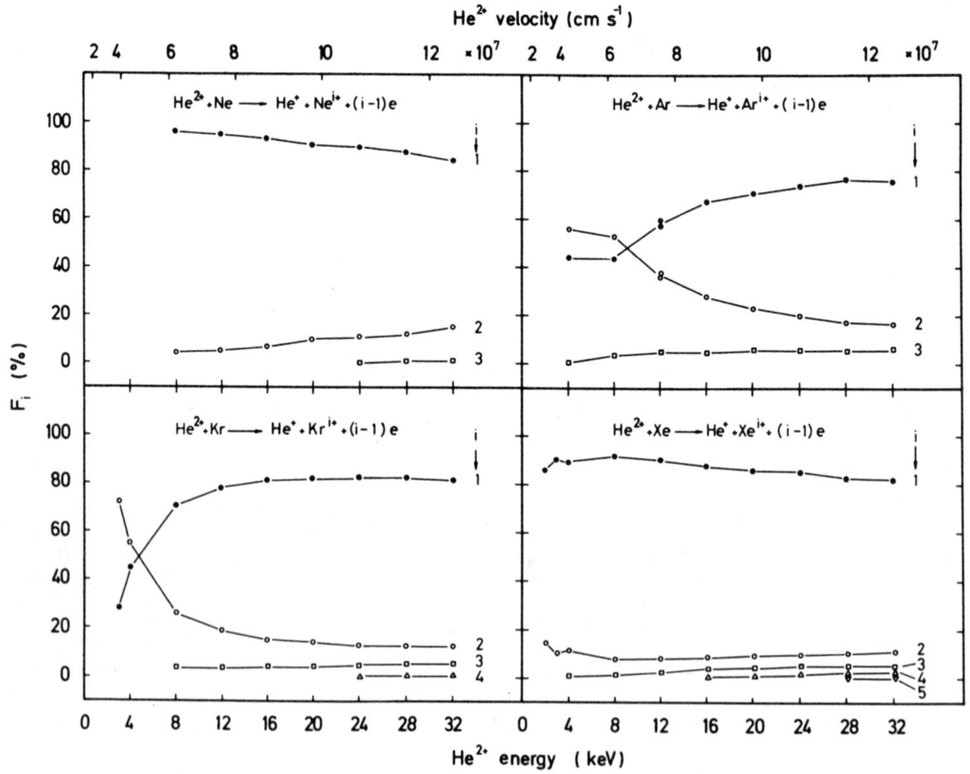

Fig. 16: Charge state fractions[23] of target ions B^{i+} produced in collisions $He^{2+} + B \rightarrow He^{+} + B^{i+} + (i-1)e$.

The behaviour of cross sections for processes I, II and III can be discussed on the basis of the corresponding potential energy curves. The potential of the incident channel which only involves polarization attraction of an ion and an atom can be regarded independent of the internuclear distance R compared to the strong repulsive Coulomb potential of the ions in the outgoing channels I, II and III. The curve crossings occur at internuclear distances $R_c = 27.2\ i \cdot a_o/Q_{(eV)}$ where i is the target ion charge state after the collision, a_o is the Bohr radius and Q is the energy defect. The cross section can then be estimated to be $\pi R_c^2\ p(R_c)$ where $p(R_c)$ describes the overlap of the wavefunctions of incoming and outgoing channels. When R_c is below 5 a_o one can assume $p(R_c) = 1$. For channel I $R_c = 0.67\ a_o$, for channel III $R_c = 3.4\ a_o$, so that TI (III) by far dominates over the pure capture of an electron into the ground state of He^+. Channel II is nearly resonant, which is usually connected with a considerable cross section, however, as the projectile energy decreases the contribution II decreases since $Q < 0$ and ultimately at low enough E_{proj} the corresponding cross section vanishes. What remains is the contribution III from TI. Similar arguments hold for the Ar target. In the case of Ne and Xe targets conditions are not favorable for an "inversion" of TI and pure capture.

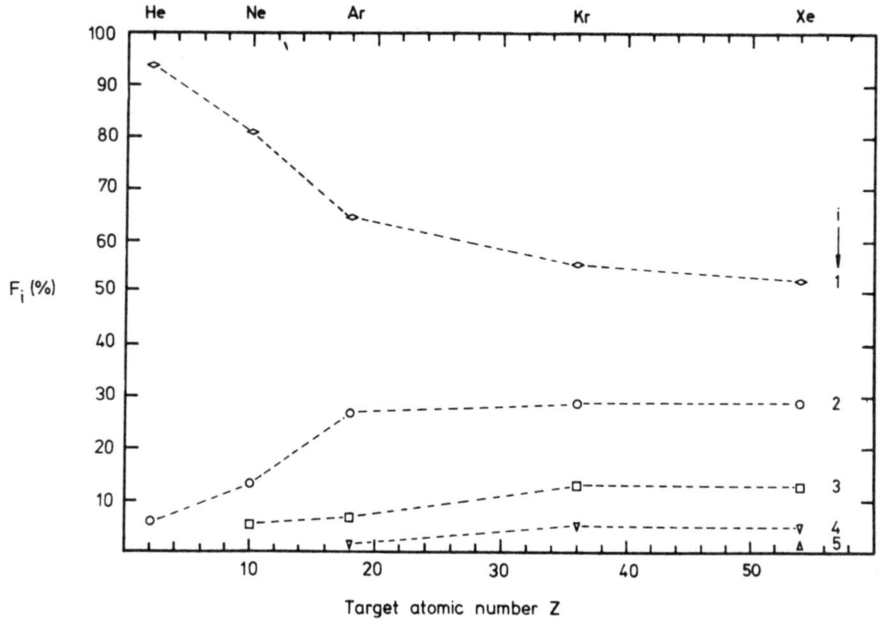

Fig. 17: Charge state fractions of target ions B^{i+} (B= He, Ne, Ar, Kr, Xe) produced in collisions $Ne^{10+} + B \rightarrow Ne^{9+} + B^{i+} + (i-1)e$ at 100 keV[24].

A general observation for TI with slow highly charged projectile ions besides the energy independence of cross sections is a threshold behaviour of the charge state fractions F_i: A target ion B^{i+} is generally only produced when the maximum potential energy available through the capture of electrons into an ion exceeds the total ionization energy of B

$$\Delta E = \sum_{j=q-k}^{q-1} I_A^{(j)} - \sum_{j=0}^{k-1} I_B^{(j)} > \sum_{j=k}^{i-1} I_B^{(j)} \tag{8}$$

Here $I_{A,B}^{(j)}$ denotes the ionization potential of the ion A^{j+} respectively B^{j+}. For the charge state fractions from Ne^{10+}+ B collisions (B = He, Ne, Ar Kr, Xe) displayed in Fig. 17 condition (8) is easily fulfilled for all data points since $I_{Ne}^{(9)}$ = 1362 eV. Condition (8) can be tested more stringently on the basis of experiments where the potential energy carried by the incident ion is smaller than in the case of Ne^{10+}. Fig. 18 shows an example of measurements where 10q keV Xe^{q+} ions have captured k = 2 electrons from Xe atoms. The fractions F_i of target ions Xe^{i+} produced are shown versus the projectile ion charge state. The arrows in Fig. 18 indicate the minimum projectile charge states for which the production of Xe^{i+} target ions becomes exoergic. Comparison with the measured threshold values of q support the assumption that the potential energy in the collision system is the key

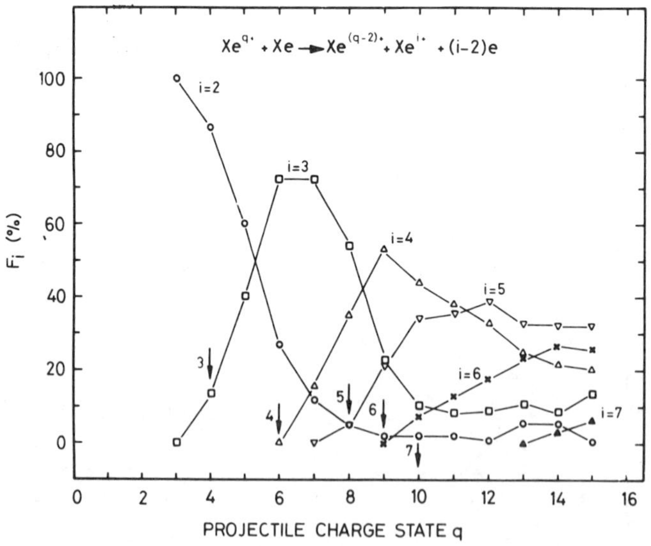

Fig. 18: Charge state fractions of target ions Xe^{i+} produced by two-electron capture of Xe^{q+} ions from Xe atoms[15]. The arrows indicate the lowest projectile charge state for which the reaction creating a recoil ion in charge state i (i = 3,4,5,6,7) is exoergic.

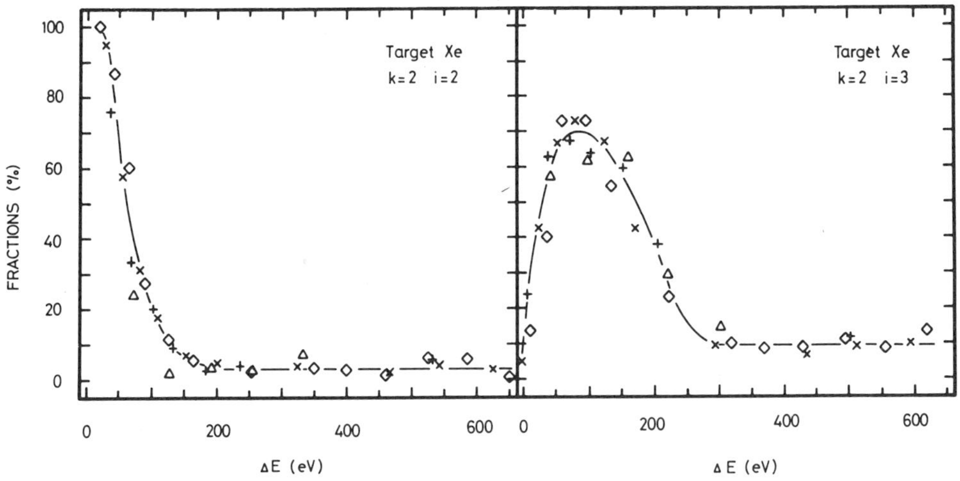

Fig. 19: Plot of experimental fractions F_2 and F_3 of Xe^{2+} and Xe^{3+} target ions produced by two-electron capture into ions A^{q+} (q = 2,3,..,15; A: Ne (Δ), Ar(+), Kr(x), Xe(◇))[22]. ΔE is the maximum potential energy defined by Eq. 8. The lines are drawn to guide the eye.

parameter which determines cross sections and charge state fractions produced in electron capture collisions. Another important evidence for the dominant role of the potential energy ΔE is given by the fact that charge state fractions of ions produced from a specific target by different projectiles in different charge states fall on common curves when plotted as a function of ΔE. Examples for this observation are shown in Fig. 19.

3.1.3. <u>Theoretical Models of TI</u> A rigorous quantum theoretical treatment of the general case of TI would be extremely difficult due to the inherent complexity of the multi-channel problem. However, simple model descriptions have been attempted with some success in the search for a scaling behaviour of TI cross sections. Because of the many different reaction channels leading to a specific set of outgoing charge states (q-k) and i in collisions (6) statistical methods appear to be adequate for the description of multiple ionization processes in charge transfer reactions.

One can assume that only such electrons which initially belong to the target atom B take part in the TI process. A simple model based on the findings of the previous section may start from the idea that the maximum potential energy ΔE is distributed among target electrons and provided ΔE is high enough one or more additional target electrons may be ionized.

Consider a potential well of depth $\langle I_B \rangle$ containing electrons in states of lowest possible energy. $\langle I_B \rangle$ is the mean binding energy for these electrons. A temperature may be defined for the electrons which is T = 0 in the initial state. By TI a maximum excess energy ΔE is distributed among the N_o remaining electrons which gain a temperature T > 0 resulting from the average energy per electron $\Delta E/N_o = 3/2\ k_B T$ (k_B is Boltzmann's constant). When the thermal energy $k_B T$ exceeds $\langle I_B \rangle$ Boltzmann statistics apply (Fermi statistics in the limit of high temperature) and the average number $\langle n \rangle$ of electrons with energy $E \geq \langle I_B \rangle$ can be calculated[30] from the normalized Boltzmann distribution function f(E)

$$f(E)\,dE = \sqrt{\frac{27 \cdot N_o^3}{2\pi\ \Delta E^3}}\ \exp\left(-\frac{3\ N_o}{2\ \Delta E}\ E\right) \sqrt{E}\ dE. \qquad (9)$$

$$\langle n \rangle = \int_{\langle I_B \rangle}^{\infty} N_o \cdot f(E)dE \qquad (10)$$

The average charge state after the transfer of the energy ΔE and the loss of high temperature electrons is

$$\langle i \rangle = k + \langle n \rangle = k + N_o\left(1 - \int_0^{\langle I_B \rangle} f(E)dE\right), \qquad (11)$$

where k is the number of electrons captured by the projectile. The number of electrons N_o among which the energy ΔE is distributed can be assumed to be the number of electrons in the outermost shell of the ion B^{k+}. For rare gas targets N_o = 8-k appears to be a reasonable number. The mean ionization potential has to be determined for each B^{k+} separately. With increasing ΔE an increasing number of electrons can be released from B^{k+}. When ΔE facilitates the ejection of n electrons but not n+1, i.e. for

$$\sum_{j=0}^{n-1} I_B^{(k+j)} \leq \Delta E < \sum_{j=0}^{n} I_B^{(k+j)} \qquad (12a)$$

one can set

$$\langle I_B \rangle = \frac{1}{n} \sum_{j=0}^{n-1} I_B^{(k+j)} \qquad (12b)$$

Now, the average target ion charge state $\langle i \rangle$ after a TI process can already be calculated. In Fig. 20 the result is compared to experimental values obtained for Xe target atoms after TI collisions with multiply charged rare gas ions. The trend and the absolute values of the data are amazingly well reproduced considering the simplicity of the above considerations.

One can even go a step further in the evaporation picture. On the basis of the theoretical work of Russek and coworkers[25] it is possible to calculate individual charge state fractions F_i. The derivation starts from a statistical distribution of infinitely small units of energy, the sum of which is ΔE, among N_o outer shell electrons of the target ion B^{k+}. One can then calculate the probability $P_n^{N_o}(\Delta E) = F_{n+k} = F_i$ that n electrons gain enough energy to overcome the mean ionization potential $\langle I_B \rangle$ and escape from B^{k+}.

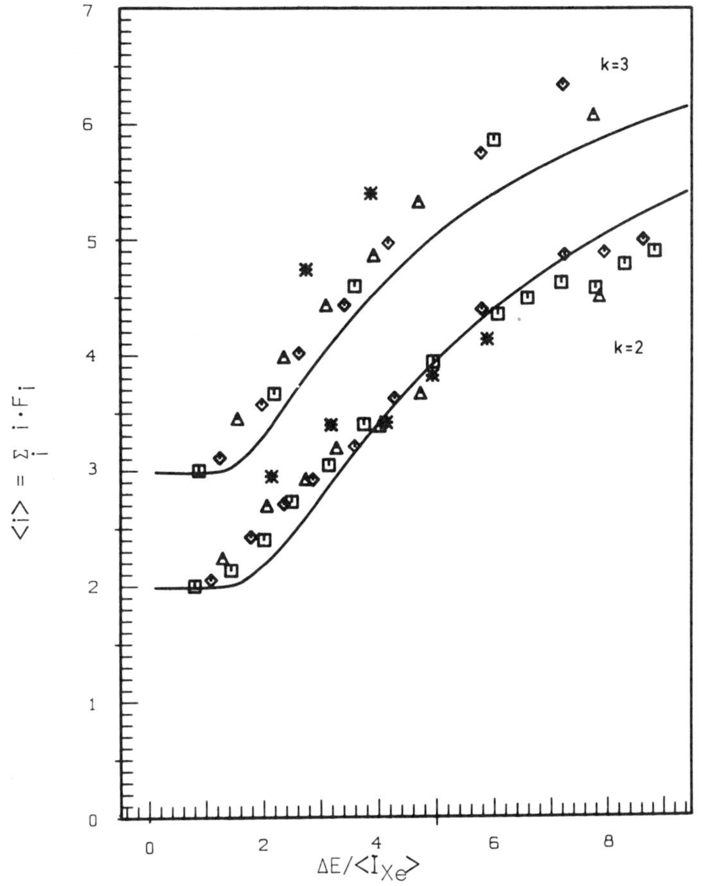

Fig. 20: Average charge states $\langle i \rangle$ of target ions Xe^{i+} produced in collisions $A^{q+} + Xe \rightarrow A^{(q-k)+} + Xe^{i+} + (i-k)e$ where k = 2 or 3 electrons are captured by the projectile[22]. The species A are: Ne(✱), Ar(Δ), Kr(◇), and Xe (□). The solid lines result from Eq. (11) and describe the thermal evaporation of electrons. $\langle I_{Xe} \rangle$ is the mean ionization potential for Xe defined by Eqs. (12).

$$P_n^{N_o}(\Delta E) = \binom{N_o}{n} \sum_{j=0}^{m} (-1)^j \binom{N_o-n}{j} \left(1-\frac{n+j}{\Delta E/<I_B>}\right)^{N_o-1} \qquad (13)$$

with m defined by $n+m \leq \Delta E/<I_B> \leq n+m+1$. For rare gas target atoms B^{k+} after the capture of k electrons the remaining number of electrons in the outer shell is $N_o = 8-k$.

Equation (13) turns out to be a very useful scaling law for TI processes involving slow multiply charged ions where k = 2 or more electrons are captured. The dominant parameter is the maximum excess energy ΔE scaled with the mean ionization potential $<I_B>$. A plot of charge state fractions F_i versus $\Delta E/<I_B>$ is shown in Fig. 21 for a large number of collision systems (6) investigated[26]. Again there is an astounding agreement between experiments and the statistical model.

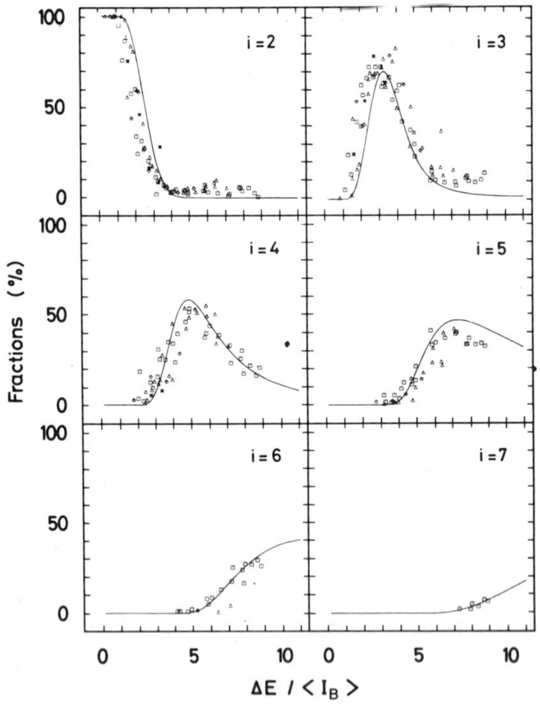

Fig. 21: Charge state fractions of target ions B^{i+} produced in reactions (6) with k=2 as a function of the reduced energy $\Delta E/<I_B>$. The figure contains all experimental results of Groh et al.[22] for different collision systems (6) investigated. Symbols for the targets are Ne(✱), Ar(Δ), Kr(◇), Xe(□). The solid lines represent the results of the statistical model[26].

Although the model of target electron evaporation is apparently successful in reproducing the experimental data, recent measurements of electron energy spectra[27] produced in TI collisions seem to indicate that most of the electrons are emitted from the projectile. The process discussed is the capture of two electrons from the target into excited states of the projectile followed by autoionization in the projectile. However, there is no conclusive picture available yet for the capture of $k = 2$ electrons in the meaning of section 2.3. Nevertheless one can try to develop a different model on the basis of the available experimental findings concerning emission of electrons from the projectile.

The basic intention of this alternative model[30] is to describe the transfer of i electrons from the target (which is left in charge state i without any excitation energy) into excited states of the projectile. By autoionization (i-k) electrons are emitted and k electrons remain attached to the projectile. The capture of electrons in this picture can be calculated by extending the classical barrier model for electron transfer[28]. The most probable mechanism of multiple electron capture is assumed to be the capture of one electron after the other at successive crossings of potential energy curves. Recent measurements on Ne^{7+} + He collisions by Roncin et al.[29] support this assumption.

Consider now that m electrons are already transferred from target atom B to the projectile A^{q+} at internuclear distances $R_x^{(1)} > R_x^{(2)} \ldots > R_x^{(m)}$. The transferred electrons occupy states with principle quantum numbers $n^{(1)}, n^{(2)}, \ldots, n^{(m)}$. The target is now B^{m+}, the projectile $A^{(q-m)+}$. As the internuclear distance R_x becomes still smaller the saddle point in the potential is further depressed and the next electron can travel across the barrier to the projectile. The energy of this electron in the target is influenced by the Coulomb potential of the projectile

$$E_B^{(m+1)*} = E_B^{(m+1)} - \frac{q-m}{R_x} \tag{14}$$

(Atomic units are used in the present derivation).
The projectile energy level into which the next electron will be captured has an energy disturbed by the presence of the target ion B^{m+}

$$E_A^{(m+1)*} = E_A^{(m+1)} - \frac{m+1}{R_x} \tag{15}$$

The resonance condition is $E_B^{(m+1)*} = E_A^{(m+1)*}$, which results in

$$E_B^{(m+1)} - E_A^{(m+1)} = \frac{q-m}{R_x^{(m+1)}} - \frac{m+1}{R_x^{(m+1)}} \tag{16}$$

$|E_B^{(m+1)}|$ is the ionization potential of the now active target electron, i.e. $E_B^{(m+1)} = -I_B^{(m)}$. The energy $E_A^{(m+1)}$ in a hydrogenic appraoch is given by $E_A^{(m+1)} = -\frac{q^2}{2(n^{(m+1)})^2}$. The already captured m electrons are in levels with higher principle quantum numbers than $n^{(m+1)}$ and thus do not screen the Coulomb potential of the projectile core. Thus the internuclear distance at which electron no. (m+1) can be transferred is

$$R_x^{(m+1)} = (q - 2m - 1)/\left(\frac{q^2}{2(n^{(m+1)})^2} - I_B^{(m)} \right) \tag{17}$$

The electron can only be transferred, when its perturbed energy $E_B^{(m+1)*}$ exceeds the height $U_{min}(R_x)$ of the potential barrier at the saddle point. This leads to another relation for R_x. The potential seen by the electron is $U(r) = -(m+1)/r - (q-m)/(R_x-r)$ where r is the radial coordinate of the electron in the frame of the target ion. Thus one gets

$$R_x \leq \frac{1}{I_B^{(m)}} \left(m-q + \frac{(m+1)}{A} + \frac{q-m}{1-A}\right) \tag{18}$$

with $A = r_m/R_x$ and r_m the radius r at which U(r) is minimum.

$$A = \left(\sqrt{(m+1)(q-m)} - (m+1)\right)/(q-2m-1) \tag{19}$$

From the two conditions for R_x one gets the principle quantum number of the level into which the electron will be captured.

$$n^{(m+1)} \leq \sqrt{\frac{q^2}{2I_B^{(m)}} / \left(1 + (q-2m-1)/(m-q + \frac{m+1}{A} + \frac{q-m}{1-A})\right)} \tag{20}$$

The cross section $\sigma_q^{o,m+1}$ for the capture of (m+1) electrons in the meaning of the present discussion can then be calculated from (17) and (20) to be

$$\sigma_q^{o,m+1} = \sum_k \sigma_{q,q-k}^{o,m+1} = \pi\left(R_x^{(m+1)}\right)^2 \tag{21}$$

This general derivation of a capture cross section (better: a target electron loss cross section) includes the case of single electron capture (m = o) and leads to the classical results of refs. 28.

The autoionization of multiply excited projectile ions after capture can be treated again with statistical methods. The probability P_{i-k}^{i} to evaporate (i-k) out of i electrons can be calculated from Eq. (13) and hence absolute partial cross sections can be determined in principle from

$$\sigma^{o,i}_{q,q-k} = \sigma^{o,i}_{q} \cdot P^{i}_{i-k} \; (E_{ex}) \tag{22}$$

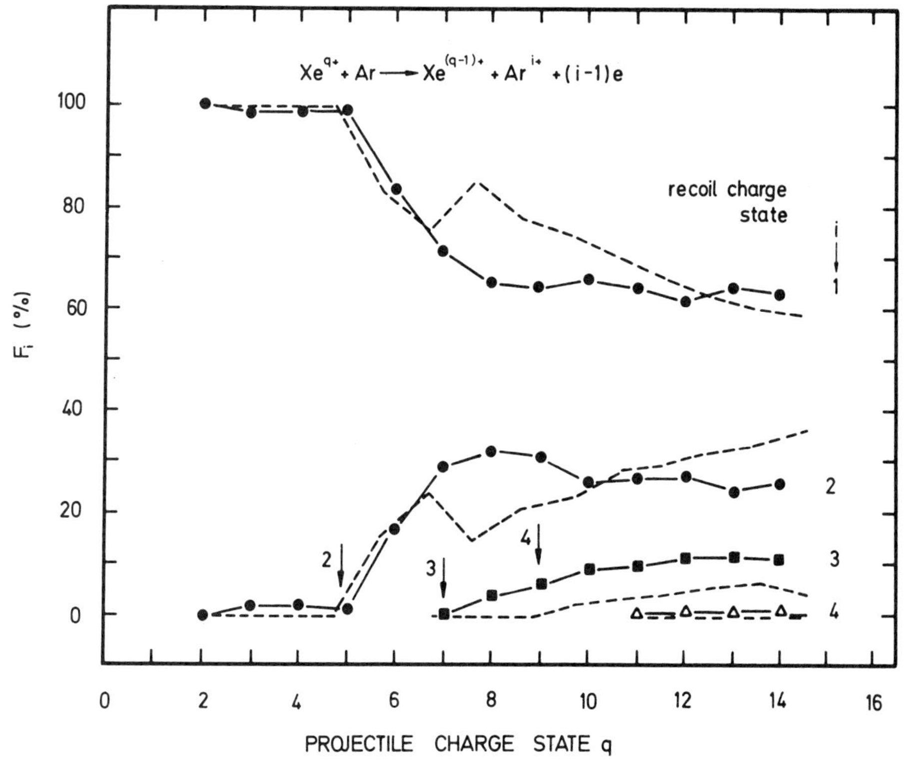

Fig. 22: Comparison of measured charge state fractions F_i with calculations (dashed lines) based on multiple electron capture and subsequent electron evaporation from the projectile[30]. The collision energy is 10q keV.

In this formula E_{ex} is still the maximum available potential energy now stored as excitation energy of the projectile. Assuming that the transfer of the electrons occured more or less resonant it is given by Eq. 8 with k replaced by $i(E_{ex} = \Delta E(i))$.

A comparison of measured and calculated charge state fractions of Ar^{i+} ions produced in collisions $Xe^{q+} + Ar \rightarrow Xe^{(q-1)+} + Ar^{i+} + (i-1)e$ at an energy of 10q keV is given in Fig. 22. In general Eq. (22) does not at all reproduce the experiments as well as Eq. (13) applied to the evaporation of target electrons. The classical model, however, has the advantage to produce absolute partial cross sections for TI which are in reasonable agreement with the global dependences.

Another statistical theory was developed by Åberg et al[31]. In their approach, two Lagrange parameters λ and λ' are introduced which have to be deduced from experimental results, λ from the mean experimental charge state $\langle i \rangle$. The second parameter λ' was set equal to zero. Calculated fractions F_i are in good agreement with experiments for k = 1, while being somewhat too evenly distributed for $k \geq 2$ (see Eq. (6)). This effect was noted and discussed in the framework of a "surprisal" theory by Åberg et al. It is interesting to note that the dominating recoil ion production mechanism (the simultaneous capture of an outer and an inner shell electron) proposed by these authors to explain the "non-statistical behaviour" of the fractions F_i also fits the physical picture given by Müller et al.[26], namely that of a target left in a state of multiple excitation from which electrons are evaporated.

Recently Bárány et al[32] and Liljeby et al.[33] have proposed a mechanistic model to describe TI. Starting from a generalization of electron capture radii in a classical barrier picture leading to similar values

$$R_x^{(i)} = \left(2(q - i + 1)^{1/2}\, i^{1/2} + i\right)/I_B^{(i-1)} \tag{23}$$

as calculated from Eqs. (17) and (20) they obtained cross sections for target electron loss

$$\sigma_q^{o,i} = \pi\left(\left(R_x^{(i)}\right)^2 - \left(R_x^{(i+1)}\right)^2\right) \tag{24}$$

which agree well with their experimental values. Eq. 24 corresponds to using $R_x^{(i+1)}$ as a non-zero minimal impact parameter, which in a sense is in conflict with a curve crossing picture (Fig. 23) and the resulting behaviour of impact-parameter dependent capture probability. When the incoming collision system follows the adiabatic potential energy curve at a crossing $R_x^{(i)}$ it does not necessarily reach another crossing within $R_x^{(i+1)}$ even when the impact parameter is small.

Therefore Liljeby et al. modify the prescription for evaluating cross sections by introducing instead the approximation:

$$\sigma_q^{o,i} = f \cdot \pi \cdot \left(R_x^{(i)}\right)^2 \tag{25}$$

and assume that f is independent of i. Through the condition

$$\sum_{i=1}^{i_{max}} \sigma_q^{o,i} = \pi \left(R_x^{(1)}\right)^2 \tag{26}$$

one arrives at $f = \left(R_x^{(1)}\right)^2 / \sum_{i=1}^{i_{max}} \left(R_x^{(i)}\right)^2$

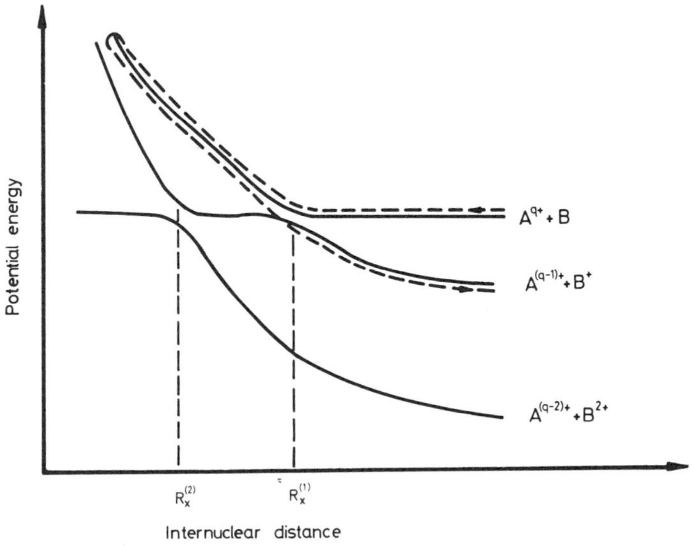

Fig. 23: Schematic potential energy curves for the capture of 1 or 2 electrons in a collision A^{q+} + B. A possible reaction path is indicated by the dashed line.

An estimate for i_{max} will be given below in connection with Eq. (31). The energy gain Q in the collision process (6) is used now to obtain information on crossing radii. The assumptions made are those described to derive Eq. (22), i.e. the excitation energy E_{ex} of the projectile after multiple electron capture determines the number of electrons emitted. Rather than assuming resonant electron capture to evaluate E_{ex} upper and lower limits are determined for the energy gain Q. The maximum energy gain for (6) is

$$Q_{max} = \sum_{j=q-k}^{q-1} I_A^{(j)} - \sum_{j=0}^{i-1} I_B^{(j)} \tag{27a}$$

The minimum energy gain is

$$Q_{min} = Q_{max} - I_A^{(q-k)} \tag{27b}$$

While Q_{max} is trivially understood Q_{min} appears somewhat arbitrary. As many results for slow projectile ions (in not too high charge states) show $\sigma^{o,i}_{q,q-k}$ can be non-zero as soon as $Q_{min} > 0$. Eq. (27b) requires that the maximum available excitation energy after capture of i electrons may not exceed the energy necessary to produce $A^{(q-k)+}$ ions from $A^{(q-i)+}$ by more than $I_A^{(q-k)}$ (otherwise the autoionization of the multiply excited ion $A^{(q-i)+*}$ would end with $A^{(q-k+1)+}$ instead of $A^{(q-k)+}$).

Anyway, under the assumptions that (27) is a reasonable boundary condition for Q and assuming capture as a one-step process at the crossing of a flat initial potential curve and the Coulomb repulsive one with a capture radius for process (6)

$$R(q,q-k,i) = (q-i)\cdot i/Q \tag{28}$$

Q_{min} and Q_{max} lead to consecutive limits R_{min} and R_{max}. The individual cross sections $\sigma^{o,i}_{q,q-k}$ are only non-zero, when $R_x^{(i)}$ (Eq. (23)) is between these limits:

$$\sigma^{o,i}_{q,q-k} = g_i\cdot\pi\cdot \mathrm{Min}\ \{(R_x^{(i)})^2, (R_{max}(q,q-k,i))^2\} \tag{29}$$

The parameter g_i follows from the condition (Eq. (25))

$$\sigma^{o,i}_{q} = \sum_{k=1}^{i} \sigma^{o,i}_{q,q-k} = f\cdot\pi\cdot(R_x^{(i)})^2 \tag{30}$$

which gives

$$g_i = f\cdot(R_x^{(i)})^2/(\sum_k \mathrm{Min}\ \{(R_x^{(i)})^2, (R_{max}(q,q-k,i))^2\}). \tag{31}$$

The summation extends over all k such that $R_x^{(i)} > R_{min}(q,q-k,i)$. The maximum i compatible with this inequality is the i_{max} value needed in Eq. (26).

Liljeby et al.[33] have compared their calculations with absolute experimental cross sections. The global features of TI are well reproduced by their model although the absolute magnitudes of cross sections sometimes agree only to within a factor of two to five. Recognizing that the model contains no free parameters and actually predicts absolute cross sections the comparisons displayed in Fig. 24 show remarkable agreement of theory and experiment.

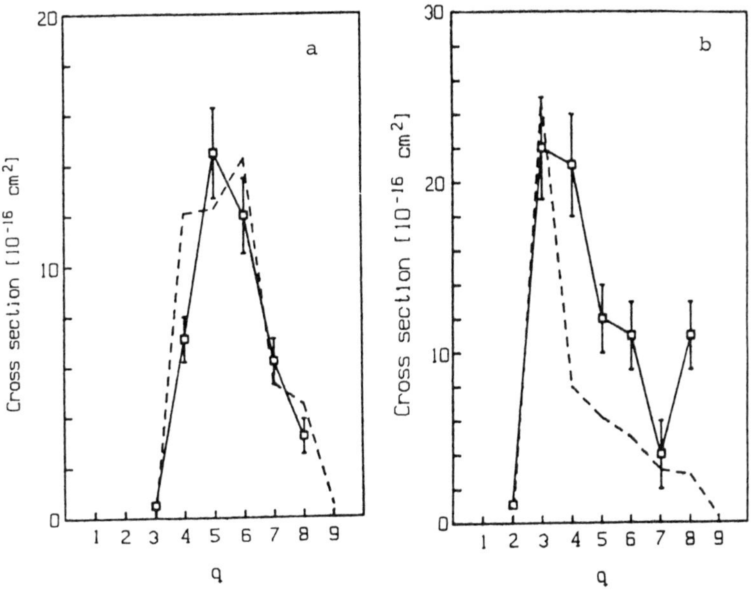

Fig. 24: Absolute cross sections $\sigma^{0,2}_{q,q-2}$ for (a) Ar^{q+}+ Ne and (b) Ar^{q+}+ Kr collisions. Experimental data (□) are compared to the model calculations presented in the text (from ref. 33).

The theoretical models for charge transfer collisions (6) of slow multiply charged ions with atoms are surprisingly successful. They describe global features of TI and partly even predict absolute cross sections. They are based on numerous assumptions and recipes rather than solid theoretical ground. Their limitations are obvious, however, they have made the problem tractable and provide a starting base for further insight into TI processes.

3.1.4. Details of TI Processes Most of chapter 3.1 dealt with global features of TI processes and in particular the multiple ionization probabilities connected with the capture of electrons by a slow highly charged ion from an atom were reviewed. A different approach to TI is the detailed investigation of isolated reaction channels in specific collision systems. Such experiments employ energy spectroscopy of the electrons emitted in TI collisions, the energy gain spectroscopy, angular differential measurements

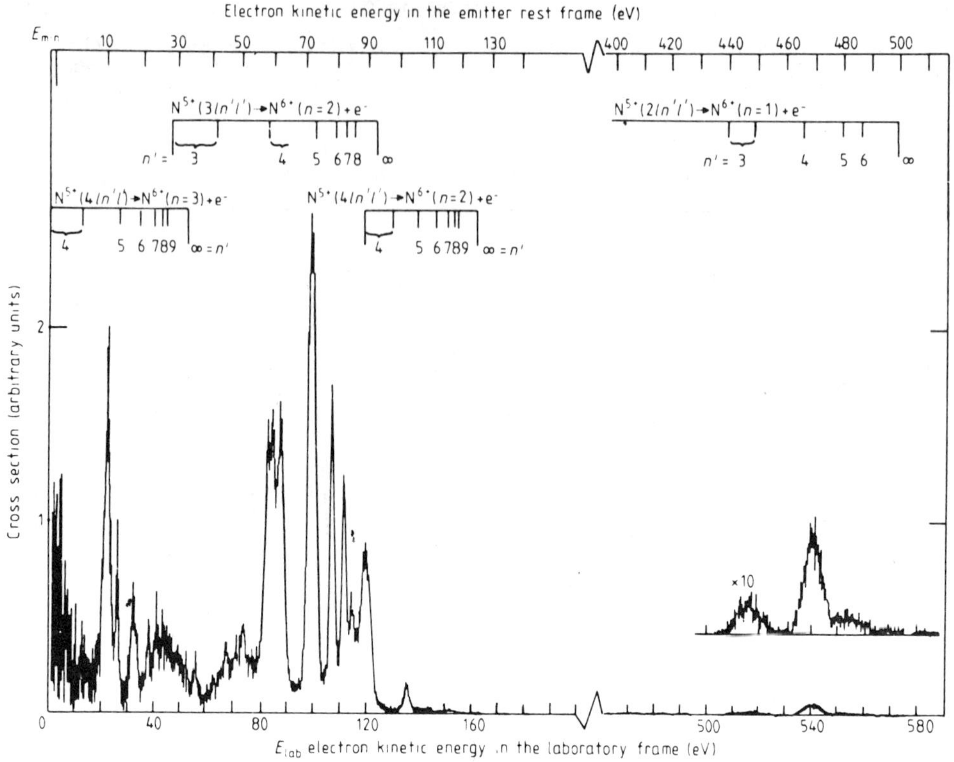

Fig. 25: Electron energy spectrum measured in N^{7+}+H_2 collisions at Θ_{lab}= 11.6° and 4.9 keV/amu. Spurious background has been subtracted. E_{min} is the minimum electron energy observable in the laboratory frame under the given experimental conditions (from ref. (27)).

of the projectile ions, and combinations thereof. Work on ions in low charge states was reviewed by Niehaus[34]. Since then some progress has been made also with multiply charged ions both in energy gain and electron spectroscopy measurements. An example for a measured electron energy spectrum is shown in Fig. 25. The collision system is 68.6 keV N^{7+}+ H_2, i.e. a completely stripped ion N^{7+} collides with a two-electron target[27]. The lines in the measured spectrum are assigned to double capture processes with subsequent emission of an autoionizing electron from the projectile. Dominant contributions appear to arise from intermediate N^{5+}(nl,n'l') states with n,n' = 3,4,5 and N^{6+}(n",l") ions left with n" = 2,3 after the autoionization process. Although the obtained information is detailed it is difficult to identify the electron lines unambiguously and follow the path of the collision system in a potential energy diagram. Apparently more work has to be done. The interpretation of experimental results can be made easier by a better definition of the experimental conditions in order to reduce the number of collision paths involved. One such approach has been made recently by Roncin et al.[29]. These authors investigated collisions of 10.5 keV Ne^{7+} on He, measured charge-state resolved energy-gain spectra of the projectiles and at the same time the angular dependence of Ne^{6+} and Ne^{5+} produced in specific excited states. From their measurements they concluded that the capture of two electrons occured via a two-step mechanism, i.e. two successive capture processes during a single collision.

Still more detailed experiments are underway in laboratories at Groningen[35] and in Giessen. In these experiments the projectiles will be charge-state analyzed (eventually even their energy can be measured), the charge state of the recoil ions will be identified and the electron energy will be determined. Coincidence measurements of all particles emerging from a TI collision are planned.

3.2. Intermediate Collision Energies ($v_{proj} \approx v_B$)

Multiply charged ions can be produced basically in two different ways. One is the ionization of thermal atoms by sufficiently many or powerful collisions such as bombardment of gas by dense electron beams or fast very highly charged heavy ions. Acceleration of the slow ions by 0.1 to about 100 kV is feasible so that typical velocities are below 10^8 cm/s. The other way leads over acceleration of ions in low charge states to high velocities and subsequent stripping of electrons in a foil target. Thus the high charge states are typically connected with ion velocities above 10^9 cm/s. It is possible to decelerate these ions or further accelerate slow highly charged ions, however, this requires considerable efforts and investment. Therefore TI data for multiply charged ions at intermediate velocities are scarce.

The experimental techniques employed in the velocity range 10^8 - 10^9 cm/s are basically the same as described in 3.1 with the exception of energy gain spectroscopy which would need too high resolution. The coincidence analysis of recoil ion charge states, however, is easily also applied here. A new feature of TI collisions is an increasing effect of the projectile velocity on the ionization efficiency and on the size of cross sections.

A compilation of data for collisions involving electron capture by protons from rare gas atoms is shown in Fig. 26. With the exception of

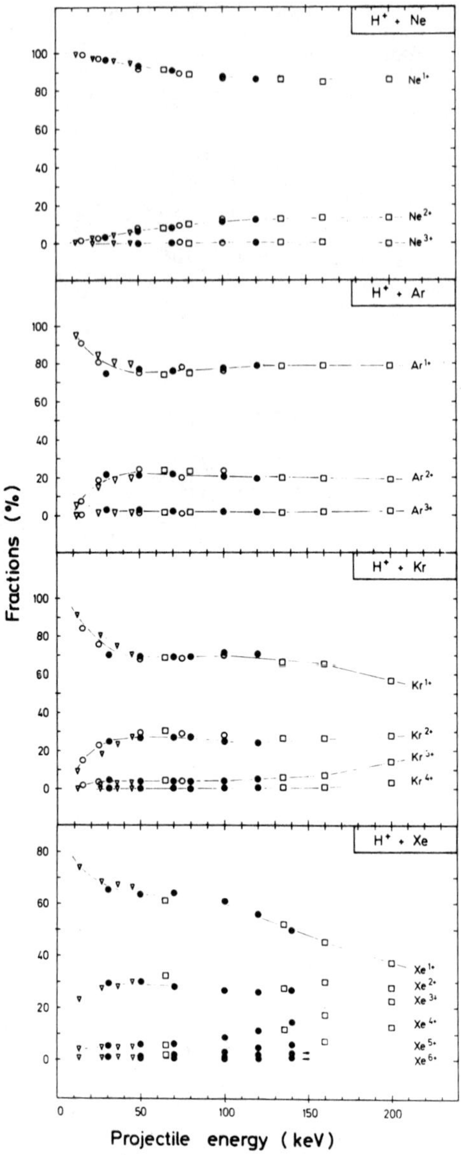

Fig. 26: Charge state fractions F_i of target ions B^{i+} (B = Ne, Ar, Kr, Xe) produced in collisions $H^+ + B \rightarrow H + B^{i+} + (i-1)e$. The data were taken by Schuch et al.[36] (●), DuBois[37] (o), Afrosimov et al.[38] (▼) and Horsdal Pedersen and Larsen[39] (□).

$H^+ + Xe \rightarrow H + Xe^{1+} + e$ all the processes investigated are endoergic. They can only proceed at the expense of kinetic projectile energy so that the increase of the fractions of multiply charged recoil ions with energy becomes understandable. In the energy range from 10 keV to 200 keV corresponding to $v_{proj} = 0.6\ v_B$ and $2.8\ v_B$, respectively, the experimental results from different experiments agree nicely, indicating that the projectile-target-ion coincidence technique is well established.

Horsdal Pedersen und Larsen[39] have studied collisions H^+ + Ne in the energy range from 50 keV to 2000 keV - the latter corresponding to $v_{proj} = 8.9\ v_B$. Their experimental fractions F_i of Ne^{i+} recoil ions are shown in Fig. 27. The production of singly charged recoil ions, i.e. pure capture

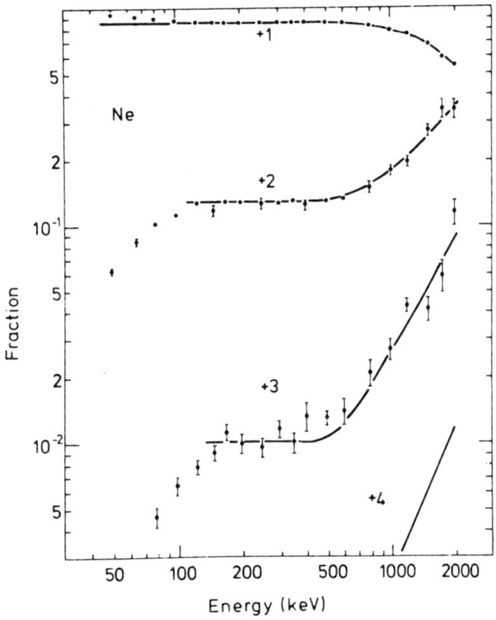

Fig. 27: Energy dependence of charge state distributions of Ne ions formed by electron capture of protons in Ne gas. The solid lines represent a calculation based on capture from inner shells with subsequent autoionization (from ref. 39)

of an electron, dominates all other channels over the entire energy range. There is, however, a characteristic dependence of F_i on the energy with a significant increase of F_i for multiply charged Ne^{i+} ions at energies above 500 keV. At low kinetic energy only Ne^{1+} ions are produced. With increasing proton energy when v_{proj} exceeds v_B double ionization becomes more and more probable where one electron is captured and removed and due to the correlation a second electron can be ionized. The ratio F_2/F_1 is similar to the one obtained from properly scaled[39] electron impact ionization experiments. Beginning from 100 to 200 keV the electron capture becomes so fast that the electron is removed from the Ne atom within 2 to $3 \cdot 10^{-17}$ s. "Sudden" removal of electrons from different shells of atoms has been investigated by Carlson et al.[40] by photoionization experiments and theoretical calculations of shake-off probabilities. Carlson et al. found for photoionzation of the Ne L-shell a final charge state distribution after the rearrangement of the Ne atomic electrons (86 % Ne^{+}, 13 % Ne^{2+}, 1 % Ne^{3+}) which agrees with the electron capture data of Horsdal Pedersen and Larsen in the energy range from 150 to 400 keV. Since the captured electron is suddenly removed from the Ne L-shell and carried away by the fast projectile there is no difference in the charge state distributions whether the electron is ionized by a photon or removed by capture.

At energies beyond 500 keV the fractions F_2 and F_3 considerably increase. The reason is that the proton velocity approaches the orbital velocity of the Ne K-electron and therefore electron capture from the K-shell becomes favourable. From the energy dependent probabilities P_L and P_K to form L- and K-shell vacancies in Ne and the charge state distributions obtained by Carlson et al.[40] after Ne K-shell photoionization (1 % Ne^{1+}, 73,6 % Ne^{2+}, 22 % Ne^{3+}, 3.4 % Ne^{4+}, 0,3 % Ne^{5+}) and Ne L-shell photoionization (see above) Horsdal Pedersen and Larsen calculated the solid lines in Fig. 27 and obtained excellent agreement with the experimental data.

A detailed study of direct ionization, capture and TI in collisions of H^{+} and He^{+} with rare gas atoms was performed by DuBois[37]. Some of his results and data of Schuch et al.[36] are displayed in Fig. 28. The investigation shows that higher order capture processes (TI) dominate over direct multiple ionization , i.e. $\sigma^{o,i}_{1,0} > \sigma^{o,i}_{1,1}$.

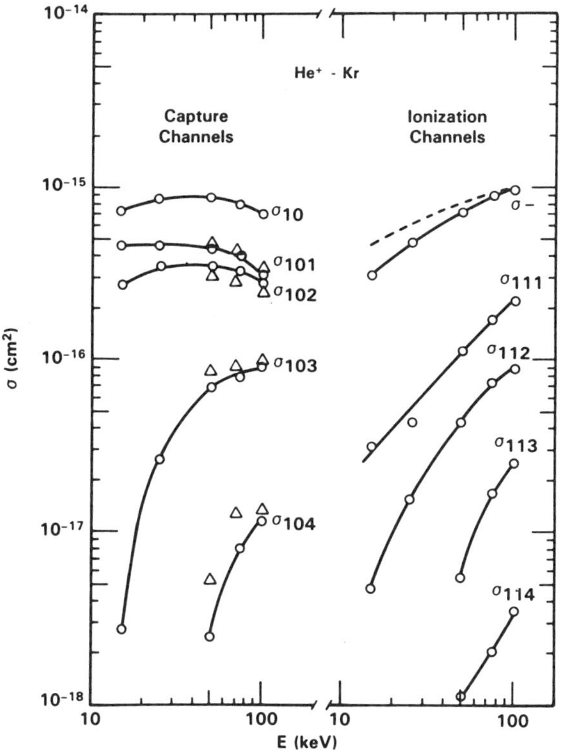

Fig. 28: Cross sections for the direct-ionization and capture channels for He^+ + Kr collisions. Cross section subscripts correspond to the present notation like $\sigma_{i,j,k} = \sigma_{i,j}^{o,k}$. The capture and direct-ionization channels are on the left-hand and right-hand side of the figure. Circles are DuBois' data[37], triangles are from Schuch et al.[36] and the dashed line is the total electron production cross section σ_- from Eckhardt and Schartner[41].

3.3. Fast Collisions of Highly Charged Ions ($v_{proj} > v_B$)

At projectile energies exceeding 1 MeV/amu (i.e. $v_{proj} = 6.3\ v_B$) the collision time is small compared to outer shell electron revolution periods, i.e. outer shells cannot adiabatically rearrange during the collision. Highly charged ions such as Fe^{20+} have radii of typically 10^{-11}m and can approximately be treated like point charges. The investigation of recoil ion charge states produced by direct ionization and electron capture with

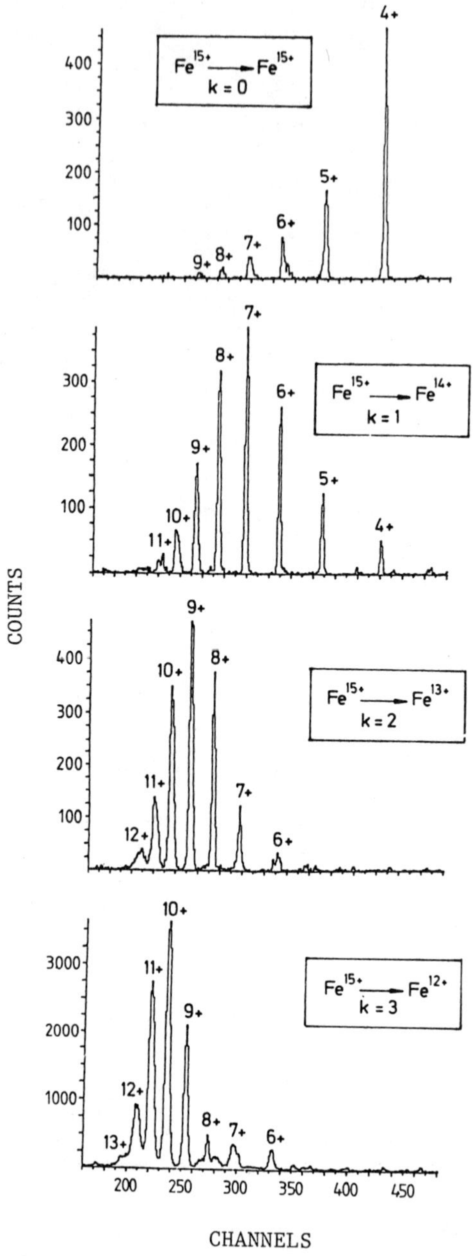

Fig. 29: Time-of- flight spectra of Ar^{i+} recoil ions produced in collisions of 1.4 MeV/amu Fe^{15+} ions on Ar atoms. The number k of target electrons found attached to the projectile after the collision increases from k = 0 (top) to k = 3 (bottom). For k = 0 the peaks produced by Ar^{1+}, Ar^{2+} and Ar^{3+} recoil ions are not displayed (from ref. 43).

ionization in collisions of fast ion with atoms was stimulated by earlier indirect observations indicating very high ionization efficiencies of highly charged ions and the possibility to use a gas target passed by an intense accelerator beam as an intense source of slow highly charged ions. Cocke[16] and Gray et al.[42] were the first to look at recoil ion production by fast ions in gases. They measured cross sections $\sigma_{q,q-k}^{o,i}$ and found, e.g. for collisions of fast completely stripped F^{9+} ions in Ne, a large shift of the average charge state $<i>$ of Ne ions from about 1.5 for direct ionization (k = 0) to about 4 for one-electron capture (k = 1) and about 7 for two-electron capture (k= 2). Such large shifts in the recoil charge states as the number of captured electrons increases is a general feature of TI at high energies. An example is shown in Fig. 29.

For a wide range of projectiles and charge states (from N^{6+} to U^{48+}) at 1.4 MeV/amu Müller et al.[43] found charge state distributions of recoil ions which nearly only depend on the target gas and the number of electrons transferred. They do not depend on the projectile as long as the initial charge state exceeds q = 10.

Cross sections for direct ionization and TI for 1.4 MeV/amu U^{44+} + Ar collisions[43] are displayed in Fig. 30. The experimental data points were obtained by normalizing measured charge state distributions to total net ionization and electron capture cross sections as they are discussed in section 2.2. The dotted lines represent the results of a theoretical estimate based on <u>c</u>lassical <u>t</u>rajectory <u>M</u>onte <u>C</u>arlo (CTMC) calculations. Although comparison with the experiment at first sight does not give a too favorable picture for the theory one has to realize the complexity of the multi-electron problem which was treated in such detail. Considering the difficulties and the fact that no free parameters enter the theory the agreement both in position and size of the maxima of the distributions and in the resulting total electron capture cross sections is remarkable. Moreover, the CTMC method presently is the only theoretical approach which can handle the problem at all.

The calculations start from the independent electron model. The probability for removing k of the K-shell electrons, l of the L-shell electrons, and m of the M-shell electrons is given by the simple product

$$P_{klm} = \binom{K}{k} P_K^k (1-P_K)^{K-k} \binom{L}{l} P_L^l (1-P_L)^{L-l} \binom{M}{m} P_M^m (1-P_M)^{M-m}. \tag{32}$$

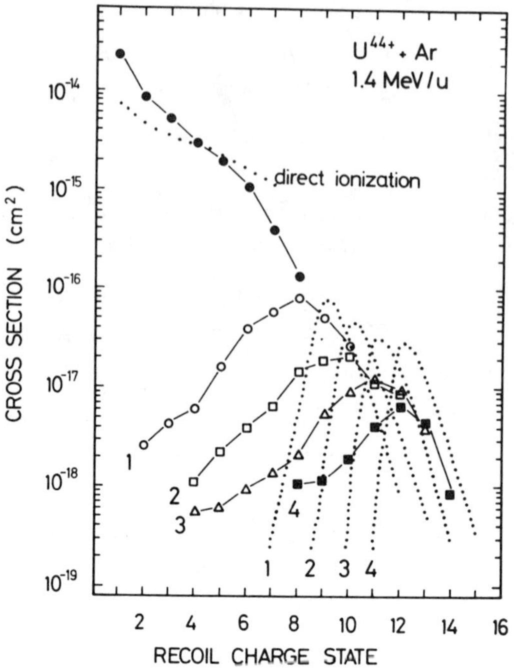

Fig. 30: Cross sections $\sigma^{o,i}_{q,q-k}$ for direct ionization (k = 0) and capture of k = 1,2,3 or 4 electrons in 1.4 MeV/amu collisions of U^{44+} ions with Ar atoms. CTMC calculations are represented by dotted lines (from ref. 43).

Here, $\binom{N}{n}$ is the binomial coefficient and P_n is the transition probability calculated as a function of impact parameter within a one-electron model. The cross section for electron capture and additional target ionization is determined by the product of two probabilities like that given in Eq. 32: the first employs the single electron capture probabilities, while the second component uses the ionization transition probabilities both calculated within the one-electron framework applying the CTMC method.

Record high charge states have been found for Xe^{i+} recoil ions produced in 15.5 MeV/amu U^{75+} + Xe collisions[44]. Kelbch et al. measured cross sections for direct ionization and total capture with ionization

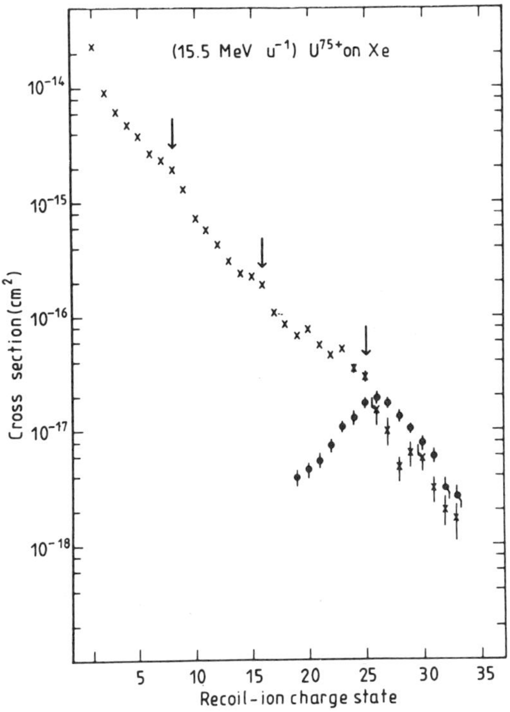

Fig. 31: Cross sections $\sigma^{o,i}_{q,q}$ and $\sum_{k\geq 1} \sigma^{o,i}_{q,q-k}$ (q = 75) for collisions $U^{75+} + Xe \rightarrow U^{(75-k)+} + Xe^{i+} + (i-k)e$ at 15.5 MeV/amu . Arrows indicate filled electron shells (from ref. 44).

and found a total production cross section for Xe^{33+} of about $5 \cdot 10^{-18}$ cm² which is dominated by TI.

Thus, once again TI appears as a process that deserves attention. The big cross sections found for TI collisions at high energies facilitate the use of recoil ion sources for the production of slow highly charged ion beams with little energy spread and thus open new fields of atomic collision studies.

REFERENCES

1. K. Dolder and B. Peart, "Experimental Aspects of Two-Body Ion-Ion Collisions" Rep.Prog.Phys. 48, 1283 - 1332 (1985)
2. K.F. Dunn, this volume
3. E. Salzborn and A. Müller in Electronic and Atomic Collisons, N. Oda and K. Takayanagi eds., p. 407 (North Holland, Amsterdam, 1980)
4. H. Knudsen in Electronic an Atomic Collisions, S. Datz ed., p. 657 (North Holland, Amsterdam 1982)
5. F.J. deHeer in Atomic and Molecular Physics of Controlled Thermonuclear Fusion, C.J. Joachain and D.E. Post eds., NATO ASI Series B: Physics, Vol. 101, p. 269 (Plenum Press, New York and London, 1983)
6. A.S. Schlachter, K.H. Berkner, H.F. Beyer, W.G. Graham, W. Groh, R. Mann, A. Müller, R.E. Olson, R.V. Pyle, J.W. Stearns, and J.A. Tanis, Physica Scripta T3, 153 (1983)
7. A. Müller and E. Salzborn, Phys. Lett. 62A, 391 (1977)
8. H. Tawara, T. Iwai, Y. Kaneko, M. Kimura, N. Kobayashi, A. Matsumoto, S. Ohtani, K. Okuno, S. Takagi, and S. Tsurubuchi, J. Phys. B18, 337 (1985)
9. A. Müller, Ch. Achenbach, and E. Salzborn, Phys. Lett. 70A, 410 (1979)
10. J.R. Macdonald and F.W. Martin, Phys. Rev. A4, 1965 (1971)
11. A.S.Schlachter, J.W. Stearns, W.G. Graham, K.H. Berkner, R.V.Pyle, and J.A. Tanis, Phys. Rev. A27, 3372 (1983)
12. H. Knudsen, H.K. Haugen, and P. Hvelplund, Phys. Rev. A23, 597 (1981)
13. A.S. Schlachter, K.H. Berkner, W.G. Graham, R.V. Pyle, P.J. Schneider, K.R. Stalder, J.W. Stearns, J.A. Tanis, and R.E. Olson, Phys. Rev. A23, 2331 (1981)
14. M. Barat, this volume
15. W. Groh, A. Müller, Ch. Achenbach, A.S. Schlachter, E. Salzborn, Phys. Lett. 85A, 77 (1981)
16. C.L. Cocke, Phys. Rev. A20, 749 (1979)
17. C.L. Cocke, R. DuBois, T.J. Gray, E. Justiniano, and C. Can, Phys. Rev. Lett. 46, 1671 (1981)
18. G. Astner, A. Bárány, H. Cederquist, H. Danared, S. Huldt, P. Hvelplund, A. Johnson, H. Knudsen, L. Liljeby, K.-G. Rensfelt, J. Phys. B17, L877 (1984)
19. K. Okuno, H. Tawara, T. Iwai, Y. Kaneko, M. Kimura, N. Kobayashi, A. Matsumoto, S. Ohtani, S. Takagi, and S. Tsurubuchi, Phys. Rev. A28, 127 (1983)
20. R.W. McCullough in Electronic and Atomic Collisions, D.C. Lorents and W.E. Meyerhof eds. (North Holland, Amsterdam 1986), in the press

21. E.H. Nielsen, L.H. Andersen, A. Bárány, H. Cederquist, J. Heinemeier, P. Hvelplund, H. Knudsen, K.B. MacAdam, J. Sørensen, J. Phys. B18, 1789 (1985)
22. W. Groh, A. Müller, A.S. Schlachter, E. Salzborn, J. Phys. B16, 1997 (1983)
23. W. Groh, A.S. Schlachter, A. Müller, E. Salzborn J. Phys. B15, L207 (1982); the data below 8 keV were taken at the Lawrence Berkeley Laboratory by A. Müller, W. Groh, J.W. Stearns, and A.S. Schlachter after publication of the measurements for $E_{proj} \geq 8$ keV
24. B. Schuch, A. Müller, E. Salzborn, S. Dousson, D. Hitz, R. Geller, Nucl. Instrum. and Meth. B9, 426 (1985)
25. A. Russek, Phys. Rev. 132, 246 (1963), A. Russek, and J. Meli, Physica 46, 22 (1970)
26. A. Müller, W. Groh, E. Salzborn, Phys. Rev. Lett. 51 107 (1983)
27. A. Bordenave-Montesquieu, A. Benoit-Cattin, A. Gleizes, A.I. Marrakchi, S. Dousson, D. Hitz, J. Phys. B17, L127 (1984)
28. N. Bohr, J. Lindhard: K. Dan. Vidensk. Selsk. Mat. Fys. Medd. 28, (1954) No. 7, H. Ryufuku, T. Watanabe, Phys. Rev. A21, 745 (1980), R. Mann, F. Folkmann, H. Beyer, J. Phys. B14, 1161 (1981)
29. P. Roncin, H. Laurent, J. Pommier, D. Hitz, S. Dousson, M. Barat in Electronic and Atomic Collisions, Abstracts of Contributed Papers of the XIV. ICPEAC, M.J. Coggiola, D.L. Huestis, R.P. Saxon eds., Palo Alto 1985, p. 677
30. A. Müller and E. Salzborn, 1983, unpublished
31. T. Åberg, A. Blomberg, J. Tulkki and O. Goscinski, Phys. Rev. Lett. 52, 1207, (1984)
32. A. Bárány, G. Astner, H. Cederquist, H. Danared, S. Huldt, P. Hvelplund A. Johnson, H. Knudsen, L. Liljeby and K.-G. Rensfelt, Nucl.Instrum. and Meth. in Phys. Res. B9, 397 (1985)
33. L. Liljeby, G. Astner, A. Bárány, H. Cederquist, H. Danared, S. Huldt, P. Hvelplund, A. Johnson, H. Knudsen and K.-G. Rensfelt, Physica Scripta, in the press.
34. A. Niehaus, "Transfer Ionization", Comm. Atom. Molec, Phys. 9, 153 (1980)
35. M. Mack, A. Drentje, A. Niehaus in Electronic and Atomic Collisions Abstracts of Contributed Papers of the XIV. ICPEAC, M.J. Coggiola, D.L. Huestis, R.P. Saxon eds., Palo Alto 1985, p. 466
36. B. Schuch, A. Müller, E. Salzborn, 1983, unpublished
37. R.D. DuBois, Phys. Rev. Lett. 52, 2348 (1984)
38. V.V. Afrosimov, Yu.A. Mamaev, M.N. Panov and V. Uroshevich, Sov. Phys.-Techn.Phys. 12, 512 (1967) and V.V. Afrosimov, Yu.A. Mamaev, M.N. Panov and N.V. Fedorenko, Sov. Phys.-Techn. Phys. 14, 109 (1969)

39. E. Horsdal Pedersen and L. Larsen, J. Phys. B 12, 4085 (1979) and J. Phys. B 12, 4099 (1979)
40. T.A. Carlson, W.E. Hunt and M.O. Krause, Phys. Rev. 151, 41 (1966)
41. K. Eckhardt and K.-H. Schartner, Z. Phys. A 312, 321 (1983)
42. T.J. Gray, C.L. Cocke and E. Justiniano, Phys. Rev. A 22, 849 (1980)
43. A. Müller, B. Schuch, W. Groh, E. Salzborn, H.F. Beyer, P.H. Mokler, R.E. Olson, Phys. Rev. A., in the press
44. S. Kelbch, J. Ullrich, R. Mann, P. Richard, H. Schmidt-Böcking, J. Phys. B 18, 323 (1985)

ELECTRON-ION AND ION-ION COLLISIONS IN ASTROPHYSICS

Jacques Dubau

Observatoire de Paris, Section d'Astrophysique
F92190 Meudon
France

INTRODUCTION

There has long been a close association between astrophysical studies and the development of atomic collision theory. This interaction took a major step forward when, some forty years ago, the real nature of the solar corona was first understood as a very hot tenuous plasma. The formulation of the so-called "coronal equilibrium" equations, (1948), pointed to the importance of excitation, ionization and recombination collisions in determining the state of the plasma, and the nature of the emitted spectra. In the earlier " local thermodynamic equilibrium" (LTE) model, the plasma state was determined only by its temperature and the laws of statistical mechanics.

EXTENDING THE WAVELENGTH RANGE

Observations made above the absorbing layers in the Earth's atmosphere have greatly enriched our knowledge of astronomy. Once one can get above the atmosphere, radiation from the Sun can be observed over a very extended range of wavelengths, out into the far UV and X-ray regions. The greater part of the solar energy output is, however, in the optical and near Infrared, and comes from the photosphere or solar "surface" which is about 6000 K. The corona, first observed during eclipses, extends out to several times the photospheric radius and the chromosphere is a transition region between the photosphere and the corona. The first definitive evidence that the corona has a high temperature $T \geqslant 10^6$K, came from the identification by Edlen that a number of lines in coronal optical spectra are due to transitions in highly-ionized atoms, such as Fe^{9+} and Fe^{13+}. Subsequent studies of the corona laid the foundations for much of our present knowledge of the physics of hot plasmas. By far the richest parts of coronal and chromosphere spectra occur, of course, at UV and X-ray wavelengths. In recent years, coronae and chromospheres have also been observed in stars other than the Sun.

For objects outside the solar system there is a strong absorption of radiation, at wavelengths $\lambda \leqslant 912$ Å, due to photoionization of interstellar atomic hydrogen, but as the cross-section fall off as λ^3, observations again become possible at X-ray wavelengths, for $\lambda \leqslant 100$ Å. In the more readily accessible ultra-violet region, $\lambda > 912$ Å, observed spectra contain many intercombination lines (allowed for electric-dipole radiation when account is taken of spin-orbit interaction) and resonance lines such as

CIV 2p-2s λ = 1548, 1551 Å. In the infrared one observes transitions between atomic fine-structure levels. Fig. 1 shows the lines of O^{2+} observed from nebulae at infrared, optical and UV wavelengths.

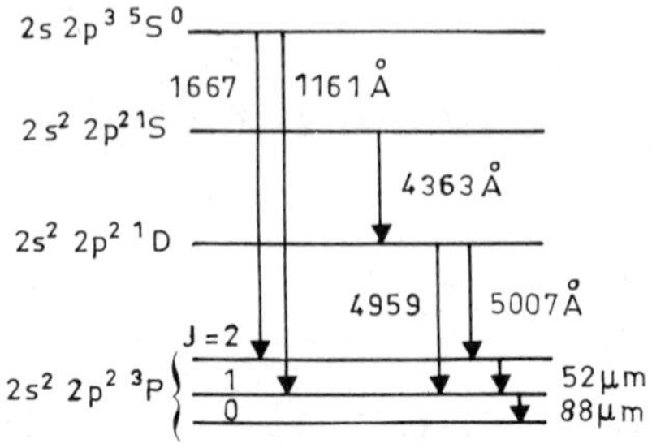

Fig. 1. Energy diagram of O^{2+} (not to scale).

O^{2+} gives strong lines in many astronomical spectra. Since the ground level terms all have the same parity, radiative transitions between them cannot occur by electric dipole radiation. Such levels are said to be "metastable" and the radiative transition to be "forbidden", but they can take place by electric-quadrupole or magnetic-dipole radiation with probabilities of the order of $1 s^{-1}$ compared with $10^9 s^{-1}$ for "allowed" lines. In radiation from laboratory sources the forbidden lines are very weak, and are rarely seen, but in the spectra of gaseous nebulae they are often the strongest features observed. We shall see later how the electron density is determined from line-intensity ratio of a forbidden and an allowed lines.

ASTROPHYSICAL PLASMAS

Star photosphere

Photospheres are optically thick and have continuous spectra with lines in absorption. They are not in thermodynamic equilibrium since radiation can eventually escape, but at each point within them it is a very good approximation to assume a local kinetic temperature. In the LTE model, thermodynamic laws are used to calculate ionization equilibrium and level populations (Boltzmann) :

$$\frac{N_i}{N_j} = \frac{\omega_i}{\omega_j} \exp[-(E_i - E_j)/k_B T_S] \tag{1}$$

where N_i, N_j are the number density of the levels i and j; T_S is the photospheric temperature; E_i, E_j are the energy of the levels i and j and ω_i, ω_j their statistical weight; k_B is the Boltzmann constant.
With the LTE assumption, quantitative spectrum analysis requires atomic data only for radiative processes. For non LTE analyses, all collisional processes must also be considered. Non LTE analyses are important in hot stars but for cooler stars LTE is generally a good approximation.

Nebulae and Coronae

Spectra of nebulae and coronae are dominated by bright emission lines, essentially because the sources are optically thin, i.e. emitted quanta can escape without re-absorption. To derive temperature and density diagnostics many electron-ion cros sections, proton-ion cross sections, are required as well as charge exchange cross-sections and photoionization cross-sections.

Nebulae are clouds of low density gas, surrounding central stars from which radiation is ionising the clouds. In planetary nebulae, the cloud is supposed to be formed by the expulsion of gaseous material from the outer layers of a dying star. Pulsating helium burning of a red giant could lead to the swift expulsion of the whole hydrogen envelope.

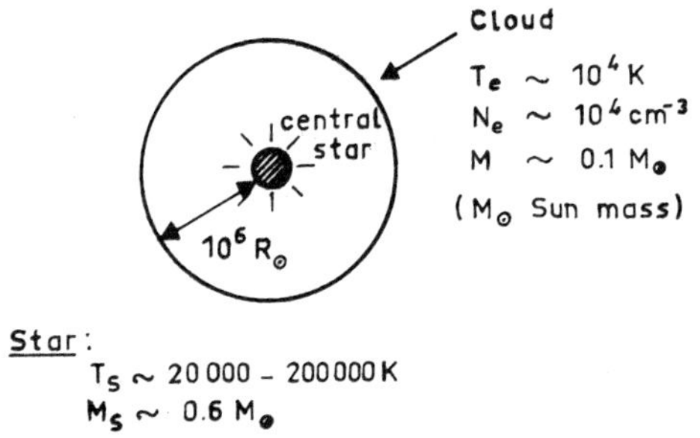

Fig. 2. Planetary nebula

The central star could eventually decay to a white dwarf. There is a rather sharp boundary between the ionized region near the star and the neutral hydrogen region further out.

LINE EMISSION FROM A LOW DENSITY PLASMA

The line emissivity $\varepsilon(\lambda)$(erg s^{-1} cm^{-3}) is given by

$$\varepsilon(\lambda_{ij}) = \frac{hc}{\lambda_{ij}} A_{ji} N_j \qquad (2)$$

where λ_{ij} is the wavelength for the transition between levels i and j; h is the Planck constant; c the speed of light; A_{ji} the spontaneous radiative transition probability and N_j the number density of the emitting level j.

The total emitted line intensity for an optically thin spectral line is

$$I(\lambda_{ij}) = \frac{1}{4\pi} \int_V \varepsilon(\lambda_{ij})dV \qquad (\text{erg s}^{-1}\text{ st}^{-1}) \qquad (3)$$

where V is the emitting volume. Of course, for an optically thick line a radiative transfer equation has to be used and equation (3) is not valid. Some of the radiation which would have been emitted towards the observer becomes re-absorbed, exciting again the upper level of the transition. The simplest case to consider is one in which the upper level can only decay to the ground, as eg. Hydrogen Lyman α. In this case, the radiation cannot be lost. However many times it is absorbed, it will always be reemitted. Although the total emitted flux is unchanged, the line profile and the direction of emission can be altered. If collisions effectively reorient the atoms before emitting, then the subsequent emission is again isotropic. The net result of multiple absorption is to transfer emitted flux from one direction into other directions. Homogeneous models of the solar chromosphere have been constructed using the observed line profiles of a number of the strongest emission lines. However these models are empirical but do predict a temperature plateau at 2 10^4 K.

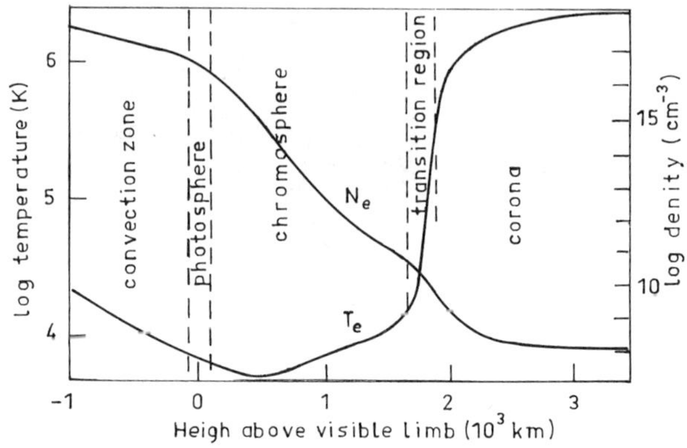

Fig. 3. Temperature and density in the solar atmosphere.

LEVEL POPULATION

The number density of the level j can be factorized :

$$N_j(X^{+m}) = \frac{N_j(X^{+m})}{N(X^{+m})} \frac{N(X^{+m})}{N(X)} \frac{N(X)}{N(H)} \frac{N(H)}{N_e} N_e \tag{4}$$

where X^{+m} denotes the m-th ionization stage of element X; $N_j(X^{+m})/N(X^{+m})$ is the population of the level j relative to the population of the ion X^{+m}; $N(X^{+m})/N(X)$ is the ionization ratio; $N(X)/N(H)$ is the elemental abundance relative to hydrogen; $N(H)/N_e$ is the hydrogen number density relative to the electron number density.

A. N_e

Unless it exists strong electric and/or magnetic fields, or a transient situation (such as runaway electrons), collisions between electrons or between electrons and ions "thermalize" rapidly the electrons and the electron velocity distribution is maxwellian, the corresponding temperature being T_e. The free electrons are produced by ionization of H and He which are the most abundant plasma elements (H $\simeq$ 90%, He $\simeq$ 10%),

$$N_e \simeq N(H^+) + N(He^+) + 2N(He^{++}) \tag{5}$$

N_e is a function of T_e but for high temperature plasmas, H and He are fully ionized and

$$N(H)/N_e \underset{T_e \text{ large}}{\simeq} \frac{0.9}{1.1} = 0.82. \qquad (5')$$

B. N(X)/N(H)

Although nuclear reactions in the centre of stars are modifying the concentration of light elements there, the outer atmosphere of a star is normally assumed to have the same classical composition as the recycled interstellar gas from which the star was first formed. There is less than 1% of heavier elements than H and He. Typical values are : $C(3.5\times10^{-4})$, $N(1.5\times10^{-4})$, $O(6\times10^{-4})$, $Ne(5\times10^{-5})$, $Mg(5\times10^{-5})$, $Si(3\times10^{-5})$, $S(1.5\times10^{-5})$, $Fe(2\times10^{-5})$. These values can still vary by some factor and it is a fundamental problem in Astrophysics to determine the relative abundance of an element to hydrogen since it has cosmological consequences. The heavier elements play a small role in influencing the physical behaviour of stars. On the other hand, these trace elements are most valuable probes when trying to measure the physical parameters (T_e, N_e, ...) by spectroscopic diagnostics. The lines emitted by trace elements are less sensitive to radiative transfer than those emitted by the most abundant elements, H and He. For example, in the Sun atmosphere where N_e is large ($N_e \geqslant 10^8$ cm^{-3}), Fe is a very valuable element. Furthermore, Fe emits line spectra for electron temperature up to 10^8 K.

C. $N(X^{+m})/N(X)$

The ionization rate is determined by the balance of ionization and recombination. In a transient situation :

$$\frac{dN(X^{+m})}{dt} = N_e(N(X^{+m-1})S_{m-1} - N(X^{+m})S_m + N(X^{+m+1})\alpha_{m+1} - N(X^{+m})\alpha_m) + (a) + (b) + (c) \qquad (6)$$

where S_m is the ionization rate of X^{+m} and α_m the recombination rate of X^{+m} (cm^{-3} s^{-1}).

(a) : photoionization plays an important role in nebulae

$$X^{+m}(i) + h\nu \rightarrow X^{+m+1}(k) + e'$$

It is very important to introduce, in the theoretical calculations, the resonances in the cross section, due to autoionizing levels, for example : the process

$$N(2s^2 2p^3\ {}^2D^o) + h\nu \rightarrow N(2s2p^4\ {}^2P) \rightarrow N^+(2s^2 2p^2\ {}^3P,\ {}^1D) + e$$

gives rise to a broad resonance in the photoionization cross section. Similar results have been obtained for photoionization of the ground ($2s^2 2p^2\ {}^3P$) and metastable ($2s^2 2p^2\ {}^1D$ and ${}^1S$) states of C.

(b) : charge exchange plays also an important role in some nebulae

$$X^{+m}(i) + H \rightleftarrows X^{+m-1}(q) + H^+$$

In nebulae, most of the hydrogen is ionized but it is a curious paradox (explained in terms of the λ^3 dependence of the hydrogen photoionization cross-section) that the fraction of neutral H can be larger for a high-excitation nebula, photoionised by a very hot star. The charge transfer is particularly important in nebulae of high excitation.

(c) : transport mechanism of the ion X^{+m} from one elementary volume dV to another one.

These transport mechanisms, related to convection and diffusion, are introduced in Tokamak plasma analysis because the plasma is very well diagnosed. But in Astrophysics the situation being more complex, transport mechanisms are not introduced, in solving equation (6) ...

We shall neglect (a) + (b) + (c) in the following discussion, for simplicity.

From equation (6), it is seen that the time taken to establish equilibrium between two ionization stages is inversely proportional to the electron density N_e. For example, in the corona where $N_e \sim 10^8$ cm^{-3}, typical time is 100 s. It is safe to assume the steady state hypothesis ($dN(X^{+m})/dt = 0$) for regions which are stable for periods longer than 100 s. Solar flares are explosive events, occuring normally in active regions and resulting in the release of a large amount of energy (up to 10^{32} ergs) over a short time of tens of minutes. Three main phases can be identified. The primary energy release : the energy source consists of the coronal magnetic field. Such a field can only release its enhanced energy by reconnection or annihilation. The released energy is manifested as beams of accelerated particles (mostly electrons) moving at energy of several hundred keV.
The impulsive phase is very short (secs). This is characterized by hard X-ray emission (photon energy > 10 keV) from very small regions. The fast electrons produced, during the first phase, are trapped by the magnetic field. When they enter the high density (chromospheric) regions near the foot points of the flare, they begin to deposit their energy by bremsstrahlung. The quasi-thermal phase is characterised by a rapid rise (1-5 mins) in soft X-ray emission followed by a decay phase. The plasma is localised in loop structures connecting regions of opposite magnetic polarity. We see that the steady state hypothesis will be justified only during the third plase when the flare is decaying.

For steady state, it is easy to show that (6) reduces to

$$N(X^{+m})S_m = N(X^{+m+1})\alpha_{m+1} \qquad (6')$$

S_m : ionisation processes include collisional ionization from the inner and outshells of the ground and metastable states - the other excited states have a negligible contribution to S_m because other processes, such as spontaneous radiative transition, are more efficient in depopulating the excited levels.

$$X^{+m}(i) + e \rightarrow X^{+m+1}(k) + e + e'$$

The collisional excitation into autoionizing levels can be very important as ionizing process through the autoionization of the intermediate levels

$$X^{+m}(i) + e \rightarrow \left\{ \begin{array}{l} (X^{+m})^{**} \rightarrow X^{+m+1}(k) + e' \\ + \\ e \end{array} \right.$$

α_{m+1} : recombination processes include radiative recombination

$$X^{+m+1}(k) + e' \rightarrow X^{+m}(i) + h\nu$$

(this process is the inverse process of photoionization) and di-electronic recombination

$$X^{+m+1}(k) + e' \rightarrow (X^{+m})^{**} \rightarrow X^{+m}(j) + h\nu$$

the excited level j decays by "cascade" to the ground level

$$X^{+m}(j) \rightarrow \ldots \rightarrow X^{+m}(g) + h\nu.$$

For low density plasmas, the bulk of the ion population is in the ground and metastable levels. Ionization processes take place from these levels. On the contrary, recombination takes place onto highly excited levels.

The ionization rates and recombination rates are almost a function of the electron temperature T_e, only. As an example, we give the ionization equilibrium theoretical calculations obtained for Fe (Jacobs, Davis, Kepple; Blaha (1977) Ap. J. 211, 605).

$$\frac{N(Z)}{N_T} \equiv \frac{N(X^{+m})}{N(X)}$$

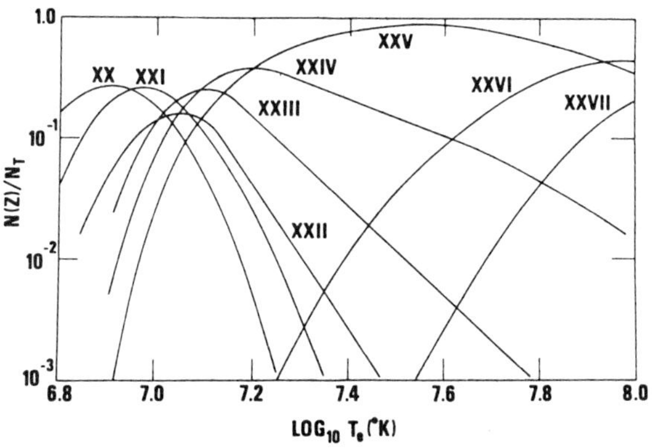

Fig. 4. The ionization equilibrium for Fe XX-Fe XXVII

The importance of dielectronic recombination in the ionization equilibrium was first recognized by Burgess (1964) in a solar corona study. Although the theory was known, since 1942, the dielectronic recombination was thought to have a negligible effect. The cases considered by Bates and Massey in 1942, were singly ionized atoms where dielectronic recombination was dominated by other processes. Burgess considered highly ionized atoms and there, the situation was completely different. Since then, Storey (1981) has shown that dielectronic recombination can also be important for low-charged ions to interpret some line intensities in nebulae, see fig. 5.

The satellite line, CIII $2p4d_o$- 2s4d 1550 Å is observed close to the parent line of CIV 2p-2s 1549 Å. (We shall see later the theory of satellite lines and their application as a diagnostics of T_e). Storey shows that the line 2297 Å, of CIII, is enhanced by cascade (411 Å). The observed anomalous intensity of 2297 Å is explained by dielectronic recombination of CIV.

D. $N_j(X^{+m}) / N(X^{+m})$

The time taken to establish an equilibrium between excited levels and their ground and metastable levels is usually very small compared to the time taken to establish an ionisation equilibrium. For example, in the corona, the typical "excitation" time is 1s compared to 100s for the "ionization-recombination" time. We can therefore assume, in most situations, that the excited levels are at any time in equilibrium with the different ground states and metastable states. As an example, on the

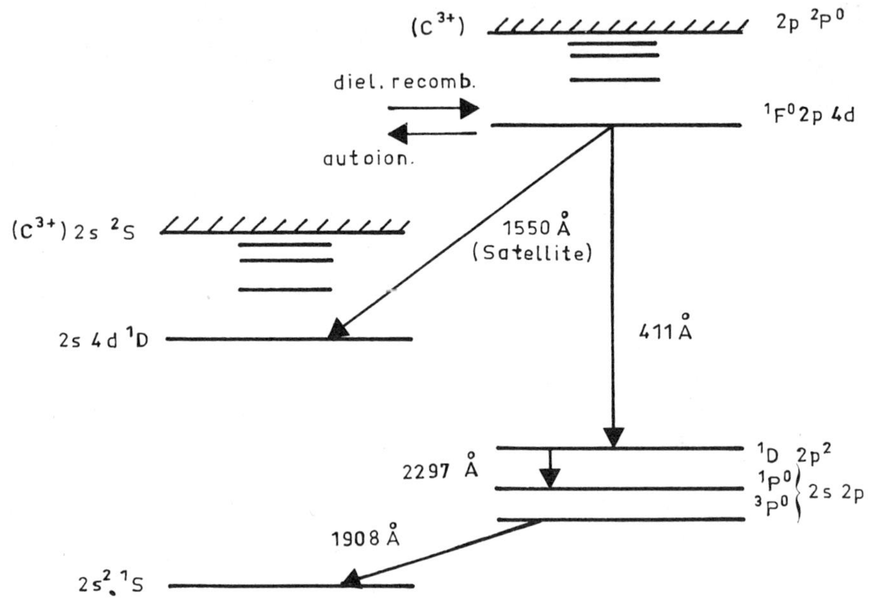

Fig. 5. CIII energy levels.

satellite Solar Maximum Mission, launched in February 1980 by NASA with the purpose to study solar active regions, the detectors of the soft X-ray spectrometers had a better resolution than 1s. Unfortunately, all the spectra recorded for less than 30s had a very poor statistics and could not be used directly for spectroscopic diagnostics. In this "steady"-state the equations satisfied by $N_j(X^{+m})$ are

$$0 = \sum_i C_{ij} N_i (X^{+m}) + N_e \{S_{m-1}(j) N(X^{+m-1}) + \alpha_{m+1}(j) N(X^{+m+1})\} \quad (7)$$

where for $i < j$:

$$C_{ij} = N_e \, q^e_{ij}(T_e) + N(H^+) q^p_{ij}(T_e) + B_{ij} \quad (8)$$

where q^e_{ij} and q^p_{ij} are the electron and proton collisional excitation rate coefficients ($cm^{-3}\ s^{-1}$); B_{ij} is the stimulated absorption rate coefficient (s^{-1}), the photon field being emitted, for example, from the central stars of the nebulae or the sun photosphere.

for $\underline{i > j}$:

$$C_{ij} = A_{ij} + N_e\, q^e_{ij}(T_e) + N(H^+) q^p_{ij}(T_e) \qquad (9)$$

where A_{ij} is the spontaneous radiative transition probability and $q^e_{ij}(T_e)$ and $q^p_{ij}(T_e)$ the electron and proton deexcitation rates which are related to their inverse excitation rate by the principle of detailed balance :

$$q_{ij}(T_e) = \frac{\omega_j}{\omega_i} q_{ji}(T_e) \exp[(E_i - E_j) / k_B T_e] \qquad (10)$$

where E_i and E_j are the energy of the (target) ionic levels i and j, respectively.

Of course, the collisional excitation rate coefficient, as for the other electron collisional processes such as ionization and recombination, are obtained by averaging the collisions cross section, multiplied by the electron velocity, over a Maxwellian velocity distribution, f(v)

$$q^e_{ij}(T_e) = \int_0^\infty Q_{ij}(E)\, f(v) v\, dv \qquad (11)$$

The electron collision strength $\Omega_{ij}(E)$, a dimensionless symmetric quantity is used in preference to the collision cross-section, where

$$Q_{ij} = \frac{\pi}{\omega_i k^2} \Omega_{ij}$$

k is the wave number of the incoming electron, $E = \hbar^2 k^2 / 2m_e$ (m_e : electron mass).

Ω_{ij}, as Q_{ij}, have resonance structures related to the presence of autoionising states. The average collision strength $< \Omega_{ij}(T_e) >$ is preferred by astrophysics because it is a smooth function of T_e - the resonances are averaged by the electron distribution (in the two following formulae k_B is Boltzmann constant)

$$< \Omega_{ij}(T_e) > = \int \Omega_{ij}(E)\, e^{-\frac{E}{k_B T_e}}\, d\left(\frac{E}{k_B T_e}\right) \qquad (12)$$

then

$$q_{ij}(T_e) = 2\sqrt{\pi}\, a^3 \tau^{-1} \frac{< \Omega_{ij}(T_e) >}{\omega_i} \left(\frac{E_H}{k_B T_e}\right)^{1/2} \exp\left(\frac{E_i - E_j}{k_B T_e}\right) \qquad (13)$$

E_H = 13.6 eV is the ionisation energy of hydrogen, a and τ are the atomic unit of length and time.

The temperature domain where $< \Omega_{ij}(T_e) >$ is interesting for plasma applications is $T_{min} \leqslant T_e \leqslant T_{max}$, where $k_B T_{max} = I(X^{+m})$, $T_{min} = T_{max}/10$ and $I(X^{+m})$ is the ionization energy of X^{+m}.

The proton collisional excitation and deexcitation processes are the only ion-ion processes important in astrophysics, since H atom is the most abundant atom in astrophysical plasmas (He^+ and He^{++} could be also considered if the cross-sections were large enough to compete with H^+).

Proton collisional excitation processes become comparable with electron collisional excitation processes for transition between fine structure levels or for very excited levels.

At last, for <u>i = j</u> : the processes are depopulating level j

$$C_{jj} = - \{ \sum_{i<j} (A_{ji} + N_e q^e_{ji} + N_p q^p_{ji}) + \sum_{j<i} (N_e q^e_{ji} + N_p q^p_{ji} + B_{ji}) \} \tag{14}$$

($N_p \equiv N(H^+)$) (For j = 1, the first RHS bracket disappears).
The stimulated absorption rate coefficient B_{ij} due to the radiation field J_ν is given by :

$$B_{ij} = B^\star_{ij} \, \delta(\nu-\nu_{ij}) \, J_\nu \tag{15}$$

where $B^\star_{ij}$ is the Einstein absorption coefficient, determined in thermodynamic condition, from A_{ji} :

$$B^\star_{ij} = \frac{c^3}{8\pi h\nu^3_{ij}} A_{ji} \frac{\omega_j}{\omega_i} \tag{16}$$

($\nu_{ij} = E_j - E_i / h$).

If we consider a star to radiate as a black body at temperature T_s. At a point P in the vicinity of the star the radiation density is (Planck law for energy density) :

$$J_\nu^{(P)} = W \, J_\nu \, (T_s) = W \frac{8\pi h\nu^3}{c^3} \left(e^{h\nu/k_B T_s} - 1\right)^{-1} \tag{17}$$

($J_\nu d_\nu$: erg cm^{-3}).

W is the geometrical "dilution" factor $W = \Omega/4\pi$ where Ω is the solid angle subtended by the star at P (see Fig. 6).

$$W = \frac{1}{2} \left[1 - \left(1 - \frac{R^2}{r^2}\right)^{1/2}\right] \tag{18}$$

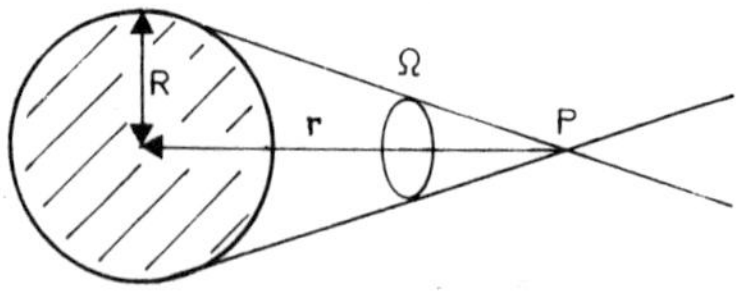

Fig. 6

In UV and X-ray wavelength range, i.e. $\lambda < 1000$ Å, the excitation processes by radiation are generally negligible compared to collisional excitation. Nevertheless, for metastable levels close to other levels by an energy difference $\Delta E < 10$ eV this is sometime not true.

The stimulated <u>emission</u> rate coefficient B_{ji} (= $B_{ij}\omega_i/\omega_j$) is always negligible compared to the spontaneous emission rate, in low density plasma.

If we come back to equation (7) : $S_{m-1}(j)$ is the ionization rate coefficient from level j - as already mention this applies essentially to the ground and metastable levels. $\alpha_{m+1}(j)$ is the recombination rate coefficient (directly) to level j - this applies to very excited levels.

The highly excited levels are populated by recombination and signifiant population inversions can occur. These lead to line intensification by maser action and, even at the very low density plasmas, collisional effects can cause line broadening. Active work on observing the "radio recombination lines" is done in Radio Astronomy. Such transitions would be difficult to detect in the laboratory, since the radii of the orbits are very large, but they have been detected by radio astronomers (the pressure in a typical gaseous nebula is of order 10^{-14} atm, that in the interstellar medium about 10^{-20} atm).

PLASMA DIAGNOSTICS

The "coronal model"

In a low density plasma and for intense allowed transitions, the coronal model can be applied. In this model, almost the entire ion population resides in the ground level (g) and excited levels (j) are populated by electron collisional excitation from (g) and depopulated by radiation decay to (g). Usually, to check the validity of the model, it is better to solve simultaneously the set of equations (7). A set of two levels satisfying the "coronal" model is very useful for spectroscopic diagnostics. Indeed, equations (7) reduce to one :

$$N_g\, N_e\, q^e_{gj} = N_j\, A_{jg} \tag{19}$$

From equations (2), (4), (13) and (19), we find

$$\varepsilon(\lambda_{gj}) = [2\sqrt{\pi}\, a^3_\tau{}^{-1} \frac{hc}{\omega_i \lambda_{gj}} \frac{N(X)}{N(H)} \frac{N(H)}{N_e}]\, A_{jg} \tag{20}$$

$$\cdot\, [\frac{N(X^{+m})}{N(X)} < \Omega_{gj}(T_e) > (\frac{E_H}{k_B T_e})^{1/2} \exp(\frac{E_g - E_j}{k_B T_e})]\, N_e^2$$

For high temperature plasma, the first RHS term is a constant (eq. 5'), the second one is a function of T_e only.

$$\varepsilon(\lambda_{gj}) = \beta\, G_{gj}\,(T_e)\, N_e^2 \tag{21}$$

From equation (3) we have :

$$I(\lambda_{gj}) = \frac{\beta}{4\pi} \int_V G_{gj}(T_e)\, N_e^2\, dV. \tag{22}$$

It is convenient to switch to a log T_e scale on which the functions $G_{gj}(T_e)$ tend to have a shape which can be approximate by a rectangular function of height $G_{gj}(T_m)$ and width Δ_{gj}

$$I(\lambda_{gj}) = \frac{\beta}{4\pi} G_{gj}(T_m) \int_{\mathbf{V}} N_e^2\, dV \tag{23}$$

the volume $\mathbf{V}$ corresponds to a plasma volume with temperature T_e such that $(\log T_m - \Delta_{gj}/2) \leqslant \log T_e \leqslant (\log T_m + \Delta_{gj}/2)$.

$\int_{\mathbf{V}} N_e^2\, dV$ is called the emission measure. If we have different lines satisfying the coronal equilibrium and corresponding to different T_m, it is possible to evaluate the energy lost by radiation, for the plasma.

Intensity ratios used to determine electron temperature

If, for the same ion X^{+m}, we have two allowed lines, which satisfy the coronal equilibrium, it is possible to evaluate T_e. We suppose they correspond to the same emitting volume V (see Fig. 7)

$$\frac{I(\lambda_{gj})}{I(\lambda_{gi})} \simeq \frac{\varepsilon(\lambda_{gj})}{\varepsilon(\lambda_{gi})} \tag{24}$$

$$\frac{\varepsilon(\lambda_{gj})}{\varepsilon(\lambda_{gi})} = \frac{\lambda_{gi}}{\lambda_{gj}} \frac{< \Omega_{gj}(T_e) >}{< \Omega_{gi}(T_e) >} \exp\left(\frac{E_i - E_j}{k_B T_e}\right) = F(T_e) \tag{25}$$

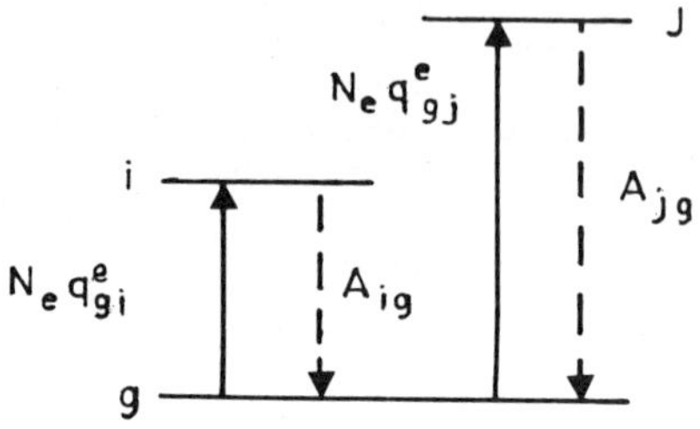

Fig. 7.

$< \Omega_{gj}(T_e) >$ and $< \Omega_{gi}(T_e) >$ have usually a similar behaviour with T_e and the variation of $F(T_e)$, as function of T_e, comes from the exponential function. This ratio is therefore sensitive to change in electron temperature if $E_j - E_i \simeq k_B T_e$.

Metastable levels; electron density diagnostic

The temperature diagnostic, described above, is model free, in the sense that we do not need to build a plasma model to deduce the electron temperature from the intensity ratio. For example, the result is not dependent of a choice of geometry. Nevertheless, to have any application, this diagnostic will have to be inserted in a model.

The next step is to try to evaluate also the electron density N_e. For some "forbidden" and "intercombination" transitions, the radiative decay rate is so small that the electron collisional deexcitation competes as a depopulating mechanism. The population of the emitting "metastable" level can become comparable with the population of the ground level (see Fig. 8). We have :

$$N_m(A_{mg} + N_e\, q^e_{mg}) = N_g\, N_e\, q^e_{gm}. \tag{26}$$

For low density N_e, $A_{mg} \gg N_e\, q^e_{mg}$ and the equation (26) reduces to (19).

For high electron density, a Boltzmann equilibrium is reached between the two levels (26) becomes (using (10))

$$\frac{N_m}{N_g} = \frac{\omega_m}{\omega_g} \exp\left[\, (E_g - E_m) \,/\, k_B T_e \,\right] \tag{27}$$

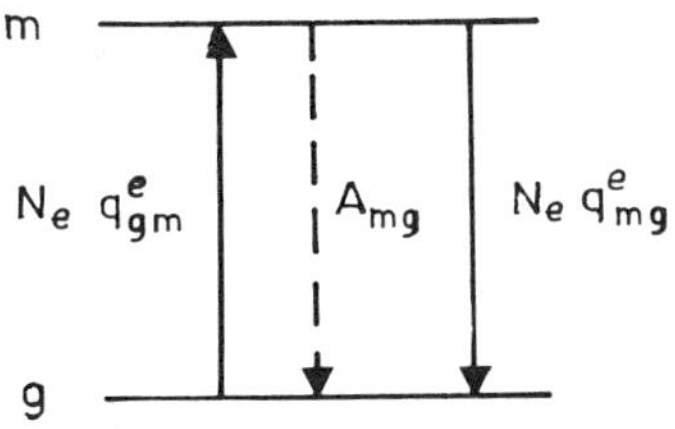

Fig. 8.

In fact, where the electron density increases other levels contribute and the situation is intricate. Indeed the population of the metastable level becomes significant, other levels can be excited from this level as well as from the ground level. For low N_e, the emissivity $\varepsilon(\lambda_{gm})$ increases as for an allowed transition until

$$N_e \simeq \frac{A_{mg}}{q^e_{mg}} = N^c_e .$$

For higher density, a part of the population of m is excited to j and therefore $\varepsilon(\lambda_{gm})$ increases less than for allowed transition

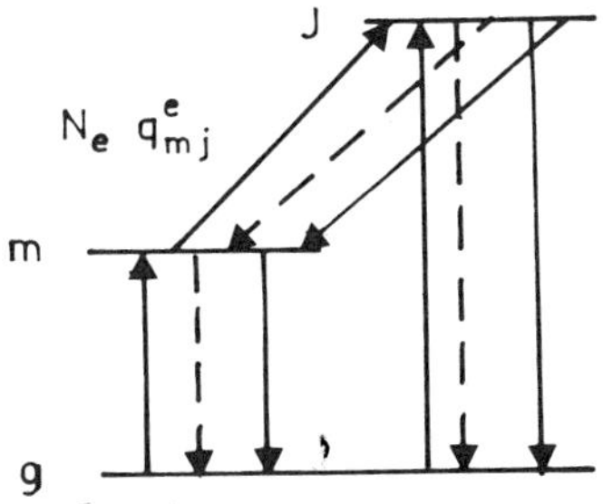

Fig. 9.

i.e. $N_e \ll N^c_e \qquad \varepsilon(\lambda_{gm}) \propto N_e^2$

$N_e \simeq N^c_e \qquad \varepsilon(\lambda_{gm}) \propto N_e^{\beta} \quad (1 < \beta < 2)$

$N_e \gg N^c_e \qquad \varepsilon(\lambda_{gm}) \propto N_e$

Three density diagnostics are used in Astrophysics described by figure 10, 11 and 12. In all of them we suppose the two lines are emitted in the same volume V so that

$$\frac{I(\lambda_{gj})}{I(\lambda_{gi})} \simeq \frac{\varepsilon(\lambda_{gj})}{\varepsilon(\lambda_{gi})} .$$

This is the main hypothesis which must be checked carefully before using such diagnostics.

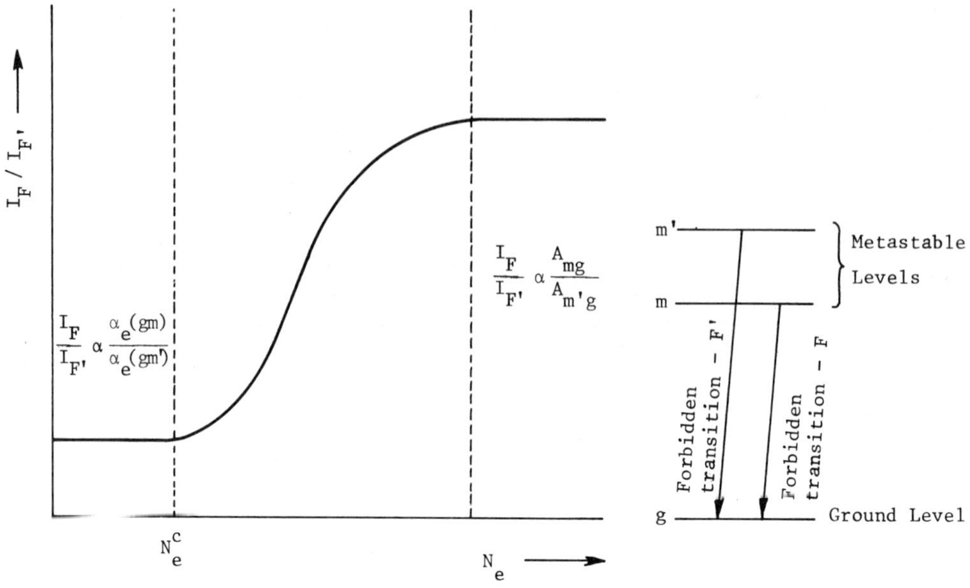

Fig. 10. The intensity ratio of two forbidden transitions from the same ion.

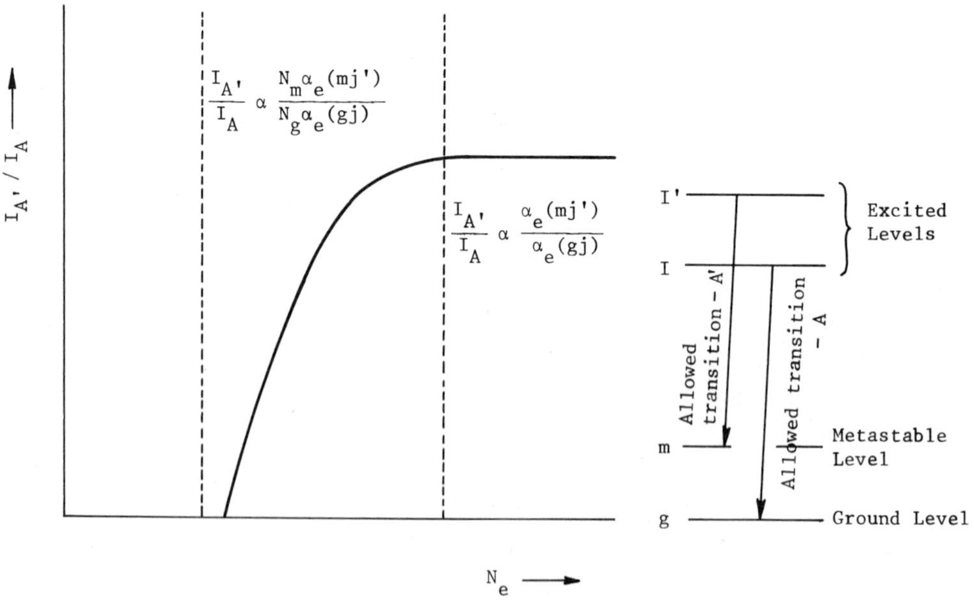

Fig. 11. The intensity ratio of two allowed transitions, one of which is excited from the fround level,the other from a "metastable" level.

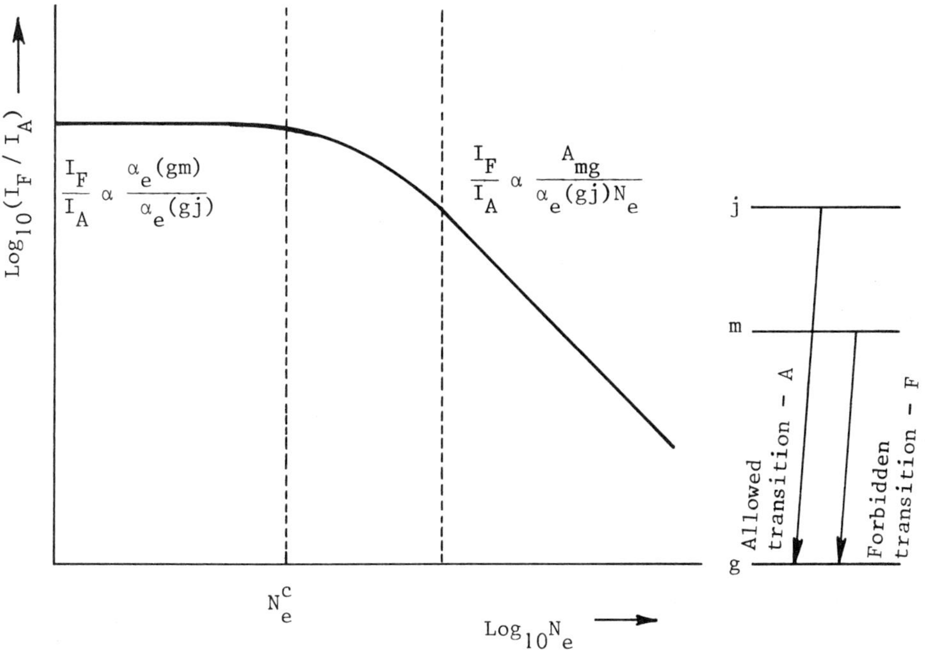

Fig. 12. The intensity ratio of an allowed to a forbidden transition.

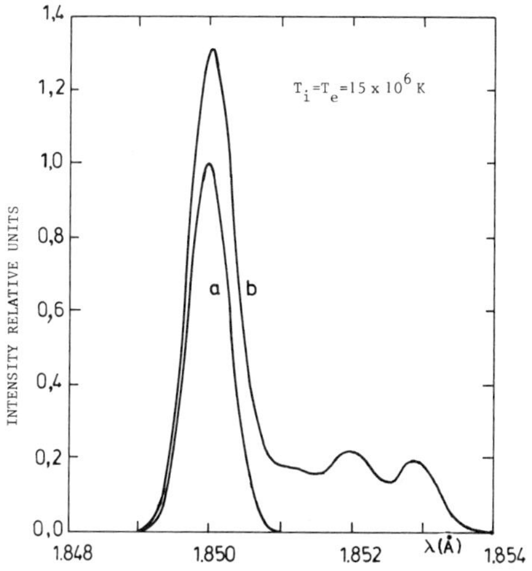

Fig. 13. Fe XXV resonance line intensity : (a) pure resonance line, (b) apparent resonance line (including unresolved satellites).

Spectra due to dielectronic recombination

For hydrogen-, helium- and neon-like ions resonance lines, satellite lines appear close to or blended with them. They have been known since the experimental work of Edlen and Tyren (1939). They correspond, for He-like, to transitions of the type :

$1s^2nl$ - $1s2pnl$ satellites to $1s^2$ - $1s2p$ (resonance) line

$1s^2nl$ - $1s3pnl$ satellites to $1s^2$ - $1s3p$ (resonance) line

etc... These lines correspond to the stabilising transitions

$$(X^{+m})^{**} \rightarrow X^{+m}(j) + h\nu$$

$$(1s2pnl \rightarrow 1s^2nl + h\nu)$$

of dielectronic recombination of helium-like system, i.e.

$$1s^2 + e \rightarrow 1s2pnl \rightarrow 1s^2nl + h\nu$$

when n = 2 these lines form a set of discrete satellites to the long wavelength side of the resonance line

$$1s^2 \rightarrow 1s2p + h\nu$$

called the parent line.

For n = 3 the lines move closer to the parent line, and the series $n \geqslant 4$ tend to merge with the resonance line, producing an asymetric broadening on its wavelength side (Fig. 13).

For n = 2 an alternative excitation mechanism exists by inner shell electron excitation of the 3-electron ion.

$$1s^22s + e \rightarrow 1s2s2p + e'$$

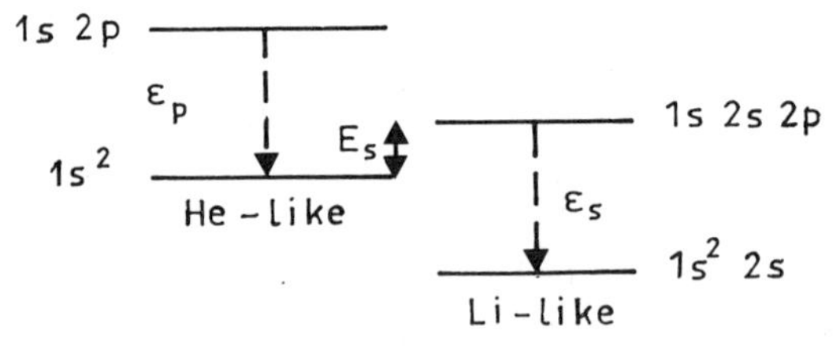

Fig. 14.

From the viewpoint of the observation of spectral lines, satellite lines are only important for highly charged systems. For simplicity, we shall consider in the next subsections only He-like satellites.
i) pure dielectronic satellite

The emissivity of a line populated only by dielectronic recombination is

$$\varepsilon_s^d = \frac{hc}{\lambda_s} A_s^r N_s^{**} \qquad (28)$$

with

$$N_s^{**} = N_e\ N(\text{He-})\ \alpha_s^d(T_e) \tag{29}$$

where λ_s, A_s^r are the wavelength and spontaneous radiative probability for the satellite line. N(He-) is the number density of the helium-like element and $\alpha_s^d(T_e)$ is the dielectronic recombination rate through the satellite line s.

$$\alpha_s^d(T_e) = 8\pi\sqrt{\pi}\ a^3 \left(\frac{E_H}{k_B T_e}\right)^{3/2} \exp\left(-\frac{E_s}{k_B T_e}\right) \frac{\omega^{**}}{2\omega_g^{He}} \frac{A_s^a}{\Sigma A^a + \Sigma A^r} \tag{30}$$

where E_s is the energy of the captured electron, i.e. the energy of the Li-like doubly excited state X_s^{**} minus the energy of the ground level of He-like (see Fig. 14); ω^{**} and ω_g^{He} are respectively the statistical weight of X_s^{**} and of the ground level g of the recombining (He-like) ion, i.e. $\omega_g^{He} = 1$, $(1s^2)$. ΣA^a, ΣA^r are the sum of all the autoionisation probabilities and radiative probabilities for the possible transitions from X_s^{**}.
The intensity ratio between the dielectronic satellite line and its parent line gives (as long as they have the same V)

$$\frac{I_p}{I_s^d} \simeq \frac{\varepsilon_p}{\varepsilon_s^d} \frac{\tau^{-1}}{4\pi} \left(\frac{k_B T_e}{E_H}\right) \exp\left(\frac{E_s - E_p}{k_B T_e}\right) \frac{2}{\omega^{**}} \left(\frac{\Sigma A^a + \Sigma A^r}{A_s^a A_s^r}\right) < \Omega_p(T_e) > \tag{31}$$

The satellite line has, by definition, a wavelength close to the parent line. Therefore $E_s \simeq E_p$ and we see as long as the variation of $< \Omega_p(T_e) >$ as function of T_e, is not very fast

$$\frac{I_p}{I_s^d} \ \alpha\ T_e\ . \tag{31'}$$

In plasma X-ray analyses, this temperature diagnostics is very much used because it is very sensitive to a small variation of T_e. It is currently used to analyse X-ray spectra of solar flares (satellites SMM (NASA), Hinotori (Japan), P78 (Navel Research Lab.), Intercosmos (USSR)). It is also used currently in Tokamaks.

ii) pure inner-shell satellite

In the case where a doubly excited level X_s^{**} has no autoionization probability A_s^a, and where it can be populated by inner-shell electron excitation of Li-like ground state, there is a second diagnostic based on the relative intensity of an inner-shell line :

$$N_s^{**} = N_e\ N(\text{Li-})\ q_s^e\ (T_e) \tag{32}$$

$$\frac{I_p}{I_s^{in-sh}} \simeq \frac{\varepsilon_p}{\varepsilon_s^{in-sh}} = \exp\left(\frac{E_s - E_p}{k_B T_e}\right) \frac{< \Omega_p(T_e) >}{< \Omega_{in-sh}(T_e) >} \frac{\omega_g^{Li}}{\omega_g^{He}} \frac{N(\text{He-})}{N(\text{Li-})} \tag{33}$$

The inner-shell excitation and the parent excitation are similar processes since it is a 1s-2p excitation and that $E_s \simeq E_p$, therefore :

$$\frac{I_p}{I_s^{in-sh}} \ \alpha\ \frac{N(\text{He-})}{N(\text{Li-})} \tag{33'}$$

This diagnostic of the relative abundance of He-like, Li-like ions of a same element is used in hot plasmas to check if the plasma is either in ionization equilibrium or in a transient situation (recombining or ionising).

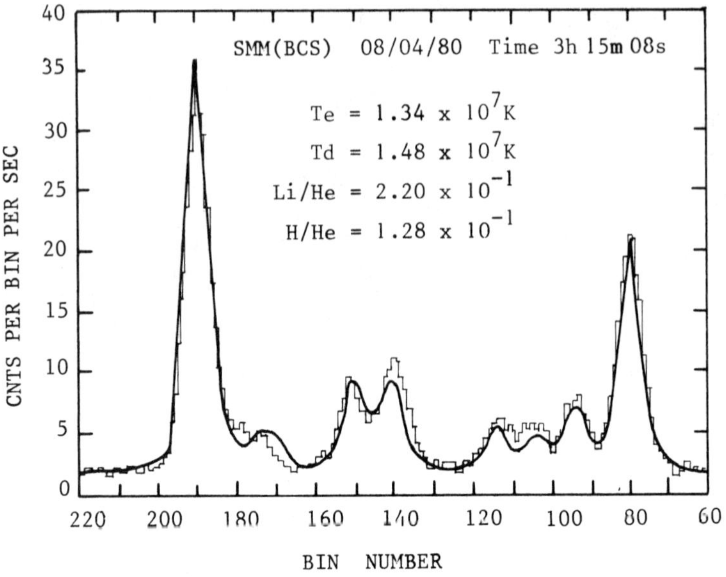

Fig. 15. Fitting of a Ca spectrum recorded by XRP in the range 3.17-3.22 Å by a synthetic spectrum.

Fig. 15 illustrates the satellite line diagnostics. It is a helium Ca^{18+} spectrum which contains 1s2p, 1s2s-1s^2 resonance (parent) and forbidden lines, as well as their satellites 1s2pnl. The spectrum was recorded by the X-ray polychromator aboard the SMM space observatory. The wavelength range is 3.17-3.22 Å. The best fit obtained includes both Ca^{18+} lines and their Ca^{17+} satellites. It corresponds to an electron temperature $T_e = 13.4 \times 10^6$K, a Doppler profile temperature $T_D = 14.8 \times 10^6$ K and relative abundances $N_{Li-}/N_{He-} = 0.22$ and $N_{H-}/N_{He-} = 0.128$.

REFERENCES

A.H. Gabriel and C. Jordan, 1972, Case studies in atomic collision Physics, Vol. 2, ed. E.W. McDaniel and M.R.C. Mc Dowell (North Holland), pp 209-91.

V.P. Myerscough and G. Peach, 1972, Case studies in atomic collision Physics, Vol. 2, pp 293-396.

C. de Michelis, M. Mattioli, 1981, "Soft X-ray spectroscopic diagnostics of Laboratory Plasmas", Nuclear Fusion, 21, pp 677-754.

A.H. Gabriel, H.E. Mason, 1982, "Solar Physics", Applied Atomic Collision Physics, Vol. 1, ed. H.S.W. Massey, B. Bederson and E.W. McDaniel, Academic Press, pp 345-397.

H.E. Mason, "Atomic Data for the interpretation of EUV Astrophysical Plasmas", 1981, ed. S. Volonté and L. Houziaux, Troisième cycle interuniversitaire en Astronomie et Astrophysique, F.N.R.S., Belgium.

J. Dubau and S. Volonté, 1980, "Dielectronic recombination and its applications in Astronomy", Reports on Progress in Physics, 43, pp 199-251.

THE ROLE OF ELECTRONIC AND IONIC COLLISIONS IN TOKAMAK DEVICES

M.F.A. Harrison

Culham Laboratory,
Abingdon, Oxon OX14 3DB, U.K.
(Euratom/UKAEA Fusion Association)

1. INTRODUCTION

The primary objective of research into controlled thermo-nuclear fusion is to develop an effective energy source based on the reaction

$$D + T \rightarrow (^4He + 3.52 \text{ MeV}) + (n + 14.06 \text{ MeV}). \qquad \text{(i)}$$

In common with all fusion reactions, the initial collision partners are charged nuclei and hence experience Coulomb repulsion. Thus the probability of fusion is small unless the colliding nuclei are energetic. The rate at which fusion energy is released is dependent upon the distribution of velocities amongst the D/T particles and the distribution of particle density throughout the reaction region. Consider, for simplicity, a homogeneous hot mixture of D and T wherein the number density $n_D = n_T$ and the distribution of particle velocities is Maxwellian. The rate K_F at which fusion collisions occur within unit volume is given by

$$K_F = n_D n_T \langle\sigma v\rangle_{DT} \qquad (1)$$

where $\langle\sigma v\rangle_{DT}$ is the rate coefficient, i.e. (cross section x velocity) averaged over the Maxwellian velocity distribution corresponding to the temperature T. The temperature dependence of this rate coefficient is such (see Fig. 4) that $kT \gtrsim 10^4$ eV is a requirement for a substantial release of fusion energy and a pre-eminent issue for a fusion reactor must therefore be the provision of a practical environment in which to house this very hot fuel. Ionization cross sections are many orders of magnitude greater than the D/T fusion cross section (i.e. σ_{DT} (max) $\sim 10^{-23}$ cm^2) so that fuel which is adequately hot for fusion will also be fully ionised. These charged particles can be confined by a magnetic field in such a way that it is possible to insulate the hot fuel from the walls of its containment vessel. A toroidal magnetic field geometry (discussed in Section 2) offers the particular advantage of providing a closed region of field which can be readily encompassed by a containment vessel. The tokamak device (described in Section 3) is a particularly successful example of toroidal confinement and many conceptual fusion reactors, e.g. INTOR[1] and NET[2], are based upon the tokamak principle.

Each fusion event produces an He^{2+} ion with an energy of 3.5 MeV and the alpha-particles can share this energy with the D^+, T^+ ions and their associated electrons. This leads to the concept of "alpha-particle ignition" whereby the energy deposited within the confined D/T plasma by alpha-particles balances the energy which is lost from the hot plasma to the walls of its containment vessel (see Section 5). Thus, after heating to an adequate temperature in the manner discussed in Sections 6 and 7, fusion reactions can become self-sustaining and the reactor could, in principle, "burn" continuously provided that sufficient D/T fuel is supplied and the non-reactive ^4He ash removed.

Energy losses from the hot D/T plasma (Section 4) are powerfully enhanced by the presence of even small quantities of impurity ions and so impurity release (Section 9) and control of impurity ingress (Section 8) are important issues.

Atomic processes relevant to magnetic confinement fall into the following broad categories:

a) Processes that are fundamental to the loss of energy from the hot D^+, T^+ and He^{2+} plasma and to the equipartition of energy amongst the electrons and ions.

b) Processes associated with the addition of energy and particles to the confined plasma, i.e. heating and fuelling.

c) Processes associated with the release of impurities from the walls of the confinement vessel, e.g. plasma-surface interactions and impurity control.

d) Processes associated with impurity elements that are present within the hot plasma and which can cause losses of energy or excessive plasma pressure.

e) Processes that are necessary for the operational control of the reactor, e.g. the maintenance of correct temperature and density and also the exhaust of "burnt" fuel.

It is the objective of this paper to provide an outline description of the impact of atomic processes upon the reactor potential of a tokamak device. The issues range widely over many fields of physics and it is not practicable to provide detailed discussion. For more comprehensive treatment of specific aspects the reader will be directed to recent review articles. The relevance of atomic processes to fusion research in general has been the subject of two previous NATO Study Institutes[3,4] and a collection of comprehensive reviews can also be found in Ref. 5.

2. PLASMA CONFINEMENT BY A CLOSED MAGNETIC FIELD

In conditions relevant to fusion research, the ionized gas (i.e. D/T in the case of a reactor and H or D in the case of present day experiments) is a "plasma" and thus has the property of being, overall, electrically neutral so that $n_e = n_i = n$, where the subscripts e and i refer respectively to electrons and ions. Another characteristic of the plasma is that the more mobile electrons screen the Coulomb field of the less mobile ions so that, at the Debye length

$$\lambda_D = (kT_e/4\pi n_e e^2)^{\frac{1}{2}} = 7.43 \times 10^2 (kT_e/n_e)^{\frac{1}{2}} \qquad [\text{cm}], \qquad (2)$$

the Coulomb field is reduced to a negligible value. In Eq.(2) and

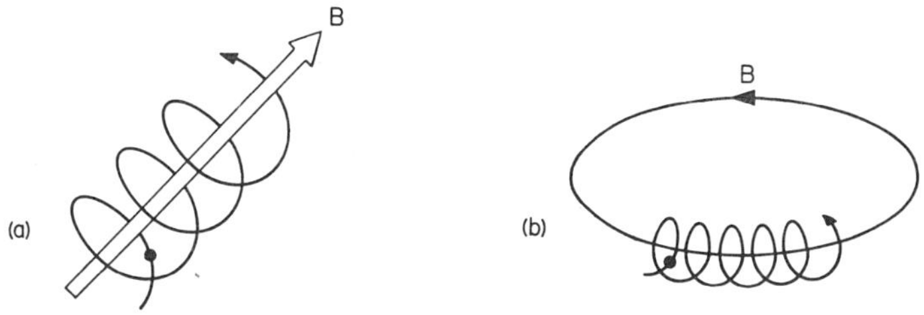

Fig. 1. (a) Trajectory of a charged particle around a magnetic field line. (b) Gyration around a closed field line.

subsequent equations*, kT is expressed in eV and n in particles/cm^3. Coulomb collisions between the charged particles give rise to equipartition of momentum. On one hand this causes scattering across the confining magnetic field and on the other it leads to equipartition of particle energy, i.e. thermalisation of the charged particles. Equipartition is dominated by a sequence of small angle scattering events rather than by a single large angle (i.e. 90°) event. The frequency of electron-electron Coulomb collisions is

$$\nu_{ee} = \frac{2.9 \times 10^{-6} n_e \ln\Lambda}{kT_e^{3/2}} \qquad [s^{-1}]. \qquad (3)$$

where $\Lambda = b_o/\lambda_D$. Here b_o is the smallest impact parameter at which small angle scattering can occur. In the regime of present interest $\ln\Lambda$ = 10 to 20.

In the presence of a magnetic field, plasma electrons and ions gyrate around the field lines as shown in Fig. 1(a). Motion across the field is thereby impeded but, if the field is uniform, the particles are free to move in the direction parallel to the field. To achieve effective confinement it is necessary that the field lines do not intercept the wall of the containment vessel and this fact has led to the evolution of closed magnetic line systems of the form indicated in Fig. 1(b). A closed line system is toroidal in shape and for stable confinement of a plasma it is necessary to consider two components of the magnetic field as shown in Fig. 2(a). These are the toroidal component (B_{tor}) which lies parallel to the axis of the plasma and the poloidal (B_{pol}) component which lies normal to the axis. The lines of the resultant field, B, are helical and twist around the torus until they close upon themselves in the manner shown by the flux tubes in Fig. 2(a). This property of rotational transform (with angle ι) can be characterized by a "safety factor" which, at the minor radius (a) of the plasma, is expressed as

$$q(a) = \frac{2\pi}{\iota} = \frac{a}{R}\frac{B_{tor}}{B_{pol}}. \qquad (4)$$

*The symbols c, e, k, etc. have their conventional meaning. The electron and ion masses are m_e and m_I, Z the atomic number and z the ion charge state.

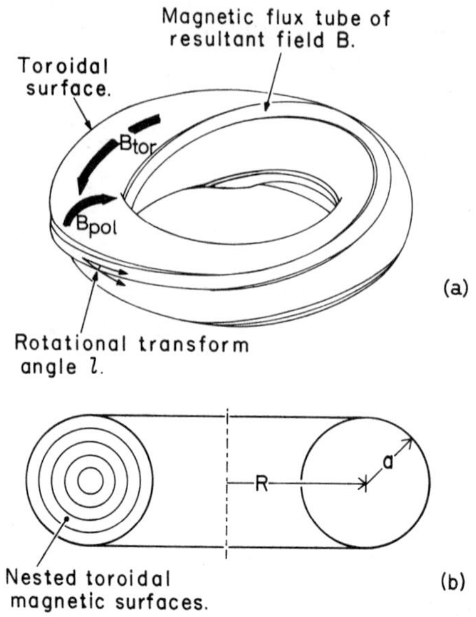

Fig. 2. Topology of a toroidal magnetic confinement device
(a) The rotational transform of the closed magnetic flux tubes defines a toroidal magnetic surface (illustrated for $q(a) = 2$).
(b) Projection of the toroidal plasma onto the poloidal plane.

Here R is the major radius of the plasma axis. The closed flux tubes map out a volume in space within which the confining magnetic field B exerts a pressure which opposes the kinetic pressure of the hot plasma, namely

$$\frac{\beta B^2(a)}{8\pi} = (n_i kT_i + n_e kT_e) \tag{5}$$

where the parameter β accounts for the efficacy of a particular toroidal magnetic configuration.

The mechanisms by which particles and energy are transported across the closed magnetic field are complex and not fully understood. Momentum exchange by scattering due to Coulomb collisions gives rise to cross field diffusion but this "classical" transport is dominated by effects which arise from the poloidally asymmetric nature of the confining magnetic field (i.e. neoclassical" processes) and from various forms of unexplained plasma transport (i.e. "anomalous" processes). Transport of plasma energy and particles across the magnetic field sets up substantial gradients of both density and temperature so that the boundary plasma (e.g. at minor radius a) is much cooler and less dense than that on the plasma axis.

3. THE TOKAMAK DEVICE

The principles of the tokamak are illustrated schematically in Fig. 3. The toroidal magnetic field (B_{tor}) is generated by external windings which are distributed around a vacuum vessel (or torus) which encircles the core of a transformer. The toroidal field windings are energized and a small

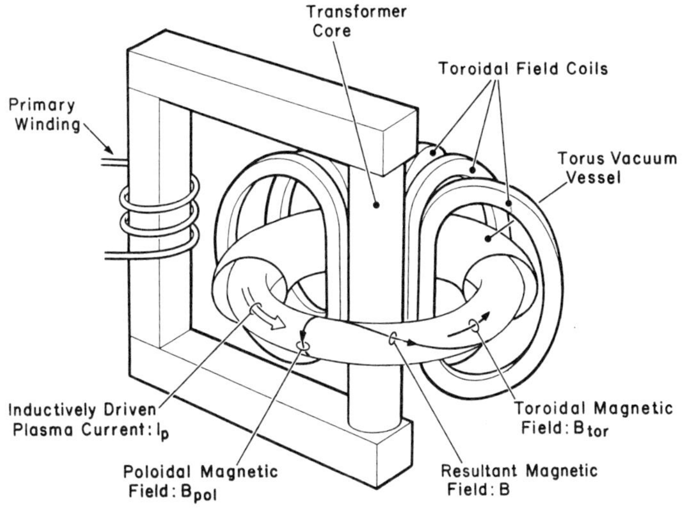

Fig. 3. Schematic illustration of the principles of a tokamak device.

amount of hydrogenous gas let into the torus. A discharge is struck within the gas and the free electrons are guided by the toroidal field so that, in effect, they form the secondary winding of the transformer. A current is now induced to flow within this conducting ring by feeding electrical energy into the primary winding of the transformer. As the induced current rises so is more gas introduced in order that both current and density may be ramped-up. This resulting powerful plasma current (I_p) generates the B_{pol} component of the confining magnetic field. As the density increases, the energy gained by the inductively driven electrons can be shared with the plasma ions and the bulk plasma is heated (i.e. Ohmic heating). The cycle of operation terminates when the energy supply to the transformer primary is exhausted.

Tokamak experiments have been particularly successful in confining hot plasma and the tokamak is regarded as the most likely form of fusion reactor. Predictions for a tokamak fusion reactor such as NET[2] indicate that q(a) is unlikely to be less than about 2, β will be only a few per cent, B (on plasma axis) about 50 kG and I_p about 10 MA. These conditions are adequate for confinement of a fusion plasma whose pressure is commensurate with $k\bar{T} \approx 10$ keV and $\bar{n} \approx 1.4 \times 10^{14}/cm^3$ where the overbars denote averaging over the minor radius (a) of the plasma. The dimensions of NET are R about 550 cm and a about 150 cm; the tokamak operational cycle is envisaged to last for $\sim 10^3$ s.

Present large tokamak experiments, e.g. JET[6], have attained (on axis) $kT_e \approx 3.5$ keV and $n \approx 3.5 \times 10^{13}/cm^3$ during a tokamak cycle which lasts for about 10 s.

4. STEADY STATE ENERGY BALANCE IN A MAGNETICALLY CONFINED FUSION PLASMA

Effects of atomic processes and plasma-surface interaction during the start-up and shutdown phases of a tokamak reactor have been considered by Harrison[7]. The following analysis is specifically related to steady state,

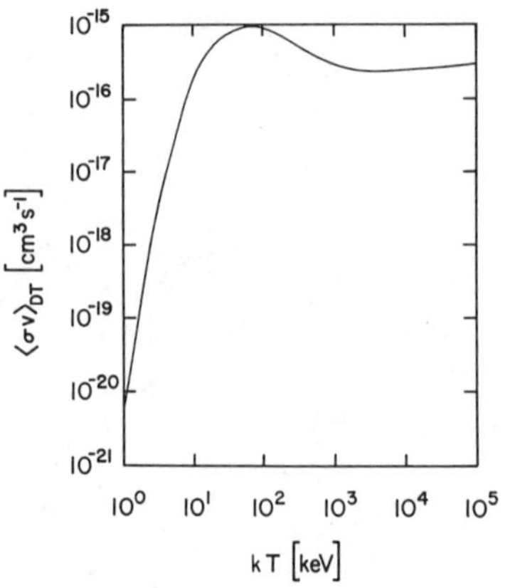

Fig.4 Rate coefficient $<\sigma v>_{DT}$ for DT fusion plotted as a function of thermal energy kT.

fusion burn conditions when the fusion power density generated in a homogeneous DT plasma wherein $n_D = n_T = n_i/2 = n_e/2$ can be expressed as

$$P_F(T_i) = \frac{n_i^2}{4} <\sigma v>_{DT} \; Q_{DT} = 8.05 \times 10^{-13} \; n_i^2 <\sigma v>_{DT} \qquad [W \; cm^{-3}]. \quad (6)$$

where Q_{DT} = 17.6 MeV per fusion event. The corresponding power density arising from fusion alpha-particles is

$$P_\alpha(T_i) = \frac{n_i^2}{4} <\sigma v>_{DT} \; Q_{DT} = 1.41 \times 10^{-13} \; n_i^2 <\sigma v>_{DT} \qquad [W \; cm^{-3}]. \quad (7)$$

where Q_α = 3,5 MeV. The temperature dependence of both $P_F(T_i)$ and $P_\alpha(T_i)$ is governed by that of the rate coefficient $<\sigma v>_{DT}$ which is shown in Fig. 4.

The alpha-particle power source must exceed all other forms of energy loss from the fusion plasma if alpha-particle ignition is to be attained. An unavoidable loss of energy arises from free-free collisions between electrons and ions. These collisions produce bremsstrahlung radiation which is emitted during acceleration of an electron in the unscreened field of an ion. For a plasma comprising fully stripped ions the power density radiated over all wavelengths is given by,

$$P_{br}(T_e) = 1.69 \times 10^{-32} \; n_e kT_e^{\frac{1}{2}} \sum (n_Z \; Z^2) \qquad [W \; cm^{-3}], \qquad (8)$$

where n_Z is the density of ions with atomic number Z. The plasma is transparent to bremsstrahlung radiation which is emitted by the hot plasma in the wavelength range below 10 Å. There is thus an unavoidable loss of energy by radiation to the wall of the containment vessel.

If it is assumed that $kT_e = kT_i = kT$, the balance of power between fusion and inherent radiative processes can be seen from the power density functions, $P_\alpha(T)/n^2$ and $P_{br}(T)/n^2$ which are plotted as functions of temperature in Fig. 5. It is apparent that the plasma temperature must be in excess of about 5 keV before energy available from the production of fusion alpha-particles exceeds that lost by bremsstrahlung radiation.

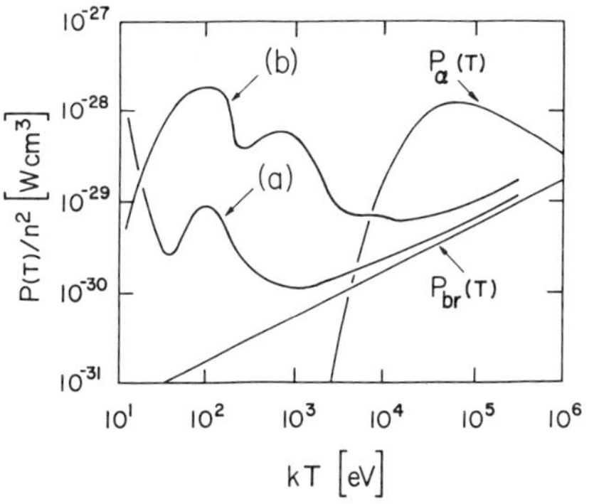

Fig. 5 The power density function $P_\alpha(T)/n^2$ and $P_{br}(T)/n^2$ plotted as a function of plasma temperature.
Also shown are the effects of small concentrations of impurity ions. Curve (a) for 1% of carbon and curve (b) for 0.1% of iron are calculated using Eq.(12).

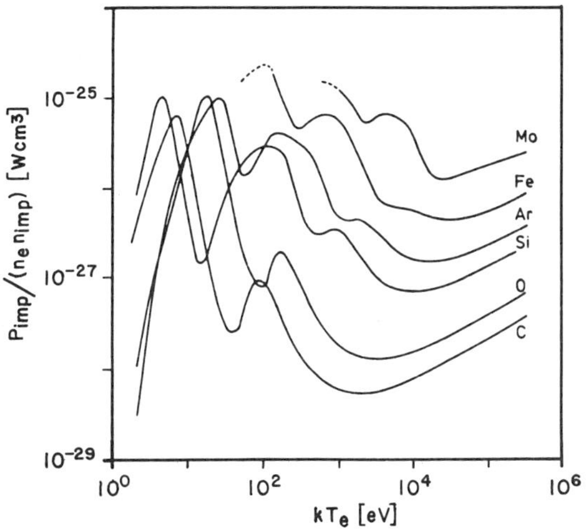

Fig. 6 Radiated power density function $P_{imp}/(n_e n_{imp})$ plotted as a function of electron thermal energy kT_e.
Data, which are taken from Summers and McWhirter[11], are based upon the assumption of local thermal equilibrium (see Section 12).

These idealised ignition conditions are appreciably degraded if the plasma contains impurity elements; not only do these enhance bremsstrahlung power losses, which are dependent upon Z^2, but the impurity ions may not be fully stripped and the consequent presence of bound electrons gives rise to atomic line-radiation losses. Details of the physics issues involved are discussed in reviews by Summers and McWhirter[8] and by Drawin[9,10].

The plasma is optically thin to the most significant components of atomic radiation from impurity ions. Figure 6 shows the total radiated power density function $P_{imp}(T_e)/(n_e n_{imp})$ for a number of likely impurities. This function depends upon the distribution of ion charge states and of excitation levels in each ion charge state at a particular plasma electron temperature. It is assumed here that conditions of local thermal equilibrium apply (see discussion in Section 12). In general, the total radiative power loss peaks in the temperature regime 10^2 to 10^3 eV (due to a maximum in the balance between the ion charge states, the electron excitation rates and the photon energies). At higher temperatures, the number of bound electrons is reduced and there is a decrease in the power loss from line-radiation; eventually bremsstrahlung becomes the dominant source of radiation and the $Z^2\sqrt{kT_e}$ dependence becomes dominant. The total radiative power losses increase with increasing atomic number of the impurity element and, because ions of the heavier elements are less readily stripped of all their electrons, so is the dominance of their line-radiation maintained throughout the temperature regime of interest.

The impact of small concentrations of impurity ions e.g. 1% of carbon and 0.1% of iron upon alpha-particle ignition conditions are shown in Fig. 5. It is apparent that whilst modest concentrations of elements such as carbon do not significantly influence the power balance of the ignited fusion plasma, they nevertheless greatly increase the fraction of energy radiated from a cold plasma and hence the power that must be fed into the plasma in order to achieve ignition. Indeed, this is a problem encountered in present day experiments. Furthermore the presence of impurity ions causes a significant and profitless increase in plasma pressure because

$$p_{imp} = \sum(n_z kT_i + z n_z kT_e) \tag{9}$$

where the summation is over the charge states of the impurity ions. The more powerful radiation losses of heavy elements extend into the higher temperature regime; this behaviour is evident from the substantial effect of 0.1% of Fe at kT ~ 10 keV.

5. ALPHA-PARTICLE IGNITION

An assessment of energy balance within a fusion plasma must account not only for the bremsstrahlung and impurity radiation losses but also for the energy input needed to sustain the thermal energy content of the hot plasma. The average energy of each particle within a pure, idealised hot plasma is 3kT/2 so that the energy required to raise the electrons and ions in a unit volume of the cold plasma to temperature T is,

$$Q_{ei} = \frac{3}{2}(n_e kT_e + n_i kT_i) = 3nkT \tag{10}$$

and this can be regarded as the energy density within the plasma. Suppose firstly that all of this thermal energy is lost from the plasma in an "energy confinement time" τ_E, and, secondly, that the heating system is capable of raising the cold plasma to a temperature kT in a time that is considerably less than τ_E. Thus the energy balance needed to sustain ignition solely by equipartition of alpha-particle energy is given by

$$\tau_E P_\alpha(T) = 3nkT + \tau_E P_r(T) \qquad (11)$$

where

$$\frac{P_r(T_e)}{n^2} = \frac{P_{br}(T_e)}{n^2} + \frac{C_{imp} P_{imp}(T_e)}{n^2} \qquad (12)$$

and $C_{imp} = (n_{imp}/n_i)$ is the concentration of impurity ions. Equation (11) can thus be expressed as

$$n\tau_E = \frac{3n^2 kT}{\left[P_\alpha(T) - P_r(T)\right]} \, . \qquad (13)$$

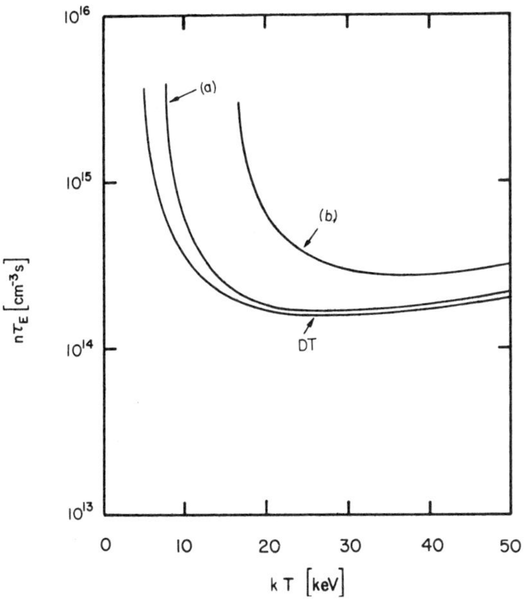

Fig. 7 The $n\tau_E$ criterion for alpha-particle ignition plotted as a function of plasma thermal energy kT.
Also shown are the effects of small concentrations of iron impurity ions; curve (a) for 10^{-3} and curve (b) for 10^{-2}.

This $n\tau_E$ criterion is shown in Fig. 7 and for pure DT it has a minimum of about 2×10^{14} cm^{-3} s in the region of kT about 1.5×10^4 eV. This simple analysis omits many important details but it does provide an indication of the necessary objectives. Accepting that $\tau_E \sim 1$ s indicates that the plasma density must be in excess of $\sim 10^{14}$ cm^{-3} in order to achieve ignition by alpha-particle heating in a pure DT plasma. The impact of impurity radiation upon the $n\tau_E$ criterion is illustrated in Fig. 7 for the case of iron ions. The maximum tolerable concentration of iron is about 0.1%.

6. OHMIC HEATING

Plasma is an excellent conductor of electricity and its resistivity can be expressed in the form

$$\eta \approx \frac{c^2 m_e^{\,2} n_i \nu_{ei}}{e^2 n_e} \tag{14}$$

where ν_{ei} is the electron-ion collision frequency which for protons is comparable to that for electron-electron collisions. Due to the temperature dependence of ν_{ei} [shown for ν_{ee} in Eq.(3)] the plasma resistivity is proportional to $kT_e^{-3/2}$. Plama current which is induced in the tokamak in order to generate the poloidal component of magnetic field is also used to heat the plasma and the power density of "ohmic heating" is thus given by

$$P_\Omega(T) = j\eta(T) \tag{15}$$

where j is the plasma current density. This ohmic heating is limited because there is an upper allowable level of Ip which, in the context of this simple discussion, is commensurate with the allowable magnitude of B_{pol}. Thus, although ohmic heating is effective at low plasma temperature, it is unlikely to be adequate to heat a tokamak plasma to ignition. The presence of impurities increases ν_{ei} and hence the ohmic power but this can only be achieved at the price of substantial radiative power losses.

7. AUXILIARY HEATING AND CURRENT DRIVE

It is envisaged that the additional heating needed to attain ignition will be provided either by feeding RF power into the plasma where it is

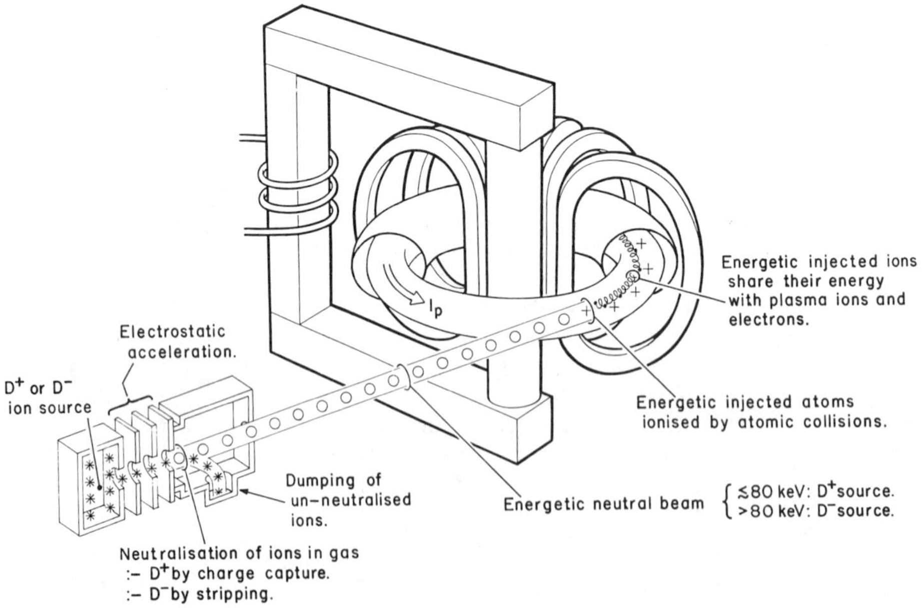

Fig. 8 Schematic illustration of the principles of neutral beam injection heating of a tokamak plasma.

absorbed by charged-particle-resonances or by injecting intense beams of energetic deuterium atoms which subsequently become ionized and share their kinetic energy with the less energetic plasma ions. Furthermore, the transformer induced plasma current in a tokamak is essentially a transient phenomenon and by using either RF power or neutral beams it is possible to impart directed velocity to the plasma particles and thereby, at least in principle, sustain the plasma current I_p and allow continuous tokamak operation. Details of various methods of current drive can be found in Ref. 12.

Atomic collisions play many important roles in neutral beam injection but they are much less significant in the case of RF systems which are not discussed further. The method of neutral beam injection is shown schematically in Fig. 8. Positive or negative ions which are produced in an external source are electrostatically accelerated and then neutralised by passage through gas. The remaining ions are deflected out of the neutral beam which then passes through the confining magnetic field and penetrates the plasma. The beam energy must lie within broad limits which are defined by the need for the atoms to penetrate into the plasma prior to ionization but the atoms must not be so energetic that they pass completely through the plasma and bombard the wall of the torus. In present experiments (e.g. JET) beams of about 80 keV D° atoms are available and the injected powers range from 5 to 25 MW. For a reactor scale tokamak, 160 keV beams of $D^{\circ}$are envisaged with injection powers of 50 to 80 MW.

Cross sections for atomic processes which ionize the energetic injected beam within the plasma are shown in Fig. 9. At beam energies relevant to a reactor, ion formation is predominantly due to plasma proton* (H_p^+) impact ionization

$$H^{o} + H_p^+ \rightarrow H^+ + e + H_p^+ \qquad \text{(ii)}$$

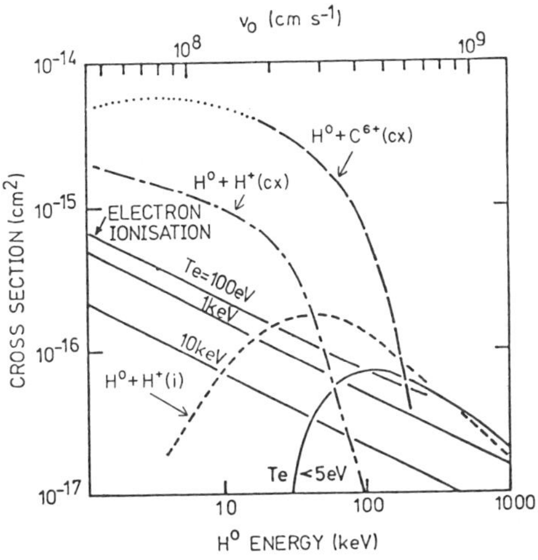

Fig. 9 Cross sections of processes which ionize the neutral beam plotted as a function of H atom velocity v_o.
The symbols cx denote charge exchange and i denotes ionization.
Electron ionization is plotted in the form $\langle \sigma_i v_e \rangle / v_o$.

*It is not necessary in this simple discussion to distinguish between the isotopes of hydrogen.

whereas, in the lower energy regime of present day experiments, charge exchange with plasma protons

$$H^o + H_p^+ \rightleftarrows H^+ + H_p^o \qquad \text{(iii)}$$

tends to be the dominant process. Electron impact ionization (shown here by a normalised form $<\sigma_i v_e>/v_o$) plays but a minor role. Cross sections for charge exchange collision with impurity ions are large and that for

$$H^o + C^{6+} \rightarrow H^+ + C^{5+}(n=4) \qquad \text{(iv)}$$

is shown as an example. Thus neutral beam penetration is reduced in the presence of impurity ions.

The efficiency of formation of a D° neutral beam from D^+ ions is dependent upon the charge exchange reaction (iii) whose cross section decreases steeply with ion energy above about 50 keV. Thus for higher energy beams it is desirable to start with D^- ions which can be stripped in reactions such as

$$D^- + D_2 \rightarrow D + e + D_2. \qquad \text{(v)}$$

Figure 10 shows the power efficiency

$$\xi^\circ = \frac{\text{power in atom beam leaving neutraliser}}{\text{power in ion beam entering}}$$

for production of D° atoms from D^- and D^+ (Ref. 13). The much greater efficiency of D^- for neutral beam formation at reactor relevant energies is evident.

The reactor requirements for current drive by neutral beam injection favour D° beam energies in an excess of 250 keV and further emphasize the importance of formation from D^- ions. Production of negative ion beams has been reviewed by Green[14].

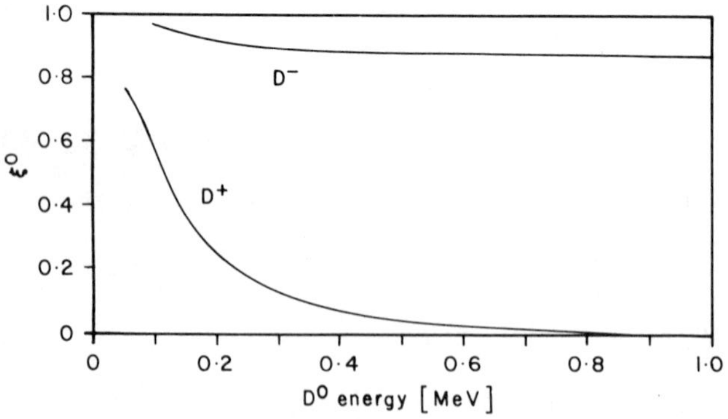

Fig. 10 The power efficiency ξ^o for formation of D atoms from D^+ and D^- ions plotted as a function of D atom energy. Data taken from Riviere[13].

8. IMPURITY CONTROL BY A LIMITER OR DIVERTOR

No device can confine plasma particles indefinitely and so there is an unavoidable flow of plasma to the vessel walls and, because ions and electrons move readily in the direction parallel to the resultant magnetic field B, the flow of escaping particles is strongly peaked wherever the magnetic flux tubes intersect the boundary of the torus vessel. Losses of plasma thermal energy exhibit the same tendency but the peaking is more pronounced because, in addition to the convective flow of energy carried by the plasma particles, there is also powerful conduction of electron thermal energy along the magnetic field. To shield the relatively delicate torus vessel wall from direct exposure to the consequences of plasma transport along the magnetic field it is desirable to insert a limiter as shown schematically in Fig. 11. The radial location of the limiter relative to the plasma axis determines the outermost extent of the "closed" region of magnetic field. Outboard of the closed region is the "scrape-off" layer where the magnetic flux tubes are open, i.e. they intersect the limiter surface or the vessel walls. Inboard of the limiter the plasma is well confined because particle motion across the field is impeded and motion along the field does not, at least in this simple picture, result in losses of particles or energy.

Plasma impurities arise as a consequence of plasma surface interactions. Examples of such processes are physical sputtering of the surface due to bombardment by energetic ions or atoms, chemical sputtering which arises when volatile compounds are formed on the surface, surface evaporation and arcing. Impurity release is most powerful wherever the incident fluxes of plasma energy and particles are most concentrated.

If the outermost surface of the closed magnetic field is defined by a limiter plate, as illustrated in Fig. 11, then impurities released at the plate can easily penetrate into the closed region of the magnetic field. This is especially true in the case of impurities (such as sputtered atoms) which are released as neutral particles and so can traverse the magnetic field until such time as they are ionized by the plasma electrons. It is possible to minimize this problem by using a magnetic configuration in which the outermost closed surface is determined by a magnetic field rather than by the material surface of a limiter.

This concept of a "magnetic limiter" or "divertor" is most simply appreciated in the case of a poloidal divertor. The poloidal field in a

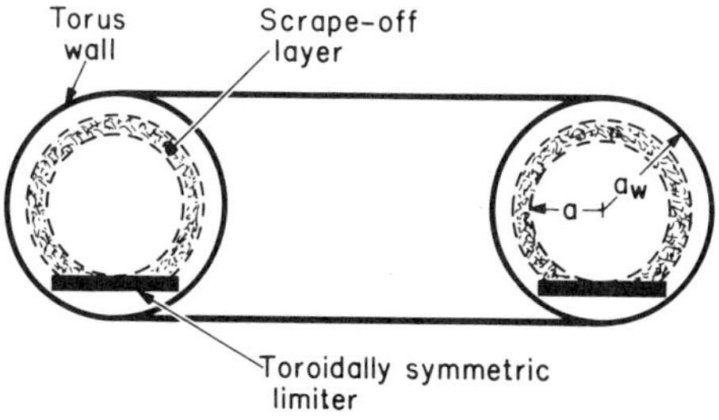

Fig. 11 Schematic illustration of a toroidally symmetric limiter at the bottom of the torus. The minor radius, a, of the limiter defines the boundary of the closed field region of the plasma whereas the vessel wall is at $a_w > a$.

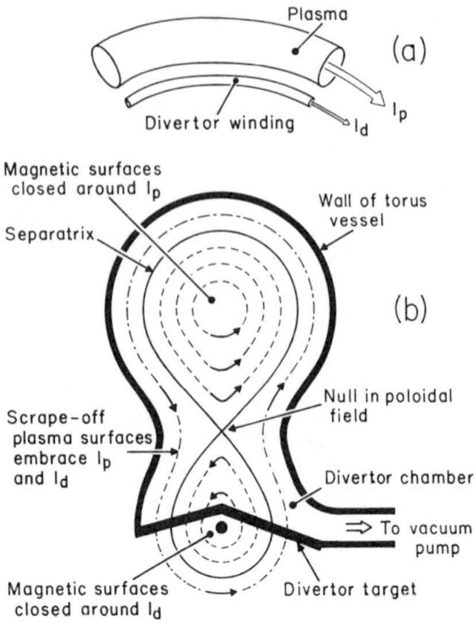

Fig. 12 Schematic illustration of the principles of a poloidal divertor. Also shown is the pumping duct through which neutral gas (e.g. helium) can be exhausted from the divertor chamber.

tokamak is produced by the plasma current, I_p, which is induced to flow within the plasma and this field can be opposed (at least over a small range of poloidal angle) when a current I_d is passed in the same direction through a nearby external conductor. The poloidal divertor relies upon such local annulment and its principles are illustrated in Fig. 12 where a single divertor winding is shown lying below but parallel to the magnetic axis of the tokamak plasma. Location of "the null" (i.e., the region where $B_{pol} = 0$) defines the "separatrix" and it is dependent upon the spacing between the winding and the plasma and also upon the relative magnitude of the currents. Inside the separatrix the magnetic surfaces arising from the individual currents I_p and I_d are discrete but those outside embrace both the plasma and the divertor winding. Because of the rotational transform of the resultant magnetic field B, plasma that diffuses across the separatrix tends to flow out of the torus and into the divertor rather than flowing to the torus wall. Thus the divertor provides a means by which the plasma collection surface (i.e. "divertor target") can be separated from the outermost closed magnetic field surface. Impurity atoms (or molecules) released from the target can thus be ionized within the divertor plasma and, because the plasma in the scrape-off layer flows into the divertor, it tends to sweep these impurity ions back to the target and thereby impede their ingress into the closed field region of the tokamak plasma.

The relative merits of the limiter and divertor in the context of impurity control of a tokamak reactor have been compared by Harrison[15].

9. RELEASE OF IMPURITIES BY PLASMA IONS

Conditions in the closed field region of the confined plasma are strongly influenced by processes which take place in the region close to the walls of the containment vessel.

Plasma-surface interactions and their impact upon tokamak performance have been reviewed by McCracken and Stott[16] and recent compilations of relevant data are available in Refs. 17 and 18. The status of theoretical and experimental knowledge of the edge plasma and of the boundary surfaces has also been the subject of a NATO Advanced Study Institute[19] and the role of atomic processes in this region has been reviewed by Harrison[20] and by Janev et al.[21].

It is convenient in this brief account to introduce the subject of impurity release by considering the energy of the T^+ (and D^+) ions of the plasma that are incident upon the boundary surface. Plasma ions in the immediate vicinity of the surface experience an electrostatic acceleration because the plasma potential is positive relative to the surface. This "sheath potential" is a characteristic property of the plasma-surface interface and it arises as a consequence of ambipolar flow (i.e. equal currents of electrons and ions) to the surface. Its magnitude, in the case of cold, singly charged ions, can be expressed as

$$U_s \approx \frac{kT_e}{2e} \ln \left(\frac{m_i}{2\pi\, m_e}\right) \tag{16}$$

so that $U_s \approx 3kT_e/e$ for a D^+/T^+ mixture of ions. A property of the plasma sheath region is that the drift velocity of the plasma (along the resultant magnetic field B) should be at least equal to the ion sound speed

$$c_s = \left(\frac{zkT_e + kT_i}{m_i}\right)^{\frac{1}{2}} . \tag{17}$$

In addition, the ions carry with them their thermal energy. The total incident energy can be expressed as

$$E_i \approx 2kT_i + zeU_s \tag{18}$$

which is about 5 kT_e/e in the case of a D^+/T^+ mixture.

Results indicate that the dominant release of impurities in tokamak experiments arises from physical sputtering whereby a surface atom of the boundary material receives sufficient energy (due to transfer of momentum from the incident D^+ or T^+ ion) that it is freed from the force which binds it to the bulk material. The sputtering yield (atoms/incident ions) is sensitive to the masses of the incident ion and the surface atoms, the binding energy of the surface atoms and the angle of incidence of the plasma ion. Suffice to state here that there is a threshold for the incident ion energy below which sputtering will not occur. Physical sputtering yields for D^+ and T^+ are shown in Fig. 13. The threshold energy for physical sputtering increases with increasing mass of the surface atoms and this dependence is opposite to that of the radiating capabilities of impurity elements (discussed in Section 4). The property offers the prospect of a compromise between impurity release and radiative power losses.

In conditions of a high degree of local recycling within a poloidal divertor (see Section 10), it is predicted[23] for a reactor such as NET that kT_s will lie in the range 20 to 30 eV. The incident ion energy is thus in the regime close to the sputtering threshold for heavy elements, such as tungsten, but significantly in excess of the threshold for lighter elements. This has prompted the selection of tungsten (or possibly molybdenum) as a divertor target material.

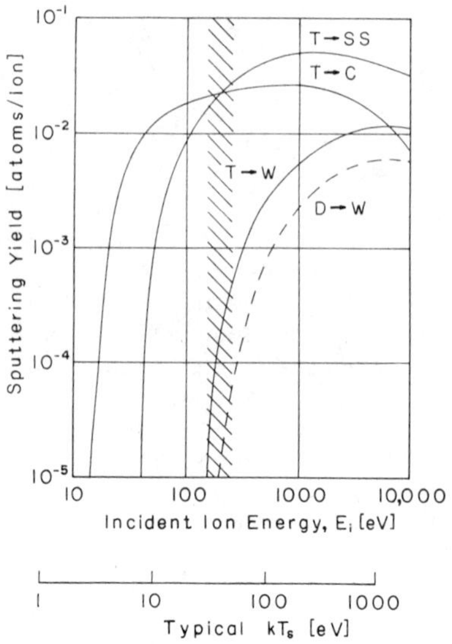

Fig. 13 Yields for physical sputtering of carbon, stainless steel and tungsten by tritons and deuterons incident normally on the surface[22].
The incident ion energy is also shown as a function of the plasma temperature kT_s at the sheath edge. The predicted operating regime of NET is shown by shading.

The presence of even small concentrations of light impurity ions in the plasma flowing to the target can significantly enhance the sputtering rate because, (a) these ions are more massive than D/T and have an intrinsically higher sputtering yield, (b) they become multiply charged whilst traversing the plasma and are thus more strongly accelerated by the sheath and (c) frictional forces exerted by the drifting D/T plasma tend to increase their energy prior to entry into the sheath (see Section 10). It is impossible to eliminate helium from a D/T fusion device but its contribution to sputtering appears to be acceptable. It is however highly desirable to minimise the concentration of oxygen ions because of the powerful chemical sputtering (i.e. release of volatile oxides) that they introduce. Small concentrations (~ 1%) of carbon in the edge plasma can probably be tolerated.

10. LOCALISED RECYCLING

In order to minimise physical sputtering it is desirable that the plasma temperature at the sheath edge be so small that the energy of the D^+ and T^+ ions incident upon the surface lies close to, or less than, the threshold. The sheath temperature can be expressed as

$$kT_s = Q_{\parallel}/(\gamma_s \Gamma_{\parallel}) \tag{19}$$

where $Q_{\parallel}$ and $\Gamma_{\parallel}$ are respectively the fluxes of energy and ion-electron pairs which flow from the plasma and through the sheath to the surface and γ_s = 5 to 7 is a coefficient which describes energy transport through the sheath. Thus a large flux of particles is advantageous in respect to lowering the plasma temperature at the sheath edge.

Bombardment of the surface by D^+ and T^+ ions saturates the underlying bulk material with D and T atoms and, after a fluence of $\sim 10^{18}$ ions/cm^2, there is negligible further absorption of incident ions. Depending on the particular combination of surface material, ion species and incident energy, a significant fraction, $R_N(E_i)$, of the incident ions are on average backscattered in the form of energetic neutral atoms. On average, these carry away from the surface a fraction $R_E(E_i)$ of the incident particle energy so that the average energy of the backscattered atoms is $E_i R_E/R_N$. The remaining fraction of neutral particles $(1 - R_N)$ tend to be molecules whose kinetic energy probably corresponds to the temperature of the surface. The coefficients R_N and R_E for carbon and tungsten are shown in Fig. 14.

The atoms and molecules which are released from the surface re-enter the adjacent plasma where they are likely to be re-ionized. The range of D/T atoms, λ^o, in a homogeneous plasma can be expressed in the form[20]

$$\lambda^o \sim \left[\frac{S_{cx}(T_i)}{3S_i(T_e)}\right]^{\frac{1}{2}} \lambda_{cx}(T_i) \tag{20}$$

where $\lambda_{cx}(T_i)$ is the mean free path for charge exchange. $S_{cx}(T_i)$ and $S_i(T_e)$ are respectively the rate coefficients for charge exchange with plasma D/T ions and for ionisation by plasma electrons and they are shown in Fig. 15. The magnitude of λ^o in a reactor such as NET is typically less than 10^{-2} m. This range is greater than the thickness of the plasma sheath (which is the order of the Debye length). The re-ionized particles tend to share their momentum and energy with the plasma which is already flowing to the surface. They thereby become entrained in the plasma flow and so return, via the ion accelerating sheath, to the surface. The cycle is repeated until such time as the neutral particles escape by being pumped from the system. The localized recycling sequence (in a divertor of the type depicted in Fig. 12) is illustrated, in a highly schematic fashion, in Fig. 16.

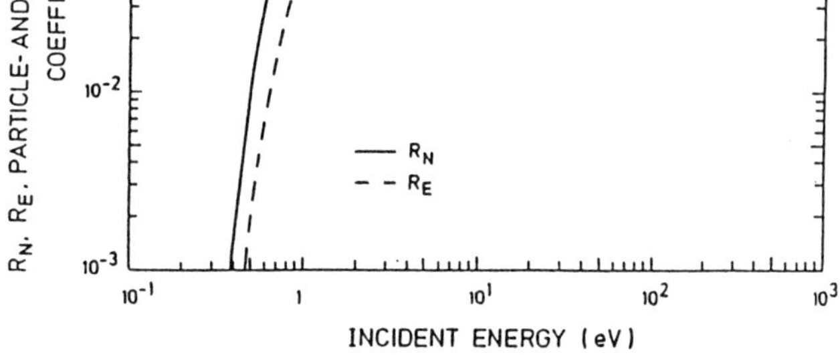

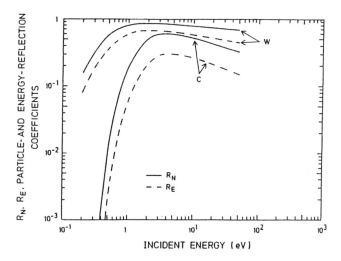

Fig. 14 Particle (R_N) and energy (R_E) reflection coefficients for carbon and tungsten surfaces.
Data are for normal incidence ions and are taken from Ref. 24.

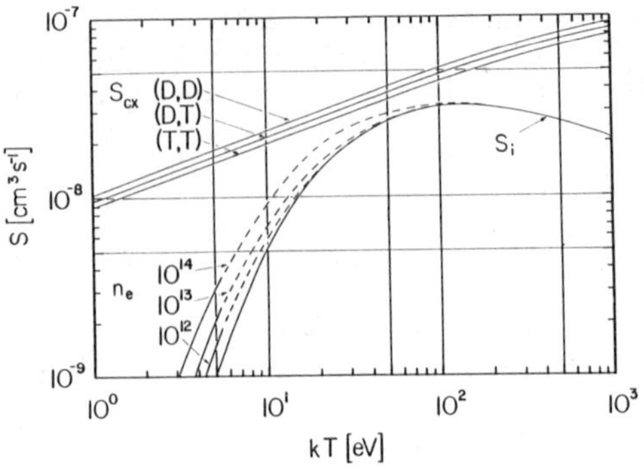

Fig. 15 Rate coefficients for charge exchange and electron ionization of D/T atoms.
The ionization rate coefficient $S_i(T_e)$ is enhanced by multi-step processes at higher electron densities (see Section 11.1) and data for $n_e = 10^{12}$, 10^{13} and $10^{14}/cm^3$ are plotted.

In the case of the open throated, single-null divertor envisaged for NET[23], most of the neutral particles are re-ionized directly whilst they are traversing the divertor plasma channel. A relatively small fraction are re-ionised indirectly after rebounding from the divertor chamber wall back into the plasma. The small flow of pumped neutrals (equivalent to a few x 10^{-3} $\Gamma_{\|}$) is adequate to exhaust the helium ash produced during burn. Plasma density at the divertor targets of present day experiments is much less than that envisaged for a reactor and so direct ionization is less likely. Nevertheless recycling of hydrogen neutrals which rebound from the divertor chamber wall is still a powerful effect.

Recycling at a limiter is inherently less localized due to the unavoidable escape of neutrals into the closed field region which lies in contact with the limiter surface (see Fig. 11).

It is predicted[23] that the power load due to electron conduction along the field B to the divertor target of a reactor is likely to be very large (typically 500 W/cm^2). However, the presence of large quantities of recycling neutral D/T helps to spread some of the incident plasma power in the form of hydrogen radiation. When the degree of localized recycling is high, this radiated power can be expressed as

$$Q^r_{DT} \approx \Gamma_{\|}\, E^r_{DT}(n_e, kT_e). \tag{21}$$

Here $E^r_{DT}(n_e, kT_e)$ is the average amount of radiated energy associated with the life history of each recycling D/T atom. The dependence of E^r_{DT} upon the local plasma electron density and temperature are discussed in Section 11.1 and are shown in Fig. 17.

When the dimensions of the divertor plasma are larger than the atom range, λ^o, most recycling occurs due to direct ionization and little energy is dissipated by atoms which escape from the plasma and reach the chamber wall. An approximate equality in power lost to the target and power lost by atomic processes then occurs when the plasma power flowing to the target is just adequate to sustain a sheath temperature of about 10 eV because at

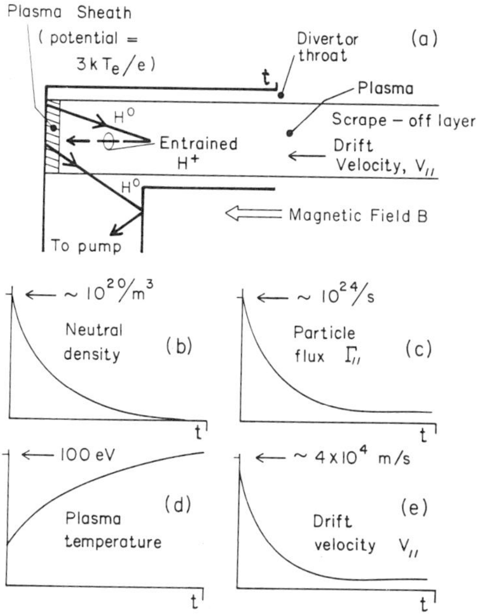

Fig. 16 Highly schematic representation of localized recycling within a divertor chamber of a reactor such as NET.
(a) Plasma environment projected in the direction of the magnetic field to the divertor throat at t;
(b) Density of neutral D/T; (c) Flux of plasma ions; (d) Plasma temperature and (e) Plasma drift velocity.

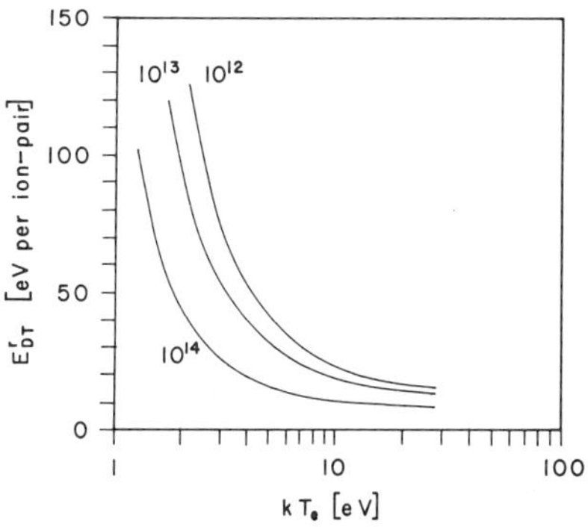

Fig. 17 The average radiation energy E^r_{DT} dissipated during the life of a recycling D/T atom.
Data from Ref. 20 are plotted for plasma densities $n_e = 10^{12}$, 10^{13} and $10^{14}/cm^3$.

this temperature the equality

$$\gamma_s kT_s \approx (\chi_i + E^r_{DT}) \tag{22}$$

is satisfied (here χ_i = 13.6 eV is the ionisation energy of H atoms). Present high recycling divertor experiments operate in the regime where the plasma power is insufficient to raise the sheath temperature above 10 eV.

The atomic and molecular processes which are dominant in hydrogen recycling are indicated schematically in Fig. 18.

11. THE RELATIONSHIP OF ATOMIC AND MOLECULAR PROCESSES TO THE PLASMA ENVIRONMENT

It is reasonable to consider three broad categories of atomic species. Firstly there is neutral hydrogen (i.e. atoms and molecules) which is present predominantly in the recycling region in the plasma edge. Two exceptional conditions are deeply penetrating, energetic D atoms arising from neutral beam injection and deeply deposited low energy hydrogen ablated from high velocity D/T pellets which may be used to fuel the hot plasma. Secondly there are higher Z species such as sputtered impurities which can be present in both the hot core and the plasma edge. To this group must be added the chemical products of plasma surface interactions, e.g. hydrocarbons, metallic oxides, etc.; these compounds are restricted to the plasma edge but their dissociation products may penetrate more deeply. Thirdly there is helium which is a special case because it is the product of fusion reactions in the hot core of this plasma and it must be exhausted as neutral gas in order that the burning plasma does not become poisened with unreactive ^4He "ash". Preceding discussion in this paper has been based largely upon considerations of plasma energy balance but atomic interactions also change the momentum balance within the plasma and thereby influence particle transport. This is particularly significant in collisions involving the ionization of neutral particles because the

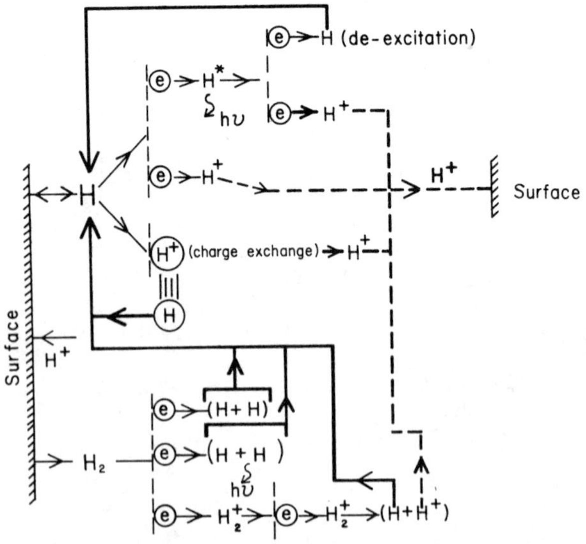

Fig. 18 Schematic representation of the dominant atomic and molecular collision processes associated with the recycling of hydrogen.

resulting ions provide a source of momentum within the plasma. Electrons contribute trivially to the momentum balance so that their role is predominantly related to energy balance. The impact of these broad issues upon the manner in which atomic data are used in modelling of fusion plasmas can be found in Ref. 25. The following account provides a summary of the salient features.

Application of basic cross section data must take account of the distribution of collision velocities within the plasma and also of the effects of multi-step collisions of the type discussed in Section 11.1. Except for analyses of injected beams, the possible presence of non-thermalised particles is neglected in order that the velocity distribution of plasma particles can be taken to be Maxwellian. The rate coefficient for an inelastic process with threshold energy χ can then be calculated from cross section data $\sigma(E)$ using

$$\langle\sigma v\rangle = \left(\frac{8kT}{\pi m}\right)^{\frac{1}{2}} \int_{\chi/kT}^{\infty} \sigma(E)\,(E/kT)\exp(-\,E/kT)\,d(E/kT). \qquad (23)$$

The influence of this integral can be appreciated by comparing the electron impact ionization cross section for H(1s) with the corresponding ground state rate coefficient S_i shown in Fig. 15. When $kT_e < \chi_i$ [say ~ 2 eV for ionisation of H(1s) atoms] the form of the velocity distribution dominates the form of the rate coefficient so that uncertainty in the atomic cross section is of minor significance. However, when $kT > \chi_i$ the rate coefficient is sensitive to the form of cross section. It is useful when considering recycling of atomic hydrogen at boundary surfaces to note that, in a low temperature regime, the variation of the ionization rate coefficient S_i with electron temperature is much greater than the possible variation of plasma parameters, e.g. $S_i(T_e)$ increases by a factor 10^4 over a temperature range kT_e = 1 to 10 eV.

11.1 Electron Collision with Hydrogen Atoms

Ionization from the ground state

$$e + H(1s) \rightarrow e + e + H^+ \qquad \text{(vi)}$$

provides a substantial sink for kinetic energy of the plasma electrons. Moreover the energy of the ejected electron is small (1 to 2 eV is typical for collisions pertinent to localized recycling) so that ejected electrons tend to dilute the energy content of the plasma.

The threshold energy χ_{pq} for excitation from a lower atomic level p to upper level q is less than the ionization threshold energy so that, in a macroscopic context, ionization by plasma electrons is always accompanied by excitation. The excitation process

$$H(p) + e \rightarrow H(q) + e \qquad \text{(vii)}$$

causes the plasma electron to lose kinetic energy. If unperturbed, the excited state q will spontaneously decay,

$$H(q) \rightarrow H(p') + h\nu, \qquad \text{(viii)}$$

to a lower level p' and the plasma is generally optically transparent to such atomic radiation so that the energy associated with reaction (viii) escapes to the walls of the vessel where it is absorbed.

The average time for collisions between plasma electrons and an atom in an excited level q is

$$\tau_q = (n_e \langle \sigma v_e \rangle_q)^{-1}. \tag{24}$$

In plasmas where $n_e > 10^{14}/cm^3$, this collision time may be appreciably less than the lifetime for spontaneous radiative decay of all but the lower excited states of hydrogen. In such conditions super-elastic collisions

$$e + H(q) \rightarrow e + \Delta E + H(p') \tag{ix}$$

become important. The collisions do not yield photons but the potential energy stored in the excited atoms H(q) is returned to the plasma electrons. These de-exciting collisions reduce the population of excited hydrogen atoms.

Ionization cross sections of excited H atoms are large. Moreover the ionization threshold energy is low so that even electrons in the low energy tail of the plasma thermal distribution are able to ionize excited atoms. Ionization of excited hydrogen

$$e + H(q) \rightarrow e + e + H^+ \tag{x}$$

provides a second route by which the population of excited H atoms is reduced. Collisional radiative (i.e. multi-step) processes involve a balance between reactions (vii), (viii), (ix) and (x) and these powerfully reduce radiative power losses from hydrogen atoms in a low temperature, high density edge plasma. Their effect can be seen in Fig. 17 and their impact upon localized recycling is considered in Refs. 20, 21 and 25. Multistep processes also enhance the ionization rate coefficient in high density plasmas in the manner shown in Fig. 15.

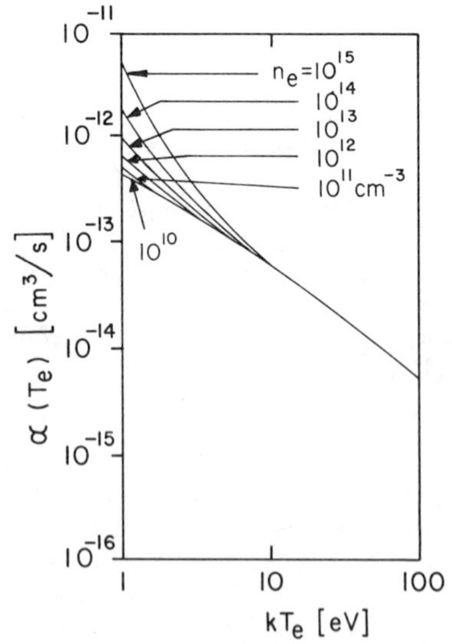

Fig. 19 Collisional radiative recombination rate coefficient for $e + H^+$ collisions.
Data are taken from Janev et al[21].

Protons can be destroyed by two-body radiative recombination with an electron

$$e + H^+ \rightarrow H(q) + h\nu. \qquad \text{(xi)}$$

The recombination rate coefficient for hydrogen $\alpha(T_e)$ is shown in Fig. 19. It is very small except at low electron temperature ($kT_e < 1$ eV) and high electron density so that the characteristic recombination time

$$\tau_\alpha = [n_e\ \alpha(T_e)]^{-1} \qquad (25)$$

is generally considerably longer than the time that the proton resides within the plasma. Two-body radiative recombination of hydrogen is therefore not likely to be significant.

Electron-proton recombination can in principle arise as a consequence of three-body collisions, but the electron density within the plasma is insufficient for this process to be significant.

11.2 Electron Collisions with Hydrogen Molecules

Collisions between plasma electrons and neutral H_2 can, in progressive order of their threshold energies, give rise to the following interactions;

$$e + H_2 \rightarrow H + H^- \qquad \text{[dissociative attachment]} \qquad \text{(xii)}$$

$$e + H_2 \rightarrow e + H + H \qquad \text{[dissociation]} \qquad \text{(xiii)}$$

$$e + H_2 \rightarrow e + H_2^* \rightarrow e + H_2 + h\nu \qquad \text{[excitation]} \qquad \text{(xiv)}$$

$$e + H_2 \rightarrow e + H_2^* \rightarrow e + H + H + h\nu \qquad \text{[dissociative excitation]} \qquad \text{(xv)}$$

$$e + H_2 \rightarrow H_2^+ + e + e \qquad \text{[ionization]} \qquad \text{(xvi)}$$

$$e + H_2 \rightarrow H^+ + H + e + e \qquad \text{[dissociative ionization]} \qquad \text{(xvii)}$$

$$e + H_2 \rightarrow H^+ + H^+ + 2e + e \qquad \text{[dissociative ionization]} \qquad \text{(xviii)}$$

Excitation of the electronic levels of the molecule together with ionization of the molecule act as energy sinks for the plasma electrons but in addition the plasma electrons dissipate energy in collisions which result in dissociation. In the context of plasma transport the molecule can be regarded as a potential source of H atom (or proton) momentum. For example when the H_2 molecule is dissociated into H + H by electron impact then two atoms each with an energy of about 2.2 eV are released for the expenditure of 8.8 eV electron energy. The dissociation products can be assumed to have a random spatial distribution within the plasma.

Rate coefficients of these reactions are shown in Fig. 20 and the most significant are (a) dissociation S_d^o [this coefficient includes contributions from reactions (xiii) and (xv)], (b) ionization S_i^o [arising from reaction (xvi)] and (c) dissociative ionization S_{di}^o [arising from reaction (xvii)].

In the case of H_2^+ the possible reactions, again in progressive ranking of threshold energy, are

$$e + H_2^+ \rightarrow H + H \qquad \text{[dissociative recombination]} \qquad \text{(xix)}$$

$$e + H_2^+ \rightarrow e + H^+ + H \qquad \text{[dissociation]} \qquad \text{(xx)}$$

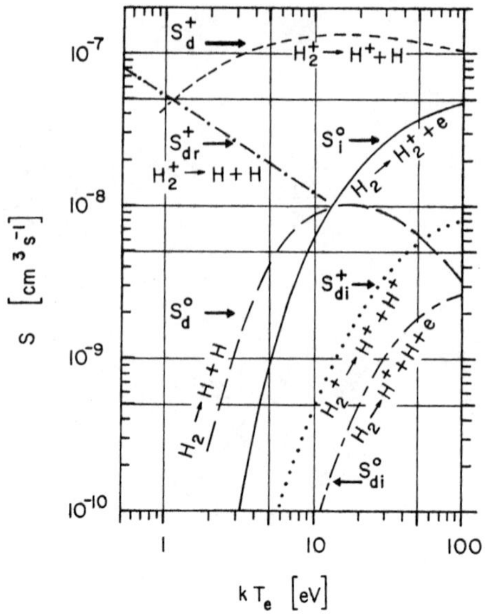

Fig. 20 Dominant rate coefficients for electron collisions with H_2 and H_2^+ plotted as a function of electron temperature. Symbols are defined in the text and taken from Ref. 20.

$$e + H_2^+ \rightarrow e + H_2^{+*} \rightarrow e + H^+ + H + h\nu \quad [\text{dissociative excitation}] \quad (xxi)$$

$$e + H_2^+ \rightarrow e + e + H^+ + H^+ \quad [\text{dissociative ionization}] \quad (xxii)$$

The rate coefficient for dissociation S_d^+ [reaction (xx) and (xxi)] is dominant whereas the rate coefficients for dissociative recombination S_{dr}^+ [reaction (xix)] has only a minor influence upon the characteristics of the boundary plasma.

11.3 Electron Collisions with Impurity Species

Collision processes similar to those in hydrogen atoms occur in the case of impurity elements but the situation is more complex because of the greater number of bound electrons. Electrons in the outermost shell of the atom (or ion) are less tightly bound than those in inner shells and so the outermost electrons participate most readily in excitation and ionization. However, in many species, there are more inner electrons so that the net contribution from inner shells may well exceed that of the outer.

The presence of many bound electrons increases the number of possible collision processes. One such example is that the plasma electron may, in a single collision, eject more than one of the bound electrons of the impurity species X^{z+}, namely

$$e + X^{z+} \rightarrow e + ae + X^{(z + a)+}. \quad (xxiii)$$

However, multiple ionization is not likely to be particularly significant because ionization by plasma electrons most probably proceeds in a stepwise manner i.e.

$$X^0 \rightarrow X^+ \rightarrow X^{2+} \text{----}\rightarrow .$$

A more significant process is the excitation of auto-ionizing states

$$e + X^{z+} \rightarrow e + X^{z+*} \rightarrow e + e + X^{(z+1)+} \qquad \text{(xxiv)}$$

and the reverse of autoionization which is dielectronic recombination

$$e + X^{z+} \rightarrow e + X^{z+*} \equiv \left(X^{(z-1)+**}\right) \rightarrow X^{(z-1)+*} + h\nu. \qquad \text{(xxv)}$$

Dielectronic recombination is akin to excitation and its contributions are significant when the electron energies lie close to the excitation threshold. The rate coefficient, illustrated here in Fig. 21 for the case of $Ne^{6+} \rightarrow Ne^{5+}$, follows somewhat the shape of an excitation coefficient and it peaks at a relatively high electron temperature. It thus differs significantly from the radiative recombination coefficient which decreases monotonically with increasing temperature.

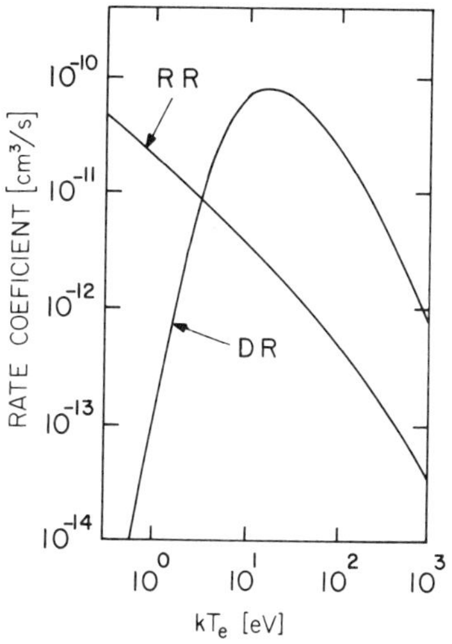

Fig. 21 Comparison of the rate coefficient for collisional dielectronic recombination compared with that for collisional radiative recombination.
Data are for $Ne^{6+} \rightarrow Ne^{5+}$ and are taken from Jacobs et al[26].

11.4 Collisions between Hydrogen and Impurity Ions

Collisions between impurity ions and their associated atoms can generally be neglected. For example the rate coefficient for the symmetrical, resonant charge exchange reaction in helium,

$$He + He^{+} \rightleftarrows He^{+} + He \qquad \text{(xxvi)}$$

is large even at low collision velocity but the effect upon He atom transport is slight because of the relatively small concentration of He^{+} in the boundary plasma.

Collisions between H atoms and impurity ions cannot be symmetric but in some cases they tend to be energy resonant. A typical example, discussed in Section 7, is

$$C^{6+} + H \rightarrow C^{5+}(n = 4) + H^{+} \tag{iv}$$

for which the cross section data are shown in Fig. 9. At relevant H atom collision energies the rate coefficient exceeds by many orders that for two body radiative recombination (i.e. $e + C^{6+}$) and so collisions of this type influence the charge state population of impurity ions in the plasma (see Section 12).

The significance of the reverse type of reaction, e.g.

$$C^{5+}_{q} + H^{+} \rightarrow C^{6+} + H, \tag{xxvii}$$

depends upon the population within the plasma of highly excited states (q) of the initial impurity ion. The collision rates could well be comparable to the electron-ion recombination rates.

12. CHARGE STATE DISTRIBUTION OF IMPURITY IONS AND RADIATIVE POWER LOSSES

It is in general reasonable to accept that ionization by electron collisions proceeds in a stepwise manner, i.e. $X^{0} \rightarrow X^{+} \rightarrow X^{2+} \dashrightarrow$ etc., so that the steady state balance of an ionization stage z can be expressed as

$$n_e n_{(z-1)} S_{CR(z-1)} - n_e n_z S_{CR,z} + n_e n_{(z+1)} \alpha_{CR(z+1)} - n_e n_z \alpha_{CR,z} - n_z/\tau_z = 0. \tag{26}$$

Here the ion density (n_z etc.) refers to the ground state and any contribution from the population of excited states are accommodated by the use of appropriate collisional radiative coefficients (denoted by subscripts CR). Details of the physics involved and of the modelling employed in determining these coefficients can be found in McWhirter and Summers[8] and also in Drawin[9]. Dielectronic recombination [reaction (xxv)] provides a powerful route for recombination of these multi-electron species and so the various contributions to recombination

$$\alpha_{CR,z} = [\alpha(\text{two body radiative}) + \alpha(\text{dielectronic}) + \alpha(\text{three-body})] \tag{27}$$

must be included.

The time τ_z in Eq. (26) is the residence time of the ions in the particular plasma region under consideration. If this region is sited deeply within the closed confinement field of the plasma then

$$\tau_z \gg [n_e \, \alpha_z(T_e)]^{-1} \tag{28}$$

and the equilibrium charge state density population tends to be governed by the electron collision rates so that

$$\frac{n_{(z+1)}}{n_z} = \frac{S_{CR,z}(T_e)}{\alpha_{CR(z+1)}(T_e)} . \tag{29}$$

This is the condition of "local thermal equilibrium" which is often referred to as "coronal equilibrium" and its characteristics are illustrated for the example of oxygen in Fig. 22. It should be noted that the population $(n_z/\Sigma n_z)$ of the charge state z is substantial when $S_{CR,z}(T_e) = \alpha_{CR(z+1)}(T_e)$ and that this equality occurs in the regime where

$kT_e < \chi_i$ (i.e. less than the energy for z to (z+1) ionization). Thus the population of ionization stages tends to be sensitive to the low temperature regime of the ionization rate coefficient.

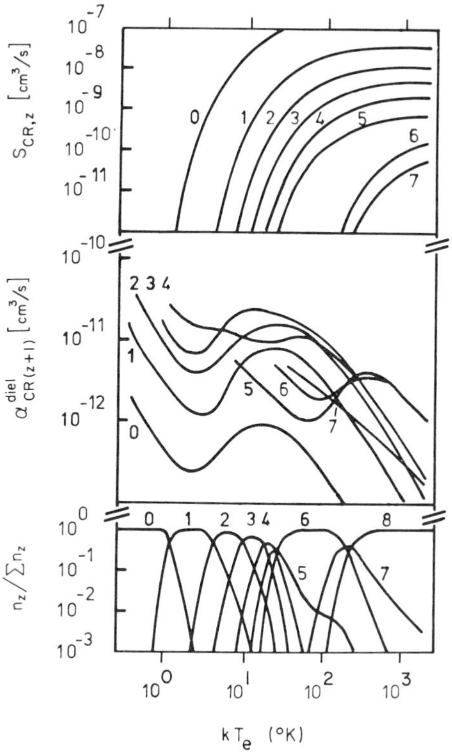

Fig. 22 Collisional radiative rates for ionization, $S_{CR,z}(T_e)$, dielectronic recombination $\alpha^{diel}_{CR(z+1)}(T_e)$ together with the charge state population $(n_z/\Sigma n_z)$ for oxygen. Data, which are taken from Summers[11], are for $n_e = 10^{12}/cm^3$. The ion charge state z is shown for each curve.

If the plasma region under consideration lies close to the wall or if the velocity of the impurity ions across the magnetic field is large then the inequality expressed in Eq.(28) is no longer valid and the ion loss rate term in Eq.(26) becomes dominant. These "non-coronal" conditions are particularly evident in the open magnetic field region of the plasma edge because here ions are lost from the system due to rapid transport along the field to the boundary surfaces so that the effects of recombination are substantially reduced. It is convenient to simplify Eq.(26) by assuming that the residence time of impurity ions is insensitive to their charge state (i.e $\tau_z = \tau_{imp}$) and the results of one such calculation making this assumption for oxygen by Abramov[27] are shown in Fig. 23. Here the average charge state

$$\bar{z} = \frac{\Sigma z n_z}{\Sigma n_z} \tag{30}$$

is plotted as a function of kT_e for various values of $n_e\tau_{imp}$; in the boundary plasma $n_e\tau_{imp}$ is typically $\sim 5 \times 10^{10}$ cm^{-3} s.

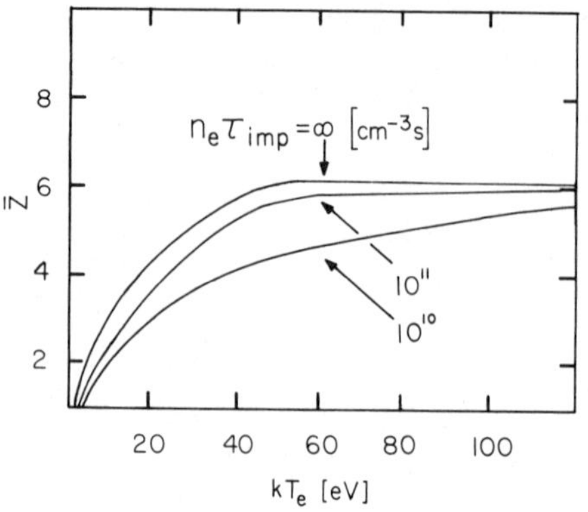

Fig. 23 The average charge state $\overline{z}$ of oxygen plotted as a function of electron temperature.
Data are taken from Abramov[27] and the condition $n_e\tau_{imp} = \infty$ is equivalent to the coronal equilibrium conditions shown in Fig. 22.

Contributions to the balance of ionization states of impurities due to charge capture from H atoms is to be expected whenever the local density of atomic hydrogen is significant and the charge state of impurity ions moderately high. Discussions of the likely significance of such effects can be found in Drawin[10] and in Hulse et al.[28]. In general, the effects of charge exchange are only likely to impact upon the ionization state population when fast atoms are injected into the plasma.

The population of excited states reaches a steady state value in a much shorter time than does the charge stage distribution. The equilibrium balance of excited states can be conveniently expressed in the form

$$\frac{n_{zq}}{n_{zo}} = \frac{n_e \, S_{oq}^{CR}}{\sum_{p<q} A_{(q\rightarrow p)}} \tag{31}$$

where n_{zq} is the density of ions (of charge state z) in excited level q, n_{zo} is the density of groundstate ions, S_{oq}^{CR} is the collisional radiative excitation coefficient for transitions $o\rightarrow q$ and $(A_{q\rightarrow p})^{-1}$ is the lifetime for spontaneous decay from q to a lower level p. The density of power radiated due to the spontaneous decay of level q to p is

$$P_{z(q\rightarrow p)} = 1.6 \times 10^{-19} \, n_{zq} \, A_{(q\rightarrow p)} \, E_{pq} \qquad [\mathrm{W/cm^3}]. \tag{32}$$

The power loss due to line radiation from ionization state z

$$P_{zl} \, (T_e) = \sum_q P_{z(q\rightarrow p)} \, (T_e) \tag{33}$$

is determined by summation over those q levels which (a) have a significant excitation rate coefficient, (b) are not depopulated by multi-step processes and (c) emit photons which carry a significant amount of energy. Details of modelling methods can be found in McWhirter and Summers[8].

The total power loss $P_{zt}(T_e)$ due to radiation associated with charge state z must also include contributions from recombination

$$P_{tz}(T_e) = [P_1(T_e) + P_\alpha(T_e) + P_{br}(T_e)]_z = n_e\, n_z\, F_z(T_e) \tag{34}$$

so that the radiated power function

$$F_z(T_e) = P_{tz}(T_e)/n_e\, n_z \qquad [\mathrm{W\ cm^3}] \tag{35}$$

can be used as a measure of the radiating efficiency of each ionisation state z. Summation of $F_z(T_e)$ over the population of ionization states yields the radiated power loss coefficient $F = P_{imp}(T_e)/n_e n_{imp}$ which is characteristic of the particular atomic species. The radiated power loss functions for the coronal equilibrium charge state distributions of likely impurity ions can be seen in Fig. 6.

The total radiated power function can be strongly sensitive to deviation of plasma conditions away from those of local thermal equilibrium. This effect can be seen in Fig. 24 where $P_{imp}/n_e n_{imp}$, for various values of $n_e\tau_{imp}$, is plotted as a function of kT_e (the associated average charge state is presented in Fig. 23). A marked sensitivity of the total radiation function to $n_e\tau_{imp}$ is evident when kT_e exceeds 30 eV. More detailed analysis which includes effects of charge exchange can be found in Carolan and Piotrowicz[29].

Detailed calculations along the lines described by McWhirter and Summers[8] yield the best available data for radiative power losses but the procedure is both complex and time consuming. Such data are therefore restricted to a relatively small number of atomic species. Furthermore, the collision data for neutral and lowly charged complex atomic species such as iron and tungsten are at present so uncertain that precise calculation of their power losses is not warranted at low plasma temperatures. Jensen et al.[30] have invoked the concept of an "average ion" model in order to simplify calculation of radiative power losses and thereby provide a wider base of data. The different ion charge states of

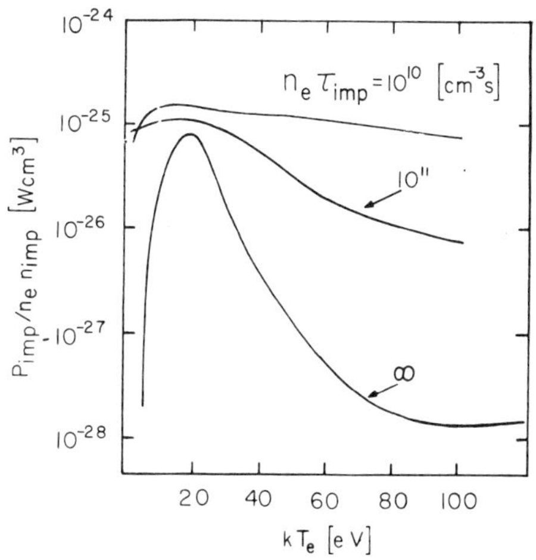

Fig. 24 Total radiated power loss functions for oxygen. Data are from Abramov[27] and apply to the average charge states shown in Fig. 23. Note coronal equilibrium conditions correspond to $n_e\tau_{imp} = \infty$.

each atomic species are replaced by a single conceptual "average ion". The populations of the ionization stages in the real plasma are statistically accounted for by assigning equivalent electron populations to the principle electron shells of the average ion. Transitions between these levels are equivalent to changes in ionization stage. Radiative power losses are related to transitions between the electron levels of this average ion. A wide range of data based upon this method has been reported by Post et al.[31]

13. CONCLUSION

The advent during the last five years of large tokamaks, exemplified by JET, has elevated fusion research from laboratory scale experiments to the status of reactor relevant studies. The results have provided further evidence of the impact of atomic and molecular processes upon the behaviour of tokamak plasmas. Atomic radiation remains the main source of energy loss from limiter bounded plasmas and this has led to the adoption of low Z coatings or liners for the torus vessel. Divertor experiments have so far been restricted to smaller sized devices, e.g. ASDEX, PDX and DOUBLET III, nevertheless they have demonstrated the effectiveness of the divertor in reducing radiation losses from the main plasma and in providing a cool recycling plasma adjacent to the divertor target. The interest in atomic and molecular processes remains strong and ranges from processes involving highly charged impurity ions to thermal energy neutral molecules of H_2 and impurities such as O_2, CO, etc. Despite the substantial advances in experimental and theoretical atomic collision physics there are many unresolved problems. Precise data for dielectronic recombination in the plasma environment is one example, excitation and ionisation data for medium charge state ions is another. Even apparently mundane processes pose problems which have not yet been adequately considered, one example is the assessment of the average energy dissipated by plasma electrons in producing H_2^+, H^o, H^+, etc. from hydrogen molecules.

In conclusion it should be noted that atomic interactions are not restricted to the hot plasma of a tokamak but extend to the detailed physics and chemistry at the surface of the vessel. There are many intellectually stimulating issues that could form topics for academic research but which fall outside the conventional disciplines of plasma, atomic and surface physics.

REFERENCES

1. "International Tokamak Reactor" - Reports of the International Tokamak Reactor Workshop.
   - Zero Phase (1979), IAEA, Vienna (1980).
   - Phase One (1980 and 1981), IAEA, Vienna (1982).
   - Phase Two A - Part 1 (1981-83), IAEA, Vienna (1983).
   - Phase Two A - Part 2 (1983-85), IAEA, Vienna (in press).
2. "Next European Torus" (Status Report, December 1985).
3. "Atomic and Molecular Processes in Controlled Thermonuclear Fusion" ed., M.R.C. McDowell and A.M. Ferendeci, NATO ASI, Bonas, France, Plenum Press, New York and London (1980).
4. "Atomic and Molecular Physics of Controlled Thermonuclear Fusion" ed., C.J. Joachain and D.E. Post, NATO ASI, Santa Flavia, Italy, Plenum Press, New York and London (1983).
5. "Applied Atomic Collision Physics" ed., H.S.W. Massey, E.W. McDaviel and B. Bederson, Vol.2, "Plasmas", ed., C.F. Barnett and M.F.A. Harrison, Academic Press, New York and London (1984).

6. A. Gibson, in "Report on the 3rd European Tokamak Programme Workshop", Plasma Physics and Controlled Fusion, 27: 801 (1985).
7. M.F.A. Harrison, "Impurity Control and Its Impact upon Start-up and Transformer Recharging in NET", in Course/Workshop on Tokamak Start-up, Int. School of Fusion Reactor Technology, Erice, 1985 (to be published by Plenum Press).
8. R.W.P. McWhirter and H.P. Summers, in Ref.5, p.51.
9. H.W. Drawin, Physica Scripta, 24: 622 (1981).
10. H.W. Drawin, in Ref.4, p.341.
11. H.P. Summers and R.W.P. McWhirter, J.Phys.B 12: 2387 (1979).
12. "Non-Inductive Current Drive in Tokamaks", Proc. IAEA Technical Committee Meeting, Culham Laboratory, 1983, ed., D. Start, Culham Laboratory Preprint, CLM-CD (1983).
13. A.C. Riviere, "Neutral Injection Heating of Toroidal Reactors", Appendix 3, ed., D.R. Sweetman, Culham Laboratory Report, CLM-R112 (1971).
14. T.S. Green, in Ref.5, p.339.
15. M.F.A. Harrison, in "Fusion Technology 1984", Proc. 13th Symposium, Varese, Italy, Pergamon Press 1984, p.81.
16. G.M. McCracken and P.E. Stott, Nucl. Fusion 19: 889 (1979).
17. "Data Compendium for Plasma-Surface Interactions", R.A. Langley, J. Bohdansky, W. Eckstein, P. Mioduszewski, J. Roth, E. Taugler, E.W. Thomas, H. Verbeek and K.L. Wilson, IAEA, Vienna, 1984.
18. "Energy Dependence of the Yields of Ion-Induced Sputtering of Monatomic Solids", N. Matsunami, Y. Yamamura, Y. Itikawa, N. Itoh, Y. Kazumata, S. Miyagawa, K. Morita, R. Shimizu and H. Tawara, Report IPPJ-AM-32 Nagoya University, Institute for Plasma Physics, Nagoya (1983).
also
"Data on the Backscattering Coefficients of Light Ions from Solids", R. Ito, T. Tabata, N. Itoh, K. Morita, T. Kato and H. Tawara, Report IPPJ-AM-41, Nagoya University, Institute for Plasma Physics, Nagoya (1985).
19. "The Physics of Plasma Wall Interactions in Controlled Fusion", eds. R. Behrisch and D.E. Post, NATO ASI, Val-Morin, Canada, Plenum Press, New York and London (in press).
20. M.F.A. Harrison, in Ref.5, p.395.
21. R.K. Janev, D.E. Post, W.D. Langer, K. Evans, D.B. Heifetz and J.C. Weisheit, J.Nucl.Mater. 121: 10, (1984).
22. J. Bohdansky, in Ref.17, p.61.
23. M.F.A. Harrison and E.S. Hotston, "Plasma Edge Physics for NET/INTOR" NET Report 50, Euratom, EUR-FU/XII-361/86/50, Brussels (1985).
also
Culham Laboratory Preprint CLM-P761 (1985).
24. R. Behrisch and W. Ekstein, "Ion Backscattering from Solid Surfaces" in Ref.19.
25. M.F.A. Harrison, "Atomic and Molecular Collisions in the Boundary Plasma", in Ref.19.
26. V.L. Jacobs, J. Davies, J.E. Rogerson and M. Blaha, Astrophys.J.230: 627 (1979).
27. V.A. Abramov, in "USSR Contributions to the INTOR Phase-Two-A Workshop", Report, Kurchatov Institute, Moscow (1982).
28. R.A. Hulse, D.E. Post and D.R. Mikkelsen, J.Phys.B, 13: 3895 (1980).
29. P.G. Carolan and V.A. Piotrowicz, Plasma Physics, 25: 1065 (1983).
30. R.V. Jensen, D.E. Post, W.H. Grasberger, C.B. Tarter and W.A. Lokke, Nucl. Fusion, 17: 1187 (1977).
31. D.E. Post, R.V. Jensen, C.B. Tarter, W.H. Grasberger and W.A. Lokke,' At. Data and Nuc. Data Tables, 20: 397 (1977).

CRYRING - A FACILITY FOR ATOMIC, MOLECULAR AND NUCLEAR PHYSICS

A. Bárány and C. J. Herrlander

Research Institute of Physics
S-104 05 Stockholm, Sweden

1. INTRODUCTION

The CRYRING project was put forward by the Research Institute of Physics (AFI), Stockholm, Sweden, as a proposal in October 1983 under the heading "A facility with CRYSIS and other ion sources connected to a synchrotron ring intended for studies of atomic, molecular and nuclear collisions, in particular in experiments with interacting beams of ions, molecules, electrons and laser photons". In September 1985 (during the present NATO ASI) funding was granted for the last subproject, comprising the actual ring structure to be completed by 1989-90. Thus started the third Swedish storage ring project, the first being the synchrotron radiation electron ring MAX at Lund[1] and the second the nuclear and particle physics light ion ring CELSIUS at Uppsala[2]. The present short contribution aims at a pedagogical (though necessarily somewhat superficial) presentation of CRYRING and its modes of operation. Some earlier presentations[3-5] can be found in the proceedings from the Workshop on Electron Cooling and Related Applications 1984 and from the 1985 Particle Accelerator Conference.

The name CRYRING should be interpreted as "CRYSIS synchrotron ring", where CRYSIS ("Cryogenic Stockholm ion source") is an electron beam ion source belonging to the CRYEBIS family developed at Orsay[6] and optimized for producing highly charged very heavy ions. A schematic outline of the CRYRING facility is given in Fig. 1. In the main operational mode, low charge state ions are fed into CRYSIS from an ion injector (isotope separator). After being stripped to the desired mean charge state in the main ion source, a pulse of highly charged ions is extracted and can be used immediately for low velocity experiments or accelerated in a linear RFQ ("Radio frequency quadrupole") accelerator[7]. Medium velocity accelerated ions can be deflected away for experiments or injected into the ring. In the storage mode experiments with a circulating beam of highly charged heavy ions interacting with photons, electrons, other ions or neutrals will be possible. The ring will also be used in a synchrotron acceleration mode to produce an extracted beam for nuclear structure and high energy atomic physics. Not shown in the schematic outline of Fig. 1, but foreseen to become quite important, is an operational mode where low velocity singly charged ions are injected from a platform and stored in the ring for studies of photon-ion interactions.

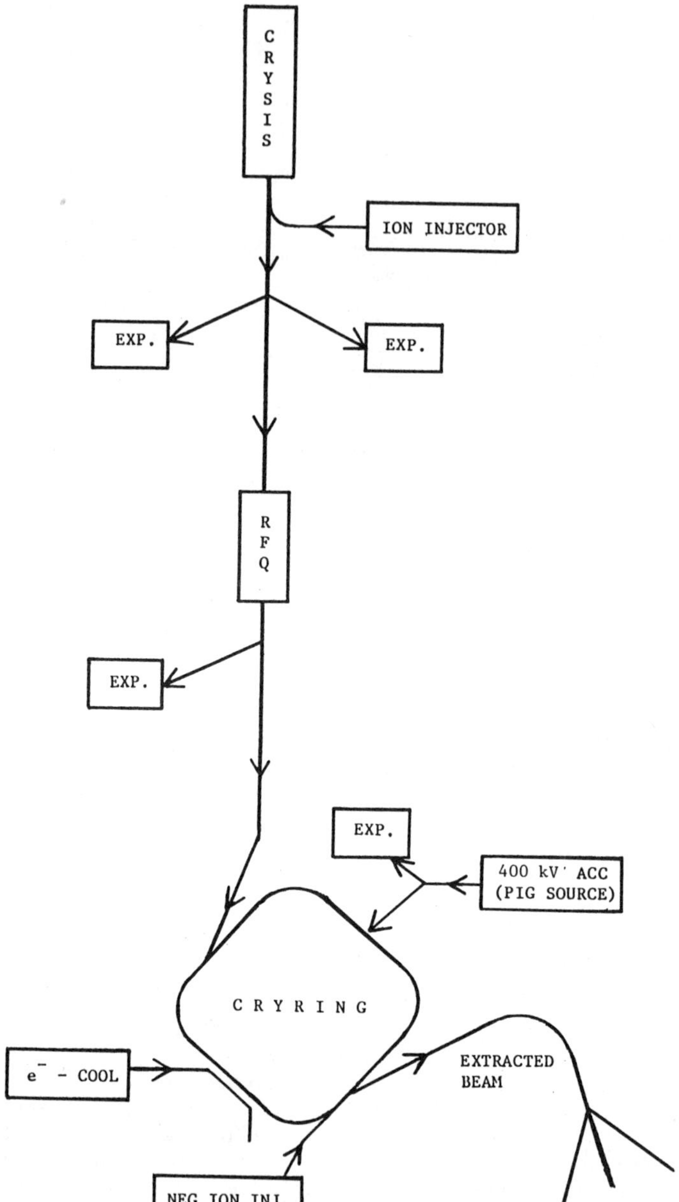

Fig. 1. Schematic outline of the CRYRING facility. Ions emanating from the isotope ion injector are to be stripped in CRYSIS and accelerated in the RFQ. After injection into the ring, the ions may be used in a storage mode for beam-beam physics or accelerated and extracted for fixed target experiments. Deflectors will make possible experiments with slow or moderately accelerated ions.

2. CRYSIS

CRYSIS, shown in Fig. 2, is an electron beam ion source with a maximum 10 keV, 2A electron beam. This beam is compressed in an 1.6 m long superconducting solenoid capable of producing an axial magnetic field of 5 T. Ions are trapped radially by the space-charge potential of the electrons and axially by a potential distribution set up in a series of cylindrical tube electrodes along the magnetic axis. Since at most around $3 \cdot 10^{11}$ electrons are in the confinement volume at the same time, space-charge neutralization shows the theoretical upper limit of the number of positive ion charges that can be trapped to be $\cong 3 \cdot 10^{11}$. Extraction from the source follows through application of a suitable potential distribution, the duration of the extracted pulse being set at 30 μs by the time it takes an ion to drift out of the ring electrode it happens to occupy at the moment of extraction.

A simple estimate of the time τ needed for stripping to a specified charge state q can be derived if one assumes that the physical mechanism is step-by-step direct electron impact ionization, as described by Lotz formula. Denoting the cross section for a one-step charge change from i-1 to i by $\sigma(i-1,i)$ and defining an effective cross section $\sigma(q)$ for production of an ion of charge state q by

$$\frac{1}{\sigma(q)} = \sum_{i=1}^{q} \frac{1}{\sigma(i-1,i)}, \qquad (1)$$

we realise that the effective cross section is dominated by the contributions from the last subshell to be ionized. For completely stripped ions or ions

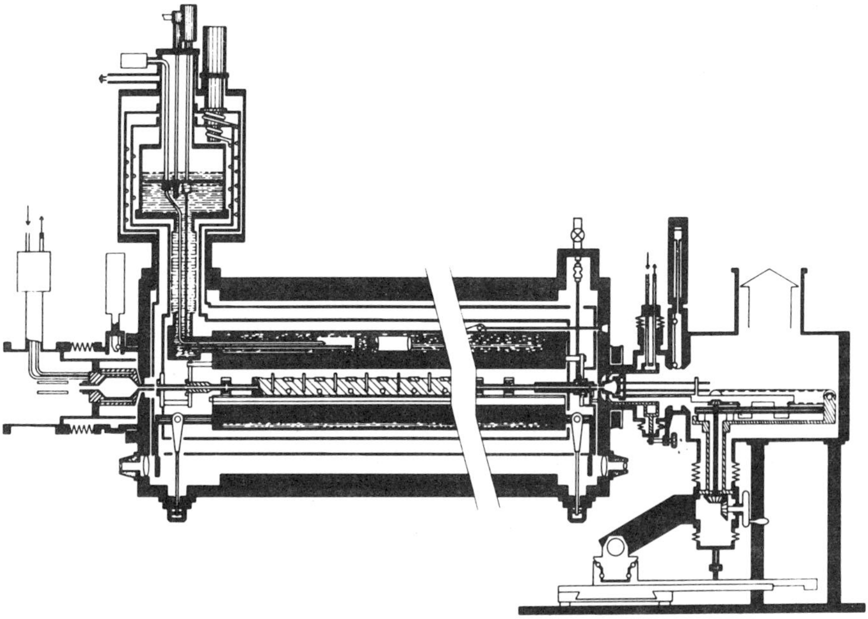

Fig. 2. CRYEBIS II, the twin of CRYSIS, showing the electron gun (right), cryostat with LHe "chimney" and electron collector with ion extraction (left).

ending up with rare-gas electronic configurations, this subshell contains two electrons and since $\sigma(q-2,q-1) \cong 2\,\sigma(q-1,q)$,

$$\sigma(q) \cong \frac{2}{3}\,\sigma(q-1,q). \tag{2}$$

With Lotz formula for the last electron in the form[8]

$$\sigma(q-1,q) = 4.5\cdot 10^{-14}\,\frac{\ln(E/I)}{E(eV)I(eV)}\ \text{cm}^2, \tag{3}$$

where E is the electron impact energy and I the ionization potential, we choose $E \cong 2I$ to be close to the cross section maximum and, using $E \cong 10$ keV, arrive at

$$\sigma(q) \cong 4\cdot 10^{-22}\ \text{cm}^2 \tag{4}$$

for fully stripped Ar (18+), He-like Kr (34+), Ne-like Xe (44+) and Ar-like Pb (64+). The condition for the confinement time τ is

$$j\ \tau\ \sigma(q) \cong 1, \tag{5}$$

where j is the electron current density of the compressed electron beam. This will be 10^3–10^4 A/cm^2 according to estimates[9], where the higher figure derives from an observed supercompression effect[10]. Settling here for $5\cdot 10^3$, we find $j = 3\cdot 10^{22}$ particles/cm$^2\cdot$s, so that

$$\tau \cong \frac{1}{12}\ \text{s} \tag{6}$$

for the particular ions considered. With a cycling frequency of 10 Hz the source will thus be able to deliver $3\cdot 10^{12}$ charges/s, corresponding to an estimated DC current of 0.5 μA. The distribution of different charge states should be quite narrow, and with 50% of the ions in the desired charge state q and an overall efficiency of 10% for the transmission through the whole CRYRING complex, an extracted beam of 25 nA is estimated for nuclear and high energy atomic physics.

CRYSIS will be installed on a 50kV platform to give the necessary ion energy for injection into the RFQ. This linear accelerator, which is a fixed velocity machine, will accept ions with a charge-to-mass ratio q/A between 0.25 and 0.5 at 10 keV/A and accelerate them to 300 keV/A. Particular care will be taken to make the quality of the accelerated beam as good as possible. Details of the energy spread and emittance properties before and after the RFQ are still largely unknown, but estimates of energy spreads $\Delta E = 50q$ eV within the source itself are found in the literature[11].

3. THE RING

The synchrotron ring (Fig. 3) is planned to become a versatile storage and acceleration ring which can be used for several physics programmes. Its circumference will be about 30 m, with bending magnets having a maximal bending power $B\rho$ of 1.35 Tm. According to

$$B\rho = 2.3\cdot 10^{-2}\,\frac{A}{q}\,\frac{v}{v_o}\ \text{Tm}, \tag{7}$$

where v_0 is the atomic unit of velocity, this corresponds, e.g., to fully stripped heavy ions travelling at about 20 v_0 (10 MeV/A). The minimal bending power $B\rho$ which will be usable is an order of magnitude smaller, so that the ring can accomodate fully stripped heavy ions at about 2 v_0 (100 keV/A) or singly charged heavy ions at a few to a few hundred keV total energy. These latter ions will be injected from a separate ion source on a platform.

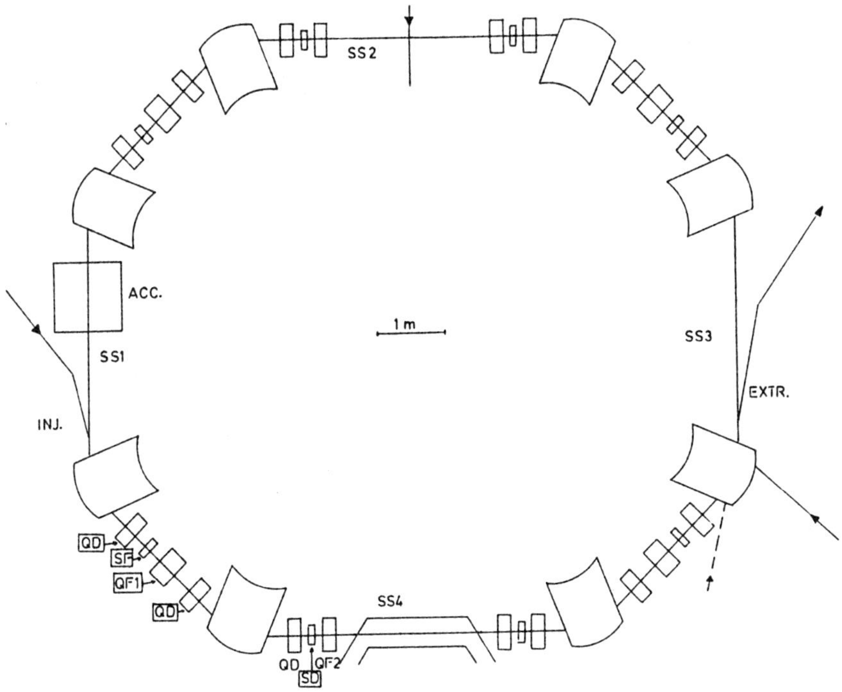

Fig. 3. Lattice of CRYRING showing dipoles, quadrupoles and sextupoles. Injection and extraction are governed by electrostatic deflectors and acceleration by an accelerating cavity. Electron cooling will be used and experiments with crossed and merged ion beams performed.

For the highly charged ions vacuum conditions become stringent, in particular at the lowest velocities. An estimated cross section for electron capture from molecular hydrogen by a slow ion of charge state q is[12]

$$\sigma(q,q-1) \cong 5 \cdot 10^{-16} q \quad cm^2. \tag{8}$$

With an ion velocity of 2 v_0, the half-life of the beam is only 3 s even for rest-gas pressures as low as 10^{-12} torr. Increasing the velocity and using an interpolated electron capture cross section (again from H_2)

$$\sigma(q,q-1) \cong 1.5 \cdot 10^{-16} \; q^3 \; (v_0/v)^7 \; cm^2, \tag{9}$$

shows that already at $3v_0$ the half-life has increased to about 60 s for the same rest-gas pressure. If v is further increased to 14 v_0 (5 MeV/A), the half-life becomes of order 60 min even if the pressure is only 10^{-11} torr. For singly charged ions at keV energies, destructive cross sections are of order 10^{-16} cm^2, giving half-lifes in the region of several minutes for an assumed pressure of 10^{-11} torr.

To fill the ring a pulse of ions is extracted from CRYSIS, accelerated by the RFQ and injected into the ring. Since the pulse length after acceleration will be around 200 m, a multiturn injection scheme has to be used. A typical CRYSIS pulse contains $5 \cdot 10^9$ ions of charge state q ≅ 30, which means that a circulating current of 5 mA will be present in the ring. (The revolution frequency is $2 \cdot 10^5$ Hz at 300 keV/A.) For the ring to be able to hold such a high current without exciting destructive instabilities the emittance and momentum spread both have to be large, $\varepsilon \cong 100\pi$ mm mrad and $\Delta p/p \cong 1\%$. With only $5 \cdot 10^5$ ions of charge state $q \cong 30$, decelerated to 100 keV/A, the limits set by instabilities are much more favourable, $\varepsilon \cong 0.1\pi$ mm mrad and $\Delta p/p \cong 0.01\%$. To actually reach this kind of beam quality some kind of mechanism reducing the phase space density of the injected beam has to be used. For CRYRING the choice has been to use electron cooling, which will be discussed in the next section.

In the synchrotron mode pulses from CRYSIS will be accelerated in the ring and extracted with a 10 Hz repetition frequency. To increase the duty cycle, a slow 3:rd order resonance extraction with pulse lengths $\cong$ 30 ms will be used (as compared to 30 μs for the CRYSIS pulse and $\cong$ 1 μs for a one-turn fast extraction). This puts rigid restrictions on the magnet design and the magnets (including power supplies) will be the most expensive part of the whole CRYRING facility.

4. ELECTRON COOLING

Ways of improving the qualities of charged-particle beams in storage rings have become of considerable interest during recent years[13]. The fundamental quality parameter which describes the beam is the phase-space density. In an idealised situation this density is conserved (Liouville´s theorem), but in actual cases it decreases due to diffusion-like heating processes such as intrabeam scattering. (This particular process tends to convert the directed translational motion of the ions into an irregular one.) It may also turn out that the phase-space density of the beam is too low already at injection and needs improvement to reduce emittance and momentum spread for certain kinds of experiments[14]. Both these reasons make cooling of the beam needed in the case of CRYRING.

The traditional way to make a high-quality ion beam is to accelerate, introduce a dispersive element and use slit collimation. After acceleration in a conservative field, the ions will be described by an anisotropic flattened velocity distribution characterized by a longitudinal temperature which has decreased due to the velocity compression effect[15]. This follows from the conservation of the energy spread during adiabatic acceleration, which implies that

$$\Delta E \cong m\, v_i\, \delta v_i \cong m\, v_f\, \delta v_f . \tag{10}$$

Applying this relation in the longitudinal direction immediately explains the velocity compression, while its application in the transverse direction shows the transverse temperature to be essentially conserved. If the ions in the source have temperature kT and the accelerated ions have energy E, the longitudinal temperature after acceleration becomes

$$kT_{long} \cong \frac{kT}{E}\, kT, \tag{11}$$

while the transverse temperature is

$$kT_{trans} \cong kT. \tag{12}$$

To reduce this temperature further in a non-destructive way, three different methods have been proposed: Stochastic cooling[14], electron cooling[13] and laser cooling[16]. While stochastic cooling is very effective for hot ion beams and laser cooling might prove useful for low charge state ions, the most powerful way of cooling the CRYRING beam seems to be electron cooling.

In electron cooling the ion beam is merged with an intense monoenergetic electron beam over a straight section of the circumference of the storage ring, as schematically shown in Fig. 4. By choosing the electron velocity equal to the mean ion velocity a situation is realised, which in the common rest frame of the two beams corresponds to a repeated mixing of the hot ion gas with a cold electron gas. This electron gas will be highly anisotropic by the velocity compression effect. Suppose, e.g., that the electrons emanate from a cathode with a temperature kT = 0.1 eV (≅ 850°C). After 50 V acceleration the longitudinal and transverse temperatures, by Eqs. (11) and (12), will be kT(long) ≅ 200 μeV and kT(trans) ≅ 0.1 eV respectively! To a first approximation (neglecting intrabeam scattering) we may consider longitudinal and transverse cooling as independent processes. It then turns out that the transverse one in general is most restrictive and we will here only consider transverse cooling.

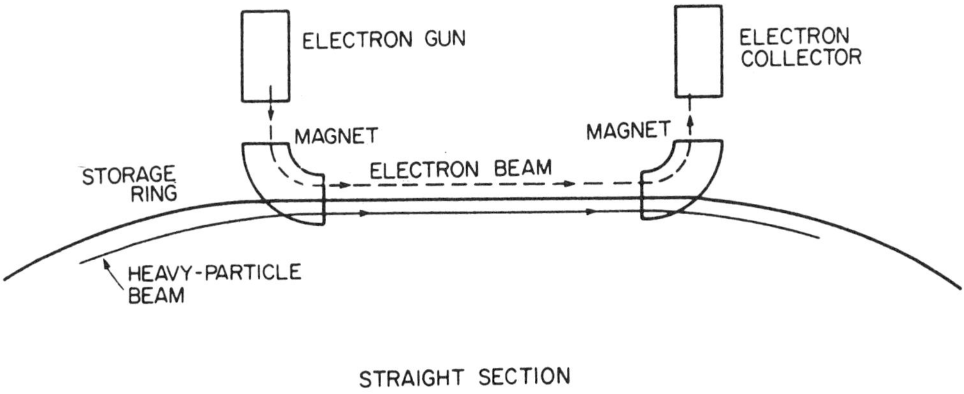

Fig. 4. Schematic principle of electron cooling. Electrons emanating from the gun are guided into the ring by a toroidal magnetic field and merged with the ion beam over a certain length. After being bent out of the ring the electrons end up at the collector. Hot ions interact repeatedly with cold electrons and are thus cooled. (From Ref. 13.)

In a simplified non-relativistic calculation of transverse cooling[17] for the important special case that the ion beam divergence θ(i) already initially is smaller than the corresponding electron beam divergence θ(e), the characteristic cooling time constant is determined as (SI units)

$$\tau_c = \frac{3e}{8(2\pi)^{1/2}} \frac{\beta^4 \theta_e^3}{\eta\, j_e L_c r_e r_i} . \tag{13}$$

Here e is the electron charge, β = v/c is the common velocity, η is the fraction of the ring circumference occupied by the cooling device, j(e) the electron current density, L(c) the Coulomb logarithm (≅ 10) and r(e), r(i) the classical electron and ion radii, respectively. Assuming a constant perveance electron gun ($I \propto V^{3/2}$) with cathode temperature kT ≅ 0.1 eV and current density j(e) = 1 A/m^2 at electron energy 50 eV, the electron beam divergence will be θ(e) = 50 mrad. For a typical ion like Kr(34+) the cooling time will then be 2 s at the equivalent ion energy 100 keV/A. Scaling properties of the cooling time includes

$$\tau_c \propto \frac{A}{q^2} \frac{1}{v^2} \tag{14}$$

showing that electron cooling is more favourable for high charge states and fast ions.

The condition that the two transverse temperatures equalize can be expressed as

$$\frac{1}{2} m_i \theta_i^2 = \frac{1}{2} m_e \theta_e^2 , \tag{15}$$

and for Kr(34+) this would lead to a final divergence θ(i) = 0.1 mrad at 100 keV/A. Knowing the focusing properties of the ring then makes it possible to estimate the ion beam emittance ε ($\propto \theta(i)^2$) as 0.01π mm mrad. But as we saw earlier (Section 3), this emittance is so small that destructive instabilities may set in, even for only $5 \cdot 10^5$ ions in the ring. A more realistic analysis shows, though, that intrabeam scattering will act as a heating mechanism on the beam, leading to an equilibrium between heating and cooling characterised by larger divergences and emittances. Typical realistic final values seem to be around θ(i) = 1 mrad and ε = 1π mm mrad for $5 \cdot 10^5$ ions at 100 keV/A [18].

A similar discussion concerning the equalization of the two longitudinal temperatures at first points at a final momentum spread $\Delta p(\text{long})/p \cong 10^{-6}$, which again might excite instabilities, but through intrabeam scattering the equilibrium value seems to be around 10^{-3} for $5 \cdot 10^5$ ions in the ring at 100 keV/A. Starting out with this realistic beam at the lowest velocity, the maximum number of ions increases while the minimum momentum spread and minimum emittance decreases at higher velocities.

5. EXPERIMENTAL PROGRAM

There will be at least five different areas of experimental activity at the CRYRING facility (Fig. 5):

5.1 Physics at CRYSIS

This program presently includes studies of electron capture in slow collisions of highly charged ions with neutral atoms [19], studies of the interaction of slow highly charged ions with surfaces [20] and spectroscopy of highly charged ions.

5.2 Physics after the RFQ

Even though the RFQ mainly will be used to accelerate ions to the energy needed for injection into the ring, also an experimental program concerning inner shell atomic collisions, beam-foil spectroscopy or solid-state interactions is foreseen.

5.3 Physics with the ion beam(s) that also will be used for crossed/merged beams

A program of beam-foil spectroscopy at the Institute [21] will be continued and extended to include also laser physics with accelerated atomic or molecular ions.

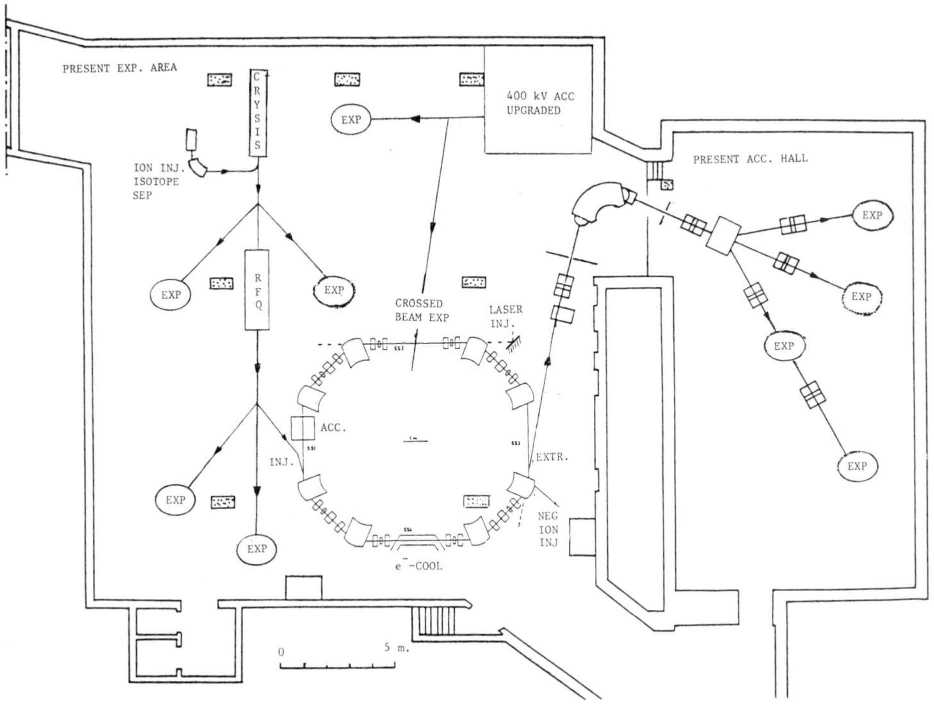

Fig. 5. Tentative layout of CRYRING facility in existing laboratory area. Experimental activities are marked.

5.4 Physics using the accelerated and extracted beam

Here a continuation and extension of the nuclear physics research presently performed at the Institute[21] will be undertaken. In particular use will be made of a newly constructed 4π detector NORDBALL. Also beam-foil spectroscopy and solid state interactions are discussed.

5.5 Physics using the stored beam

At present mainly atomic physics experiments are planned with the stored ion beam. Generally the storage ring can be used to:

Increase the current of rare or expensive ions by re-circulation.

Increase the beam quality by phase-space manipulations such as electron cooling.

Increase the beam purity by letting unwanted metastable excitations decay.

Increase the interaction time with a particular ion in a non-destructive situation (such as laser spectroscopy) by having the ion return again and again.

Out of a long list of possible experimental programs it seems today that electron-ion interactions will be studied in the electron cooler, photon-ion interactions in a colinear geometry using a c.w. dye laser and ion-ion interactions in crossed or merged beams. Details of these programs are at present being worked out.

REFERENCES

1. M. Eriksson, Nucl. Instr. Meth. 196, 331 (1982).
2. D. Reistad and A. Johansson in "Nuclear Physics with Stored, Cooled Beams", P. Schwandt and H. O. Meyer, eds. (American Institute of Physics, New York, 1985).
3. C. J. Herrlander and A. Bárány in "Proceedings of the Workshop on Electron Cooling and Related Applications", H. Poth, ed., Kernforschungszentrum Karlsruhe, Report KfK 3846 (1984).
4. C. J. Herrlander, L. Bagge, A. Bárány, S. Borg, H. Danared, P. Heikkinen, S. Hultberg, L. Liljeby, and Th. Lindblad, IEEE Trans. Nucl. Sci. NS-32, 2718 (1985).
5. P. Heikkinen, IEEE Trans. Nucl. Sci. NS-32, 2715 (1985).
6. J. Arianer and R. Geller, Ann. Rev. Nucl. Part. Sci. 31, 19 (1981).
7. H. Klein, IEEE Trans. Nucl. Sci. NS-30, 3313 (1983).
8. C. F. Barnett in "Physics Vade Mecum", H. L. Andersson, ed. (American Institute of Physics, New York, 1981).
9. J. Arianer, C. Collart, Ch. Goldstein, H. Laurent, and M. Malard, Physica Scripta T3, 35 (1983).
10. I. G. Brown and B. Feinberg, Nucl. Instr. Meth. 220, 251 (1984).
11. J. Arianer, Nucl. Instr. Meth. B9, 516 (1985).
12. R. K. Janev, L. P. Presnyakov and V. P. Shevelko, "Physics of Highly Charged Ions", (Springer-Verlag, Berlin-Heidelberg, 1985).
13. F. T. Cole and F. Mills, Ann. Rev. Nucl. Part. Sci. 31, 295 (1981).
14. S. van der Meer, Science 230, 900 (1985).
15. S. L. Kaufman, Opt. Comm. 17, 309 (1976).
16. O. Poulsen in "Proceedings of the 6:th General Conference of the European Physical Society", J. Janta and J. Pantoflicek, eds. (European Physical Society, Prague, 1984).
17. L. Spitzer, "Physics of Fully Ionized Gases", (Wiley, New York, 1962).
18. P. Heikkinen, private communication.
19. G. Astner, A. Bárány, H. Cederquist, H. Danared, S. Huldt, P. Hvelplund, A. Johnson, H. Knudsen, L. Liljeby and K. G. Rensfelt, J. Phys. B: Atom. Molec. Phys. 17, L877 (1984).
20. T. Fried and M. at Ugglas, "SMILE: Surface-Multicharged ions Interactions at Low Energies", Research Institute of Physics, Stockholm, Report RIPS 83:4 (1983).
21. "AFI Annual Report 1982-83", (Research Institute of Physics, Stockholm, 1985).

ELECTRON IMPACT IONISATION OF ATOMIC HYDROGEN AND HELIUM

Bernard Piraux

Blackett Laboratory, Imperial College
London SW7 2BZ
England

INTRODUCTION

The electron-impact ionisation of atoms is one of the most important and challenging areas of research in the field of atomic collisions. It presents to the theorist the difficulties of a many-body problem coupled with several delicate features related to the infinite range of the Coulomb interaction. It is of practical significance also in fields such as plasma physics and astrophysics where a deep understanding of the reaction mechanisms together with a knowledge of the ionisation rates for many atoms is required.

The experimental work concerning this process was restricted for a long time to the analysis of the energy dependence of the total cross section; the systematic study of (e^-, $2e^-$) processes started in 1969 with the first measurements by Ehrhardt et al[1] of the triple differential cross section (T.D.C.S.) in the case of the electron impact ionisation of helium. Such coïncidence experiments in which the kinematics is fully determined provide the most detailed information available about the (e^-,$2e^-$) reactions. Since that time, many experiments of the same kind have been carried out on various atomic targets for a wide range of incident energies. This large amount of data, in contrast to electron impact ionisation of ions for which differential cross sections are extremely difficult to measure[2],stimulated the development of very sophisticated theoretical models. Two different energy regimes have been particularly explored : the very low energy region just above threshold and the high energy domain.

In this contribution we shall review the main points of these theories and attempt to extract the most important mechanisms which govern the dynamics of these (e^-,$2e^-$) processes, confining ourselves to the analysis of the electron impact ionisation of atomic hydrogen and helium.

Section 1 is devoted to a brief discussion of the general theory of the ionisation of atoms by electron impact.

In section 2, we analyse the threshold laws. Two different approaches are considered : the classical Wannier theory[3] and the quantum mechanical approach of Temkin[4]. Their results are compared with the experimental data.

In section 3 we study the high energy domain (corresponding to incident energies greater than ten times the ionization potential). In particular, we consider the T.D.C.S. in the case of coplanar symmetric and asymmetric kinematics for which important progress have been made recently[5].

1. BRIEF REVIEW OF THE GENERAL THEORY OF ELECTRON IMPACT IONISATION OF ATOM

1.1. The triple differential cross section

The most sensitive probe of the theory of single ionization by electron impact is provided by the measurements of the triple differential cross section (T.D.C.S.) in a $(e^-,2e^-)$ coïncidence experiment. The T.D.C.S. is a measure of the probability that in a single ionization reaction an incident electron of energy E_0 and momentum $\underset{\sim}{k}_0$ will produce on collision with the target two electrons having energies E_A and E_B and momenta $\underset{\sim}{k}_A$ and $\underset{\sim}{k}_B$ emitted respectively into the solid angles $d\Omega_A$ and $d\Omega_B$ centred about the directions (θ_A,ϕ_A) and (θ_B,ϕ_B) as shown on fig. 1.

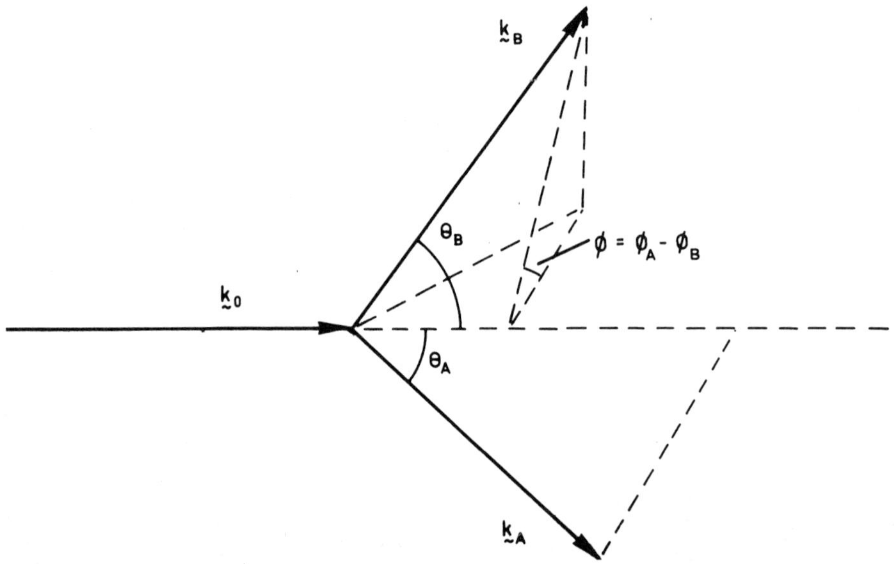

Fig. 1. The kinematics of a $(e^-,2e^-)$ reaction. k_0 is the incident electron momentum and the outgoing momenta are $\underset{\sim}{k}_A$ and $\underset{\sim}{k}_B$ also shown are the angles θ_A,θ_B and $\phi = \phi_A-\phi_B$ (from[5]).

For convenience, one usually call the electron A, the "scattered" electron with $k_A \geqslant k_B$ while electron B is called the "ejected electron". The momentum and energy conservation laws can be expressed respectively as follows :

$$\underset{\sim}{k}_0 = \underset{\sim}{k}_A + \underset{\sim}{k}_B + \underset{\sim}{Q} \qquad (1)$$

$$E_0 - (W_f^{A^+} - W_i^A) = E_A + E_B + T \qquad (2)$$

where $\underset{\sim}{Q}$ is the recoil momentum of the ion A^+; W_i^A, W_f^{A+} are respectively the internal energy of the target atom A in the initial state and of the ion A^+ in the final state. $T = Q^2/2M$ is the (small) kinetic energy of the recoiling ion of mass M.

For an unpolarized e^--H system, the T.D.C.S. is given by[6a,7] :

$$\frac{d^3}{d\Omega_A d\Omega_B dE} = \frac{k_A k_B}{k_o} \{\frac{1}{2}\ |f|^2 + \frac{1}{2}|g|^2 + \frac{1}{2}|f-g|^2\} \qquad (3)$$

$$= \frac{k_A k_B}{k_o} \{|f|^2 + |g|^2 - \mathrm{Re}(f^\star g)\} \qquad (4)$$

where $E \equiv E_A$ or E_B and f,g are respectively the direct and the exchange amplitudes. It is important to note that we do not take into account the spin interaction so that the exchange effects are a consequence of the Pauli principle. The exchange amplitude g describes a process in which the outgoing electrons are exchanged in the final channel. This amplitude is related to the direct amplitude f by the Peterkop theorem[8]

$$g(\underset{\sim}{k}_A \underset{\sim}{k}_B) = f(\underset{\sim}{k}_B \underset{\sim}{k}_A) \qquad (5)$$

which is not a surprising result since there are no real distinction between the direct and the exchange process as is the case for the elastic and inelastic collisions[9].

When the incident and the target electrons are polarized and the spin of both escaping electrons observed, one can measure the quantity $|f|^2 + |g|^2$ when the spins of both the incident and the target electrons are antiparallel and $|f-g|^2$ if they are parallel. However, in contrast to elastic or inelastic collisions[10], it is impossible to "isolate" a pure exchange or a pure direct process so that $|f|$, $|g|$ and the relative phase Φ between f and g cannot be known simultaneously[11].

Since it is possible to measure separately both the triplet and the singlet ionization T.D.C.S. it is interesting to introduce the so-called spin asymmetry parameter A defined as follows

$$A = \frac{(|f|^2+|g|^2)-|f-g|^2}{|f|^2+|g|^2+|f-g|^2} \qquad (6)$$

$$= \frac{\mathrm{Re}[f^\star g]}{\frac{1}{2}(|f|^2+|g|^2+|f-g|^2)} \qquad (7)$$

This parameter A provides information on the relative phase Φ and allows stringent tests of the theoretical models. However, because of the extreme complexity of its measurement, one usually defines A in terms of triplet and singlet total cross section[12]. A discussion of the results will be presented in section 2.

In the case of complex targets (more than one target electron) we need to consider a capture amplitude w describing a process in which the incident electron is captured into a bound state while two initially bound electrons are ejected[13]. In general, the amplitude w is very small compared to both amplitude f and g[14] so that equation (3) still holds.

1.2. Basic equations for the amplitudes

Let $\underset{\sim}{r}_1$, $\underset{\sim}{r}_2$ be respectively the coordinates of the incident and the ejected electron and let us denote by x the coordinates of the other bound electrons. According to Peterkop[15], Rudge and Seaton[13,16], an integral representation for the direct scattering amplitude corresponding to the electron impact ionisation of the target A is given by :

$$f(\mathbf{k}_A,\mathbf{k}_B) = -(2\pi)^2 \exp(i\Delta(\mathbf{k}_A,\mathbf{k}_B)) < \phi_f(\mathbf{r}_1,\mathbf{r}_2)\phi_{A^+}(x)|H-E_S|\Psi_i^{(+)}(\mathbf{r}_1,\mathbf{r}_2,x) > \tag{8}$$

H is the Hamiltonian of the e^--A system and E_S $(= E_o + W_i^A)$ its total energy; $\Psi_i^{(+)}(\mathbf{r}_1,\mathbf{r}_2,x)$ is an exact scattering wave function with outgoing (+) spherical wave boundary conditions. The wave function $\phi_f(\mathbf{r}_1,\mathbf{r}_2)$ describes in the asymptotic region the outgoing electrons and $\phi_{A^+}(x)$ is the wave function associated with the residual ion A^+; $\Delta(\mathbf{k}_A,\mathbf{k}_B)$ is a logarithmic phase due to the long range of the Coulomb potentials.

In many calculations, $\Psi_i^{(+)}(\mathbf{r}_1,\mathbf{r}_2,x)$ is replaced by its first Born approximation :

$$\Psi_i^{(+)}(\mathbf{r}_1,\mathbf{r}_2,x) \cong (2\pi)^{-3/2} \exp(i\mathbf{k}_o\cdot\mathbf{r}_1)\phi_A(\mathbf{r}_2,x) \tag{9}$$

where $\phi_A(\mathbf{r}_2,x)$ is the initial target state wave function. This treatment can be improved by replacing the incident plane wave by a distorted wave $\chi_{\mathbf{k}_o}(\mathbf{r}_1)$ which takes into account the complicated interaction of the incident electron with the target[17-21]. Recently Byron et al[22] evaluated $\Psi_i^{(+)}(\mathbf{r}_1,\mathbf{r}_2,x)$ by means of its second Born approximation in the case of the electron impact ionization of atomic hydrogen and helium; this more sophisticated model will be discussed in section 3.

In the final channel, one usually assumes that electron A moves in a Coulomb field $-Z_A/r_1$ and electron B in a Coulomb field $-Z_B/r_2$, Z_A and Z_B being effective charges depending on both $\mathbf{k}_A$ and $\mathbf{k}_B$[13,15,16]. Hence, we have :

$$\phi_f(\mathbf{r}_1,\mathbf{r}_2) = \psi^{(-)}_{c,\mathbf{k}_A}(Z_A,\mathbf{r}_1)\;\psi^{(-)}_{c,\mathbf{k}_B}(Z_B,\mathbf{r}_2) \tag{10}$$

where $\psi^{(-)}_{c,\mathbf{k}}(Z,\mathbf{r})$ is a Coulomb wave function with incoming spherical wave behaviour corresponding to momentum $\mathbf{k}$ and charge Z. The effective charges satisfy the asymptotic relation

$$\frac{Z_A(\mathbf{k}_A,\mathbf{k}_B)}{k_A} + \frac{Z_B(\mathbf{k}_A,\mathbf{k}_B)}{k_B} = \frac{1}{k_A} + \frac{1}{k_B} - \frac{1}{|\mathbf{k}_A-\mathbf{k}_B|} \tag{11}$$

and the phase is given by

$$\Delta(\mathbf{k}_A,\mathbf{k}_B) = \frac{Z_A(\mathbf{k}_A,\mathbf{k}_B)}{k_A}\,\ell n\left[\frac{k_A^2}{k_A^2+k_B^2}\right] + \frac{Z_B(\mathbf{k}_A,\mathbf{k}_B)}{k_B}\,\ell n\left[\frac{k_B^2}{k_A^2+k_B^2}\right] \tag{12}$$

It must be emphasized that this approach does not solve the delicate problem of the determination of the wave function $\phi_f(\mathbf{r}_1,\mathbf{r}_2)$ since we have only equation (11) to evaluate the effective charges Z_A and Z_B. Different choices of this charges will be discussed in section 3. Let us note that this ambiguity concerning Z_A and Z_B also occurs in the case of electron impact ionization of multicharged ions[23] but is less crucial since in that case, the interaction between the ion and the escaping electrons is dominant.

If we take the exchange effects into account, the problem of the determination of Z_A, Z_B and the phase Δ casts doubt on the validity of the relation (5) which is true for the exact amplitudes only. The procedure usually adopted consists to "relax" the condition (5), apply the Peterkop theorem to the moduli of the amplitudes and introduce a relative phase $\tau(\underset{\sim}{k}_A,\underset{\sim}{k}_B)$ such that

$$g(\underset{\sim}{k}_A,\underset{\sim}{k}_B) = \exp[i\tau(\underset{\sim}{k}_A,\underset{\sim}{k}_B)]f(\underset{\sim}{k}_B,\underset{\sim}{k}_A). \tag{13}$$

The relative phase $\tau(\underset{\sim}{k}_A,\underset{\sim}{k}_B)$ is totally arbitrary; it is usually chosen (without physical justification) so that the interference term present in the right hand side of equation (4) is maximum; this is done after integrating the T.D.C.S. over all the angles and the energy of one of the electrons[23].

2. THRESHOLD LAWS

2.1. Preliminary remarks

In the case of electron impact ionization of atom, the threshold laws or the energy dependence of the cross-sections just above ionization threshold provide information on the dynamics of two electrons escaping in the field of a positive ion. Wigner[24] pointed out that the threshold behaviour is a characteristic of the escape configuration and does not involve details on what goes on in the "reaction zone" where the particles are close together and interact strongly. This consideration leads to two important conclusions :

(i) the threshold laws can be worked out in the absence of a full quantum mechanical solution inside the reaction zone. It requires however some assumptions about the phase-space distribution of the system at the exit of the reaction zone.

(ii) the fact that the details of the dynamics of the system in the reaction zone is irrelevant in determining the threshold behaviour implies that the threshold law is the same for any process leading to a two electron escape in the field of a positive ion. The same law is therefore applicable to two electron photoionization of negative ions and atoms as well as electron impact ionization of neutrals and positive ions.

We shall now present the classical Wannier theory and the Temkin approach. We shall focus on the main points of these theoretical treatments avoiding the complicated mathematical details given for example in[4,25-29].

2.2. Wannier theory

The concept of dynamic screening

Let us start this section by outlining the interesting discussion of the dynamic screening as given by Rau[3,30]. At any instant of the escape process, the energy partition between the electrons determines their mutual screening which in turn leads to energy and angular momentum exchanges and thereby affects the configuration at the next instant. In other words, the effective charges "seen" by each escaping electron is now dependent on the radial distances of the electrons; this consideration leads to two important consequences :

(i) the dynamic screening gives rise to an inherent instability of the double escape process. Let us suppose that at a given instant, $E_A \neq E_B$; then the slower electron will stay closer to the ion and screen more of its charge as seen by the other. The faster electron will therefore gain energy at the expense of the other.

Thus, any initial discrepancy in the energies will be enhanced as the escape proceeds and given the long time involved for escape, the probability that the slower electron ends up in a bound state is important.

(ii) the static picture which consists to define two effective charges Z_A and Z_B independent of the radial distances of the electrons and to describe both electrons by a product of Coulomb wave functions (see equations 10 and 11) over the entire escape process is totally inadequate. However, this model can be used to derive the threshold behaviour of the total cross section σ in two limiting cases; when there is no screening (Z_A=Z_B=Z the charge of the residual ion) we get by means of equation 8 (see also[31]) $E_T = E_o - (W_f^{A^+} - W_i^A)$

$$\sigma \underset{E_T \to o}{\sim} E_T \tag{14a}$$

in the case of full screening (Z_A=0, Z_B=Z) we have

$$\sigma \underset{E_T \to o}{\sim} E_T^{3/2} \tag{14b}$$

In order to understand more deeply how the dynamic screening affects the radial and angular correlation, it is useful to introduce the hyperspherical coordinates $\rho, \alpha, \theta_{12}$ ($o \leq \alpha \leq \pi/2$, $o \leq \theta_{12} \leq 2\pi$) defined as follows :

$$\rho = (r_1^2 + r_2^2)^{1/2}, \tag{15}$$

$$\alpha = \tan^{-1} (r_2/r_1), \tag{16}$$

$$\theta_{12} = \cos^{-1} (\hat{r}_1 \cdot \hat{r}_2); \tag{17}$$

ρ describes the "size" of the system and α, θ_{12} respectively the radial and the angular correlations[32]. In this system of coordinates, the potential "felt" by the electrons in the field of the ion of charge Z takes the following form (in atomic units)

$$V = - \frac{Z}{r_1} - \frac{Z}{r_2} + \frac{1}{r_{12}} = (- \frac{Z}{\rho})\, C(\alpha, \theta_{12}) \tag{18}$$

Figure 2 shows the potential term C as a function of α and θ_{12}.

This well-known potential surface exhibits a saddle point around $\alpha = \pi/4$ and $\theta_{12} = \pi$; this reflects :

(i) the inherent instability associated to the radial correlation : α must stay close to $\pi/4$ ($r_2/r_1 \sim 1$) for a large range of values of ρ in other for double escape to result otherwise, one electron could fall in one of the two deep valleys (α=o or $\pi/2$) and stay close to the nucleus. This unstable behaviour therefore acts to reduce the ionisation cross section close to threshold.

(ii) the inherent stability associated to the angular correlation. This correlation restricts the angle θ_{12} to a cone around $\theta_{12}=\pi$ because of the electrostatic repulsion but do not act to suppress any configuration leading to double escape.

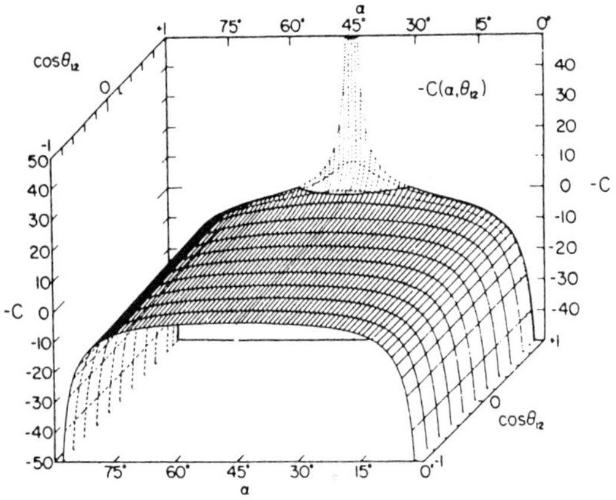

Fig. 2. Dependence of the potential term C on the hyperspherical coordinates α and θ_{12}. One half of the range of variation of θ_{12} (from o to π) is shown (from[32]).

Basic arguments of the Wannier theory

The next step of the Wannier analysis consists to divide the ρ-space in three zones : the reaction zone, the Coulomb zone and the outer zone as shown on Fig. 3. The radius ρ_R of the reaction zone is essentially independent of the total energy E_T of the system and is assumed to be of the order of magnitude of the Bohr radius.

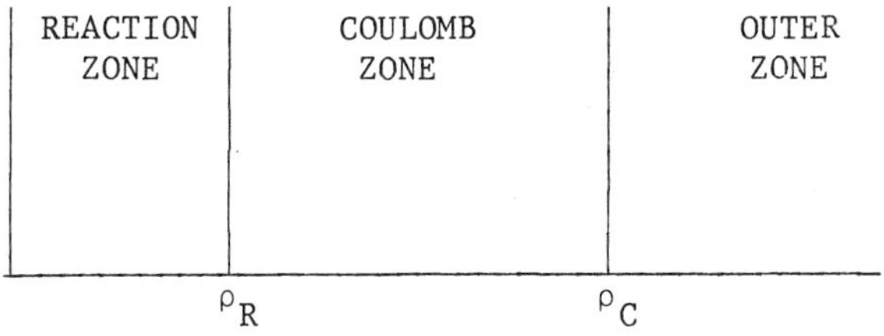

Fig. 3. Division of the ρ space into the reaction, Coulomb and free zones (from[33]).

In the Coulomb zone, the potential energy of the system stays less than the combined kinetic energy of both electrons untill the outer boundary ρ_C where they are of the same magnitude; ρ_C is therefore dependent on the energy E_T and tends to infinity when E_T tends to zero. In the outer zone the electrons move independently of each other. It is important to note that this subdivision is meaningfull only if α is not close to o or π/2 when one electron stays far behind the other one.

Let us now recall the basic assumptionsof the Wannier theory :

1. the total angular momentum L as well as the total spin S of the electrons are assumed to be zero. It can be shown[3,33,34-38] that the inclusion of non zero LS components does not affect the threshold law except when the Pauli principle imposes restrictions.
2. the motion of the escaping electrons is essentially classical outside the reaction zone even if E_T tends to zero. In other words, the wave length of each electron is slowly varying at large values of ρ so

that wave packets can be constructed for each electron and classical mechanics employed. This point is discussed in detail in[25,33].

3. the quasi-ergodic hypothesis : Wannier assumed that the distribution in phase space of both electrons is approximately uniform when they enter the Coulomb zone[25].

Threshold behaviour of the total and differential ionisation cross sections

The threshold behaviour of the total cross section can be derived as follows : first, one expands the potential term $C(\alpha,\theta_{12})$ around the saddle point and solves around this point, the classical equation of motion; then by analysing the dependence on energy of the form of the trajectories in the Coulomb zone, one gets the well known Wannier law[39]

$$\sigma \sim E_T^n \tag{19}$$

where

$$n(Z) = \frac{1}{4}\left[\left(\frac{100Z-9}{4Z-1}\right)^{1/2} -1\right] \tag{20}$$

The Z-dependence on n reflects the fact that the escape process is actually a continuum competition between the attractive force of the ion which is Z-dependent and the inter-electronic repulsion which is independent of Z^3. In the case of electron impact ionization of atoms (Z=1) we get the well known value n=1.127.

We now turn our attention to the threshold behaviour of the differential cross sections starting with the energy partitioning of the electrons. The earlier numerical and theoretical studies of energy partitioning based on classical mechanics and W.K.B. theory[29,40-42] showed that the distribution of the available energy E_T between E_A and E_B is uniform. However, recent numerical calculations by Read[33,43] based on the classical Wannier theory revealed that the distribution in the interval $E_A=0$ to $E_A=E_T$ exhibits a maximum around $E_A/E_T=0.5$ showing that an equipartition of the energies is the most probable (see fig. 7). As pointed out by Read[33], a possible explanation of the non uniformity lies in the definition of ρ_C which is proportional to E_T^{-1} but independent of E_A/E_B (or r_1/r_2); we can expect that this definition is valid over most of the interval $E_A=0$ to $E_A=E_T$ but not near the ends of this interval. If for example $E_B \ll E_T$, the value of r_2 at $\rho=\rho_C$ may be small enough to invalidate the assumption of classical motion. Read[33] showed that when one choses ρ_C larger if E_B/E_T is small, the qualitative effect is a reduction of the ionisation yield which might explain the reduction found in his new results.

Let us now analyse the angular distribution of the electrons. By using a W.K.B. -like treatment, Rau[27,41] obtained a Gaussian distribution

$$\sigma(\theta_{12}) = \exp\left[-a(\pi-\theta_{12})^2 / \sqrt{2E_T}\right] \tag{21}$$

with the full width at half maximum (FWHM) given by

$$\theta_{1/2} = b\, E_T^{1/4}; \tag{22}$$

a and b are Z-dependent; in the case of electron impact ionization of atoms (Z=1), Rau found a=0.34 and, b=3.38 (provided that the energy E_T is expressed in atomic unit). As pointed out by Fano[44], the energy behaviour of the F.W.H.M. reflects tne fact that the angular momenta of both electrons behave as $E_T^{-1/4}$. Other treatments based on W.K.B. theory and classical mechanics exist, all of them agree on the energy dependence of $\theta_{1/2}$ but yield rather different angular distributions; these results are discussed in detail in[33].

To finish this brief review of the Wannier theory, let us mention how the Pauli principle can affect the threshold behaviour. If one applies the Wannier treatment to final states of the electrons characterized by more general values of the quantum numbers S, L and π (the parity) it can be shown[34-37] that in all the cases, except $^3S^e$ and $^1P^e$ the threshold behaviour of the total cross section is the same as for $^1S^e$. In the cases $^3S^e$ and $^1P^e$, the exponent n in equation (19) is larger : instead of 1.127. For electron impact ionization of atoms, one finds 3.881. The origin of this behaviour comes from the fact that the assumption $\alpha \sim \pi/4$ (meaning that $r_1/r_2 \sim 1$) together with the Pauli principle which constrains the electron wave function to be spatially antisymmetric demand that these wave functions have a node at $\alpha=\pi/4$, reducing the number of configurations leading to double escape. This effect is specific to the Wannier theory; unfortunately, it is very difficult to find out an experimental set-up leading to final state $^3S^e$ or $^1P^e$ even in the case of the double detachment by photon impact[3].

2.3. Temkin's theory

The quantum mechanical approach of Temkin[4,45-48] is based on the following assumption : the threshold law is dominated by the part of the interaction in which the inner electron sees the charge of the residual ion directly whereas the outer electron sees the dipole potential caused by the combined effect of the ion and the inner electron so that the full interaction potential given by

$$V = -\frac{1}{r_1} - \frac{1}{r_2} + \frac{1}{r_{12}} \qquad (23)$$

becomes

$$V \simeq -\frac{1}{r_2} + r_2 \cos\theta_{12} \ / \ r_1^2 \qquad (24)$$

assuming that the high order terms do not alter the functional dependence on energy of the total cross section. In addition, Temkin made two simplifications : the first is the restriction to the total S wave and the second is the use of a two dimensional model.

Taking the assumption into account, let us briefly describe Temkin's method to obtain the threshold behaviour; we start with the transition matrix in its "post" form[6b,47]

$$T = \langle \Psi_f^{(-)} | V_i | \phi_i \rangle \qquad (25)$$

where V_i reduces to

$$V_i = r_2/r_1^2 \qquad (26)$$

In the case of electron impact ionisation of atomic hydrogen initially in the 1s state $|\phi_{1s}\rangle$, the unperturbed initial state wave function ϕ_i is given by :

$$\phi_i(r_1,r_2) = j_o(k_o r_1)\phi_{1s}(r_2) \qquad (27)$$

where j_o is the usual spherical Bessel function. The final state wave function $\Psi_f^{(-)}$ is approximated (when $r_1 > r_2$) by

$$\Psi_f^{(-)} \simeq F^{(C)}(r_2)\, F^{(d)}(r_1) \tag{28}$$

where $F^{(c)}(r_2)$ is the zero energy Coulomb wave describing the inner electron and $F^{(d)}(r_2)$ a dipole wave describing the outer electron. After performing the integration of $|T|^2$ on the energy E_B from 0 to E_T, one gets near threshold a modulated linear law :

$$\sigma \sim E_T\ [1 + C \sin(\alpha \ell n E_T + \mu)] \tag{29}$$

where c, α and μ are adjustable parameters. It is important to note that this behaviour is actually the result of a tunneling effect; as pointed out by Temkin[48], there are many more orbits of double escape than can arise from a classical description and it is the increase of the number of these orbits with energy that determines the threshold law.

2.4. Experimental test

Total cross section

The first experimental results[49-51] about the threshold behaviour of the total cross section for electron impact ionization of atom indicated, when fitted by a power law E_T^n, that the exponent n was greater than unity, the most accurate value being[51] 1.16±0.03 consistent with the Wannier law. Later, Cvejanovic and Read[52] obtained very accurate experimental results in the case of electon impact ionization of helium. In a preliminary experiment, they found that the energy distribution of the escape electrons is in good approximation uniform and then measured the yield of electrons having an energy between 0 and 20 meV as a function of the incident energy E_0. Their results are presented on Fig. 4.

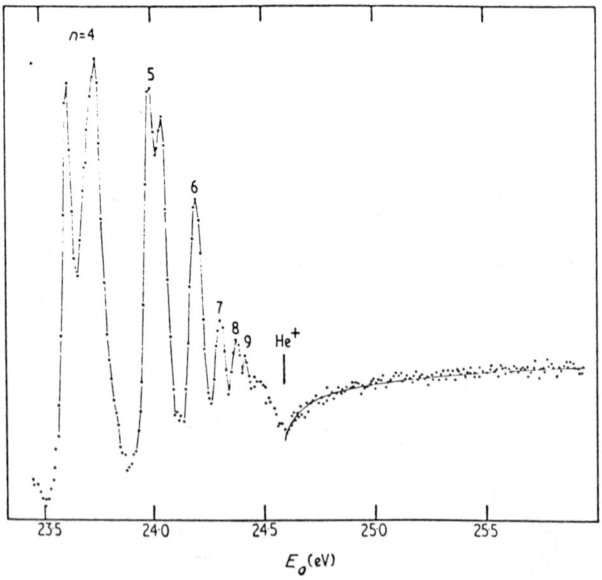

Fig. 4. The yield of very low energy ($\leqslant$ 20 meV) electrons following the electron impact ionization of Helium. The solid curve above the ionization threshold at 24.6 eV is proportional to $E_T^{0.127}$ (from[52]).

In addition, they showed that this partial yield is consistent in the range of E_0 from 0.2 eV to 1.7 eV with a total ionisation cross section having an energy dependence E^n with $n = 1.131 \pm 0.019$. Very recently, Donahue et al.[53,54] measured the total cross section of two electron-photoionization of H^- above the threshold which occurs at a photon energy of 14.352 eV. Their results are presented on Fig. 5 and have been fitted by a power law of the form

$$\sigma \sim A\, E_T^n + B \tag{30}$$

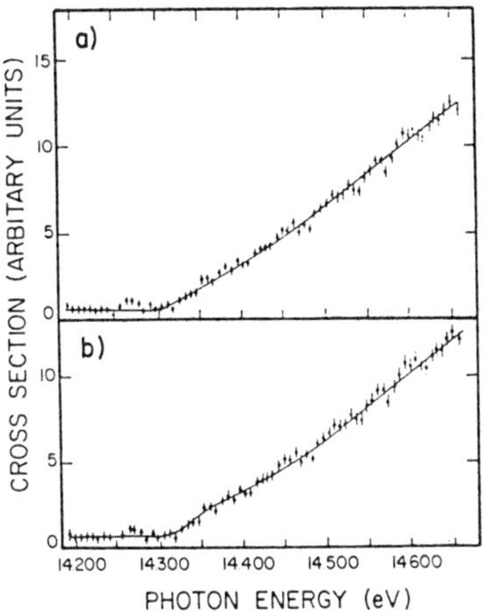

Fig. 5. Cross section for double photodetachment of H^-.
a. The solid curve is the best fit to the power law (30) with $n = 1.15 \pm 0.04$.
b. The solid curve is the best fit to the modulated linear law (31) (from [53]).

with $n = 1.15 \pm 0.04$ and by the Temkin modulated linear law

$$\sigma \sim E_T\,[A + D\,\sin(c\,\ell n E_T + F)] + B \tag{31}$$

unfortunately, because of field ionization background, the results are not accurate enough to distinguish between both laws; let us note in passing that such a distinction requires extremely accurate results since, it is always possible, over a limited range of values of E_T to ajust the three parameter in equation (29) so that both laws are in very close accord[3].

Before finishing this section, let us turn our attention to the asymmetry parameter A defined in terms of total cross sections; this parameter has been measured by using polarized electrons incident on polarized atomic targets[11,12,55-57]. As mentioned before, the Wannier theory predicts the same threshold behaviour for both singlet and triplet mode in most cases so that A must be independent of E_T when E_T tends to zero. This result has been confirmed by all the experiments. We present on Fig. 6 the results obtained by Baum et al.[11] in the case of electron-impact of Li.

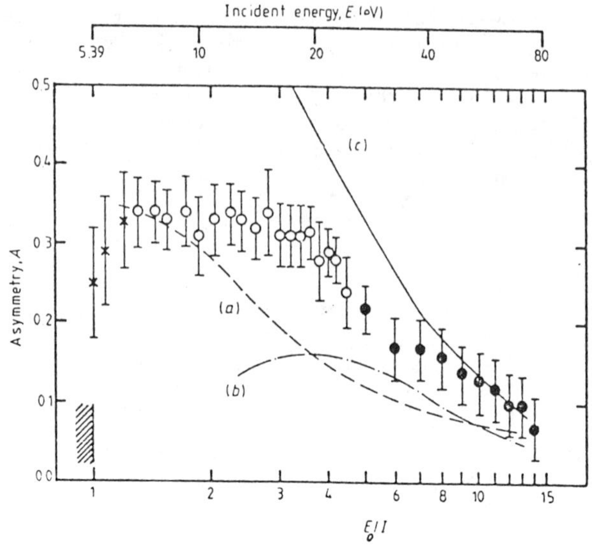

Fig. 6. Asymmetry parameter A in term of total cross section (see section 1.1) for electron impact ionization of Li. The curves are from various non threshold theoretical calculations (from[11]).

Differential cross sections

Let us first examine the energy partitioning of the electrons. The first experiment has been performed by Cvejanovic and Read[52]. As mentioned before, they found that the energy distribution is uniform to within approximately 5%. The next experiments by Pichou et al[58] confirmed the previous result. Recently however, Hammond et al[59] reexamined more accurately the energy partitioning. Their results are presented on Fig. 7 : the function f which is plotted versus E/E_T (E is the energy of one of the escaping electron) is related to the differential cross section $\sigma_{E_T}(E)$ as follows[33]

$$\sigma_{E_T}(E) = E_T^{n-1} \; f(E/E_T) \tag{32}$$

Where n is given by equation (20), these results reveal that the energy

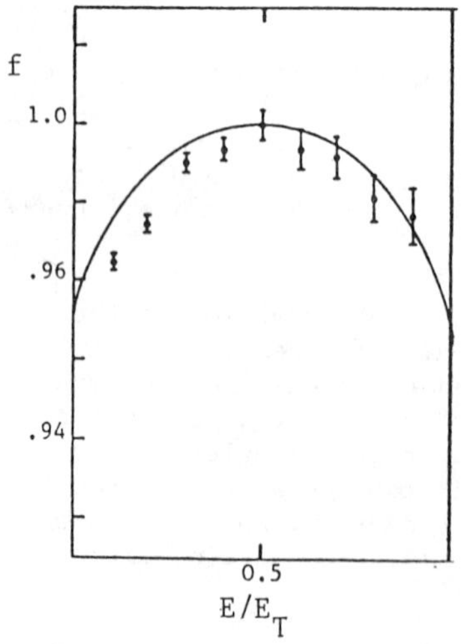

Fig. 7. Experimental and theoretical results for the function f (related to the differential cross section $\sigma_{E_T}(E)$ by equation (32)) as a function of E/E_T. The experimental data have been obtained by Hammond et al[59] and the theoretical data by Read[33] (from[59]).

distribution exhibits a maximum around $E/E_T = 0.5$ in perfect agreement with the theoretical results obtained recently by Read[33].

An other stringent test of the Wannier theory is the study of the T.D.C.S. close to threshold. All the experimental results so far concern the electron impact ionization of helium. The first experiment of this type has been done by Ehrhardt et al[60] and later by Schubert et al[61]. They selected a coplanar kinematics and measured the angular distribution in the range from 30.5 eV to 6 eV above threshold. By comparing their results with the W.K.B. treatment of Rau[27], they concluded that the threshold regime is not yet reached at their lowest energy. In the energy range between 0.2 eV and 3 eV Cvejanović and Read[52] detected both electrons in a plane perpendicular to the incident direction and for two values of their mutual angle (θ_{12}=180° and 150°); their results are consistent with the predictions of the W.K.B.-like theory of Rau for the angular distribution of the escaping electrons except for the angular width which is smaller by a factor two than the theoretical one. Very recently, Fournier-Lagarde et al[62] measured the angular distribution in the energy range from 6 eV to 1 eV above threshold and for a coplanar kinematics. Their results at E_T=1eV and $E_A = E_B = 0.5$ eV are shown in Fig. 8 for six different values of the angle θ_A. It is clear that the L=0 W.K.B. treatment of Rau cannot account for all the experimental features but, as pointed out by Fournier-Lagarde[62], two points should be noted : (i) one observes for θ_A=90° a good agreement with the Rau treatment and (ii) there is a global localization of the second electron around θ_{12}=180° supporting one of the basic assumptions of the Wannier theory. Further investigations also showed[62] that the experimental results presented in Fig. 8 are consistent with the Rau treatment when generalized to include the mixing of (L S π) states up to L=2.

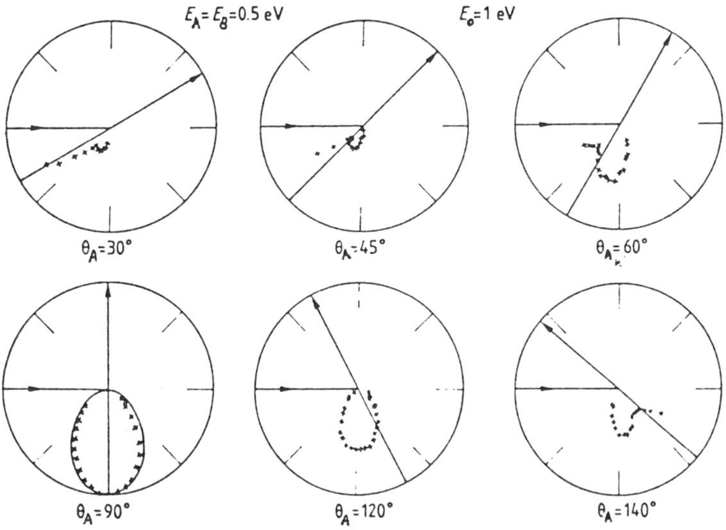

Fig. 8. Polar graph of the T.D.C.S. for E_A=E_B=0.5 eV and six angles θ_A; in each case, θ_B varies from 0° to 360°. The data are normalized together and the radius of each circle represents the maximum observed cross section which is here at θ_A=90° and θ_B=90°. At θ_A=90°, the prediction of the L=0 theory of Rau[27] (solid curve) is presented for comparison with the experimental data (from[62]).

In conclusion it is clear from the previous discussion that all the experimental data are compatible with the Wannier theory. In the case of electron impact ionisation of atoms, it is true if we include the contribution of several (L S π) states. The next stringent test will be a detailed analysis of each (L S π) contribution to the T.D.C.S.

3. HIGH ENERGY REGIME

3.1. Various types of kinematics

The dynamical aspects of the single ionization processes are strongly related to the kinematics selected in a given $(e^-,2e^-)$ coincidence experiment. It is therefore useful to distinguish between several kinematical regimes. In coplanar geometries the momenta $\underset{\sim}{k}_0$, $\underset{\sim}{k}_A$ and $\underset{\sim}{k}_B$ are in the same plane (so that $\phi_A=0$ and $\phi_B=0$ or π) while in non coplanar geometries the vector $\underset{\sim}{k}_B$ is out of the $(\underset{\sim}{k}_0 \underset{\sim}{k}_A)$ reference plane. Another important distinction can be made between asymmetric and symmetric geometries. In asymmetric geometries, for a given (high) incident electron energy E_0, a fast electron A is detected in coïncidence with a slow electron B, both electrons being analysed with respect to energy. As already mentioned, the most extensive series of experiments of this kind have been performed by Ehrhardt et al (see for example[1,63-66] and further references therein) in helium and other noble gases. Ehrhardt et al used typical incident energies E_0 of a few hundred eV and selected a coplanar geometry in which the scattering θ_A of the fast electron is fixed and small (typically < 20°) while the angle θ_B is varied. We shall refer to this kinematical arrangement as the Ehrhardt asymmetric geometry. It is important to note that in this kinematical regime the magnitude of the momentum transfer denoted by $\underset{\sim}{K} = \underset{\sim}{k}_0 - \underset{\sim}{k}_A$ is small. Measurements of the T.D.C.S. for this kinematics have also been made in helium and argon by Lahmam-Bennani et al[67-69] using high incident energy ($E_0 \sim 8$ keV) and in helium by Martino et al[70] in the energy range 500-2000 eV. Very recently, coplanar asymmetric $(e^-,2e^-)$ coïncidence experiments have been carried out in atomic hydrogen by Lohmann et al[71] and by Ehrhardt et al[72]. In the later case, the experimental data are absolute so that they provide the most detailed experimental information to date about the dynamics of $(e^-,2e^-)$ reactions. These results will be analysed in detail in the next section.

Symmetric geometries are such that $\theta_A=\theta_B(=\theta)$ and $E_A=E_B$. The first $(e^-,2e^-)$ coïncidence experiments of this type were performed in 1969 by Amaldi et al[73]. Since then, a large amount of measurements of the same kind have been made, in particular by Camilloni et al[74,75], Weigold et al[76-78], de Heer et al[79-81] and Pochat et al[82]. In contrast to the Ehrhardt asymmetric geometry, the magnitude of the momentum transfer $\underset{\sim}{K}$ is never small in the symmetric case. The magnitude of the vector Q the recoil momentum of the ion ($Q = |2k_A\cos\theta - k_0|$) remains small or moderate for scattering angle $\theta \lesssim 70°$ (with $Q = 0$ when $\theta = 45°$); this angular domain may be called the $(e^-,2e^-)$ spectroscopy region since in this case the momentum density distribution of the target electron can be obtained [74-78] and [83]. For $\theta \geqslant 70°$, both K and Q are large; $(e^-,2e^-)$ coincidence experiments in this angular domain are extremely difficult and have only been performed recently in helium[82]. The results will be discussed in detail in section 3.3.

3.2. Coplanar asymmetric $(e^-,2e^-)$ reactions

In this section we shall discuss the theory of coplanar asymmetric $(e^-,2e^-)$ reactions in the particular case of atomic hydrogen and helium targets. The main features emerging from the experimental data obtained during the last fifteen years[1,63-66] are illustrated in Fig. 9 which shows the recent absolute T.D.C.S. measured by Ehrhardt et al.[72] for the (e^-,2e^-) reaction in atomic hydrogen for the case E_0=250 eV, E_B=5 eV and

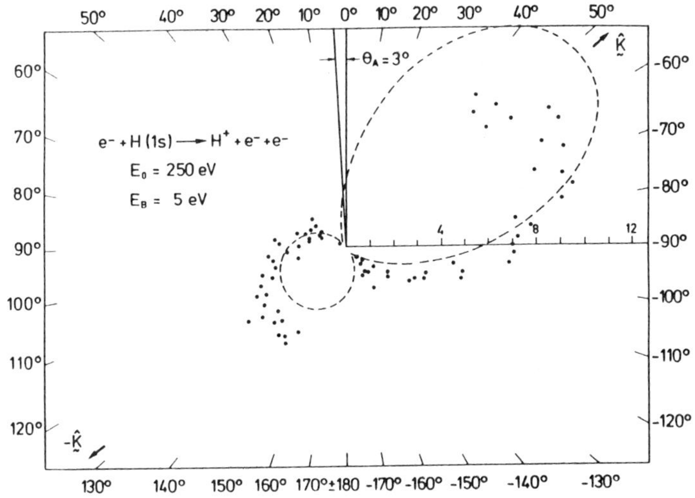

Fig. 9. Polar graph of the T.D.C.S. (in a.u.) for electron impact ionization of atomic hydrogen in its ground state for the case E_0=250 eV, E_B=5 eV and θ_A=3°. --- : First Born approximation, (•) absolute experimental data of Ehrhardt et al[72]. Also shown on the figure are the directions of the vectors $\underset{\sim}{K}$ and $-\underset{\sim}{K}$ which give respectively the angular positions of the binary and recoil peak maxima as predicted by the First Born approximation. Following a frequently used convention, the ejected electron angles (θ_B, ϕ_B=180°) have been denoted by $-|\theta_B|$ and the angles (θ_B, ϕ_B=0) have been written as $|\theta_B|$ (from[5]).

θ_A=3° as seen in Fig.9, there is a strong angular correlation between the scattered and ejected electrons characterized by two peaks : the forward or "binary" peak and the backward or "recoil" peak. According to the first Born approximation. the forward peak should occur in the direction of the momentum transfer $\underset{\sim}{K}$ and that of the backward peak in the direction $-\underset{\sim}{K}$. However it is clear from fig. 9 that the positions of both peaks are shifted towards larger angles with respect to the first Born prediction; in addition, the magnitude of the forward peak is reduced and that of the recoil peak considerably enhanced when compared with the first Born results. It is important to note that in the case of electron impact ionization of atomic hydrogen the wave function associated with both the initial and the final state are known exactly so that the comparison with absolute experimental data is unambiguous. In the case of (e^-,$2e^-$) reactions in helium, theoretical difficulties arise in describing accurately the initial helium state and especially its final continuum state. This problem introduces additional complications in the interpretation of data on T.D.C.S. In particular, a precise evaluation of the first Born amplitude is in that case very difficult explaining the substantial differences between the presently available first Born calculations of the T.D.C.S.[84-89]. Nevertheless, the experimental data clearly show the same features as in the case of atomic hydrogen : there is a shift of both peaks towards larger angles when compared with the first Born prediction and the relative magnitude of the forward and recoil peaks is never given correctly by the first Born approximation. This is illustrated in Fig. 10 where recent absolute measurements are compared with first Born calculations[90] for the case E_0=250 eV, E_B=3 eV and θ_A=10°. The experimental work carried out since 1969 stimulated a lot of theoretical treatments. In addition to the first Born calculations these

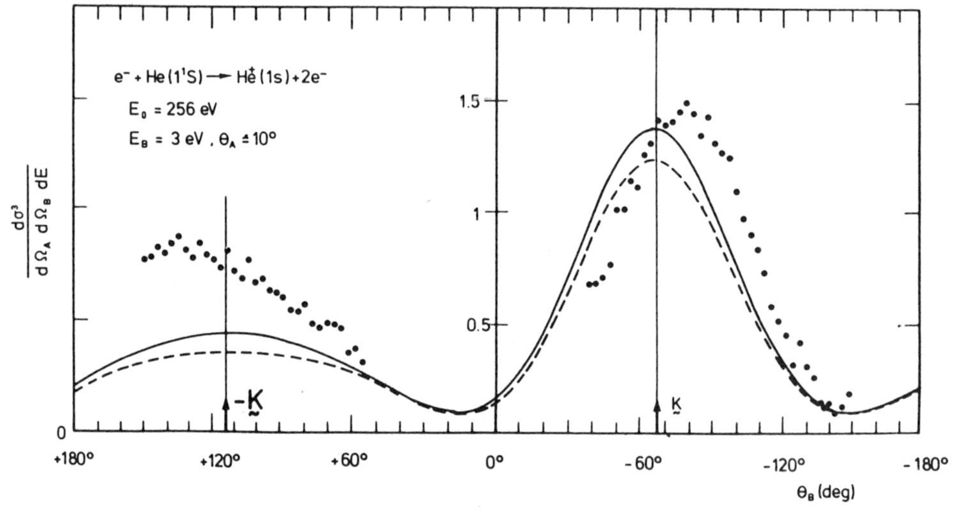

Fig. 10. The D.C.S. (in a.u.) for the electron impact ionization of helium for the case E_0=256 eV, E_B=3 eV and θ_A=10°. --- First Born values of Jacobs[90], —— first Born values of Jurig et al[93], (•) absolute experimental data of Müller-Fiedler et al[90]. Also indicated are the directions of the vectors $\underset{\sim}{K}$ and $-\underset{\sim}{K}$ (from[5]).

include Coulomb projected Born calculations[84,91-94], and distorted-wave Born approximation treatments[17-21] which are both of first order in the electron-electron interaction. Although these treatments are improvements over the first Born approximation and were able to describe relatively well the binary peak, severe discrepancies remained concerning the size, shape and position of the recoil peak.

The first theoretical treatment in which all the main features of the experiments of Ehrhardt et al where reproduced was a second Born calculation performed by Byron et al[94] in atomic hydrogen within the framework of the Eikonal-Born-Series (E.B.S.) method[6c,95,96]. According to the E.B.S. method, the direct amplitude is given by

$$f_{EBS} = f_{B1} + f_{B2} + f_{G3} \tag{33}$$

where f_{B1} is the first Born amplitude, f_{B2} the second Born term and f_{G3} the third term of the Glauber series obtained by expanding the Glauber amplitude[97] in power of the direct projectile-target interaction V_d. In the case of the ionisation reaction

$$e^- + H(1s) \rightarrow H^+ + e^- + e^- \tag{34}$$

the terms f_{B1} and f_{B2} are given by[22]

$$f_{B1}(\underset{\sim}{k}_A, \underset{\sim}{k}_B) = -(2\pi)^{-1} < \exp(i\underset{\sim}{k}_A \cdot \underset{\sim}{r}_1)\Psi^{(-)}_{c,\underset{\sim}{k}_B}(Z_B=1, \underset{\sim}{r}_2) \left| \frac{1}{r_{12}} - \frac{1}{r_1} \right| \exp(i\underset{\sim}{k}_A \cdot \underset{\sim}{r}_1)\phi_{1s}(r_2) > \tag{35}$$

$$f_{B2}(\underset{\sim}{k}_A, \underset{\sim}{k}_B) = (8\pi^4)^{-1} \sum_n \int d\underset{\sim}{q} \frac{1}{q^2 - k_n^2 - i\varepsilon}$$

$$\times < \exp(i\underset{\sim}{k}_A \cdot \underset{\sim}{r}_1)\Psi^{(-)}_{c,\underset{\sim}{k}_B}(Z_B=1, \underset{\sim}{r}_2) \left| \frac{1}{r_{12}} - \frac{1}{r_1} \right| \exp(i\underset{\sim}{q} \cdot \underset{\sim}{r}_1)\phi_n(\underset{\sim}{r}_2) >$$

$$\times < \exp(i\underset{\sim}{q} \cdot \underset{\sim}{r}_1)\phi_n(\underset{\sim}{r}_2) \left| \frac{1}{r_{12}} - \frac{1}{r_1} \right| \exp(i\underset{\sim}{k}_o \cdot \underset{\sim}{r}_1)\phi_{1s}(r_2) > \quad \varepsilon \rightarrow o^+ \tag{36}$$

where $k_n^2 = k_o^2 - 2(W_n - W_{1s})$, W_n being the eigenvalue of the hydrogen atom Hamiltonian corresponding to the eigen-function ϕ_n; the summation on n runs over both the discrete and continuum spectrum. A schematic diagram representing the contribution of an intermediate target state $|n>$ to the second Born term f_{B2} is shown in Fig. 11.

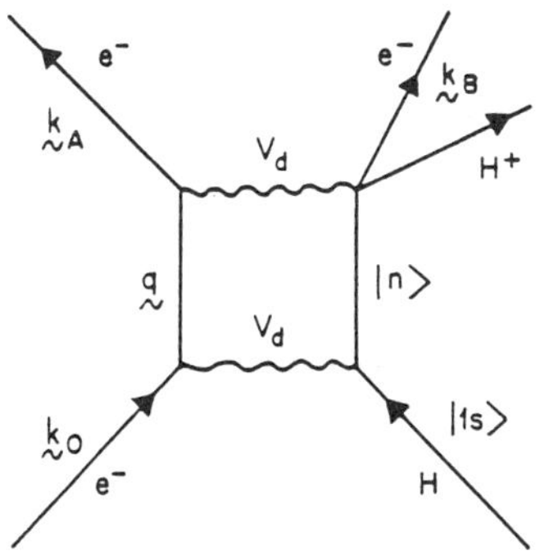

Fig. 11. Diagram representing the contribution of the intermediate target state $|n>$ to the second Born term f_{B2} for the electron impact ionization of atomic hydrogen in its ground state; the potential $V_d(=1/r_{12}-1/r_1)$ is the direct interaction between the incident electron and the target hydrogen atom (from[22]).

In the high energy region where an infinite number of channels are open, we expect that continuum intermediate states will play a significant role. In order to take these states into account Byron et al[22,98-99] evaluated $\bar{f}_{B2}$ by using the closure approximation which consists to replace the target energy difference (w_n-w_{1s}) in (36) by an average excited energy $\bar{w}$ so that the summation over n can be performed. This approximation is justified in the case of the Ehrhardt kinematics where E_B is very small[100]. In addition, Byron et al included in the previous calculation the exact contribution of the intermediate bound states 1s, 2s, 2p in order to insure the stability of the results with the choice of $\bar{w}$. Let us note that amongs these states, it is the 2p intermediate state which gives the most significant contribution to $\bar{f}_{B2}$ since the 2p state as well as the other p states are responsible for the polarization effects.

In the case of the Ehrhardt kinematics it turns out that the term $\bar{f}_{G3}$ gives always a very small contribution in equation 33. A detailed analysis of this term can be found in[96,100]

In the kinematical regime considered here where on electron emerges with a high velocity and the other one with a slow velocity we expect exchange effects to be small. Byron et al[22] therefore treated the exchange amplitude at lowest order of perturbation theory using the familiar Ochkur expression[101]

$$g_{och} = \frac{K^2}{k_o^2} f_{B_1} \qquad (37)$$

Since K^2 is typically less than one atomic unit in the cases considered in this section and since k_o^2 is large, we see that as expected, the exchange amplitude is small compared with the first Born amplitude f_{B1}; a detailed discussion of expression (37) is given in[22].

In figure 12, the E.B.S. results obtained by Byron et al[22] for the T.D.C.S. in atomic hydrogen using equations (4), (33) and (37) are compared with the experimental T.D.C.S. of Lohmann et al[71] and Ehrhardt et al[72] for the case E_o=250 eV, E_B=5 eV and θ_A=3°.
Also shown in figure 12 are the first Born results; it is clear that the E.B.S. results are in excellent agreement with the experimental data and that the first Born result are clearly deficient. Similar conclusions can be drawn from figure 13 which illustrates the case E_o=250 eV. E_B=5 eV and θ_A=8°. Consequently, the previous discussion shows unambiguously that for the Ehrhardt kinematical regime the second order dynamical effects are very important in (e^-, $2e^-$) processes.

Let us now turn our attention to the (e^-, $2e^-$) reaction

$$e^- + He(1^1S) \rightarrow He^+(1s) + e^- + e^- \qquad (38)$$

According to the previous discussion, Byron et al[65,102,103] evaluated the T.D.C.S. as follows

$$\frac{d\sigma^3}{d\Omega_A d\Omega_B dE} \simeq | f_{B1} + f_{B2} |^2 \qquad (39)$$

using the closure approximation to calculate the second Born term f_{B2}. The initial ground state of helium system was chosen to be the Byron-Joachain analytical fit to the Hartree-Fock wave function[104] while the final state was taken to be a symmetrized product of the $He^+(1s)$ wave function times a Coulomb wave corresponding to a charge Z=1, orthogonalized to the ground state Hartree-Fock orbital. Fig. 14 shows the theoretical results obtained from equation (39) compared with the

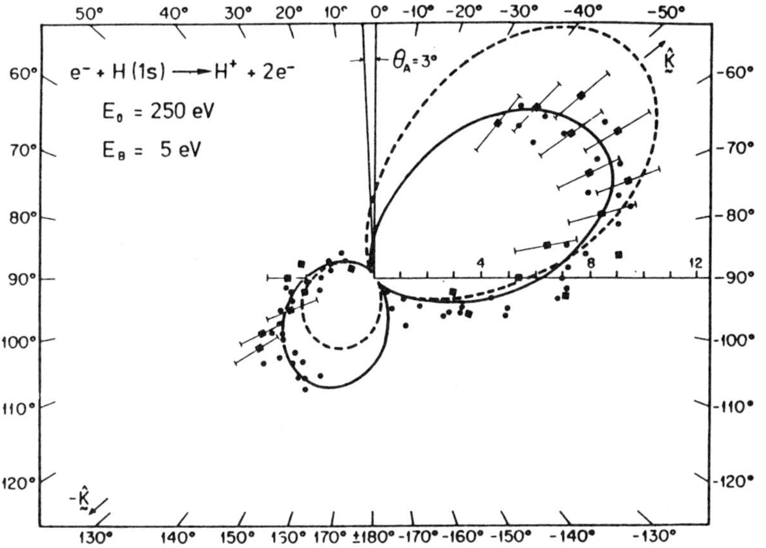

Fig. 12. Polar graph of the T.D.C.S. (in a.u.) for the electron-impact ionization of atomic hydrogen in its ground state for the case E_O=250 eV, E_B=5 eV and θ_A=3°.
(——) E.B.S. results of Byron et al[22] (in this calculation, $\bar{w}$ = 0.5 a.u.)
(---) first Born approximation, (•) absolute experimental data of Ehrhardt et al[72], (■) relative measurements of Lohmann et al[71], normalized to the E.B.S. results at θ_B=-70° (from[22]).

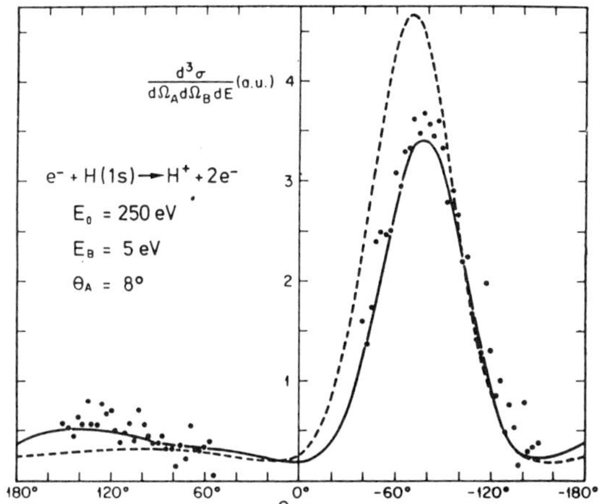

Fig. 13. The T.D.C.S. (in a.u.) for electron impact ionization of atomic hydrogen in its ground state for the case E_O=250 eV E_B=5 eV and θ_A=8° as a function of the ejected electron angle θ_B.
(——) E.B.S. results of Byron et al[22] (in this calculation, $\bar{w}$ = 0.5 a.u.), (---) first Born approximation, (·) absolute experimental data of Ehrhardt et al.[72] (from[22]).

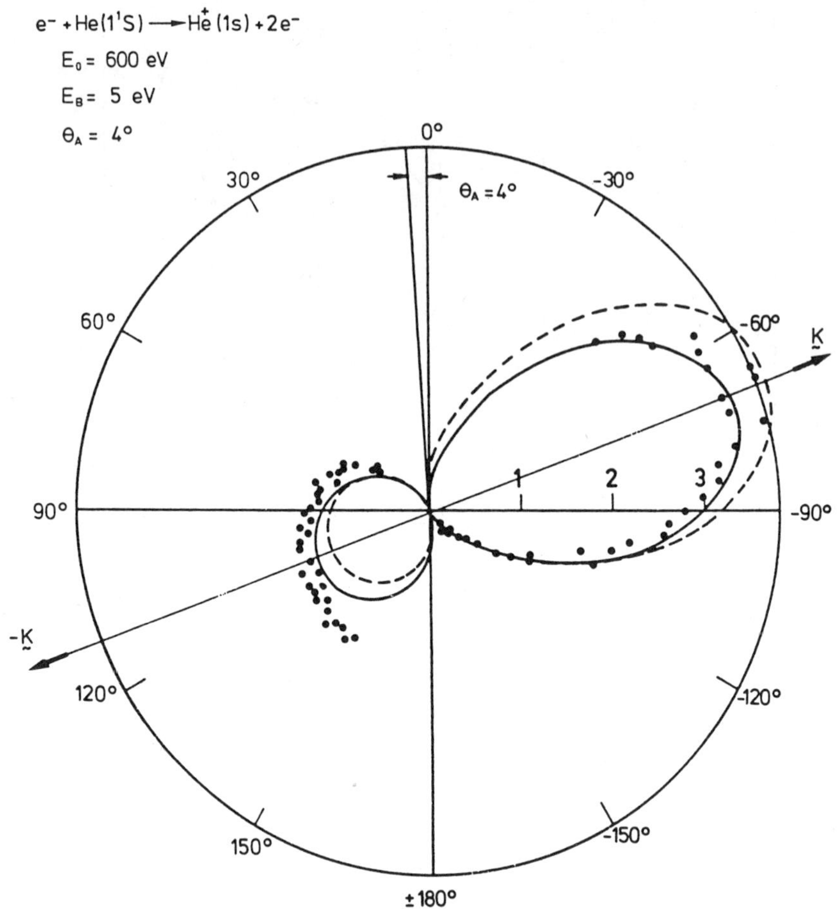

Fig. 14. Polar graph of the T.D.C.S. (in u.a.) for the electron impact ionisation of helium in its ground state for the case E_0=600 eV, E_B=5 eV, θ_A=4°. (——) Second Born results of Byron et al[102,103] (the value of $\bar{w}$ was chosen to be 0.9 a.u.), (---) first Born approximation[103], (●) absolute experimental data of Ehrhardt et al.[105] (from[5]).

recent absolute experimental data of Ehrhardt et al.[105] for the case E_0=600 eV, E_B=5 eV and θ_A=4° while Fig. 15 displays the angular position θ_B^{Rec} of the recoil peak as a function of θ_A for E_0=500 eV and E_B=10 eV. It is clear that the second Born calculations represent a marked improvement over first order results and that second order dynamical effects must be included in order to account for the magnitude and the angular position of both peaks.

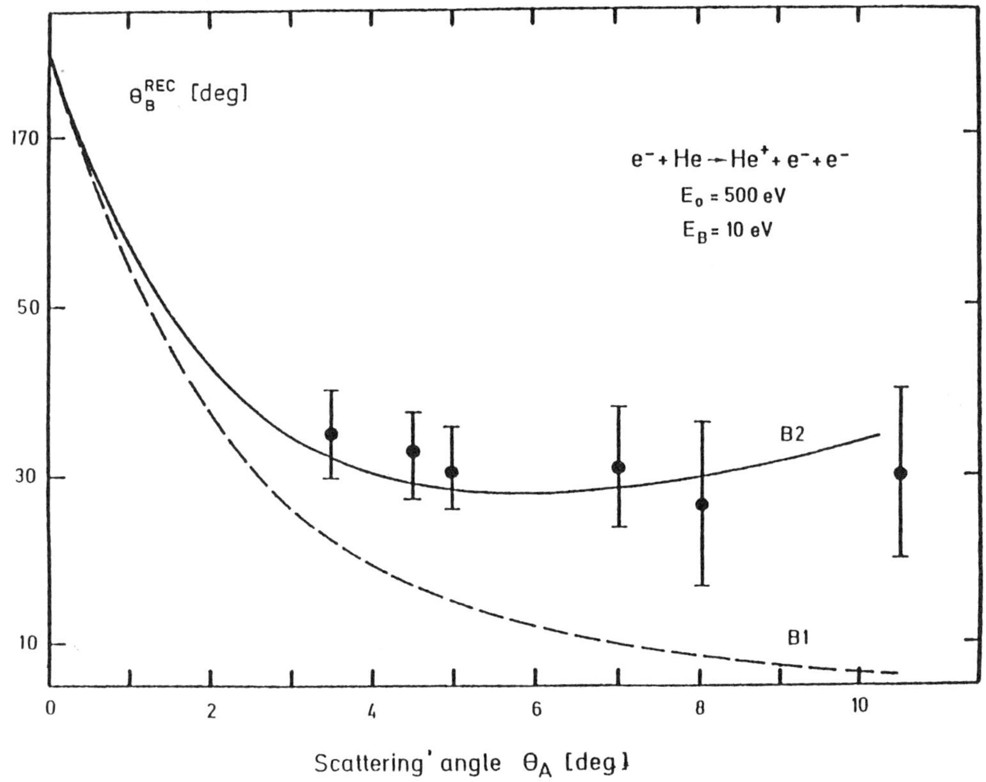

Fig. 15. The angular position θ_A^{Rec} of the recoil peak as a function of the angle θ_A for the electron impact ionization of helium in its ground state for the case E_0=500 eV, E_B=10 eV. The experimental data of Ehrhardt (•) are compared with the first Born approximation (---) and the second Born approximation (——) (the value of $\bar{w}$ was chosen to be 0.9 a.u.) (from[65]).

3.3. Coplanar symmetric (e^-, $2e^-$) reactions

As seen in section 3.1 one may distinguish at high energy two different coplanar symmetric kinematical regime depending on the angle θ. In this section we will focuse our attention on the wide-angle domain ($\theta \gtrsim 70°$) for which both K and Q are large. According to previous studies concerning inelastic scattering by electron impact[106], the second Born term should be important and dominated by the contributions of the initial and final target states acting as intermediate states. This is confirmed by recent calculations performed by Byron et al[107] in the case of atomic hydrogen. They showed that for large k_o and large θ, the first Born term f_{B1} is of order k_o^{-6} while the contribution of the intermediate 1s target state of f_{B2}, represented by the diagram of Fig. 16 is given by

$$f_{B2}(1s) = -\frac{8\sqrt{2}}{\pi}\frac{1}{k_o^5}\frac{1}{(1-\sqrt{2}\cos\theta)^2}\frac{1}{k_o\cos(2\theta)-2\sqrt{2}\,i\cos\theta} \tag{40}$$

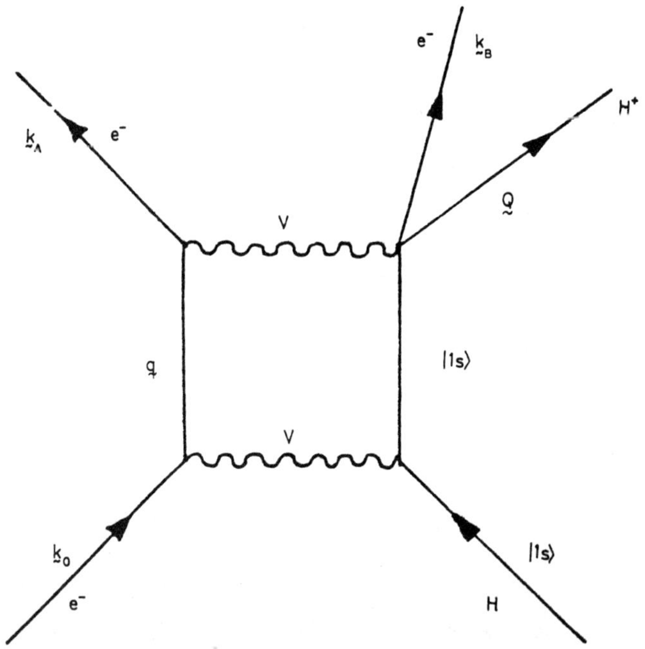

Fig. 16. Diagram representing $f_{B2}(1s)$ namely the contribution of the intermediate $|1s>$ state to the second Born term f_{B2} for the electron impact ionization of atomic hydrogen in its ground state. The potential $V_d(=1/r_{12}-1/r_1)$ is the direct interaction between the incident electron and the target hydrogen atom (from[107])

It is clear that $f_{B2}(1s)$ behaves like k_o^{-6} except around θ=135° when it is of order k_o^{-5}. The contribution f_{B2} (cont) of the final (continuum) target state to f_{B2} has been also estimated by Byron et al and has been found to be of order k_o^{-6}. Consequently, second order effects due to the term $f_{B2}(1s)$ should be important in the large θ region, especially near θ=135°.

This analysis is confirmed by an exact (numerical) calculation of the quantity $f_{B2}(1s)$. Fig. 17 shows the T.D.C.S.

$$\frac{d^3\sigma}{d\Omega_A d\Omega_B dE} = \frac{k_A k_B}{k_0} |f|^2 \qquad (41)$$

as a function of θ, for an incident energy E_0=250 eV. In writing equation (41), it has been taken into account the fact that the scattering occurs only in the singlet mode in the present case. The broken curve, B1, corresponds to the first Born approximation in which f is replaced by f_{B1} and the full curve, B2(1s) is obtained by replacing f by $f_{B1}+f_{B2}(1s)$. We note the dramatic second order effects in the large angle region, in particular the local minimum around $\theta=90°$ and the maximum near $\theta=130°$; at this maximum, the second order effect is actually fifty times larger than the first Born result. A similar behaviour of the theoretical T.D.C.S.

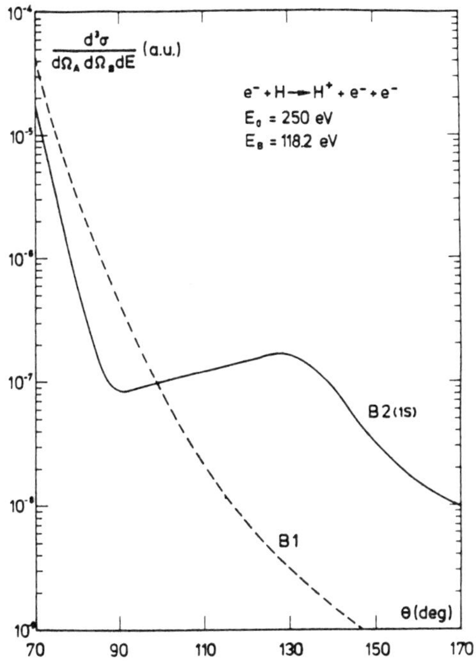

Fig. 17. T.D.C.S. (in a.u.) for the electron impact ionization of atomic hydrogen in its ground state calculated from equation (41) for the case of a coplanar symmetric geometry with E_0=250 eV, $E_A=E_B$=118.2 eV as a function of the scattering angle $\theta=\theta_A=\theta_B$. The full curve, B2(1s) corresponds to a second Born calculation in which $f=f_{B1}+f_{B2}(1s)$[107], the broken curve, B1, refers to the first Born approximation (from[107]).

due to second order effects arises for large angle symmetric $(e^-,2e^-)$ reactions in helium and has recently been observed by Pochat et al[82]. As seen from Fig. 18 the agreement between theory and experiment is considerably improved at large θ when the contribution $f_{B2}(1^1S)$ of the initial (1^1S) target state to the second Born term f_{B2} is included in the scattering amplitude. The physical interpretation of this second order effect can be understood as follows : the system undergoes a large-angle ionization transition by first making a large-angle elastic (off shell) scattering, followed by a small angle ionization (of shell) transition (see fig. 16 in the case of reaction (34)). During the first elastic scattering event the projectile can feel the singular Coulomb potential of the target nucleus (which is very important in this angular region) while in first Born approximation, the electron-target interaction is missing since the initial and final state wave functions are orthogonal.

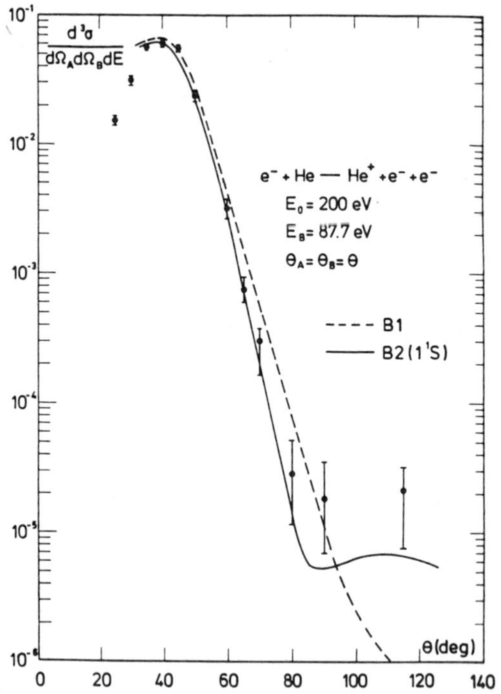

Fig. 18. The T.D.C.S. (in a.u.) for the electron impact ionization of helium in its ground state from equation (41) for the case of a coplanar symmetric geometry with E_0=200 eV, $E_A=E_B$=87.7 eV as a function of the scattering angle $\theta=\theta_A=\theta_B$. The full curve, $B_2(1^1S)$ corresponds to a second Born calculation[82] in which $f=f_{B1}+f_{B2}(1^1S)$. The broken curve, B1, refers to the first Born approximation (from[107]).

CONCLUSION

We have attempted in this contribution to analyse the most important mechanisms which govern the dynamics of the (e^-,$2e^-$) processes and considered in particular the case of atomic hydrogen and helium.

In the energy domain around threshold, we saw that all the predictions of the Wannier classical theory are so far consistent with all the experimental results. We also discussed the quantum mechanical approach of Temkin which gives a different threshold behaviour of the total cross-section; unfortunately, the experimental data are still not accurate enough to distinguish between this threshold law and the Wannier law.

In the high energy domain, we saw that the inclusion of the second order effects is crucial in order to understand the dynamics of the electron impact ionisation processes in the case of the asymmetric and large angle symmetric coplanar kinematics. In the later case however, the contributions of the intermediate continuum state to the second Born term as well as those due to high order should be investigated.

In the intermediate energy domain, a very few results exist. We must mention however, the work of Pochat et al[108,109] who studied how the autoionizing states affects, in the case of helium the double and the triple differential cross-sections.

REFERENCES

1. H. Ehrhardt, M. Schultz, T. Tekaat, K. Willmann, Phys. Rev. Lett. 22, 89 (1969).
2. P. Defrance, Electron Impact Excitation and Ionization of Ions. Experimental Methods. This Institute.
3. A.R.P. Rau, Phys. Reports 110, 369 (1984).
4. A. Temkin, Phys. Rev. Lett. 49, 365 (1982).
5. C.J. Joachain, B. Piraux, Comments At. Mol. Phys. 17, 261 (1986).
6. C.J. Joachain, "Quantum Collision Theory" 3rd edn (Amsterdam, North-Holland, 1983) - a) p. 541; b) p. 359, p. 410; c) p. 558.
7. The T.D.C.S. is sometimes called the five-fold differential cross section and denoted by $d^5\sigma/d\Omega_A d\Omega_B dE$ since the directions (θ_A,ϕ_A) and (θ_B,ϕ_B) involve the four angles $\theta_A,\theta_B,\phi_A,\phi_B$.
8. R.K. Peterkop, Proc. Phys. Soc. (London) 77, 1220 (1961).
9. B.H. Bransden, "Atomic Collision Theory" 2nd edn (New York, Benjamin 1983) p. 297.
10. J. Kessler "Polarized electrons", (Berlin, Springer Verlag 1976) p. 89.
11. G. Baum, E. Kisker, W. Raith, W. Schröder, U. Sillmen, D. Zenses, J. Phys. B : At. Mol. Phys. 14, 4377 (1981).
12. M.J. Alguard, V.W. Hughes, M.S. Lubell, P.F. Wainwright, Phys. Rev. Lett. 39, 334 (1977).
13. M.R.H. Rudge, Rev. Mod. Phys. 40, 564 (1968).
14. D.H. Phillips, M.R.C. McDowell, J. Phys. B 6, L165 (1973).
15. R.K. Peterkop, Opt. Spectrosc. 13, 87 (1962).
16. M.R.H. Rudge, M.J. Seaton, Proc. Roy. Soc. A 283, 262 (1965).
17. K.L. Baluja, H.S. Taylor, J. Phys. B : At. Mol. Phys. 9, 829 (1976).
18. D.H. Madison, R.V. Calhoun, W.N. Shelton, Phys. Rev. A 16, 552 (1977).
19. B.H. Bransden, J.J. Smith, K.H. Winters, J. Phys. B : At. Mol. Phys. 11, 3095 (1978).
20. J.J. Smith, K.H. Winters, B.H. Bransden, J. Phys. B : At. Mol. Phys. 12, 1723 (1979).
21. R.J. Tweed, J. Phys. B : At. Mol. Phys. 13, 4467 (1980).
22. F.W. Byron, Jr., C.J. Joachain, B. Piraux, J. Phys. B : At. Mol. Phys. 18, 3203 (1985).
23. J. Jakubowić, D.L. Moores, Comments At. Mol. Phys. 9, 55 (1980).
24. E.P. Wigner, Phys. Rev. 73, 1002 (1948).
25. G.H. Wannier, Phys. Rev. 90, 817 (1953).
26. R. Peterkop, J. Phys. B. : At. Mol. Phys. 4, 513 (1971).
27. A.R.P. Rau, Phys. Rev. A 4, 207 (1971).

28. R. Peterkop, "Theory of Ionization of Atoms by Electron Impact" (English edition)(Boulder, Colorado : Associated University Press 1977) chap. VII.
29. R. Peterkop, A. Liepinsh, J. Phys. B : At. Mol. Phys. 14, 4125 (1981).
30. A.R.P. Rau, Comments At. Mol. Phys. 14, 285 (1984).
31. M.R.H. Rudge, M.J. Seaton, Proc. Phys. Soc. 83, 680 (1964).
32. U. Fano, Phys. Today 29, 32 (September 1976).
33. F.H. Read in Electron Impact Ionization, ed. T.D. Märk and G.H. Dunn, (Springer Verlag, Wien, New York, 1985) p. 42.
34. H. Klar, W. Schlecht, J. Phys. B : At. Mol. Phys. 14, 3255 (1981).
35. C.H. Greene, A.R.P. Rau, Phys. Rev. Lett. 48, 533 (1982).
36. C.H. Greene, A.R.P. Rau, J. Phys. B : At. Mol. Phys. 16, 99 (1983).
37. A.D. Stauffer, Phys. Lett. 91A, 114 (1982).
38. J.M. Feagin, J. Phys. B : At. Mol. Phys. 17, 2433 (1984).
39. This result has been reproduced by several authors by using W.K.B.-like treatments see for example[26,27,29].
40. S. Cvejanović, P. Grujić, J. Phys. B : At. Mol. Phys. 8, L305 (1975).
41. A.R.P. Rau, J. Phys. B : At. Mol. Phys. 9, L283 (1976).
42. M.S. Yurev, Opt. Spectrosc. 42, 594 (1977).
43. F.H. Read, J. Phys. B : At. Mol. Phys. 17, 3965 (1984).
44. U. Fano, J. Phys. B : At. Mol. Phys. 7, L401 (1974).
45. A. Temkin, J. Phys. B 7, L450 (1974).
46. A. Temkin, Y. Hahn, Phys. Rev. A9, 708 (1974).
47. A. Temkin in "Recent developments in electron-atom and electron-molecule collision processes", Proceedings of the Daresbury Study Weekend, March (1982) p. 44.
48. A. Temkin, Comments At. Mol. Phys. 11, 287 (1982).
49. J.W. McGowan, E. Clark, Phys. Rev. 167, 43 (1968).
50. C.E. Brion, G.E. Thomas, Phys. Rev. Lett. 20, 241 (1968).
51. P. Marchand, C. Paquet, P. Marmet, Phys. Rev. 180, 123 (1969).
52. S. Cvejanović, F.H. Read, J. Phys. B : At. Mol. Phys. 7, 1841 (1974).
53. J.B. Donahue, P.A.M. Gram, M.V. Hynes, R.W. Hamm, C.A. Frost, H.C. Bryant, K.B. Butterfield, D.A. Clark, W.W. Smith, Phys. Rev. Lett. 48, 1538 (1982)
54. Phys. Rev. Lett. 52, 164 (1984).
55. T.J. Gay, F.D. Fletcher, M.J. Alguard, V.W. Hughes, P.F. Wainwright, M.S. Lubell, Phys. Rev. A 26, 3663 (1982).
56. D. Hils, W. Jitschin, H. Kleinpoppen, J. Phys. B 15, 3347 (1982).
57. M.H. Kelley, W.T. Rogers, R.J. Celotta, S.R. Mielczarek, Phys. Rev. Lett. 51, 2191 (1983).
58. F. Pichou, A. Huetz, G. Joyez, M. Landau, J. Phys. B : At. Mol. Phys. 11, 3683 (1978).
59. P. Hammond, F.H. Read, S. Cvejanović, G.C. King in the book of Abstracts of the Second European Conference on Atomic and Molecular Physics (Amsterdam, April 1985) p. 158.
60. H. Ehrhardt, K.H. Hesselbacher, K. Jung, K. Willmann, J. Phys. B : At. Mol. Phys. 5, 1559 (1972).
61. E. Schubert, K. Jung, H. Ehrhardt, J. Phys. B : At. Mol. Phys. 14, 3267 (1981).
62. P. Founier-Lagarde, J. Mazeau, A. Huetz, J. Phys. B : At. Mol. Phys. 17, L591 (1984).
63. H. Ehrhardt, K.H. Hesselbacher, K. Jung, K. Willmann in Case Studies in Atomic Physics (North-Holland, Amsterdam), 2, 159 (1971).
64. H. Ehrhardt, M. Fischer, K. Jung, Zeit. Phys. A 304, 119 (1982).
65. H. Ehrhardt, M. Fischer, K. Jung, F.W. Byron, Jr., C.J. Joachain, B. Piraux, Phys. Rev. Lett. 48, 1807 (1982).
66. H. Ehrhardt, Comments At. Mol. Phys. 13, 115 (1983).
67. A. Lahmam-Bennani, H.F. Wellenstein, A. Duguet, M. Rouault, J. Phys. B. : At. Mol. Phys. 16, 121 (1983).
At. Mol. Phys. 16, 4089 (1983).

68. A. Lahmam-Bennani, H.F. Wellenstein, C. Dal Cappello, M. Rouault, A. Duguet, J. Phys. B : At. Mol. Phys. 16, 2219 (1983).
69. A. Lahmam-Bennani, H.F. Wellenstein, C. Dal Cappello, A. Duguet, J. Phys. B : At. Mol. Phys. 17, 3159 (1984).
70. V.D.I. Martino, R. Fantoni, A. Giardini-Guidoni and R. Tiribelli, Phys. Lett. 103 A, 45 (1984).
71. B. Lohmann, I.E. McCarthy, A.T. Stelbovics, E. Weigold, Phys. Rev. A 30, 758 (1984).
72. Ehrhardt, private communication.
73. U. Amaldi, Jr., A. Egidi, R. Marconero, G. Pizzella, Rev. Sci. Instr. 40, 1001 (1969).
74. R. Camilloni, A. Giardini-Guidoni, R. Tiribelli and G. Stephani, Phys. Rev. Lett. 29, 618 (1972).
75. A. Giardini-Guidoni, R. Fantoni, R. Camilloni, G. Stephani, Comments At. Mol. Phys. 10, 107 (1981).
76. E. Weigold, S.T. Hood, P.J.O. Teubner, Phys. Rev. Lett. 30, 475 (1973).
77. I.E. McCarthy, E. Weigold, Phys. Reports 27, 275 (1976).
78. E. Weigold, I.E. McCarthy, Adv. At. Mol. Phys. 14, 127 (1978).
79. B. van Wingerden, J.T. Kimman, M. van Tilburg, E. Weigold, C.J. Joachain, B. Piraux, F.J. de Heer, J. Phys. B : At. Mol. Phys. 12, L627 (1979).
80. V. van Wingerden, J.T. Kimman, M. van Tilburg, F.J. de Heer, J. Phys. B : At. Mol. Phys. 14, 2475 (1981).
81. J.T.N. Kimman, Pan Guang-Yan, C.W. McCardy, F.J. de Heer, J. Phys. B : At. Mol. Phys. 16, 4203 (1983).
82. A. Pochat, R.J. Tweed, J. Peresse, C.J. Joachain, B. Piraux, F.W. Byron, Jr., J. Phys. B : At. Mol. Phys. 16, L775 (1983).
83. E. Weigold, Comments At. Mol. Phys. 15, 223 (1984).
84. M. Schultz, J. Phys. B : At. Mol. Phys. 6, 2580 (1973).
85. V.L. Jacobs, Phys. Rev. A 10, 499 (1974).
86. W.D. Robb, S.P. Rountree, T. Burnett, Phys. Rev. A 11, 1193 (1975).
87. M. Zarcone, D.L. Moores, M.R.C. Mc Dowell, J. Phys. B : At. Mol. Phys. 16, L11 (1983).
88. K. Jung, R. Müller-Fiedler, P. Schlemmer, H. Ehrhardt, H. Klar, to be published.
89. H. Klar, K. Jung, H. Ehrhardt, Phys. Rev. A 29, 405 (1984).
90. R. Müller-Fiedler, P. Schlemmer, K. Jung, H. Ehrhardt, to be published.
91. S. Geltman, M.B. Hidalgo, J. Phys. B : At. Mol. Phys. 7, 831 (1974).
92. S. Geltman, J. Phys. B : At. Mol. Phys. 7, 1994 (1974).
93. The Coulomb-projected Born approximation for the direct amplitude is obtained from expression (8) in which relation (9) is used and $Z_A=Z_B=1$.
94. F.W. Byron, Jr., C.J. Joachain, B. Piraux, J. Phys. B : At. Mol. Phys. 13, L673 (1980).
95. F.W. Byron, Jr., C.J. Joachain, Phys. Rev. A 8, 1267 (1973).
96. F.W. Byron, Jr., C.J. Joachain, Phys. Reports 34, 233 (1977).
97. R.J. Glauber, "Lectures in Theoretical Physics", Vol. 1, ed. W.E. Brittin (New York : interscience) p. 315 (1959).
98. F.W. Byron, Jr., C.J. Joachain, B. Piraux, Phys. Lett. 99A, 427 (1983).
99. F.W. Byron, Jr., C.J. Joachain, B. Piraux, Phys. Lett. 106A, 289 (1984).
100. B. Piraux, Thèse de doctorat, Université Catholique de Louvain (1983).
101. V.I. Ochkur, Zh. Eksp. Teor. Fiz 45, 734 (1963); [English translation Sov. Phys. JETP 18, 503 (1984)].
102. F.W. Byron, Jr., C.J. Joachain, B. Piraux, J. Phys. B : At. Mol. Phys. 15, L293 (1982).
103. F.W. Byron, Jr., C.J. Joachain, B. Piraux, to be published.
104. F.W. Byron, Jr., C.J. Joachain, Phys. Rev. 146, 1 (1966).
105. H. Ehrhardt, private communication.
106. C.J. Joachain, Comments At. Mol. Phys. 6, 69 (1977).
107. F.W. Byron, Jr., C.J. Joachain, B. Piraux, J. Phys. B : At. Mol. Phys. L769 (1983).

108. A. Pochat, R.J. Tweed, M. Doritch, J. Peresse, J. Phys. B : At. Mol. Phys. 15, 2269 (1982).
109. A. Pochat, M. Doritch, J. Peresse, Phys. Lett. A 90, 354 (1982).

INDEX